# Virus-Induced Immunosuppression

Edited by

**Steven Specter**
*University of South Florida*
*Tampa, Florida*

**Mauro Bendinelli**
*University of Pisa*
*Pisa, Italy*

and

**Herman Friedman**
*University of South Florida*
*Tampa, Florida*

Plenum Press • New York and London

Library of Congress Cataloging in Publication Data

Virus-induced immunosuppression / edited by Steven Specter, Mauro Bendinelli, and
Herman Friedman.
    p.    cm. — (Infectious agents and pathogenesis)
    Includes bibliographies and index.
    ISBN-13: 978-1-4684-5585-4    e-ISBN-13: 978-1-4684-5583-0
    DOI: 10.1007/ 978-1-4684-5583-0
    1. Virus-induced immunosuppression. 2. Immune response — Regulation. I. Specter,
Steven. II. Bendinelli, Mauro. III. Friedman, Herman, 1931-    . IV. Series.
    [DNLM: 1. Immunosuppression. 2. Virus Diseases — immunology. 3. Viruses —
immunology. QW 920 V821]
    QR188.46.V57   1988
    616.07′9 — dc19
    DNLM/DLC
    for Library of Congress                    88-31629
                                              CIP

© 1989 Plenum Press, New York
Softcover reprint of the hardcover 1st edition 1989
A Division of Plenum Publishing Corporation
233 Spring Street, New York, N.Y. 10013

# Contributors

WARREN A. ANDIMAN • Departments of Pediatrics and Epidemiology and Public Health, Yale University School of Medicine, New Haven, Connecticut 06510

LAURE AURELIAN • Departments of Pharmacology and Experimental Therapeutics and Microbiology, University of Maryland School of Medicine, Baltimore, Maryland 21201; and Divisions of Comparative Medicine and Biophysics, The Johns Hopkins Medical Institutions, Baltimore, Maryland 21205

LORNE A. BABIUK • Department of Veterinary Microbiology and Veterinary Infectious Disease Organization, University of Saskatchewan, Saskatoon, Saskatchewan, Canada S7N 0W0

GIUSEPPE BARBANTI-BRODANO • Institute of Microbiology, School of Medicine, University of Ferrara, I-44100 Ferrara, Italy

FULVIO BASOLO • Institute of Pathological Anatomy, University of Pisa, I-56100 Pisa, Italy

MAURO BENDINELLI • Department of Biomedicine, University of Pisa, I-56100 Pisa, Italy

HENRY R. BOSE, JR. • Department of Microbiology, University of Texas at Austin, Austin, Texas 78712-1095

PAOLO CASALI • Laboratory of Oral Medicine, National Institute of Dental Research, National Institutes of Health, Bethesda, Maryland 20892

UMESH C. CHATURVEDI • Department of Microbiology, King George's Medical College, Lucknow, India 226 003

FRANK DOMURAT • Infectious Diseases Unit, Department of Medicine, University of Rochester School of Medicine, Rochester, New York 14642

J. M. DUPUY • Immunology Research Center, Institute Armand-Frappier, University of Quebec, Quebec, Canada. *Present address:* 5 Boulevard des Belges, Lyon 6906, France.

v

MARIO R. ESCOBAR • Department of Pathology, Medical College of
  Virginia, Richmond, Virginia 23298-0001
ROBERT FINBERG • Laboratory of Infectious Diseases, Dana Farber
  Cancer Institute, Boston, Massachusetts 02115
HERMAN FRIEDMAN • Department of Medical Microbiology and
  Immunology, College of Medicine, University of South Florida, Tampa,
  Florida 33612-4799
CARLO GARZELLI • Institute of Microbiology, University of Pisa, 56100
  Pisa, Italy
P. GRIEBEL • Department of Veterinary Microbiology and Veterinary
  Infectious Disease Organization, University of Saskatchewan, Saskatoon,
  Saskatchewan, Canada  S7N 0W0
HELEN L. GRIERSON • Departments of Pathology and Microbiology, and
  Pediatrics, and the Eppley Institute for Research in Cancer and Allied
  Diseases, University of Nebraska Medical Center, Omaha, Nebraska
  68105-1065
JOHN HADDEN • Departments of Internal Medicine and Medical
  Microbiology and Immunology, College of Medicine, University of
  South Florida, Tampa, Florida 33612-4799
TIMO HYYPIÄ • Department of Virology, University of Turku, SF-20520
  Turku, Finland
L. LAMONTAGNE • Immunology Research Center, Institute Armand-
  Frappier, University of Quebec, Quebec, Canada.  *Present address:*
  Department of Biological Sciences, University of Quebec at Montreal,
  Montreal, Quebec, Canada H3C 3P8
M. J. P. LAWMAN • Department of Veterinary Microbiology and
  Veterinary Infectious Disease Organization, University of Saskatchewan,
  Saskatoon, Saskatchewan, Canada S7N 0W0
MAYRA LOPEZ-CEPERO • Department of Medical Microbiology and
  Immunology, College of Medicine, University of South Florida, Tampa,
  Florida 33612-4799
MICHAEL B. McCHESNEY • Scripps Clinic and Research Foundation, La
  Jolla, California 92037
DONATELLA MATTEUCCI • Institute of Hygiene, University of Pisa,
  56100 Pisa, Italy
MINORU NAKAMURA • Laboratory of Oral Medicine, National Institute
  of Dental Research, National Institutes of Health, Bethesda, Maryland
  20892
TAKASHI ONODERA • Laboratory of Immunology, National Institute of
  Animal Health, Kodaira, Tokyo 187, Japan
BELLAR S. PRABHAKAR • Laboratory of Oral Medicine, National
  Institute of Dental Research, National Institutes of Health, Bethesda,
  Maryland 20205
DAVID T. PURTILO • Departments of Pathology and Microbiology, and
  Pediatrics, and the Eppley Institute for Research in Cancer and Allied
  Diseases, University of Nebraska Medical Center, Omaha, Nebraska
  68105-1065

WILLIAM E. RAWLS • Department of Pathology, McMaster University, Hamilton, Ontario, Canada L8N 3Z5

MARIE F. ROBERT • Departments of Pediatrics and Epidemiology and Public Health, Yale University School of Medicine, New Haven, Connecticut 06510

NORBERT J. ROBERTS, JR. • Infectious Diseases Unit, Department of Medicine, University of Rochester School of Medicine, Rochester, New York 14642

JAGDEV M. SHARMA • Regional Poultry Research Laboratory, U.S. Department of Agriculture, Agricultural Research Service, East Lansing, Michigan 48823. *Present address:* Department of Veterinary Pathobiology, College of Veterinary Medicine, University of Minnesota, St. Paul, Minnesota 55108

STEVEN SPECTER • Department of Medical Microbiology and Immunology, College of Medicine, University of South Florida, Tampa, Florida 33612-4799

ROBERT W. STORMS • Department of Microbiology, University of Texas at Austin, Austin, Texas 78712-1095

DAVID S. STRAYER • Department of Pathology and Laboratory Medicine, University of Texas Health Science Center, Houston, Texas 77030

MAN-SUN SY • Department of Pathology, Harvard Medical School, Boston, Massachusetts 02115

ANTONIO TONIOLO • Institute of Microbiology, University of Sassari, I-07100 Sassari, Italy

RAIJA VAINIONPÄÄ • Department of Virology, University of Turku, SF-20520 Turku, Finland

JOSEPH L. WANER • Department of Pediatrics, University of Oklahoma, Health Sciences Center, Oklahoma City, Oklahoma 73190; and Virology Laboratory, Oklahoma Children's Memorial Hospital, Oklahoma City, Oklahoma 73104

KATHRYN E. WRIGHT • Scripps Clinic and Research Foundation, La Jolla, California 92037. *Present address:* Department of Microbiology and Immunology, University of Ottawa, Ottawa, Ontario, Canada K1H 8M5

# Preface to the Series

The mechanisms of disease production by infectious agents are presently the focus of an unprecedented flowering of studies. The field has undoubtedly received impetus from the considerable advances recently made in the understanding of the structure, biochemistry, and biology of viruses, bacteria, fungi, and other parasites. Another contributing factor is our improved knowledge of immune responses and other adaptive or constitutive mechanisms by which hosts react to infection. Furthermore, recombinant DNA technology, monoclonal antibodies, and other newer methodologies have provided the technical tools for examining questions previously considered too complex to be successfully tackled. The most important incentive of all is probably the regenerated idea that infection might be the initiating event in many clinical entities presently classified as idiopathic or of uncertain origin.

Infectious pathogenesis research holds great promise. As more information is uncovered, it is becoming increasingly apparent that our present knowledge of the pathogenic potential of infectious agents is often limited to the most noticeable effects, which sometimes represent only the tip of the iceberg. For example, it is now well appreciated that pathologic processes caused by infectious agents may emerge clinically after an incubation of decades and may result from genetic, immunologic, and other indirect routes more than from the infecting agent in itself. Thus, there is a general expectation that continued investigation will lead to the isolation of new agents of infection, the identification of hitherto unsuspected etiologic correlations, and, eventually, more effective approaches to prevention and therapy.

Studies on the mechanisms of disease caused by infectious agents demand a breadth of understanding across many specialized areas, as well as much cooperation between clinicians and experimentalists. The series *Infectious Agents and Pathogenesis* is intended not only to document the state of the art in this fascinating and challenging field but also to help lay bridges among diverse areas and people.

M. Bendinelli
H. Friedman

ix

# Preface

It is now widely acknowledged that at the beginning of this century Claude von Pirquet first pointed out that a viral disease, i.e., measles, resulted in an anergy or depression of preexisting immune response, namely, delayed continuous hypersensitivity to PPD derived from *Mycobacterium tuberculosis*. Thereafter observations that viral infections may result in immunosuppression have been recorded by many clinicians and infectious disease investigators for six or seven decades. Nevertheless, despite sporadic reports that infectious diseases caused by viruses may result in either transient or prolonged immunodepression, investigation of this phenomenon languished until the mid-1960s, when it was pointed out that a number of experimental retroviral infections of mice with tumor viruses may result in marked immunosuppression. However, it was not until the recognition of the new epidemic of acquired immunodeficiency syndrome (AIDS) caused by the human immunodeficiency virus and related viruses that acquired immunodeficiencies associated with virus infection became general knowledge among biomedical investigators as well as the lay public.

A number of reviews published during the past decade or so pointed out that numerous viruses may affect humoral and cellular immune responses. Furthermore, expanding knowledge about the nature and mechanisms of both humoral and cellular immunity and pathogenesis of viral infections has provided clinical and experimental models for investigating in depth how and why viruses of man and animals profoundly affect immune responses.

The realization that viruses, including both those which obviously infect lymphoid cells and those which have nonlymphoid cells as their primary target tissues, may either directly or indirectly influence host immune responses has led to many investigations concerning the mechanisms of these host–parasite relationships. This volume, therefore, presents reviews of major human pathogenic viruses as well as several animal viruses which influence immunity, mainly in a negative manner.

Virus-induced immune suppression is accomplished by a wide variety of mechanisms. Some viruses may induce dysregulation of immune responses, including an increase in one arm of the response and a decrease in another arm, or a hyporeactivity of both arms of the immune system. More common is a transient immunosuppression associated with virus infection. This can be seen from the wide variety of viruses reviewed herein.

There is now sufficient information pertaining to the mechanisms involved in virus-induced immune suppression that rational approaches to therapy for viral infection and controlling immunoderegulation induced by such viruses can be undertaken. A number of laboratories and clinical investigators are in the process of developing such therapeutic strategies. Thus, the AIDS problem has not only electrified the biomedical community to study the mechanisms whereby viruses affect immunity and how such immunodeficiencies can be reversed, but also reminded the biomedical community that many viruses other than retroviruses can affect the immune system. It is anticipated by the editors of this volume that publication of these reviews will provide a foundation of established and newer information which can be utilized, through both experimental and clinical investigation, to determine precisely how viruses affect immunity and cause immunosuppression.

S. Specter
H. Friedman
M. Bendinelli

# Contents

## 3. Papovaviruses

GIUSEPPE BARBANTI-BRODANO

## 4. Adenoviruses

WARREN A. ANDIMAN and MARIE F. ROBERT

7. Epstein–Barr Virus-Induced Immune Deficiency

DAVID T. PURTILO and HELEN L. GRIERSON

8. Immunosuppression by Bovine Herpesvirus 1 and Other Selected Herpesviruses

LORNE A. BABIUK, M. J. P. LAWMAN, and P. GRIEBEL

## 9. Poxviruses

DAVID S. STRAYER

## 10. Reovirus-Induced Immunosuppression

CARLO GARZELLI and TAKASHI ONODERA

## 14. Togavirus-Induced Immunosuppression

UMESH C. CHATURVEDI

## 15. Rhabdoviruses: Effect of Vesicular Stomatitis Virus Infection on the Development and Regulation of Cell-Mediated and Humoral Immune Responses

MAN-SUN SY and ROBERT FINBERG

16. Virus-Induced Immunosuppression: Influenza Virus

NORBERT J. ROBERTS, JR., and FRANK DOMURAT

17. Paramyxoviruses

RAIJA VAINIONPÄÄ and TIMO HYYPIÄ

18. Immunosuppression by Measles Virus

PAOLO CASALI, MINORU NAKAMURA, and
MICHAEL B. McCHESNEY

19. Avian Retroviruses

ROBERT W. STORMS and HENRY R. BOSE, JR.

20. Nonhuman Mammalian Retroviruses

STEVEN SPECTER, MAURO BENDINELLI, and HERMAN FRIEDMAN

21. Implications for Immunotherapy of Viral Infections

MAYRA LOPEZ-CEPERO, STEVEN SPECTER, and JOHN HADDEN

22. Conclusions and Prospects

MAURO BENDINELLI, STEVEN SPECTER, and HERMAN FRIEDMAN

# Viruses and Immunosuppression

## General Comments

STEVEN SPECTER, MAURO BENDINELLI,
and HERMAN FRIEDMAN

## 1. HISTORY

The recognition that viruses are able to compromise immunity dates back to the observation by von Pirquet in 1908 that measles infection resulted in a reduced delayed hypersensitivity response in patients who would normally respond to tubercle bacillus antigens. Thus, von Pirquet was the first to suggest an immunologic explanation for the increased susceptibility to superinfection observed in patients with viral diseases. This was followed a decade later by a report in 1919 that influenza virus could also suppress tuberculin reactivity. The investigation of viruses and their effects on immunity then went unreported for 40 years. Beginning about 1960, oncogenic viruses were given serious consideration as immunosuppressive agents. This was first alluded to by Old and colleagues, and a few years later, Good and co-workers presented the first systematic evaluation of suppression of antibody responses by murine leukemia viruses.[1,2] During the late 1960s and early 1970s, there was a flurry of activity in this field. Numerous reports supported the concept that oncogenic viruses suppress immunity. Both humoral and cellular immunity were shown to be depressed. Concomitant to studies with oncogenic viruses, similar studies with many nononcogenic viruses also resulted in findings of immu-

STEVEN SPECTER and HERMAN FRIEDMAN • Department of Medical Microbiology and Immunology, University of South Florida, College of Medicine, Tampa, Florida 33612-4799.    MAURO BENDINELLI • Department of Biomedicine, University of Pisa, I-56100 Pisa, Italy.

nosuppressive activity.[3,4] Many investigators considered virus-induced immunosuppression important to the establishment of persistent infections that lead to chronic diseases or tumor formation. However, during the mid-1970s, the emphasis in virus biology moved away from this field and the number of studies in this area decreased. These studies have been revived again during the 1980s. Most recent studies have focused on understanding the molecular mechanisms involved in virus-induced immunosuppression. Thus, the "science" of interactions between viruses and immunity is not recent, albeit activity in this field has been generated to a frenzied pace during the past half-decade.

The outbreak of acquired immune deficiency syndrome (AIDS) in the United States and elsewhere is the obvious cause for these research efforts. The need to control AIDS has been a powerful force in generating support for research that will lead to a "cure for AIDS." A major prerequisite to achieving a cure is to gain a full understanding of the mechanism(s) of the virus interactions with host defenses, especially virus-induced immunosuppression. In this regard, a strong interest has been generated for understanding immune suppression induced by all viruses, as any one may serve as a model to help us delineate some of the underlying mechanisms involved in AIDS. Obviously, the discovery of the etiologic agent has been a major breakthrough in the understanding of AIDS. This virus has been designated human immunodeficiency virus (HIV), both to describe it in the proper taxonomic manner and to quiet the dispute about the original names for the virus[5], i.e., human T lymphotropic virus III (HTLV-III), lymphadenopathy-associated virus (LAV), and AIDS-related virus (ARV). Once the virus was described and demonstrated to affect the helper T lymphocyte (Th), a better understanding of pathogenesis was established.[6] Yet we are far from solving the problems of HIV infection and the development of AIDS. In fact, the recent description of a second virus, HIV2, that causes AIDS or a similar disease, presents an additional threat to control of infectious disease. Although HIV can be cytopathic for T lymphocytes *in vitro*, the mechanisms whereby immune alterations are induced by this virus *in vivo* are not fully understood. The long latency and other aspects of the disease have repeatedly led to suggestions of modes of action more complex than direct cytotoxicity.

It has been estimated that upward of one million people in the United States alone have been exposed to HIV, yet "only" about 60,000 cases of full-blown AIDS have been reported to date. While many of those infected will develop AIDS over the next several years, it is likely that a large percentage of these individuals will not, even though judging from pathology observed in retrovirus-infected animals, other forms of disease (e.g., autoimmune, neoplastic) caused by HIV persistence are likely to occur. This is not a unique property of HIV. Most viruses capable of causing severe life-threatening disease usually do so only in a relatively small percentage of infected persons. Most of these infections are self-limiting. For example, before the advent of vaccination, the proportion of poliovirus-infected persons in whom paralytic poliomyelitis developed was on the order of 0.1–1%, and many fewer succumbed to the disease. The (co)factors that determine the difference in the outcome of poliovirus infection have never been clearly understood. Thus, it is not surprising that we still do not know those cofactors that influence HIV infection. It is reasonable, however, to suppose that, if the outcome of infections in general is influenced by factors, this is certainly more likely for long latency diseases such as AIDS.

Attention is frequently focused on the immunosuppressive effects of human retroviruses. However, as discussed in the chapters of this volume, a variable degree of immunodeficiency is associated with most, if not all, viral infections. Interestingly, even those viruses that are capable of causing severe depression of host immune responses, at least by the criteria presently available, seem to do so only transiently. In certain cases, the clinical importance of such effects is well appreciated. In others, the immunosuppressive action does not appear to contribute significantly to the clinical features of the infection. The field warrants more intense consideration, however. The proportion of subjects who, for varied reasons (e.g., old age, cancer, iatrogenic immunosuppression), have a reduced level of immune competence has considerably increased in recent years and will likely grow steadily in the future. It seems likely that viral infections, which do not cause clinically relevant immunodepression in normal subjects, may do so and worsen the immunologic status in such subjects.

## 2. ANTIVIRAL DEFENSES

When viruses interact with an animal host, a variety of specific and nonspecific components are involved in the defense against the virus. Similarly, the virus acts to subvert one or more of these components to prolong its survival. This volume is filled with numerous examples of these interactions. This chapter briefly introduces the major elements of host-immune defenses against viruses.

### 2.1. Antibodies

Generally, during viral infections antibodies are generated by plasma cells, the end stage of the B lymphocyte lineage, to viral antigens (usually proteins). A variety of antibodies specific for different viral antigens are produced and can be measured by a number of immune assays.[7,8] However, only those antibodies effective in the neutralization test provide protective immunity directly. Antibody to viral antigens can also contribute to host defenses by facilitating viral clearance due to virion opsonization and agglutination and by participating in complement- and cell-mediated lysis of virus-infected cells. An inhibition of viral synthesis by antibodies interacting with viral antigens on the cell surface has also been described in certain models of viral infection *in vitro*.

In some cases, the production of non-neutralizing and noncytotoxic antiviral antibody may lead to the formation of immune complexes, which may become involved in damage to host tissues. This is most notable when such complexes are deposited on the glomerular basement membrane, resulting in glomerulonephritis.[9]

The role of antibodies in recovery from a primary infection is often uncertain, although there is no doubt that they provide protection from reinfection with the same or antigenically related viruses. In fact, when one looks at persistent viral infection, either with active virus replication or latency, antibody is often detected in host serum, sometimes in excess amounts. Thus, this antibody does not appear to be sufficient to clear the host of the infecting virus. This has contributed to the emphasis on the function of cell-mediated immune responses as the critical factor in controlling many primary viral infections. It is clear, however, that antibodies are frequently crucial in containing viral spread to critical target tissues during acute infection. They may also be a key element in determining the latent state of many persistent infections.

## 2.2. Cellular Immunity

### 2.2.1. T Lymphocytes

The thymus-derived (T) lymphocytes are effectors of important specific antiviral responses. At least three subclasses of T cells are involved. These are defined as Th, which provide for the expansion of both humoral and cellular responses; cytotoxic T cells (Tc or CTL), which are involved in the destruction of virus-infected cells; and suppressor cells (Ts), which downregulate immune responses and are often implicated as being responsible for virus-induced immunosuppression. Numerous experiments with many of the major pathogenic viruses for humans and animals have demonstrated the need for T cells for recovery from virus infections.[10]

An important function of T lymphocytes is the ability of CTL to destroy virus-infected cells. Not only do they remove effete cells and virus but often they may have a detrimental effect, causing immunopathologic damage, destroying tissues vital to the host. In this context, Ts and other suppressor cells may provide an important contribution, by limiting such damaging responses. All these lymphocyte functions are under genetic restriction controlled within the major histocompatibility (MHC) locus. This implies that viral antigens are seen by T cells only when expressed on the cell surface.[10]

### 2.2.2. Monocytes/Macrophages

The blood monocytes and tissue macrophages are another vital cell in host defenses. They function in several manners in antiviral immunity.[11] These cells are among the first to encounter foreign particles and thus are often the initial cells to interact with an infecting virus. The virus may be digested and destroyed by the macrophages, or it may replicate and be transmitted to other cells and tissues. Under both circumstances, especially the former, virus antigenic components may be presented on the surface of the macrophage (accessory cells), which may be instrumental in triggering T and/or B lymphocytes, leading to an immune response against the virus. In addition, macrophages can be stimulated to produce interleukin-1 (IL-1) and additional soluble factors that can recruit and activate other cells, especially resting T cells.[12] Thus, the macrophages may act as effector cells, antigen-processing/presenting cells, or helper cells. A fourth function of the macrophages is that of a nonspecific suppressor cell, regulating a variety of lymphocyte functions.[13]

### 2.2.3. Natural Killer Cells

Natural killer (NK) cells were first described as antitumor cells that could kill their target cells without prior sensitization, as required for CTL.[14] However, there is a specificity to this killing, since only certain targets are lysed by the NK cell. While this is a much broader specificity than we consider for B and T lymphocytes, it is a controlled response. The basis of selectivity is not known but is most probably related to recognition structures on the surface of the effector and target cells.

Subsequently, it was recognized that this heterogeneous cell population of large granular lymphocytes (LGL) was also capable of lysing virus-infected cells.[15] NK activity is enhanced within the first days of many viral infections, an effect mediated by interferon (IFN) or directly by viral molecules. Thus, NK cells may also be considered a first line of defense against viruses.

### 2.2.4. Antibody-Dependent Cellular Cytotoxicity

Antibody-dependent cellular cytotoxicity (ADCC) is a combination of humoral and cellular immunity. It requires both an effector cell and an antibody of the immunoglobulin G (IgG) class, acting as a ligand between the effector and target cell. This response may be carried out by a few cell types, including macrophages, polymorphonuclear leukocytes, and non-T, non-B lymphocytes. Classically, it is killing due to a cell in the LGL population (K cell) that we speak about in ADCC. The precise relationship between the K and NK cells is uncertain, as is the in vivo function of K cells.[16] However, in vitro studies demonstrate that K cells can destroy virus-infected cells in the presence of virus-specific antibodies. This process is considerably more effective than complement-mediated cytolysis.

## 2.3. Nonspecific Soluble Factors

Antiviral defense mechanisms are also mediated or amplified by soluble factors produced by cells of the immune system or other tissues.

### 2.3.1. Complement

Antibody responses are often aided by the lytic or opsonic components of the complement system. Complement may therefore be involved with cytotoxic antibody in virus particle lysis or the lysis of virus-infected cells as well as virus neutralization.[10] These effects may be due to complement activation by the classic pathway, triggered by the binding of specific antibody to target antigens. Enveloped viruses and cells infected with these viruses may also activate the alternative complement pathway. This may cause lysis of virions or of infected cells, before antiviral antibody has been produced. In addition, complement component C3b may be involved in the destruction of virus particles by polymorphs in the absence of antibody.[10,17] It has also been suggested that binding of complement may be involved in clumping of virions, which may facilitate their clearance by a nonimmune mechanism.[18] Complement activation also leads to inflammation and accumulation of leukocytes at the site of viral replication.

### 2.3.2. Interleukins

Soluble substances that facilitate communication between leukocytes are called interleukins. Several of these substances are described in the literature, but only a few are fully accepted.[12,19,20] IL-1 is produced by macrophages and other cells and activates resting T lymphocytes.[12] This also generates a second interleukin, designated IL-2. IL-2 has been demonstrated to enhance both T-cell growth[19] and T-cell differentiation.[21,22] While the importance of IL-2 has not been fully recognized, its participation in antiviral and antitumor immunity has been clearly established.[23,24] IL-1 is also pyrogenic, and high body temperatures may hamper viral replication. Several additional factors, designated IL-3–IL-6, have been added to the growing list of substances involved in regulating leukocyte growth and differentiation.

### 2.3.3. Interferons

Interferon was first described in 1957 by Isaacs and Lindenmann as an antiviral substance produced by many types of cells in response to viral infection. Its function was described as being directed against viral replication, mainly at the translational level. While this early observation was correct, we have learned that this was only the tip of the iceberg. Three classes of IFNs are described: leukocyte (IFN$_\alpha$), produced in response to infections by viruses and bacteria; fibroblast (IFN$_\beta$), generated by viral infections of other cell types; and immune (IFN$_\gamma$),produced by T lymphocytes in response to antigen or mitogen stimulation. The IFNs not only have direct antiviral effects but immunomodulatory activities as well. They may enhance or depress T lymphocyte, macrophage, and NK cell antiviral functions, depending on timing, dose, and route of administration. Thus, their production by leukocytes, as well as by other cells, is vital to host-immune defenses.[25]

## 3. MECHANISMS OF IMMUNE SUPPRESSION BY VIRUSES

While we have indicated that virtually all viruses are capable of suppressing immune functions, the extent to which they do so and the means by which this is accomplished are quite varied. This will be apparent in the chapters that follow. To provide a general framework, we outline below the various mechanisms that may be involved.

### 3.1. Infection of Lymphoid Cells

The most obvious way that a virus may suppress immune function is by replicating in the cells which carry out that function (Tables I and II; Fig. 1). This mechanism has been suggested for a number of viruses capable of invading B lymphocytes, T lymphocytes, and/or macrophages. Infection of lymphoid or monocytic cells may lead to the direct destruction of such cells. Beyond a certain threshold, the reduced number of immune cells will ultimately lead to decreased immune function. Even if only one subpopulation of cells is selectively infected, as Th cells appear to be in HIV infection, this may cause an imbalance in immunity. When Th cells are destroyed, this might also lead to a relative excess of Ts cells and suppression; when Ts cells are infected, the result may be overactive B cells or T effector cells, and this might exacerbate immunopathologic processes.

In the absence of overt lytic infection, the invasion of lymphoid cells may result in the establishment of a persistent infection, thereby permitting viral replication without acute destruction of the host cell. Actually the lymphoid system seems to be one of the preferential sites for viral persistence. Viruses that have been shown to replicate in lymphocytes and macrophages are numerous. Also, highly cytopathic viruses, such as adenoviruses and enteroviruses appear to be able to persist in these cells by establishing a carrier-culture equilibrium with the cell population. The consequences of these persistent infections may be normally functioning infected cells, cells that cease to function, or cells that function abnormally. One such abnormal function could be the enhanced expression of suppressor cell activity. In addition, the infection of lymphocytes may alter their ability to migrate normally. This may result in a selective increase or decrease of a lymphoid cell population or subpopulation in lymphoid organs or the bloodstream. Selective depletion of specific lymphoid zones has been observed in certain viral infections. The relative change in the microenvironment of the organ affected due to a defect in homing may contribute to immunodepression. Whether similar mechanisms are responsible for the generalized involution of lymphoid organs observed in a number of viral infections of animals and humans, most notably Lassa fever, remains to be established.

### 3.2. Activation of Suppressor Cells

Enhanced suppressor cell activity can be generated during viral infections, most commonly with retroviruses. Infecting viruses, unlike most antigens used experimentally, are self-replicating immunogens. In addition, they not only express their antigens on the surface of many different cell types but also induce the neoexpression of cellular antigens. If neoantigens are MHC antigens, this may create the conditions for antigen presentation by cells that do not usually do so. Thus, viruses are bound to overload the immune system and

**TABLE I**

**Potential Interactions between Viruses and Lymphocytes**[a]

| | Type of interaction | Examples | Immune functions of cells |
|---|---|---|---|
| | Surface interaction without internalization | Avian and mammalian retroviruses + Hu PBL | Normal to slightly impaired |
| | Adsorption with internalization | Inactivated viruses (measles, influenza, reo-3) + Hu PBL | Moderately impaired |
| | Abortive cycle (frequent in resting cells) | HSV, influenza, measles + Hu PBL | Moderately to severely impaired |

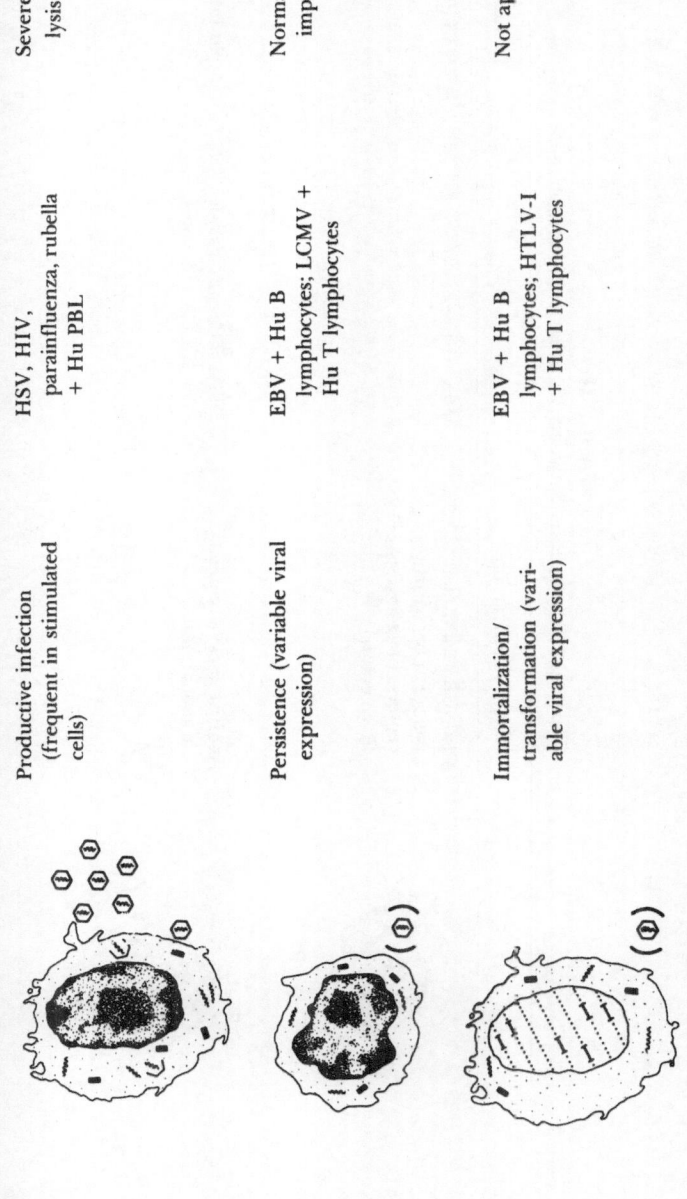

| | Productive infection (frequent in stimulated cells) | HSV, HIV, parainfluenza, rubella + Hu PBL | Severely impaired to cell lysis |
| | Persistence (variable viral expression) | EBV + Hu B lymphocytes; LCMV + Hu T lymphocytes | Normal to moderately impaired |
| | Immortalization/transformation (variable viral expression) | EBV + Hu B lymphocytes; HTLV-I + Hu T lymphocytes | Not applicable |

[a]EBV, Epstein–Barr virus; HIV, human immune deficiency virus; HTLV-I, human T lymphotropic virus-I; Hu, human; PBL, peripheral blood lymphocytes.

## TABLE II
### Potential Interactions between Viruses and Macrophages[a]

| | Type of interaction | Examples | Immune functions of cells |
|---|---|---|---|
| | Adsorption with internalization (via virus receptors, phagocytosis, Fc receptors) | Most viruses | No information |
| | Abortive cycle (frequent in monocytes) | HSV, influenza, measles + Hu PB monocytes | No information |

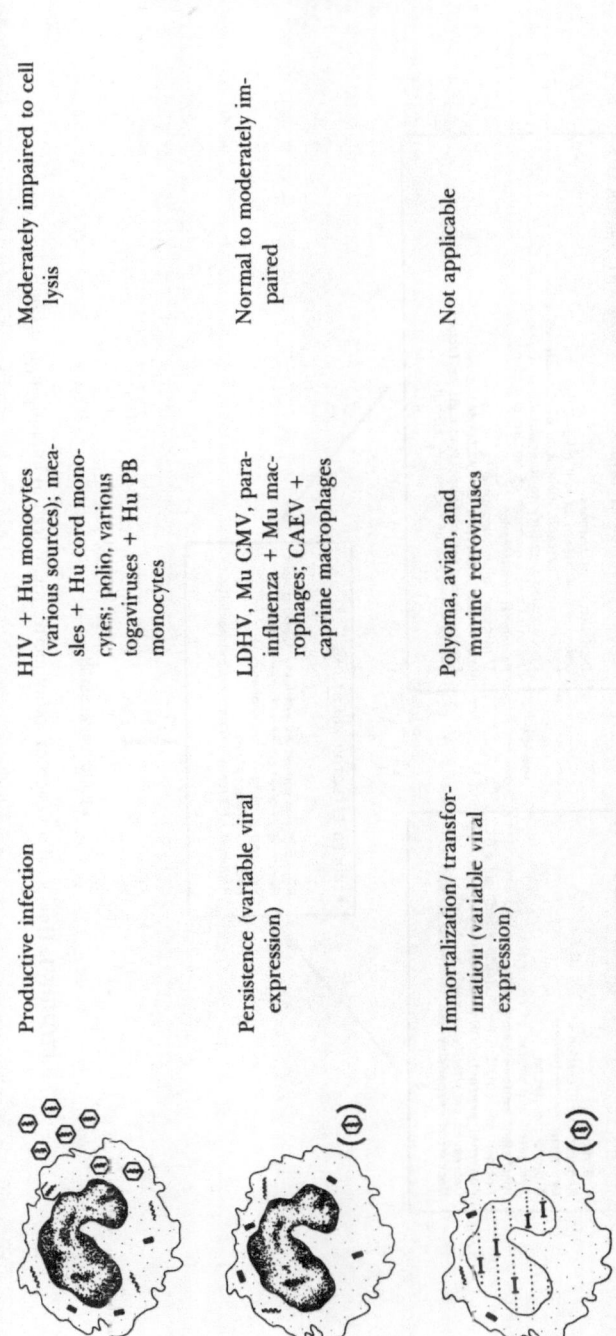

| | | |
|---|---|---|
| Productive infection | HIV + Hu monocytes (various sources); measles + Hu cord monocytes; polio, various togaviruses + Hu PB monocytes | Moderately impaired to cell lysis |
| Persistence (variable viral expression) | LDHV, Mu CMV, parainfluenza + Mu macrophages; CAEV + caprine macrophages | Normal to moderately impaired |
| Immortalization/ transformation (variable viral expression) | Polyoma, avian, and murine retroviruses | Not applicable |

[a]CAEV, caprine arthritis encephalitis virus; CMV, cytomegalovirus; HIV, human immune deficiency virus; HSV, herpes simplex virus; Hu, human; LDHV, lactic dehydrogenase elevating virus; Mu, murine; PB, peripheral blood.

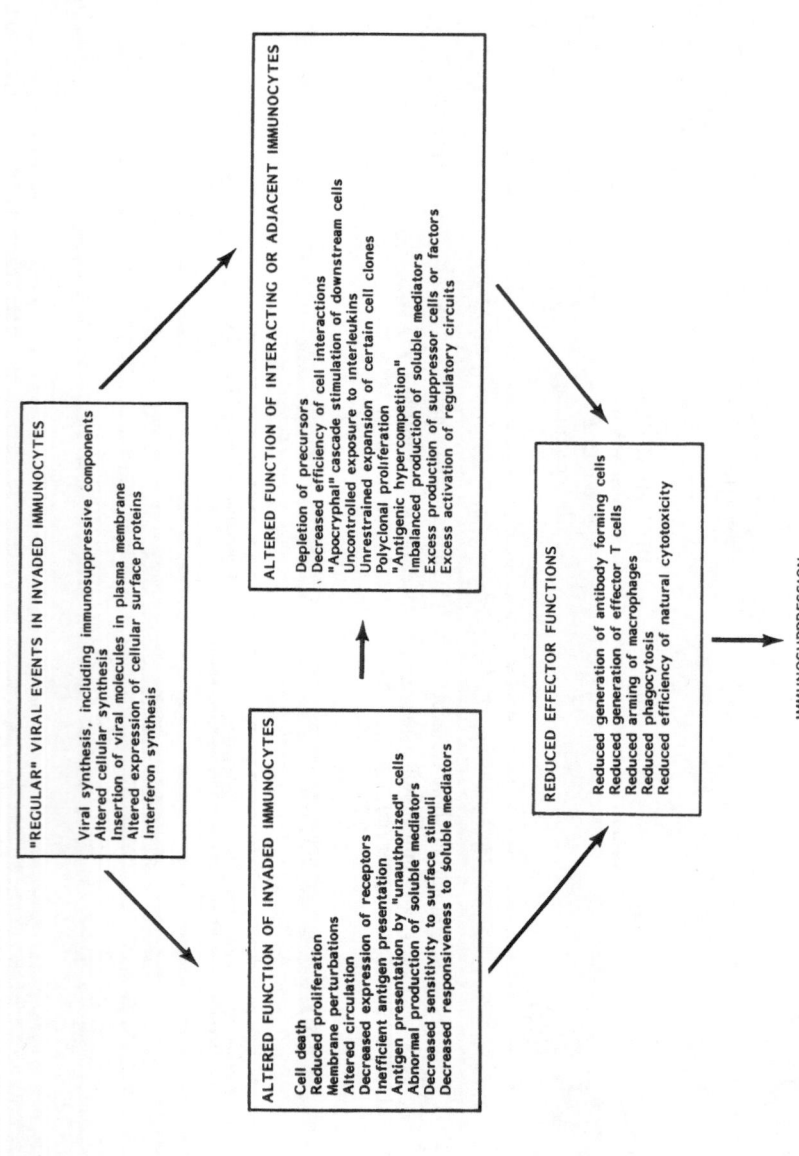

**FIGURE 1.** How viral invasion of immunocytes might lead to immunosuppression.

to activate it through improper pathways. It is therefore likely that excess stimulation of immunoregulatory circuits contributes to the generation of suppressor cell activation and/or recruitment. Most suppressor cells described in viral infections are either Ts or macrophages, although B cells and NK cells are known to exhibit suppressor activity. The suppressor cells may exert their activity through either direct cellular interactions or suppressor factors.

## 3.3. Suppressor Factors

### 3.3.1. Host Derived

Suppressor factors may be elaborated by lymphocytes, macrophages, or infected cells other than those of the immune system. A wide spectrum of factors has been described from lymphoid cells and macrophages displaying a broad range of molecular weights and activities. These factors may act specifically to inhibit only antiviral responses or they may have nonspecific activity. In some infections, several factors may be detected with both specific and nonspecific components present. Suppressor factors are also frequently produced by virus-transformed tumor cells. Frequently when suppression is due to a soluble factor produced by macrophages, this can be attributed to prostaglandin $E_2$ ($PGE_2$). This suggests that infection may lead to an overproduction of normal host factors, resulting in immunodepression. Whether this is the case for many of the factors thought to participate in virus-induced immune suppression remains to be determined. Another example of a natural factor that can inhibit immune function during infection is IFN. While IFN has already been mentioned as immunostimulatory and directly antiviral, under the appropriate circumstances it also can be immunosuppressive.

### 3.3.2. Virion Components

The virion per se may be immunosuppressive due to components that are toxic to lymphocytes and have the ability to turn off cell functions or to induce suppressor cells. By using inactivated virus or purified viral components, it is sometimes possible to suppress immunity in the absence of viral replication. Under such circumstances, any attempt at reversing immune suppression may prove fruitless, unless one is also successful in removing viral proteins as well as in neutralizing infectious virus.

## 3.4. Other Mechanisms Effecting Virus-Induced Immune Suppression

The damage of nonlymphoid tissue by viruses can also lead to suppression of the immune response by indirect means. For example, infection of mouse viruses that damage pancreatic tissue results in an influx of excess enzymes into the circulation, which might ultimately cause immune suppression. Likewise, viral attack on the adrenal cortex can cause corticosteroid imbalances that result in immune suppression. While it is difficult to assess the effect of virus-induced stress on the establishment of immunosuppression, it has been shown to be important. In a few instances, adrenalectomy prior to viral infection has prevented or alleviated the development of immune abnormalities, but in others has not. The extent to which other physiopathologic changes may alter the host's immunity in viral infections is still being investigated.

## 4. FACTORS THAT MIGHT INFLUENCE VIRUS-INDUCED IMMUNE SUPPRESSION

A variety of host factors may influence the development of viral infections, directly or more often through an action on the immune system. It seems reasonable that such factors may also influence the occurrence and degree of immune suppression. Most notable among these are the host's age, genetics, presence of other infections or diseases, and the environment. Furthermore, hosts whose immune system is already functionally defective for physiologic or pathologic reasons could be affected by virus-induced immunosuppression more severely than hosts with a fully functioning immune system.

### 4.1. Age

It has been clearly established that very young animals are often more susceptible to infection than are adults. This has been linked to the immune competence of the host, especially T-cell function. Likewise, elderly persons are more susceptible to serious viral infections that can also be linked to their immune status. That young subjects are more susceptible to the immunodepressive properties of most viruses has clearly been shown in several systems, including retroviruses. However, if the pathologic processes induced by the virus are immunologically mediated, their reduced immune competence may be advantageous. This is exemplified by infection with lymphocytic choriomeningitis virus (LCMV) in the mouse. The lethal effects of the virus are dependent on functioning T lymphocytes. In the newborn mouse, in which these cells are not yet fully developed, the virus replicates to high titer, but no acute disease is observed. These mice can be shown to have developed a T-cell tolerance to LCMV. This virus also induces manifestations of generalized immunodepression.

### 4.2. Genetics

Resistance to viral infection has repeatedly been demonstrated to be controlled at the genetic level. These genes may be associated with the MHC or may be located elsewhere in the genome. In some cases, this genetic control may require the concomitant action of multiple genes, with loci within the MHC and outside involved. This genetic effect has frequently been demonstrated to involve control of macrophage or NK cell functions. Regardless of the cells affected, one can demonstrate that virus-resistant mice often exhibit little or no suppression of immune reactivity during the aborted infection.

### 4.3. Presence of Other Infections and Diseases

Immunocompromised individuals have no greater likelihood of being infected by a virus than do normal individuals but, once infected, they have a higher morbidity and mortality. This may be the key factor as to why only certain HIV-infected people develop AIDS. More than 95% of AIDS victims are either homosexual, drug addicts, hemophiliacs, transfusion recipients, or a combination of these. None of these persons is truly in a normal state of health. For example, each has a much higher than average incidence of other viral infections, most notably the hepatitis and herpes group viruses. These agents are capable of severely altering the functioning of the immune system, which may leave these persons susceptible to the most severe consequences of HIV infection.

Alternatively, these viral infections may stimulate immunocompetent cells so that they become better host cells for HIV. It has been reported that HIV replicates well in activated Th cells, but not in resting cells. Thus, any immunologic stimulus may increase susceptibility to infection. While these two theories seem contradictory, they are actually complementary. In most viral infections in which there is transient suppression, this is usually followed by an active, often heightened immune response. Thus, the immunosuppressed state of the host attributable to other causes may permit replication of the virus in a small percentage of cells when the host is incapable of combating a small number of infecting particles. The enhanced proliferation phase may then provide an abundance of host cells for the virus before the development of protective immunity. By this time, the virus could provide its own suppressive activity to prevent maturation of a response to the virus.

### 4.4. Environment

Environment may be defined as the environment in which the host lives or the microenvironment for the virus within the host. In the former, we refer to environmental factors that may affect infectivity of the virus, immune status of the host or other undefined phenomena. It is still unclear why Epstein–Barr virus (EBV) infection causes a self-limiting disease (infectious mononucleosis) in the Western world but progressive neoplastic diseases in Equatorial Africa (Burkitt lymphoma) and China (nasopharyngeal carcinoma). However, there seem to be contributions by environmental factors, age, and perhaps genetics (see Chapter 7). In addition, life-style factors have been implicated as determinants of progression in HIV-infected hosts.

The microenvironment affecting the virus may depend on the route of infection. In experimental models, in which the route can be readily controlled, both the ability to modulate immune responses and the outcome of infection can be shown to depend on this factor.

## 5. VIRUS-SPECIFIC VERSUS GENERALIZED IMMUNOSUPPRESSION

Regardless of the mechanism by which suppression is induced, the nature and extent of the decrease in immunity may vary, depending on the inducing virus. This may include viruses that suppress only one aspect of immunity, most commonly cell-mediated immunity, or those that selectively suppress several but not all responses. Occasionally, a virus may cause a generalized immunosuppression in which virtually all immune function decreases. This can be observed during the later stages of many retrovirus infections.

Interestingly, one may see a decrease in responses to mitogens or nonviral antigens, while antiviral responses seem to be quite normal. The reasons for this are not clear but are possibly related to a selective effect on a particular lymphoid cell subpopulation. The inverse situation, in which antiviral responses are depressed but nonviral responses seem to be intact, is also reported. In this case, cells previously committed immunologically may be unaffected, whereas antiviral effector cells or their precursors might be suppressed. In this regard, Nash[26] recently reviewed the concepts of virus-induced immunologic tolerance and suppression as mechanisms by which viruses can alter immune reactivity. The induction of tolerance generally affects only immune responses to viral antigens; it may be T-cell specific (class II MHC antigen specific), or both B and T cell responses may be suppressed. By contrast, in infections in which suppressor cells are induced suppression also affects heterologous antigens and/or mitogen responses. In any case, the relationships of viral specific to generalized immunosuppression are not clear. For example, it seems possible that the latter is essential for the former to become established.

## 6. HOW IMMUNOSUPPRESSION CAN AFFECT PATHOGENESIS

The complexity of the interactions between viruses and the immune system is only beginning to be understood. However, there are some areas in which we have developed a reasonable understanding of how these relationships help tip the balance to favor either host or virus. Clearly, this is most apparent in infections in which the virus can cause the destruction of large numbers of immunocompetent cells. Ultimately, a generalized immunodeficiency develops that enhances the probability that the host will develop opportunistic infections of neoplastic disease. In the compromised state of the host, these secondary infections are usually fatal. This seems to be the story of HIV, a vicious catch-22. Immunosuppressive superinfections, viral or otherwise, may be cofactors that enhance susceptibility to AIDS in HIV-infected individuals; HIV in turn further suppresses the system, allowing even more severe consequences attributable to opportunistic infections, which would normally not be pathogenic.

In certain circumstances, the ability of the virus to compromise the immune system appears to have little effect on the pathogenesis of the virus. This is most evident with some live attenuated virus vaccines that transiently suppress immunity with no apparent consequence in immunocompetent individuals. In the immunocompromised host, however, these vaccines can have devastating effects. The failure of immune suppression to generate severe pathology is somewhat puzzling but is most likely related to the fact that the immune functions affected do not include important effectors of antimicrobial defenses, such as CTL. Alternatively, it may be that the suppressive effects induced by the virus are too short-lived to cause problems for the host.

Finally, there may be the positive aspect of immune suppression, from the point of view of the host. Virus-specific immune responses can have pathologic consequences. By suppressing these responses, such consequences can be avoided. One must be wary that the loss of these immune reactions that prevent acute pathogenesis may contribute to the establishment of viral persistence; this may have severe consequences for the host in the form of chronic diseases.

Thus, the balance of immune responsiveness and virus-induced immune suppression is a delicate state. We must be guided by this realization when trying to develop effective therapies to counteract viral infections by manipulation of the immune responsiveness (see Chapter 21). Strong immune reactivity, if not properly balanced, is not necessarily a benefit.

This volume is dedicated to the studies of all the major pathogenic virus groups and their ability to downregulate the immune response in their natural host and/or experimental models; it also examines the consequences of the immunosuppressive activity. These studies provide examples of virtually all the possible modes by which the immune response can be suppressed.

## REFERENCES

1. Old, L., B. Benacerraf, D. A. Clark, and H. Goldsmith, The reticuloendothelial system and the neoplastic process, *Ann. NY Acad. Sci.* **88:**264–280 (1960).
2. Peterson, R. D. A., R. Hendrickson, and R. A. Good, Reduced antibody forming capacity during the incubation period of passage A leukemia in C3H mice, *Proc. Soc. Exp. Biol. Med.* **114:**517–520 (1960).
3. Specter, S., and H. Friedman, Viruses and the immune response, *Pharmacol. Ther. A* **2:**595–622 (1978).
4. Szentivanyi, A. and H. Friedman (eds.), *Viruses, Immunity and Immunodeficiency*, Plenum, New York (1986).
5. Editor, What to call the AIDS virus, *Nature (Lond.)* **321:**10 (1986).
6. Popovic, M., M. G. Sarngadharan, E. Read, and R. C. Gallo, Detection, isolation and continuous production of cytopathic retroviruses (HTLV-III) from patients with AIDS and pre AIDS, *Science* **224:**497–500 (1984).
7. Rose, N. R., H. Friedman, and J. L. Fahey (eds.), *Manual of Clinical Laboratory Immunology*, 3rd ed., American Society of Microbiology, Washington, D. C. (1986).
8. Specter, S., and G. J. Lancz (eds.), *Clinical Virology Manual*, Elsevier, New York (1986).
9. Escobar, M. R., and P. D. Swenson, Mechanisms of viral immunopathology, in: *The Reticuloendothelial System: A Comprehensive Treatise*, Vol. 4: *Immunopathology* (N. R. Rose and B. V. Siegel, eds.), pp. 201–254, Plenum, New York (1983).
10. Sissons, J. G. P., and M. B. A. Oldstone, Host response to viral infections, in: *Fundamental Virology* (B. N. Fields and D. M. Knipe, eds.), pp. 265–279, Raven, New York (1986).

11. Mogensen, S. C., Macrophages and genetically determined natural resistance to virus infections, in: *Viruses, Immunity and Immunodeficiency* (A. Szentivanyi and H. Friedman, eds.), pp. 13–24, Plenum, New York (1986).

12. Dinarello, C. A., Interleukin-1, *Rev. Infect. Dis.* **6**:52–57 (1984).

13. Varesio, L., Suppressor cells and cancer: Inhibition of immune functions of macrophages, in: *The Reticuloendothelial System: A Comprehensive Treatise*, Vol. 5: *Cancer* (R. B. Herberman and H. Friedman, eds.), pp. 217–252, Plenum, New York (1983).

14. Herberman, R. B., M. E. Nunn, H. T. Holden, and D. H. Lavrin, Natural cytotoxic reactivity of mouse lymphoid cells against syngeneic and allogeneic tumors. II. Characterization of effector cells, *Int. J. Cancer* **16**:230–239 (1975).

15. Lopez, C., Natural resistance mechanisms in herpes virus infections, in: *Viruses, Immunity and Immunodeficiency* (A. Szentivanyi and H. Friedman, eds.), pp. 3–11, Plenum, New York (1986).

16. Herberman, R. B., Natural killer cell activity and antibody-dependent cell-mediated cytotoxicity, in: *Manual of Clinical Laboratory Immunology*, 3rd ed. (N. R. Rose, H. Friedman, and J. L. Fahey, eds.), pp. 308–314, American Society of Microbiology, Washington, D. C. (1986).

17. Hirsch, R. L., The complement system: Its importance in the host response to viral infection, *Microbiol. Rev.* **46**:71–85 (1982).

18. Dulbecco, R., and H. S. Ginsberg, *Virology*, Harper & Row, Hagerstown, Maryland (1985).

19. Watson, J. D., R. L. Prestridge, D. Y. Mochizuki, and S. Gillis, Interleukin 2, in: *Recognition and Regulation in Cell-Mediated Immunity* (J. D. Watson and J. Marbrook, eds.), pp. 265–290, Dekker, New York (1985).

20. Ihle, J. N. and Y. Weinstein, Interleukin 3: Regulation of a lineage of lymphocytes characterized by the expression of 20 αSDH, in: *Recognition and Regulation in Cell-Mediated Immunity* (J. D. Watson and J. Marbrook, eds.), pp. 291–324, Dekker, New York (1985).

21. Ceredig, R., J. W. Lowenthal, M. Nahholz, and H. R. MacDonald, Expression of interleukin 2 receptors as a differentiation marker on intrathymic stem cells, *Nature (Lond.)* **314**:101–103 (1985).

22. Hadden, J. W., S. Specter, and E. Hadden, Effects of T cell growth factor/interleukin 2 on prothymocytes, *Lymphokine Res.* **5**:549–554 (1986).

23. Matis, L. A., S. Shu, E. S. Groves, S. Zinn, T. Chou, A. M. Kruisbeek, M. Rosenstein, and S. A. Rosenberg, Adoptive immunotherapy of a syngeneic murine leukemia with a tumor-specific cytotoxic T cell clone and recombinant human interleukin 2: Correlation with clonal IL-2 receptor expression, *J. Immunol.* **136**:3496–3501 (1986).

24. Conlon, P. J., T. L. Washkewicz, D. Y. Mochizuki, K. L. Urdal, S. Gillis, and C. S. Henney, The treatment of induced immunodeficiency with interleukin 2, *Immunol. Lett.* **10**:307–314 (1985).

25. Sehgal, P. B., L. M. Pfeffer, and I. Tamm, Interferon and its inducers, in: *Chemotherapy of Viral Infections* (P. E. Came and L. A. Caliguiri, eds.), pp. 205–312, Springer-Verlag, Berlin (1982).

26. Nash, A. A., Tolerance and suppression in virus diseases, *Br. Med. Bull.* **41**:41–45 (1985).

# 2

# Immunomodulation by Hepatitis B and Related Viruses

## MARIO R. ESCOBAR

## 1. INTRODUCTION

Immunodeficiency disorders may occur as a result of one or more defects within the immune system spanning stem cell deficiencies, through immunoregulatory dysfunction, to a restricted failure to recognize or mount an immune response against certain antigens. The identification of these defects is complicated by their secondary effects, or by opportunistic infections in the immunocompromised host. Although the extent to which viral infections can initiate these deficits remains to be evaluated in most cases, the active role of many viruses in inducing immunosuppression is well established.

The conventional view held for many years, that hepatitis B virus (HBV) was strictly hepatotrophic has recently been re-evaluated. A number of reports have been published documenting the presence of hepadnaviral genomes in extrahepatic sites, particularly T and B lymphocytes from humans, chimpanzees, Peking ducks, and woodchucks.[1-7] The infection of cells of the immune system by HBV is of special relevance to the subject of viral immunomodulation because this is how HBV may exert an effect on their functions.

The recent finding that the basic replication cycle of hepadnaviruses involves the synthesis of viral DNA by reverse transcription of an RNA template, a step previously thought to be unique to retroviruses, is also crucial.[8] Accordingly, viral replication appears to be indicated only when RNA : DNA hybrid molecules are present in the productively infected cell. In addition to duck

MARIO R. ESCOBAR • Department of Pathology, Medical College of Virginia, Richmond, Virginia 23298-0001.

HBV (DHBV), hepadnavirus infection of peripheral blood lymphocytes of chimpanzees and woodchucks *in vivo* has also been demonstrated.[9]

Despite the remarkable paucity of data available in the literature dealing with the immunosuppressive role of HBV, this review focuses on related immunologic events and potential mechanisms of HBV immunosuppression. Studies of the effects of HBV on the immune system mostly have been limited to investigations of the immunopathogenesis of chronic viral hepatitis. Nevertheless, there is already adequate justification for delving further into this exciting area.

## 2. BASIC PROPERTIES OF HUMAN HEPATITIS B AND RELATED VIRUSES

Human HBV has the distinction of being the prototype member of a relatively new and expanding family of animal viruses known as the *Hepadnaviridae*.[10,11] The other members of this family are viruses closely related to HBV, which infect woodchucks (WHV),[12] beechy ground squirrels (GSHV),[13] and Peking ducks (DHBV).[14] These viruses have several basic properties in common, including morphology, antigenic composition (except for DHBV), DNA size and structure, DNA polymerase, a tropism for the liver and an association with persistent infection and, for WHV and HBV, primary hepatocellular carcinoma. With the use of monoclonal antibodies, hepatitis B surface antigen (HBsAg) and the surface antigens of WHV and GSHV have been demonstrated to be antigenically related via a common determinant.[15]

The Dane particle, which is the infectious noncytopathic form of HBV, is a 42-nm spherical virus consisting of a 7-nm outer shell and 27-nm inner core that encloses the viral DNA and DNA polymerase.[16] The outer surface component contains HBsAg and the inner core component contains hepatitis B core antigen (HBcAg). In addition to Dane particles, numerous HBsAg 22-nm spherical particles and tubular structures of the same diameter and variable length circulate in the blood of HBV-infected individuals accumulating to high levels ($10^{13}$/m).[17] Four major subtypes have been reported (*adw, adr, ayw,* and *ayr*) based on the presence of the common group-reactive determinant *a* and the mutually exclusive antigenic subspecificities *d* or *y*[18] and *w* or *r*.[19]

The HBcAg consists of a major polypeptide, p25, and its glycosylated form, gp28. Additional polypeptides of high molecular weight (p39/gp42 and p33/gp36) have been associated preferentially with the intact virion and the filamentous form of 22-nm particles.[20,21] The p25 polypeptide is encoded by the S gene beginning from the third possible translational initiation site of a larger open reading frame (ORF) and is preceded in phase by 174 codons (adw subtype) designated the pre-S region.[22] The large ORF for HBsAg terminates in a single stop codon but can initiate at three possible translational start codons, which define the pre-S(1), pre-S(2), and S regions, yielding p39, p33, and p25, respectively. All three polypeptides share the 226 amino acid residues of the S region (p25): p33 consists of the p25 sequence plus an amino-terminal 55 residues [pre-S(2)],[23,24] and p39 consists of the p33 sequence plus an amino-terminal 119 residues [pre-S(1)].[20] Consequently, three different envelope polypeptides are expressed by the variable use of initiation codons in one ORF, and HBsAg-containing virions or particles may vary in composition relative to these polypeptides. It should be noted, therefore, that HBsAg/p39 particles contain all three polypeptides.

HBcAg, in contrast to HBsAg, is not readily detectable in the serum, but removal of the outer HBsAg component of the Dane particle with nonionic detergents results in the release of serologically reactive HBcAg.[16] Solubilization of either the inner HBcAg component of Dane particles or core components from infected liver with sodium dodecyl sulfate (SDS) and 2-mercaptoethanol results in the appearance of hepatitis e antigen (HBeAg) and the disappearance of HBcAg reactivity, suggesting that HBeAg is a structural component of the inner core of the Dane particle.[25,26] HBeAg can also be detected as a soluble protein in the serum of many HBV-infected individuals. Three subspecificities of HBeAg, termed $e_1$, $e_2$, and $e_3$, have been identified using immunodiffusion,[27,28] and two HBeAg determinants have been demonstrated more recently using monoclonal antibodies.[29] During HBV infection, HBsAg, HBcAg, and HBeAg can induce an antibody response in the host with the development of anti-HBs, anti-HBc, and anti-HBe, respectively. The inner core component of the Dane particle consists of an endogenous DNA polymerase and a unique viral genome, ~3000 nucleotides in length, is circular and partly single stranded, containing an incomplete plus (+) strand. The incomplete plus strand is complexed with the DNA polymerase in the virion. Under appropriate conditions, the polymerase can elongate the plus strand, using the complete minus (−) strand as the template. These morphologic and structural features distinguish HBV from all known classes of DNA-containing viruses. The replication cycle of HBV is strikingly different from other DNA-containing viruses and suggests a possible relationship with the RNA-containing retroviruses.[30]

## 3. IMMUNOPATHOLOGIC SEQUELAE OF HEPATITIS B VIRUS INFECTION

The natural course and immunopathologic sequelae of HBV infection are illustrated diagrammatically in Fig. 1. They may be influenced by pathogenetic determinants, which include both viral and host factors (e.g., HBV subtype, dose, mode of transmission, and host age, sex, physiologic state). The persistence of viral synthesis in patients without liver injury (i.e., silent carriers of HBsAg) suggests that HBV itself is not directly cytopathic for hepatocytes. Nonetheless, viral persistence is frequently incriminated in chronic hepatitis indicating that the virus is at least necessary, if not a causative factor, in the disease process.[31]

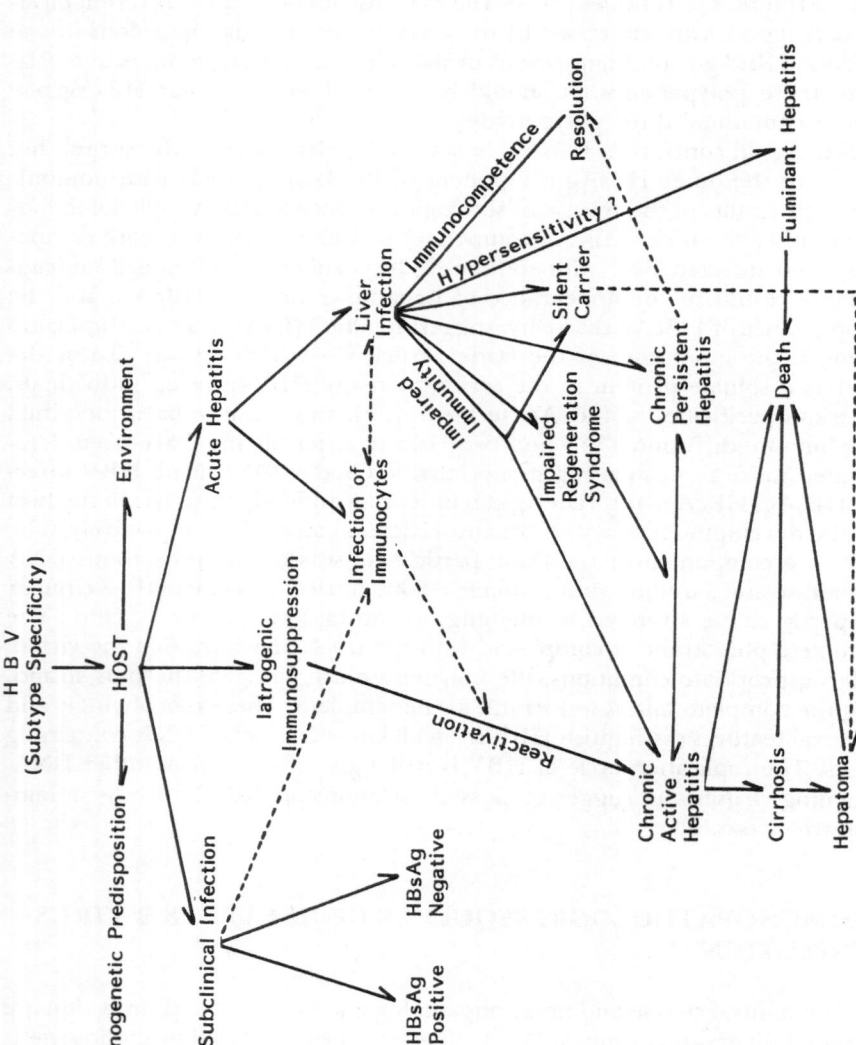

**FIGURE 1.** Immunopathologic sequelae of human hepatitis B virus infection.

Blumberg[32] postulated that HBV has the characteristics of a genetic polymorphic trait on its surface (similar to those of inherited erythrocyte antigens and enzymes, serum proteins, and others) and that putative hosts are also polymorphic for related antigens that may or may not match the antigens present on the virus. Thus, the fate of the virus in the host as well as the clinical sequelae of infection and the host's immunologic responses may be determined by the nature of the match between the virus and the putative host. On the other hand, genetic factors may potentially control the immune response to HBV which determines who recovers from HBV infection and who is left with either persistent infection or chronic liver disease, or both. Although no HLA markers have been associated conclusively with chronicity of HBV infection, nonresponders to hepatitis B vaccine have a high frequency of HLA-D7 and lack HLA-D1.[33] Similarly, H-2 loci appear to control T-dependent humoral immune responsiveness to HBsAg in inbred strains of mice.[34-36] Paradoxically, however, nonresponse of humans to the HBsAg in hepatitis B vaccine does not correlate with an altered clinical response to HBV. Among nonresponders to hepatitis B vaccine, the likelihood of remaining chronically infected after natural HBV infection remains unchanged.[37] Although there is evidence that the level of HBV replication correlates inversely with the degree of HLA protein expression on hepatocytes,[38] to date no role for genetic influences has been documented in chronic active hepatitis B.[39]

Since males have a greater likelihood of remaining chronically infected than do females, the influence of sex-linked factors on immune responsiveness to HBV is possible. It has been observed that renal transplant recipients who are anti-HBs positive are more likely than anti-HBs-negative recipients to reject a male than a female donor kidney. This also suggests that the immune response to HBV antigens is affected differently by gene products of males versus females. Nonimmune clearance of HBsAg was found to be diminished in male mice compared with female mice, and male mice with diminished clearance of HBsAg also mounted lower anti-HBs responses.[40] However, a clear effect of gender on measurable indicators of immunologic responsiveness to HBV antigens has not been demonstrated.[39]

There are numerous reports of clinical observations, in vitro investigations, and hypotheses to suggest that cell-mediated immunity (CMI) plays a causative role in the production of liver damage in hepatitis B. This follows directly from the belief that HBV is not a cytopathic agent and that the mononuclear cells adjacent to dying hepatocytes in hepatic infiltrates are responsible for the death of those cells. Furthermore, there are many clinical studies that indicate that CMI mechanisms are responsible for the liver injury observed in patients with chronic active hepatis.[41] Nevertheless, despite this abundance of data to support the major role of CMI in liver damage and the apparent liver specificity of CMI responses demonstrated by Smith and co-workers in vitro, these responses may be the results of the liver injury and not to have triggered its pathogenesis.[42,43] Therefore, even the best in vitro assay of CMI cannot discriminate between primary pathologic events and epiphenomena.[41]

## 4. INTERACTION OF HEPATITIS B VIRUS WITH THE IMMUNE SYSTEM

### 4.1. Antibody Production

Most persons infected with HBV develop acute hepatitis followed by clearance of the virus and resolution of their disease, as indicated by disappearance of HBsAg from their blood and, ultimately, the development of anti-HBc and anti-HBs. However, a small percentage of adults infected with HBV are unable to eliminate the virus. Their serum remains positive for HBsAg and frequently HBeAg, thus becoming chronic HBV carriers. These carriers do not develop detectable anti-HBs, although they do produce high titers of anti-HBc and anti-HBe when HBeAg is lost. The persistence of HBsAg in the absence of anti-HBs is strong evidence that these individuals have a specific immunologic nonresponsiveness to this antigen, therefore, explaining their inability to eradicate HBV. The chronic carrier state is particularly common among individuals with conditions or disease associated with significant immunologic hyporesponsiveness, such as newborns, patients who are iatrogenically immunosuppressed, or those who are undergoing renal dialysis and patients with Down syndrome, Hodgkin disease, acute leukemia, and lepromatous leprosy.[44–46] Since the vast majority of chronic HBsAg carriers are otherwise normal with no generalized perceptible immunologic deficiency,[47] one would have to expect that the carrier state is due to a specific immune defect, such as a deficiency in recognition, processing, or response to one of the HBV antigens.[48] A very elegant transgenic mouse model of the chronic HBsAg carrier state has been described by Chisari and his group,[49] which along with numerous human *in vitro* experiments,[48,50–53] have attempted to address this important issue.

Investigation of the murine humoral and cellular immune response to the S region of HBsAg has revealed the influence of at least two H-2-linked immune response (Ir) genes and identified HBsAg/p25 high-responder (H-2$^d$ and H-2$^q$) and nonresponder (H-2$^f$ and H-2$^s$) phenotypes.[34] Later, studies from the same laboratory showed that (1) the pre-S(2) region was significantly more immunogenic than the S region of HBsAg; (2) the immune responses to the pre-S(2) region are regulated by H-2 linked genes distinct from those that regulate the response to the S region; and (3) immunization of an S-region nonresponder strain with HBV envelope particles that contain both the pre-S(2) and the S regions can circumvent nonresponsiveness to the S region.[54] More recently, this analysis has been extended to the pre-S(1) region indicating that (1) it is immunogenic at the T and B cell levels; (2) specific antibody production is regulated by H-2-linked genes and can be independent of anti-S and anti-pre-S(2) antibody production; (3) immunization of H-2$^f$ strains with HBsAg/p39 particles containing the pre-S(1) region can bypass nonresponsiveness to the S and pre-S(2) regions in terms of antibody production; (4) two synthetic peptides, p32-53 and p94-117, define murine and human antibody binding sites on the pre-S(1) region, and pl-21 and p12-32 define additional human antibody binding sites; (5) pre-S(1)-specific T cells can be elicited in S and pre S(2) region nonresponder mice (H-2$^f$) and provide functional T cell help for S-pre-S(2)-, and pre-S(1)-specific antibody production; and (6) a T-cell recognition site in the pre-S(1) region, p12-32, was identified. These findings were considered of importance to HBV vaccine development, and possibly to viral clearance mechanisms, since the higher-molecular-weight polypeptides are preferentially expressed on intact virions.[55]

It is quite clear that the immune response to HBsAg in patients who recover from HBV infection and in vaccinees has a major function in the immunity against HBV infection.[21,56]−[58] There is evidence from *in vitro* and animal experiments that, as with most soluble and particulate antigens, synthesis of anti-HBs is regulated by T cells[59]−[61] and that T cells appear to be involved both in the immunity and immunopathology of hepatitis B.[39,58]

The mechanisms underlying the inability of chronic HBs carriers to produce anti-HBs antibodies have yet to be identified with certainty. Actually, approximately one third of these carriers do have low titers of serum anti-HBs. Studies have shown, however, that this anti-HBs is heterotypic; i.e., it is directed against subdeterminants of HBsAg that are not present in that serum.[62] It has been postulated by Kerlin and co-workers that the persistence of HBsAg in these individuals may be related to a defect in T-cell function promoting continuous *in vivo* production of injurious autoantibodies, or deficient suppressor T-cell (Ts) function with insufficient modulation of the antibody response.[52] These investigators found that T-cell regulation of pokeweed mitogen (PWM)-stimulated total IgG and IgM synthesis *in vitro* failed to distinguish between chronic HBV-infected patients and controls with circulating anti-HBs. Alternatively, a lack of helper T-cell (Th) function or excessive Ts activity in patients with chronic HBV infection could explain the inability to mount a primary antibody response to the virus. In this regard, studies of greater relevance to our understanding of immunoregulation of anti-HBs synthesis during HBV infection were carried out by Dusheiko et al.[48] These workers demonstrated that 33% of carriers in their study appeared to have a defect in Th augmentation of anti-HBs antibody production, and 41% of carriers had enhanced Ts activity that decreased such production. The presence of these abnormalities did not correlate with the clinical status or degree of chronic hepatitis in their patients. Although these experiments, in contrast to those by Kerlin et al.[52] measured anti-HBs antibody specifically, Th activation was assessed by PWM stimulation rather than by HBsAg. It would have been more valuable to have used specific antigen, since it has been reported that PWM activates B cells in a nonspecific manner.[63] These studies may have been potentially flawed by allogeneic differences and the independent effect of suppressor monocytes in the co-culture experiments. They were able, however, to demonstrate an absence of circulating B cells capable of producing anti-HBs antibody despite the elimination of Ts activity and under conditions of maximal Th function.

The number of anti-HBs antibody-producing lymphocytes in the circulation or their state of activity has not been measured. This information would help determine whether these cells are still present in the body, but at different sites (e.g., liver and spleen). The possibility remains also that these cells may have been tolerized or that their *in vitro* proliferation and ability to synthesize anti-HBs (up to 500 μg/ml) also may quickly absorb available homotypic anti-HBs present in chronic HBsAg carriers. This possibility is consistent with the presence of immune complexes of HBsAg/anti-HBs in the serum and tissues of these individuals. However, this explanation for the lack of detectable serum anti-HBs antibody in chronic HBV carriers has been questioned[48] because of the technical limitation of the tests employed in these studies.

A more recent and particularly attractive hypothesis to explain the failure of chronic HBsAg carriers to develop anti-HBs antibody is based on the earlier finding that HBV as well as HBsAg can interact with host serum albumin (HSA). The recruitment of HSA is achieved by the presence of receptor structures for polymerized HSA (pHSA) on the outer coat of HBV, coded for by the pre-S region of the HBV genome.[64,65] The extraordinary capacity of HBV to bind this "self" component and its relation to tolerance induction is an important immunological aspect, which needs to be further investigated. Recent studies[66] have shown that antibodies against the receptor for pHSA, determined by radioimmunoassay (RIA), were present in 1% (4 of 358 sera) of asymptomatic carriers of HBsAg containing antibodies to HBeAg; in none of 67 sera containing HBeAg; in 74% (111 of 150 sera) of blood donors, who presumably had acquired anti-HBs following natural infection; and in none of 77 successful vaccinees against hepatitis B. These observations may be relevant to our understanding of the immune mechanisms of HBV infection. All these studies are based on the observation that the pre-S(2) sequences of HBV and HBsAg react with glutaraldehyde-polymerized HSA (GA-HSA). However, naturally occurring pHSA, unlike GA-HSA, does not bind significantly to HBsAg,[67] casting doubt on the validity of the postulate. More recently, Neurath *et al.*[68] located within the HBV envelope proteins a sequence mediating the attachment of HBV to human hepatoma HepG2 cells. A synthetic peptide analogue is recognized by both cell receptors and anti-HBV antibodies and elicits antibodies reacting with native HBV.

Anti-HBs antibody produced after exposure to HBV has been reported to consist primarily of anti-*a* specificity.[69] The development of antibody to the common *a* determinant appears to confer protection against reinfection with HBV, regardless of the subtype.[56,70] However, up to 2.8% of HBV-infected individuals fail to develop antibodies to such antigen.[69] This situation was illustrated by the report of a case in a patient in whom acute hepatitis B developed despite the presence of pre-existent anti-HBs, which was detected during an episode of acute non-A, non-B (NANB) hepatitis. This marker persisted up to the onset of acute hepatitis B, disappeared for about 1 month, and subsequently reappeared. Since this patient had no clinical history of viral hepatitis before the acute NANB hepatitis, it was presumed that the anti-HBs antibody present during the acute phase of NANB hepatitis was a residuum from a previous subclinical HBV infection. The HBsAg antibody present during this patient's second HBV infection was subtyped as *ayw* by a blocking RIA procedure. Using a similar technique, the anti-HBs antibody detected before the period of HBs antigenemia appeared to be almost exclusively of anti-*d* specificity, since only HBsAg subtypes with $d$ specificity blocked the anti-HBs activity significantly, while that detected after HBsAg positivity appeared to have both anti-$d$ and anti-$y$ specificity. A later serum sample collected 7 months after onset of acute hepatitis B clearly demonstrated an increase in the anti-$y$ activity with the anti-$d$ activity remaining constant. The source of this patient's second HBV infection may have been a known sexual partner whose anti-HBs antibody was broadly reactive with blocking by all eight subtypes. However, the sexual contact appeared to have significantly more anti-$y$ than anti-$d$ activity, consistent with the hypothesis that he was the source of the patient's HBV infection.[71]

A rather novel hypothesis to explain the suppression of production of anti-HBs antibody and the maintenance of the carrier state of HBsAg has been based on the idiotype-antiidiotype network.[72] An IgM-specific anti [anti-HBs] antibody was found only in sera of HBV-infected patients, both acute and chronic. However, not all HBsAg-positive patients exhibited this reaction, and activity was correlated with the presence of HBeAg. Approximately 93% of the sera with antiidiotype activity also contained HBeAg. Conversely, 70% of the sera positive for HBeAg reacted in the IgM assay. No correlation was observed between the presence of antiidiotype and rheumatoid factor or elevated serum alanine aminotransferase (ALT; formerly SGPT) levels.

Despite the early appearance of anti-HBc antibody in acute viral hepatitis B, it is the anti-HBs that has been considered protective against reinfection, hence the use of purified HBsAg in vaccination. It is important to re-evaluate this concept in view of two recent reports. Tabor and Gerety[73] found that chimpanzees immunized with purified HBcAg became resistant to virus challenge. This observation may clarify the immune status of individuals who are anti-HBc positive alone. Goudeau and Dubois[74] obtained some unexpected results when immunizing anti-HBc-positive volunteers during a clinical trial of hepatitis B vaccine in hospital personnel. Of these, 14 who were anti-HBc positive alone received three doses of 5 µg HBsAg alum vaccine (Hevac B, Institut Pasteur) spaced 1 month apart. The anti-HBs seroconversion rate was 85%, significantly lower than the 96% observed in seronegative staff in the same settings.[75] A true anamnestic response (rapid increase in anti-HBs after the first injection) was observed in only one subject, all the others having a primary type of response. There was no obvious relationship between this relatively low response and either increasing (two cases), decreasing (three cases), or stable anti-HBc titers, or the presence of anti-HBc IgM. These investigators suggested that a degree of anergy to HBsAg may be one characteristic of individuals who do not acquire anti-HBs after HBV infection or who lose it quickly. Some of them remain chronically infected, with low-level synthesis of HBsAg undetectable by conventional procedures.[76,77]

There is usually no difficulty in detecting antibodies to other HBV antigens (e.g., anti-HBc and anti-HBe antibodies) by conventional assays. It is interesting to note that in patients with chronic active hepatitis B and in silent HBsAg carriers, whose immunologic tolerance to HBsAg may be induced and maintained by the HBsAg-HSA complex, sensitization of immunocompetent cells and establishment of effector-cell functions toward HBcAg is altered.[78] That is, monomeric (7-8S) anti-HBc IgM is produced, in contrast to the 19S anti-HBc IgM detected during acute viral hepatitis B.[79,80] Membrane-bound IgG antibodies with specificity for HBcAg have been demonstrated on liver cells from these patients. Eluates obtained after urea treatment of isolated hepatocytes contained HSA, HBsAg, and IgG, when analyzed for protein content.[81] These antibodies may participate in the maintenance of immunologic tolerance to HBV-infected hepatocytes by local activation of Ts[82] or by antigen-receptor blockade through cross-reactivity with HSA in association with HBV antigens expressed on the hepatocyte, or steric hindrance.[78] It should be reiterated that although numerous studies have been reported in favor of the role of HBcAg in hepatocellular injury, playing down the importance of HBsAg in this connection, there are both clinical and laboratory observations that cannot be reconciled with this hypothesis. Nevertheless, this hypothesis is still viable and merits further consideration. If indeed HBcAg is the target for cytolytic T cells (CTL), other factors (e.g., target cell properties and immunoregulatory function) may explain differences in outcomes among individuals infected with HBV.[41]

## 4.2. T Lymphocytes

Lymphocyte subpopulations in patients with acute and chronic hepatitis have been investigated in an effort to elucidate their regulatory and effector functions in liver damage. Early studies involved the enumeration of circulating B and T lymphocytes and yielded conflicting findings from which no conclusions could be drawn. Nouri-aria et al.[83] recently investigated the relationship between histologic classification and the control of B-cell differentiation in HBV infection, including 68 patients with HBs antigenemia and a spectrum of liver damage in addition to 25 controls. Spontaneous immunoglobulin G (IgG) production was related to the degree of inflammation with significantly elevated IgG levels only in patients with chronic active hepatitis. Concanavalin A (Con A)-induced suppressor cell regulation of IgG producing cells was also impaired in these patients. The degree of impairment was directly proportional to the severity of portal tract inflammation. These abnormalities were also found in patients with acute viral hepatitis, however, but were transient in those with self-limited disease. The advent of monoclonal antibodies against helper-inducer (T4) and cytotoxic-suppressor (T8) T-cell subsets led to the analysis of T-cell subpopulations. In acute hepatitis B, the T4/T8 ratio was found to be minimally[84] or significantly[85] reduced, whereas in chronic active hepatitis B the ratio was variable[84] or low.[85] The decreased ratio in acute hepatitis B returns to normal upon resolution of the disease. This finding is similar to what has been reported for other viral infections, such as Epstein–Barr virus (EBV) and cytomegalovirus (CMV).[86]

Of great interest, however, was the fact that the T4/T8 ratio in silent carriers of HBsAg was consistently subnormal and independent of viral replication,[84] perhaps reflecting an increase in the suppressive immunoregulatory forces limiting immunologic attack on infected hepatocytes.[39,86] In contrast to these findings in certain aspects were those of Thomas et al.[85] These workers found that in chronic active hepatitis B, T4/T8 ratios depended on the level of viral replication, independent of disease activity. That is, patients with HBeAg and DNA polymerase had low T4/T8 ratios, while those with anti-HBe actually had mild increases. They also found that silent carriers of HBsAg had normal T4/T8 ratios.

Additional disagreements in this area have also been reported by others,[87,88] including the lack of correlation between Ts phenotype and function. Barnaba et al.[87] investigated the relationship between T-cell subsets, as defined by monoclonal antibodies, and suppressor activity, using a short-lived suppressor cell assay. Their studies showed that patients with chronic HBV infection had an absolute reduction in the T4-positive subset and a significantly decreased T4/T8 ratio, as compared with healthy controls. Conversely, Pignata et al.[89] reported that in children with HBsAg-positive chronic active hepatitis the T4/T8 ratio was lower than in age-matched controls, which was mainly due to an increase of T8-positive cells. The number of T4-positive cells was even lower in severe chronic hepatitis patients, thereby reducing still further the T4/T8 ratio.

Alexander et al.[90] recently investigated the relationship of suppressor cell activity, viral replication, and the histologic type of disease to T-cell cytotoxicity against autologous hepatocytes in 42 consecutive HBsAg carriers undergoing a liver biopsy. The proportion of suppressor/cytotoxic lymphocytes directly correlated with T-cell cytotoxicity to autologous hepatocytes, and both were higher in those with HBeAg in serum than in those with anti-HBe antibody or in patients on corticosteroid therapy. There was no relationship to underlying histologic classification. By contrast, suppressor cell regulation of IgG-producing cells was unrelated to the proportion of suppressor/cytotoxic lymphocytes in peripheral blood or HBeAg status, but impaired function was associated with chronic hepatitis, particularly chronic active hepatitis. These investigators suggest that the increased proportion of suppressor/cytotoxic lymphocytes in these patients represents an increase in the cytotoxic and not the suppressor cell subset.

One of the problems in cell analysis is the inability to distinguish between Ts and CTL, owing to the lack of specific monoclonal antibodies needed to characterize the components of these lymphocyte subsets. Complicating this already major problem is the detection of T4-positive lymphocytes that can be induced to suppress the function of other Th.[86] It is also possible that the current methodology for evaluation of immunoregulatory phenotypes or function is not sufficiently sensitive to detect changes in minor antigen-specific regulatory or effector cell subpopulations. In addition, it is impossible at this time to determine whether these changes represent primary pathogenetically important events, or secondary phenomena. In this regard, it is known that infection with a variety of viruses, even in the absence of abnormal protracted clinical outcomes, leads to altered T4/T8 ratios,[91] suggesting that the immunosuppression observed results from viral infection and does not necessarily contribute to chronicity. Furthermore, since a substantial number of patients with hepatitis B being studied are homosexual men, some of the abnormalities in T4/T8 ratios may reflect an underlying immunosuppressed state, possibly associated with multiple frequent infections, rather than HBV infection alone. In the final analysis, these studies of immunoregulatory T-cell subsets, like studies of immunoregulatory and effector function, have not yet provided us with definitive information to identify the immunologic abnormalities involved with HBV infection.

### 4.3. Macrophages

The role of macrophages in murine viral hepatitis is relatively well defined.[92] By contrast, their role in HBV infection has yet to be determined. The extrahepatic sites in humans and chimpanzees where HBV DNA, either alone or as an RNA : DNA hybrid, has been detected include T and B lymphocytes.[9] Although HBV DNA could not be found in macrophages by these investigators, the protective function of the cells of the reticuloendothelial system cannot be ruled out; i.e., HBV might persist in these cells for only short periods of time. It is possible that these cells are responsible for the removal of infectious virus particles from the circulation before they can gain access into the hepatocyte. Whether this function is achieved may depend on the size of the virus inoculum. The reticuloendothelial system might be overwhelmed by too large a virus inoculum, which could render the macrophages unable to process the virus adequately, thus paving the way to hepatocyte infection and disease.[31] Macrophage functions other than phagocytosis may also be involved in the immunopathology of hepatitis B infection, but to date no pertinent studies have been reported.

### 4.4. Natural Killer Cells

The cytotoxic capacity of peripheral blood mononuclear cells (PBMC) from patients with acute and chronic hepatitis has been assessed *in vitro* against a variety of target cell lines that do not express known liver and HBV antigens. In other studies, defined liver or HBV antigens such as liver-specific protein or HBsAg were bound artificially to red blood cells (RBGs) or cultured cells and were used as targets for cell-mediated cytotoxicity by PBMC from patients with acute and chronic hepatitis B. Although many of the results from these *in vitro* studies were interpreted as reflecting the mechanism of liver cell damage in hepatitis B, the experiments were often flawed by the lack of appropriate controls or by technical pitfalls.[39] In addition, the literature is replete with conflicting reports of investigations using the same target cell line, PLC/PRF/5. Thus, even when relatively similar methodologies and the identical cell line were employed by different investigators, divergent observations emerged. In short, the role of natural killer (NK) cells in the pathogenesis of liver injury associated with HBV infection remains obscure.[39]

## 4.5. Soluble Factors

Several reviews[39,47,49,86,93,94] have dealt with the immunomodulatory role of factors found in the serum of patients with viral hepatitis. In addition to the products of impaired catabolism by the damaged liver, serum proteins, such as very-low-density lipoproteins (VLDL), two classes of immunoregulatory proteins synthesized in patients with viral hepatitis have been described. One is an abnormal low-density lipoprotein that inhibits the ability of T cells to form rosettes with sheep RBCs and has been named rosette inhibitory factor (RIF);[49] the other inhibits lymphocyte proliferation and proliferation-dependent functions and is termed serum inhibitory or immunosuppressive factor (SIF).[95,96] Although these serum factors had both been proposed to contribute to the immunopathogenesis of HBV-induced liver injury, evidence now indicates that RIF, and even more frequently, SIF, are also associated with nonviral liver diseases. In any case, RIF has been shown to inhibit primarily the rosetting activity of T cells expressing Fc receptors for IgM, a property that has been associated with Th.[97] It has been postulated that RIF exerts an immunomodulatory effect by interfering with the cellular–humoral interactions required for the production of anti-HBs antibody. Several findings support this concept: (1) RIF levels are higher in HBsAg carriers without anti-HBs antibody; (2) RIF appears to interfere with Th; and (3) RIF inhibits the Ig secretory response of lymphocytes to stimulation with PWM.[98]

The biochemical properties of SIF are different from those of RIF and other liver-derived immunoregulatory factors; it is a low-molecular-weight albumin-associated lipid or lipophilic peptide, which can even be detected, at low levels of activity, in albumin fractions of normal serum.[99] In the latter investigation, SIF did not appear to correlate as well with liver cell injury as RIF, a finding at odds with the detection by others of RIF in asymptomatic HBsAg carriers.[98]

The role of interferon (IFN), which has both antiviral and immunoregulatory functions, has also been investigated, as it relates to the immunopathogenesis of hepatitis B. It has been found, however, that HBV is not an efficient IFN inducer and that the PBMC of patients produce less, not more, IFN than controls.[100–103] Several studies have recently dealt with $IFN_\alpha$ (virus-induced, a product predominantly of macrophages and NK cells but also of lymphocytes) and $INF_\gamma$ (immune IFN, a product of T cells). In contrast to earlier reports, Levin and Hahn found normal $IFN_\alpha$ and $IFN_\gamma$ in the serum of patients with uncomplicated acute viral hepatitis (A, B, and NANB), but depressed levels in those with fulminant viral hepatitis.[101] This study suggested that the short-term outcome of acute viral hepatitis might essentially be determined by the endogenous IFN production. That is, in self-limited acute viral hepatitis there was a rising IFN serum level and the mononuclear cells were in an antiviral state, whereas in patients with fulminant disease, the IFN system appeared grossly defective. It was speculated that this deficiency shown by the failure to detect IFN in the serum might be the cause of the fulminant course;

treatment with $IFN_\alpha$ was followed by rapid activation of the IFN system and an uncomplicated recovery in some of these patients. These data[101] have far-reaching implications with regard to our understanding of the different forms of outcome and a potential treatment for acute viral hepatitis but they await confirmation.

More recently, Pirovino et al.[104] measured IFN activity in the serum of patients with acute and chronic viral hepatitis B and compared the results with those in patients with nonviral liver disease, in patients with influenza, and in healthy controls. In all patients with acute and chronic viral hepatitis, no serum IFN could be detected, confirming data from earlier studies of acute viral hepatitis in which no circulating IFN was found. These investigators concluded that their results disprove the view that the amounts of serum IFN, detected at the time of the acute clinical illness, may be a determinant of outcome.

Kato et al.[105] found that peripheral lymphocytes produced normal amounts of $IFN_\alpha$ and $IFN_\beta$ (type I) in asymptomatic HBsAg carriers but decreased levels in patients with chronic active hepatitis. This report agrees with that by Davis et al.,[106] which showed that the ability of lymphocytes from patients with chronic hepatitis B, NANB, and delta (D) to produce $INF_\alpha$ was diminished but that $IFN_\gamma$ production was normal. This finding was interpreted as reflecting a defect of B-lymphocyte function. Finally, Yousefi et al.[103] evaluated the $IFN_\gamma$ response of peripheral lymphocytes from patients with acute and chronic hepatitis following incubation with partially purified HBsAg. Whole lymphocytes from patients with acute hepatitis B, resolved hepatitis B, and chronic hepatitis B responded by producing $IFN_\gamma$, while lymphocytes from asymptomatic healthy carriers did not. The significance of these observations remains to be elucidated.

## 5. SIGNIFICANCE OF HEPATITIS B VIRUS-INDUCED IMMUNOMODULATION IN PATHOGENESIS

The mechanisms by which viruses in general induce immunosuppression are slowly being unraveled. Evidence accumulated over many years suggests that virus-induced immunomodulation may often be, as expected, the result of infection of cells directly involved in the immune response. Cytolysis of these target cells may be produced by certain viruses but, for HBV, this is most likely not the case because HBV is not a cytopathic virus. Destruction of the hepatocytes occurs instead, according to most experts in the field, by CMI directed against viral antigens expressed on the surface of the infected hepatocyte. The liver is the source of a number of immunoregulatory molecules; therefore, liver injury may, directly or indirectly, alter the overall immunocompetency of the host. Although this should be an important consideration, the notion that HBV is not strictly a hepatotropic agent has opened new avenues of investigation.

There is now enough evidence to show that the HBV genome can be detected in other tissues, particularly T and B cells, but not in macrophages.[107–109] Frequently, most cells involved in the immune response are poorly permissive for viral replication; however, the virus may persist in a latent form or integrated into their genome. Both replicative (HBV RNA : DNA hybrids) and nonreplicative (HBV DNA alone) forms of HBV have been found in lymphocytes. Obviously, these observations are important to the tenet that HBV can exert a functional influence on cells participating in the immune response.

Another noteworthy aspect is the role of HBV in the acquired immune deficiency syndrome (AIDS). Before the discovery of the etiologic role of the human T lymphotropic virus III/lymphadenopathy virus (HTLV-III/LAV) in AIDS, a number of other agents were incriminated as the etiologic agents of this syndrome, including HBV. It is interesting that the possible role of HBV as a cofactor in AIDS is being re-evaluated in the laboratory of the investigators who discovered HTLV-III/LAV.[3] DNA sequences of HBV were found in fresh and cultured lymphocytes from patients with AIDS or pre-AIDS, even in the absence of conventional HBV serologic markers. Furthermore, the restriction DNA pattern was consistent with the integration of the HBV DNA. Two recent studies[110,111] have shown that HBV-infected individuals (e.g., homosexuals, drug users) are more prone to become HBV carriers with liver disease if they are anti-HIV positive. However, anti-HIV positive individuals with past or current HBV infection were not more likely to progress to AIDS than those who had no evidence of HBV infection.

# REFERENCES

1. Blum, H. E., L. Stowring, A. Figus, C. K. Montgomery, A. T. Haase, and G. N. Vyas. Detection of hepatitis B virus DNA in hepatocytes, bile duct epithelium, and vascular elements by in situ hybridization, Proc. Natl. Acad. Sci. USA **80**:6682–6685 (1983).
2. Chong-Jin, O. and D. Jenk-Ling, Hepatitis B virus in lymphocytes of seronegative carriers, Lancet **1**:395–396 (1984).
3. Laure, F., D. Zagury, A. G. Saimot, R. C. Gallo, B. H. Hahn, and C. Brechot, Hepatitis B virus DNA sequences in lymphoid cells from patients with AIDS and AIDS-related complex, Science **229**:561–563 (1985).
4. Lie-Injo, L. E., M. Balasegaram, C. G. Lopez, and A. R. Herrera, Hepatitis B viral DNA in liver and white blood cells of patients with hepatoma, DNA **2**:301–308 (1983).
5. Pontisso, P., M. C. Poon, P. Tiollais, and C. Brechot, Detection of hepatitis B virus DNA in mononuclear blood cells, Br. Med. J. **288**:1563–1566 (1984).
6. Romet-Lemonne, J. L., E. Elfassi, W. Haseltine, and M. Essex, Infection of bone marrow cells by hepatitis B virus, Lancet **2**:732 (1983).
7. Romet-Lemonne, J. L., M. F. McLane, E. Elfassi, W. Haseltine, J. Azocar, and M. Essex, Hepatitis B virus infection in cultured human lymphoblastoid cells, Science **221**:667–669 (1983).
8. Summers, J. and W. Mason, Replication of the genome of a hepatitis B-like virus by reverse transcriptase of an RNA intermediate, Cell **29**:403–415 (1982).
9. Korba, B. E., F. Wells, B. C. Tennat, G. H. Yoakum, R. H. Purcell and J. L. Gerin, Hepadnavirus infection of peripheral blood lymphocytes in vivo: Woodchuck and chimpanzee models of viral hepatitis, J. Virol. **58**:1–8 (1986).

10. Melnick, J. L., Classification of hepatitis A virus as enterovirus type 72 and hepatitis B as a hepadnavirus type 1, *Intervirology* **18**:105–106 (1982).

11. Robinson, W. S., P. Marion, M. Feitelson, and A. Siddiqui, The hepadnavirus group: Hepatitis B and related viruses, in: *Viral Hepatitis, 1981 International Symposium* (W. Szmuness, H. J. Alter, and J. E. Maynard, eds.), pp. 57–68, Franklin Institute Press, Philadelphia (1982).

12. Summers, J., J. Smolec, and R. Snyder, A virus similar to hepatitis B associated with hepatitis and hepatoma in woodchucks, *Proc. Natl. Acad. Sci. USA* **75**:4523–4537 (1978).

13. Marion, P. L., L. S. Oshiro, D. C. Regnery, G. H. Scullard, and W. S. Robinson, A virus in Beechy ground squirrels that is related to hepatitis B virus of humans, *Proc. Natl. Acad. Sci. USA* **77**:2941–2945 (1980).

14. Mason, W. S., G. Seal, and J. Summers, Virus of Pekin ducks with structural and biological relatedness to human hepatitis B virus, *J. Virol.* **35**:829–836 (1980).

15. Cote, P. J., Jr., G. M. Dapolito, J. W-K. Shin, and J. L. Gerin, Surface antigenic determinants of mammalian "hepadnavirus" defined by group and class-specific monoclonal antibodies, *J. Virol.* **3**:135–142 (1982).

16. Howard, C. R., The nature of the hepatitis B virus and its mode of replication, *Springer Semin. Immunopathol.* **3**:397–419 (1981).

17. Chernesky, M. A., M. R. Escobar, P. D. Swenson, and S. Specter, Laboratory diagnosis of hepatitis viruses, *Cumitech* **18**:1–12 (1984).

18. LeBouvier, G. L., The heterogeneity of Australia antigen, *J. Infect. Dis.* **123**:671–675 (1971).

19. Bancroft, W. A., F. F. Mundon, and P. K. Russel, Detection of additional antigenic determinants of hepatitis B antigen, *J. Immunol.* **109**:842–848 (1972).

20. Heermann, K. H., U. Goldman, W. Schwartz, T. Seyffarth, H. Baumgarten, and W. H. Gerlich, Large surface proteins of hepatitis B virus containing the pre-S sequence, *J. Virol.* **52**:396–401 (1984).

21. Stibbe, W. and W. H. Gerlich, Variable protein composition of hepatitis B surface antigen from different donors, *Virology* **123**:436–442 (1982).

22. Tiollais, P, P. Charnay, and G. N. Vyas, Biology of hepatitis B virus, *Science* **213**:406–411 (1981).

23. Machida, A., S. Kishimoto, H. Ohnuma, K. Baba, Y. Ito, H. Miyamoto, G. Funatsu, K. Oda, S. Usuda, S. Togami, T. Nakamura, M. Miyakawa, and M. Mayumi, A polypeptide containing 55 amino acid residues coded by the pre-S region of hepatitis B virus deoxyribonucleic acid bears the receptor for polymerized human as well as chimpanzee albumins, *Gastroenterology* **86**:910–918 (1984).

24. Stibbe, W. and W. H. Gerlich, Structural relationships between minor and major proteins of hepatitis B surface antigen, *J. Virol.* **46**:626–628 (1983).

25. Takahashi, K., Y. Akahane, T. Gotanda, T. Mishiro, M. Imai, Y. Miyakawa, and N. Mayumi, Demonstration of hepatitis B e antigen in the core of Dane particles, *J. Immunol.* **122**:275–279 (1979).

26. Yoshizawa, H., Y. Itoh, J. P. Simonetti, T. Takahashi, A. Machida, Y. Miyakawa, and M. Mayumi, Demonstration of hepatitis B e antigen in hepatitis B core particles obtained from the nucleus of hepatocytes infected with hepatitis B virus, *J. Gen. Virol.* **42**:513–519 (1979).

27. Murphy, B., E. Tabor, V. McAuliffe, A. Williams, J. Maynard, R. Gerety, and R. Purcell, Third component, HBeAg/3, of hepatitis B e antigen system identified by the different double diffusion techniques, *J. Clin. Microbiol.* **8**:349–350 (1978).

28. Williams, A. and G. L. LeBouvier, Heterogeneity and thermolability of "e" antigen, *Bibl. Haematol.* **42**:71–75 (1976).

29. Imai, M., M. Nomura, T. Gotanda, T. Sano, K. Tachibana, H. Miyamoto, K. Takahashi, S. Toyama, Y. Miyakawa, and M. Mayumi, Demonstration of two distinct antigenic

determinants on hepatitis B e antigen by monoclonal antibodies, *J. Immunol.* **128**:69–72 (1982).

30. Summers, J., Replication of hepatitis B viruses, in *Viral Hepatitis and Liver Disease* (G. N. Vyas, J. L. Dienstag, and J. H. Hoofnagle, eds.), pp. 87–96, Grune & Stratton, Orlando, Florida (1984).

31. Escobar, M. R. and P. D. Swenson, Mechanisms of viral immunopathology with special reference to viral hepatitis, in *Immunomodulation by Microbial Products and Related Synthetic Compounds* (Y. Yamamura, S. Kotani, I. Azuma, A. Koda, and T. Shiba, eds.), pp. 134–147, Excerpta Medica, Amsterdam (1982).

32. Blumberg, B. S., Comments on the biology of HBV, in *Viral Hepatitis* (G. N. Vyas, S. N. Cohen, and R. Schmid, eds.), pp. 591–592, The Franklin Institute Press, Philadelphia (1978).

33. Walker, M. E., W. Szmuness, C. Stevens, and P. Rubenstein, Genetics of anti-HBs responsiveness. I. HLA-DR7 and nonresponsiveness to hepatitis vaccination, *Transfusion* **21**:601 (1981) (abst.).

34. Milich, D. R., H. Alexander, and F. V. Chisari, Genetic regulation of the immune response to hepatitis B surface antigen (HBsAg). III. Circumvention of nonresponsiveness in mice bearing HBsAg nonresponder haplotypes, *J. Immunol.* **130**:1401–1407 (1983).

35. Milich, D. R. and F. V. Chisari, Genetic regulation of the immune response to hepatitis B surface antigen (HBsAg). I. H-2 restriction of the murine humoral immune response to *a* and *d* determinants of HBsAg, *J. Immunol.* **129**:320–325 (1982).

36. Milich, D. R., G. G. Leroux-Roels, and F. V. Chisari, Genetic regulation of the immune response to hepatitis B surface antigen (HBsAg). II. Qualitative characteristics of the humoral immune response to the *a*, *d*, and *y* determinants of HBsAg, *J. Immunol.* **130**:1395–1400 (1983).

37. Szmuness, W., C. E. Stevens, E. A. Zang, E. J. Harley, and A. Kellner. A control clinical trial of the efficacy of the hepatitis B vaccine (Heptavax B): A final report, *Hepatology* **1**:377–385 (1981).

38. Montano, L., G. C. Miescher, A. H. Goodall, K. H. Wiedmann, G. Janossy, and H. C. Thomas, Hepatitis B virus and HLA antigen display in the liver during chronic hepatitis B virus infection, *Hepatology* **2**:557–561 (1982).

39. Dienstag, J. L., Immunologic mechanisms in chronic viral hepatitis, in *Viral Hepatitis and Liver Disease* (G. N. Vyas, J. L. Dienstag, and J. H. Hoofnagle, eds.), pp. 135–166, Grune & Stratton, Orlando (1984).

40. Craxi, A., L. Montano, A. Goodall, and H. C. Thomas, Genetic and sex-linked factors influencing HBs antigen clearance. I. Nonimmune clearance in inbred strains of mice, *J. Med. Virol.* **9**:117–123 (1982).

41. Klingenstein, R. J. and J. L. Dienstag, Immunopathogenesis of acute and chronic hepatitis B, in *Hepatitis B* (R. J. Gerety, ed.), pp. 221–245, Academic, Orlando (1985).

42. Smith, C. I., W. G. E. Cooksley, and L. W. Powell, Cell-mediated immunity to liver antigen in toxic liver injury. I. Occurrence and specificity, *Clin. Exp. Immunol.* **39**:607–617 (1980).

43. Smith, C. I., W. G. E. Cooksley, and L. W. Powell, Cell-mediated immunity to liver antigen in toxic liver injury. II. Role in pathogenesis of liver damage, *Clin. Exp. Immunol.* **39**:618–625 (1980).

44. Blumberg, B. S., H. I. Sutnick, and W. T. London, Australia antigen as a hepatitis virus. Variation in host response. *Am. J. Med.* **48**:1–8 (1970).

45. Dudley, F. J., R. A. Fox, and S. Sherlock, Cellular immunity and hepatitis-associated Australia antigen in liver disease, *Lancet* **1**:723–726 (1972).

46. Serjeantson, S., and D. G. Woodfield, Immune response of leprosy patients to hepatitis B virus, *Am. J. Epidemiol.* **107**:321–327 (1978).

47. Chisari, F. V., J. A. Routenberg, D. S. Anderson, and T. S. Edgington, Cellular immune reactivity in HBV-induced disease, in: *Viral Hepatitis* (G. N. Vyas, S. N. Cohen, and R. Schmid, eds.), pp. 245–266, Franklin Institute Press, Philadelphia (1978).
48. Dusheiko, G. M., J. H. Hoofnagle, W. G. Cooksley, S. P. James, and E. A. Jones, Synthesis of antibodies to hepatitis B virus by cultured lymphocytes from chronic hepatitis B surface antigen carriers, *J. Clin. Invest.* **71:**1104–1113 (1983).
49. Chisari, F. V., and T. S. Edgington, Lymphocyte E rosette inhibitory factor: A regulatory serum lipoprotein, *J. Exp. Med.* **142:**1092–1107 (1975).
50. Celis, E., P. C. Kung, and T. W. Chang, Hepatitis B virus-reactive human T lymphocyte clones: Antigen specificity and helper function for antibody synthesis, *J. Immunol.* **132:**1511–1516 (1984).
51. Hellstrom, U., and S. Sylvan, Regulation of the immune response to hepatitis B virus and human serum albumin. I. Hepatitis B surface antigen-induced secretion of antibodies with specificity for human serum albumin in hepatitis B-immune donors *in vitro, Scand. J. Immunol.* **23:**545–553 (1986).
52. Kerlin, P., K. M. Nies, and M. J. Tong, Unimpaired B-cell function and T-cell regulation of immunoglobulin synthesis in patients with chronic hepatitis B virus infection, *Clin. Immunol. Immunopathol.* **25:**149–156 (1982).
53. Vento, S., J. E. Hegarty, A. Alberti, C. J. O'Brien, G. J. M. Alexander, A. L. W. F. Eddleston, and R. Williams, T-lymphocyte sensitization to HBcAg and T-cell mediated unresponsiveness to HBsAg in hepatitis B virus-related chronic liver disease, *Hepatology* **5:**192–197 (1985).
54. Milich, D. R., G. B. Thornton, A. R. Neurath, S. B. Kent, M -L. Michel, P. Tiollais, and F. V. Chisari, Enhanced immunogenicity of the pre-S region of hepatitis B surface antigen, *Science* **228:**1195–1198 (1985).
55. Milich, D. R., A. McLachlan, F. V. Chisari, S. B. H. Kent, and G. B. Thornton, Immune response to the pre-S(1) region of the hepatitis B surface antigen (HBsAg): A pre-S(1) specific T cell response can bypass nonresponsiveness to the pre-S(2) and S regions of HBsAg, *J. Immunol.* **137:**315–322 (1986).
56. McAuliffe, V. J., R. H. Purcell, and J. L. Gerin, Type B hepatitis: A review of current prospects for a safe and effective vaccine, *Rev. Infect. Dis.* **2:**470–492 (1980).
57. Melnick, J. L., G. R. Dreesman, and F. B. Hollinger, Approaching the control of viral hepatitis type B, *J. Infect. Dis.* **133:**210–229 (1976).
58. Thomas, H. C., and D. P. Jewell, Acute and chronic viral hepatitis, in: *Clinical Gastrointestinal Immunology* (H. C. Thomas and D. P. Jewell, eds.), pp. 164–210, Blackwell Scientific Publications, London (1979).
59. Fathman, C. G., and F. W. Fitch (eds.), *Isolation, Characterization and Utilization of T Lymphocyte Clones,* Academic, New York (1982).
60. Pernis, B. and H. J. Vogel, *Regulatory T. Lymphocytes,* Academic, New York (1980).
61. Roberts, I. M., C. C. Bernard, G. N. Vyas, and I. R. MacKay, T-cell dependence of immune response to hepatitis B antigen in mice, *Nature (Lond.)* **254:**606–607 (1975).
62. Tabor, E., R. J. Gerety, L. A. Smallwood, and L. F. Barker, Coincident hepatitis B surface antigen and antibodies of different subtypes in human serum, *J. Immunol.* **118:**369–370 (1977).
63. Lane, H. C., G. Whalen, and A. S. Fauci, Dichotomy between antigen and mitogen-induced T cell help in human B cell activation, *Clin. Res.* **30:**252 (1982) (abst.).
64. Machida, A., S. Kishimoto, H. Ohnuma, H. Miyamoto, K. Baba, K. Oda, T. Nakamura, Y. Miyakawa, and M. Mayumi, A hepatitis B surface antigen polypeptide (p31) with the receptor for polymerized human as well as chimpanzee albumins, *Gastroenterology* **85:**268–274 (1983).
65. Okamoto, H., M. Imai, S. Usuda, E. Tanaka, K. Tachibana, S. Mishiro, A. Machida, T. Nakamura, Y. Miyakawa, and M. Mayumi, Hemagglutination assay of polypeptide coded by the pre-S region of hepatitis B virus DNA with monoclonal antibody: Correla-

tion of pre-S polypeptide with the receptor for polymerized human albumin in serums containing hepatitis B antigens, *J. Immunol.* **134**:1212–1216 (1985).

66. Okamoto, H. S. Usuda, M. Imai, K. Tachibana, E. Tanaka, T. Kumakura, M. Itabashi, E. Takai, F. Tsuda, T. Nakamura, Y. Miyakawa, and M. Mayumi, Antibody to the receptor for polymerized human serum albumin in acute and persistent infection with hepatitis B virus, *Hepatology* **6**:354–359 (1986).

67. Yu, M. W., J. S. Finlayson, and J. W-K. Shih, Interaction between various polymerized human albumins and hepatitis B surface antigen, *J. Virol.* **55**:736–743 (1985).

68. Neurath, A. R., S. B. H. Kent, N. Strick, and K. Parker, Identification and chemical synthesis of a host cell receptor binding site on hepatitis B virus, *Cell* **46**:429–436 (1986).

69. Gold, J. W. M., H. J. Alter, P. V. Holland, J. L. Gerin, and R. H. Purcell, Passive hemagglutination assay for antibody to subtypes of hepatitis B surface antigen, *J. Immunol.* **112**:1100–1106 (1974).

70. Markenson, J. A., R. J. Gerety, J. H. Hoofnagle, and L. F. Barker, Effect of a cyclophosphamide on hepatitis B virus infection and challenge on chimpanzees, *J. Infect. Dis.* **131**:79–87 (1975).

71. Swenson, P. D., M. R. Escobar, R. L. Carithers, and T. J. Sobieski III, Failure of preexistent antibody against hepatitis B surface antigen to prevent subsequent hepatitis B infection, *J. Clin. Microbiol.* **18**:305–309 (1983).

72. Troisi, C. L. and F. B. Hollinger, Detection of an IgM antiidiotype directed against anti-HBs in hepatitis B patients, *Hepatology* **5**:758–762 (1985).

73. Tabor, E. and R. J. Gerety, Possible role of immune responses to hepatitis B core antigen in protection against hepatitis B infection. (Letter.), *Lancet* **1**:172 (1984).

74. Goudeau, A. and F. Dubois, Immune status of anti-HBc positive individuals, *Lancet* **1**:396 (1984).

75. Goudeau, A., F. Dubois, F. Barin, M. D. Dubois, and P. Coursaget, Hepatitis B vaccine: Clinical trials in high-risk settings in France. (September 1975–September 1982), *Dev. Biol. Std.* **54**:267–284 (1983).

76. Brechot, C., M. Hadchouel, J. Scotto, M. Fonck, F. Potet, G. N. Vyas, and P. Tiollais, State of hepatitis B virus DNA in hepatocytes of patients with hepatitis B surface antigen-positive and -negative liver diseases, *Proc. Natl. Acad. Sci. USA* **78**:3906–3910 (1981).

77. Katchaki, J. N., R. Brouwer, and T. H. Siem, Anti-HBc and blood, *N. Engl. J. Med.*, **298**:1421–1422 (1978).

78. Hellstrom, U. and S. Sylvan, Editorial Reviews: Human serum albumin and the enigma of chronic hepatitis type B, *Scand. J. Immunol.* **23**:523–527 (1986).

79. Fakunle, Y. M., F. Aranguibel, D. De Villiers, H. C. Thomas, and S. Sherlock, Monomeric (7S) IgM in chronic liver disease, *Clin. Exp. Immunol.* **38**:204–210 (1979).

80. Sjogren, M. and J. H. Hoofnagle, Immunoglobulin M antibody to hepatitis B core antigen in patients with chronic type B hepatitis, *Gastroenterology* **89**:252–258 (1985).

81. Trevisan, A., G. Realdi, A. Alberti, G. Ongaro, E. Pornaro, and R. Meliconi, Core antigen-specific immunoglobulin G bound to the liver cell membrane in chronic hepatitis B, *Gastroenterology* **82**:218–222 (1982).

82. Pape, G. R., E. P. Rieber, J. Eisenburg, R. Hoffman, C. M. Balch, G. Baumgartner, and G. Riethmuller, Involvement of the cytotoxic/suppressor T-cell subset in liver diseases, *Gastroenterology* **85**:657–662 (1983).

83. Nouri-Aria, K. T., G. J. M. Alexander, B. Portmann, D. Vergani, A. L. W. F. Eddleston, and R. Williams, *In vitro* study of IgG production and concanavalin A induced suppressor cell function in acute and chronic hepatitis B virus infection, *Clin. Exp. Immunol.* **64**:50–58 (1986).

84. Klingenstein, R. J., A. M. Savarese, J. L. Dienstag, R. H. Rubin, and A. K. Bahn, Immunoregulatory T cell subsets in acute and chronic hepatitis, *Hepatology* **1**:523 (1981) (abst.).

85. Thomas, H. C., D. Brown, G. Routhier, G. Janossy, P. C. Kung, G. Goldstein, and S.

Sherlock, Inducer and suppressor T-cells in hepatitis B virus-induced liver disease, *Hepatology* **2**:202–204 (1982).

86. Dienstag, J. L. Immunopathogenesis of acute and chronic hepatitis B, in: *Hepatitis B* (R. J. Gerety, ed.), pp. 221–245, Academic, Orlando, Florida (1985).

87. Barnaba, V., A. Musca, C. Cordova, M. Levrero, G. Ruocco, V. Albertini-Petroni, and F. Balsano, Relationship between T cell subsets and suppressor cell activity in chronic hepatitis B virus (HBV) infection, *Clin. Exp. Immunol.* **53**:281–288 (1983).

88. Hodgson, H. J. F., J. R. Wands, and K. J. Isselbacher, Alteration in suppressor cell activity in chronic active hepatitis, *Proc. Natl. Acad. Sci. USA* **75**:1549–1553 (1978).

89. Pignata, C., P. Vajro, R. Troncone, G. Monaco, and M. Ciriaco, Immunoregulatory T subsets in chronic active viral hepatitis: Characterization by monoclonal antibodies, *J. Pediatr. Gastroenterol. Nutr.* **2**:229–233 (1983).

90. Alexander, G. J. M., M. Mondelli, N. V. Naumov, K. T. Nouri-Aria, D. Vergani, D. Lowe, A. L. W. F. Eddleston, and R. Williams, Functional characterization of peripheral blood lymphocytes in chronic HBsAg carriers, *Clin. Exp. Immunol.* **63**:498–507 (1986).

91. Rubin, R. H., W. P. Carney, R. T. Schooley, R. B. Colbin, R. C. Burton, R. A. Hoffman, W. P. Harrsen, A. B. Cosimi, P. S. Russell, and M. S. Hirsch, The effect of infection on T-lymphocyte subpopulations: A preliminary report, *Int. J. Immunopharmacol.* **3**:307–312 (1981).

92. Mallucci, L. and B. Edwards, Influence of the cytoskeleton on the expression of a mouse hepatitis virus (MHV-3) in peritoneal macrophages: Acute and persistent infection, *J. Gen. Virol.* **63**:217–221 (1982).

93. Chisari, F. V. Regulation of lymphocyte function and viral transformation by hepatic bioregulatory molecules, *Hepatology* **2**:97–106s (1982).

94. Levy, G. A., and F. V. Chisari, The immunopathogenesis of chronic HBV induced liver disease, *Springer Semin. Immunopathol.* **3**:439–459 (1981).

95. Berg, P. A., N. W. Brattig, and W. Grauer, Immunoregulatory serum factors in acute and chronic hepatitis, *Liver* **2**:275–278 (1982).

96. Brattig, N. W., and P. A. Berg, Serum inhibitory factors (SIF) in patients with acute and chronic hepatitis and their clinical significance, *Clin. Exp. Immunol.* **25**:40–49 (1976).

97. Sanders, G. and R. P. Perillo, Rosette inhibitory factor: T-lymphocyte subpopulation specificity and potential immunoregulatory role in hepatitis B virus infection, *Hepatology* **2**:547–552 (1982).

98. Grauer, W., N. W. Brattig, H. Schomerus, G. Frosner, and P. A. Berg, Immunosuppressive serum factors in viral hepatitis. III. Prognostic relevance of rosette inhibitory factor and serum inhibitory factor in acute and chronic hepatitis, *Hepatology* **4**:15–19 (1984).

99. Brattig, N. W., G. E. Schrempf-Dekker, C. W. Brockl, and P. A. Berg, Immunosuppressive serum factors in viral hepatitis. II. Further characterization of serum inhibition factor as an albumin-associated molecule, *Hepatology* **3**:647–655 (1983).

100. Hill, D. A., J. H. Walsh, and R. H. Purcell, Failure to demonstrate circulating interferon during incubation period and acute stage of transfusion-associated hepatitis, *Proc. Soc. Exp. Biol. Med.* **136**:853–856 (1971).

101. Levin, S., and T. Hahn, Interferon system in acute viral hepatitis, *Lancet* **1**:592–594 (1982).

102. Tolentino, P., F. Dianzani, M. Zucca, and R. Giacchino, Decreased interferon response by lymphocytes from children with chronic hepatitis, *J. Infect. Dis.* **132**:459–461 (1975).

103. Yousefi, S., M. R. Escobar, and R. L. Carithers, Immune interferon induction: A measure of cellular immune deficiency in hepatitis B carriers, *Hepatology* **3**:826 (1983) (abst.).

104. Pirovino, M., M. Aguet, M. Huber, J. Altorfer, and M. Schmid, Absence of detectable serum interferon in acute and chronic viral hepatitis, *Hepatology* **6**:645–647 (1986).

105. Kato, Y., H. Nakgawa, K. Kobayashi, N. Hattori, and K. Hatano, Interferon production

by peripheral lymphocytes in HBsAg-positive liver diseases, *Hepatology* **2**:789–790 (1982).

106. Davis G. L., J. L. Jicha, and J. H. Hoofnagle, Alpha and gamma interferon in patients with chronic type B, non-A, non-B, and delta hepatitis: Serum levels and *in vitro* production by lymphocytes, *Gastroenterology* **86**:1315 (1984) (abst.).

107. Anderson, R. E., W. Winkelstein, Jr., H. E. Blum, and G. N. Vyas, Hepatitis B virus infection in the acquired immunodeficiency syndrome (AIDS), in: *Viral Hepatitis and Liver Disease* (G. N. Vyas, J. L. Dienstag, and J. H. Hoofnagle, eds.), pp. 339–343, Grune & Stratton, Orlando, Florida (1984).

108. McDonald, M. I., J. D. Hamilton, and D. T. Durack, Hepatitis B surface antigen could harbor the infectious agent of AIDS, *Lancet* **2**:882–884 (1983).

109. Ravenholt, R. T., Role of hepatitis B virus in acquired immunodeficiency syndrome, *Lancet* **2**:885–886 (1983).

110. Taylor, P. E., C. E. Stevens, S. Rodriguez de Cordoba, and P. Rubinstein, Hepatitis B virus and human immunodeficiency virus: Possible interaction, in: *Viral Hepatitis and Liver Disease* (A. J. Zuckerman, ed.), pp. 198–200, Alan R. Liss, Inc., New York (1988).

111. Govindarajan, S., V. M. Edwards, M. L. Stuart, E. A. Operskalski, J. W. Mosley, and The Transfusion Safety Study Group, Influence of human immunodeficiency virus infection on expression of chronic hepatitis B and D virus infections, in: *Viral Hepatitis and Liver Disease* (A. J. Zuckerman ed.), pp. 201–204, Alan R. Liss, Inc., New York (1988).

# 3

# Papovaviruses

## GIUSEPPE BARBANTI-BRODANO

### 1. GENERAL CHARACTERISTICS OF THE VIRUSES

Viruses of the papovavirus group belong taxonomically to the family Papovaviridae.[1,2] The family is subdivided into two genera, *Papillomavirus* and *Polyomavirus*, whose members are widely distributed among several mammalian species. However, knowledge of immunologic aspects is limited to a few viruses of the group, the object of this chapter.

Papovavirus virions are naked icosahedral particles that contain a double-stranded circular DNA genome. In contrast to these common characteristics, viruses of the *Papillomavirus* and *Polyomavirus* genus are different in virion and genome size as well as in sedimentation coefficient of the viral particle (Table I). Some members of the polyomavirus group (polyoma, BK, and JC virus) possess a virion-associated hemagglutinin that reacts with neuraminidase-sensitive receptors present on guinea pig and human red blood cells (RBCs). Each virus species has distinct surface antigens that do not cross-react with other members of the family. All members of each genus, however, share a common internal antigen shown by disrupting the virions. While viruses of the *Polyomavirus* genus are readily grown *in vitro*, no tissue-culture systems are available to propagate papillomaviruses. Papovaviruses can infect cells lytically *in vitro* or transform them to a neoplastic phenotype.

All papovaviruses are oncogenic. Papillomaviruses generally produce tumors in the species of origin. Polyomaviruses seem to be naturally nononcogenic or weakly oncogenic in the species of origin, whereas they are highly oncogenic when inoculated experimentally in rodents. While the genome organization and mechanisms of transformation by papillomaviruses are still poorly understood, convincing evidence[3] has been accumulated that polyomaviruses

---

GIUSEPPE BARBANTI-BRODANO • Institute of Microbiology, School of Medicine, University of Ferrara, I-44100 Ferrara, Italy.

**TABLE I**

**Properties of Viruses of the Family Papovaviridae**

| Genus | Capsid diameter (nm) | DNA (daltons) | Sedimentation coefficient | Major species | Host |
|---|---|---|---|---|---|
| Papillomavirus | 55 | $5 \times 10^6$ | 300 $S_{20w}$ | Rabbit papillomavirus | Rabbit |
| | | | | Bovine papillomavirus (five types) | Cattle |
| | | | | Human papillomavirus (~40 types) | Humans |
| Polyomavirus | 45 | $3 \times 10^6$ | 240 $S_{20w}$ | Polyomavirus | Mouse |
| | | | | Simian virus 40 | Monkey |
| | | | | BK virus | Humans |
| | | | | JC virus | Humans |
| | | | | B lymphotropic papovavirus | Monkey, humans (?) |

transform cells by the expression of specific proteins (tumor or T antigens) encoded in the early region of the genome. In transformed cells, viral DNA integrates into cellular chromosomes or may remain free, in an episomal state. Most papovaviruses are ubiquitous in their host species and generally produce an inapparent, latent infection that is affected by the immunologic status of the host.

## 2. INTERACTIONS WITH THE IMMUNE SYSTEM

### 2.1. Papillomaviruses

Excellent reviews on general problems concerning papillomaviruses have recently been published.[4,5] Before considering the immunology of human papillomaviruses (HPV), it is worthwhile to examine the role of the immune system in animal papillomavirus infections. In fact, rabbit papillomavirus (RPV) and bovine papillomavirus (BPV) represent useful models to elucidate results with HPV.

### 2.1.1. Rabbit Papillomavirus

Rabbits respond to papillomavirus infection with both humoral and cell-mediated immunity. Neutralizing antibodies are detected that protect against reinfection. Antibodies, however, do not seem to be responsible for regression of papillomas because they are produced by animals with both regressing and persisting papillomas.[6,7] Cellular immune mechanisms seem to represent the main factor in papilloma regression, as suggested by the frequent presence of mononuclear cell infiltrates in regressing tumors.[8] Interestingly, reduction of cell proliferation begins in the upper layers of the tumor, whereas leukocyte infiltrates are mostly evident at the epithelial basement membrane, suggesting the involvement of soluble factors and lymphokines.

Cell-mediated immunity was also demonstrated by *in vitro* experiments.[9] Lymph node cells from rabbits with both regressing and persisting papillomas inhibited colony formation by cells derived from papillomas or carcinomas. Sera from nonregressor rabbits, however, blocked the inhibitory effect induced by lymphocytes, indicating that humoral factors can play an important role in preventing regression. Further evidence that cell-mediated immune responses contribute to the outcome of RPV infection comes from tumor-bearing animals treated with methylprednisolone.[10] Although steroid treatment did not influence the length of the latency period, papilloma growth rate or malignant conversion, treated animals showed a high incidence of secondary papillomas and a very low rate of regressions: in 47% of untreated animals, papillomas regressed, whereas only 2.5% of the steroid-treated animals showed signs of regression.

### 2.1.2. Bovine Papillomaviruses

The host immune response seems to be an important factor in controlling the effects of BPV infection. BPV type 4 causes alimentary tract papillomas in cattle.[11,12] Progression of papillomas to carcinomas occurs with relatively high frequency in cows of the Scottish highlands feeding on bracken fern.[11,12] Malignant transition may occur because this plant contains radiomimetic substances[11] that display immunosuppressive and cocarcinogenic activities when present in a long-term diet.

### 2.1.3. Human Papillomaviruses

*2.1.3.a. Humoral Immunity.* Antibodies to HPV structural proteins were detected in wart patients by immune electron microscopy (IEM)[13,14] and immunoelectrophoresis.[15] By using a solid-phase radioimmunoassay (RIA), the percentage of persons producing antibodies to HPV-1 in an unselected population was found to reach a maximum of about 50% at 20 years and to decrease in the following age groups to a value of about 35%.[16] Since warts are typically a disease of youth and their incidence drops in the age groups beyond 20–30 years, it is tempting to speculate that humoral immunity is protective. This suggestion was supported by the observation that at the onset of regression all patients had virus-specific immunoglobulin M (IgM) antibodies, whereas only 12% of patients with nonregressive warts were IgM positive.[17,18]

Concurrent with wart regression, patients also develop complement-fixing IgG antibodies.[19–22] No significant differences in virus-specific IgG antibody frequency and titer were found, however, between patients with regressing and persisting warts.[23] In addition, patients have been described with occasionally high levels of IgG antibodies but without any signs of wart regression.[24] Finally, in wart patients, the anti-HPV antibody titers are usually low,[15,16,25] suggesting a limited antigenic stimulation by virus-specific antigens. Thus, the protective effect of humoral immunity in HPV infection and its role in wart regression remains uncertain. Even though the appearance of anti-HPV IgG antibodies may simply represent an epiphenomenon of regression, it may be a useful indicator to monitor wart evolution.

*2.1.3.b. Cell-Mediated Immunity.* Several epidemiologic observations indicate a crucial role for cell-mediated immune mechanisms in protection against HPV infection. Patients treated with immunosuppressive drugs after organ transplantation often develop disseminated persisting warts.[26–28] Likewise, patients with cell-mediated immune deficiencies, particulary those affected by malignant conditions such as Hodgkin disease and chronic lymphatic leukemia, suffer from severe papillomavirus infections more frequently than do patients with humoral immune deficiencies.[29–32] These data are clearly in accordance with the observation that regression of flat and plantar warts shows many characteristics of a cell-mediated immune reaction. Macrophages and lymphocytes accumulate in the dermis at the basal membrane and invade the basal layers of epidermal cells, generally without diffuse infiltration of the degenerating wart tissue.[33—35]

Further support for the role of cell-mediated immunity in wart regression is given by epidermodysplasia verruciformis.[36] This rare disease with a frequent familial occurrence is associated with genetic disorders and immunologic defects. It is characterized by a lifelong generalized eruption of flat, often confluent, warts that cover extensive areas of the skin. Patients have a high risk for malignant transition of warts to carcinomas in skin sites exposed to light, since as many as 30% may develop skin cancer over the areas of verrucosis.

In epidermodysplasia verruciformis, several parameters of cell-mediated immunity are impaired.[37—40] Patients show a reduction in the frequency and index of 1-nitro-2,4-dichlorobenzene skin sensitization, percentage of E-rosette-forming lymphocytes, lymphocyte blastogenesis, and leukocyte migration inhibition. By contrast, the virus-specific humoral immune response is normally preserved.[41] Although the impairment of cell-mediated immunity seems to be less pronounced in cases of common warts, reduced responsiveness in lymphocyte transformation tests[42] and decreased number of T cells[43] are frequently observed in wart patients, particularly in those with persistent or long-lasting warts. Patients with genital papillomas show similar immunological defects.[40,44] In this connection, it has been reported recently that in immunosuppressed women with HPV-associated lower genital intraepithelial neoplasia, an altered T-helper/T-suppressor ratio and a deficient response to mitogenic stimulation correlated with persistent and recurrent tumors as well as with progression to invasive epidermoid carcinoma.[45]

*2.1.3.c. Modulation of Immunity by HPV.* Several lines of evidence suggest that HPV may modulate the host-immune response or even display immunosuppressive activity. These effects seem to be related to different HPV types.[37,38,40] Cell-mediated immunity was well preserved in patients with HPV-1, HPV-4, and HPV-7-induced skin warts, whereas it was considerably altered during infections by HPV-5 and HPV-8. The latter HPV types most commonly are associated with epidermodysplasia verruciformis. Moreover, in cases of persisting warts, cell-mediated immunity improved after surgical removal or spontaneous regression[31,46] suggesting that immune depression was, at least in part, specifically determined by HPV infection. There are also reports on soluble factors present in wart extracts that may be able to protect persistent warts from immune attack by blocking the local expression of cellular immunity.[47]

*2.1.3.d. Malignant Conversion of Warts and Genital Papillomas.* As in the RPV system,[10] malignant conversion does not seem to depend on the immune status of the patient, but rather to be related to the carcinogenic potential of some HPV types. Indeed, cell-mediated immunity was impaired to the same extent in 13 cases of epidermodysplasia verruciformis with lesions induced by HPV-3 or HPV-5, but only the seven patients infected with HPV-5 developed skin carcinomas.[38] Similarly, malignant transition of genital papillomas seem to be related to infection by HPV-16 and HPV-18,[48] whereas HPV-6 and HPV-11 are mostly associated with benign condylomata[49,50] with a relatively high rate of regression.

*2.1.3.e. Vaccination.* The evidence that wart virus infection can be controlled by immune mechanisms suggests that vaccination is a possible approach to protection against HPV. Stimulation of humoral immunity could be of protective value in preventing HPV infection, while an increase in cell-mediated immunity could have a therapeutic significance by inducing regression of the lesions. Recent knowledge about the role of HPV infection in human diseases may prompt efforts toward protective vaccination, since papillomaviruses seem to be related not only to superficial skin and genital warts, but also to carcinomas of the lower genital tract as well as papillomas and carcinomas of the larynx, lung, oral cavity, and esophagus.[48]

## 2.2. Polyomaviruses

### 2.2.1. Polyomavirus

Polyomavirus infection is common in colonies of wild and laboratory mice.[51] Under natural circumstances, infant mice are infected by the virus when partially protected by maternal antiviral antibodies and the infection is completely asymptomatic.[52] After acute infection, polyomavirus persists for life in a latent state in the kidneys and salivary glands.[53,54] Although the mechanisms of persistence are still poorly understood, evidence obtained from natural and experimental situations indicates that virus latency is controlled by the immunologic status of the host.

*2.2.1.a. Effect of Cell-Mediated Immunity in Polyomavirus Latent Infection and Tumor Formation.* During the latent state, the kidneys probably harbor only viral DNA or a negligible quantity of complete virions.[55] However, since latently infected animals produce virus-specific antibodies,[56] it is assumed that small amounts of capsid proteins are produced and shed from persistently infected cells to stimulate an immunologic response. The low production of virus-specific antigens may represent a critical mechanism of latency, because expression of viral antigens may be absent or the density of antigenic molecules may be so low on the cell surface to render the infected cells insusceptible to T-cell lysis or to antibody-dependent cellular cytotoxicity (ADCC) mediated by complement and K cells.

Virus infectivity is usually not detected at 12 weeks postinfection or later if susceptible cell-culture monolayers are exposed to kidney homogenates from persistently infected, immunocompetent mice.[55] In such animals, infectious virus can only be reactivated in explant cultures of the latently infected kidney tissue,[55] even though viral DNA can be directly detected by DNA–DNA hybridization of total kidney DNA with $^{32}$P-labelled polyomavirus DNA probes.[57] On the contrary, polyomavirus infection of unprotected infant mice, thymectomized at birth, or reactivation of the latent infection in adult mice by the administration of antithymocyte serum and hydrocortisone lead to the isolation of high-titer infectious virus from kidney homogenates and urine.[56] In these animals, multiple tumors, formed by polyomavirus-transformed cells, appear in several organs.[58] Tumors are very rarely observed in immunocompetent mice under conditions of natural infection.[58] A neurologic disease with wasting and paralysis has also been described in T-cell-deficient nude mice inoculated as adults with polyomavirus.[59]

2.2.1.b. Reactivation of Polyomavirus Latent Infection in Pregnancy. In pathogen-free mice experimentally infected during pregnancy, polyomavirus can cross the placenta and infect fetuses with consequences which vary in severity with the stage of gestation.[60] Moreover, polyomavirus latent infection is reactivated late in pregnancy of persistently infected mice. In the latter case, infectious virus can readily be detected in kidneys, but not in other organs and fetuses.[55,56] Transplacental transmission does not occur in persistently infected mice probably because the viremia is neutralized by antivirus antibodies in the serum.

Although the factors underlying reactivation have not been clearly identified, the depression of cell-mediated immunity and the peculiar hormonal state that accompany pregnancy are probably important. Indeed, polyomavirus replication was enhanced both in infected mice and in tissue culture cells after treatment with estradiol benzoate and progesterone.[55] It seems likely that antiviral antibodies protect mice in acute polyomavirus infection by neutralizing the virus but that the latent infection and tumor induction are controlled by T lymphocytes and other effectors of cell-mediated immune responses.

### 2.2.2. Simian Virus 40

Simian virus 40 (SV40) produces a persistent asymptomatic infection in several species of monkeys. The main site of persistence is the kidney, and virus is shed in urine.[61] Indeed, it is common to isolate SV40 from tissue culture cells derived from rhesus or African green monkey kidneys. No other hosts have been described for SV40, except for humans who live in close contact with monkeys[62,63] or have been given polio virus vaccine contaminated by live SV40.[64,65] The virus is apparently not pathogenic in its immunocompetent natural host and does not cause tumors under conditions of natural infection, although it is highly oncogenic in experimentally inoculated rodents.[66]

Very little is known about the immunologic response of monkeys to SV40. Nevertheless, it seems that the immune system plays an important role in controlling SV40 persistence. Immunosuppressed monkeys developed progressive multifocal leukoencephalopathy, a demyelinating disease first described in humans[67] where it has been linked to JC virus infection. Virions with papovavirus morphology were detected by electron microscopy in glial cells, and SV40 was isolated from the brains of these monkeys.[68,69]

### 2.2.3. BK and JC Human Polyomaviruses

The two human polyomaviruses, BK virus (BKV) and JC virus (JCV), show a very similar epidemiology.[70] They have a worldwide distribution and are ubiquitous in normal human populations, since 60–80% of adults have antiviral serum antibodies.[71–74] Primary infection occurs in childhood and is generally inapparent. It is followed by a persistent, latent infection which is reactivated under conditions of impaired immunologic response. BKV acute infection is sometimes the cause of upper respiratory[75,76] or urinary tract disease,[77–79] and it has been related also to cases of Guillain-Barré syndrome.[80] No overt disease has yet been associated with JCV primary infection. The main site of persistence of both BKV and JCV is the kidney[56,81–83] and both viruses can be isolated from urine. BKV is latent also in tonsils,[76] whereas both viruses have not been found to be latent in the brain.[56,82,83]

2.2.3.a. *Participation of Lymphoid Cells in BKV and JCV Primary and Latent Infections.* The interactions of human polyomaviruses with the immune system are not fully understood. In particular, while many serologic surveys have been performed and an abundance of data is available on the humoral immune response,[71–74] almost no investigations have been carried out on cell-mediated immunity. Only recently cell-mediated immune responses to BKV have been studied in normal individuals,[84] opening the way to the analysis of specific cellular immunity in patients with acute infections or reactivation of the latent infection by human papovaviruses.

Some recent results suggest that lymphoid cells may be involved in the pathogenesis of human polyomavirus primary and latent infection. Viruses with papovavirus morphology have been observed by electron microscopy in lymphocytes of measles patients during the phase of the disease characterized by generalized depression of the cell-mediated immunity.[85] Moreover, direct experimental evidence suggests that BKV is able to infect human B and T lymphocytes *in vitro*.[86,87] BKV replication is somehow restricted in lymphocytes, since these cells produce about 100-fold less virus than do fibroblasts and eventually develop a persistent infection.[87]

Since BKV probably enters the body by the respiratory route, and the oropharynx is the initial site of primary infection,[75,76] lymphocytes persistently infected in the tonsils could carry the virus to the bloodstream, allowing its transport to other organs. Only lymphocytes from seropositive persons were found to bear virus receptors on the cell membrane and could be productively infected.[88] This suggests that individuals in the 20–40% of the general population that do not exhibit antibodies may be refractory to infection because of a lack of virus receptors on their lymphocytes. By contrast, monocytes are resistant to BKV infection,[87] although they possess surface receptors for virus adsorption and were shown by electron microscopy to bind and ingest BKV. Resistance to BKV was maintained, even when monocytes were infected after their differentiation into macrophages.[87] Therefore, it is likely that during the course of natural infection, irrespective of their state of differentiation, monocytes are involved in the degradation rather than in the replication of engulfed BKV particles.

2.2.3.b. *Role of Cell-Mediated Immunity in BKV and JCV Latent Infection.* Immunologic surveillance seems to be of paramount importance in controlling BKV and JCV latency. Virus reactivation is observed in immune-deficient or immunosuppressed patients.[70,89] This typically occurs in patients with malignant diseases of the lymphoid tissue, congenital immune deficiencies or severe iatrogenic immunosuppression following organ transplantation or antiblastic therapy for neoplasia; a series of conditions characterized by a profound impairment of cell-mediated immunity. The humoral response is preserved and high titers of antiviral antibodies are occasionally detected during reactivation.[70,89] The role of cell-mediated immunity in papovirus latency and reactivation is supported by the observation that SV40-specific cytotoxic T lymphocytes abrogate the virus lytic cycle by recognizing SV40 T antigen on the surface of infected cells.[114]

Reactivation of BKV latent infection in the urinary tract after renal transplantation may cause rejection of the grafted kidney, due to inflammation of the ureteric tissues at the junction site, followed by ureteric stenosis and obstruction.[90,91] In this situation, treatment with corticosteroids, with the aim of blocking rejection, may obtain the opposite result by further preventing the host immunologic response and thereby favoring the local infectious process.[91] Reactivation of JCV latent infection may cause progressive multifocal leukoencephalopathy (PML),[67] a demyelinating neuropathy that is invariably fatal. Demyelination is caused by JCV replication in oligodendrocytes the nuclei of which are greatly enlarged and full of viral particles.[92] Most lesions also contain giant astrocytes with bizarre hyperchromatic nuclei that resemble the malignant astrocytes of pleomorphic glioblastomas.[92]

The pathogenesis of the disease is not completely clear. Since JCV is not found latent in normal brains,[82,83] it is postulated that the virus reaches the brain under conditions of immunosuppression, when the cell-mediated immune surveillance becomes inefficient. Situations of heavy immunosuppression are rather common in medical practice, while PML is a very rare disease. This suggests that some other yet unknown factor, perhaps producing a selective reactivation of JCV latent infection in some patients, participates in the pathogenesis of PML. A marked deficiency of specific cell-mediated immunity to JCV antigens has been documented by *in vitro* tests on peripheral blood lymphocytes of PML patients.[93] This could represent the critical mechanism allowing spread of the virus to the brain. It is also unclear why only JCV is associated with PML, since JCV and BKV have very similar biologic properties and the latent infection by both viruses is reactivated by immunosuppression. The reason may be the marked tropism of JCV for the nervous system.

*2.2.3.c. Reactivation of BKV and JCV Latent Infection in Pregnancy.* Like polyomavirus infection, BKV and JCV latent infections are reactivated during pregnancy,[89] but so far there is no evidence of virus transmission to the fetus.[94] Recent studies suggest that BKV reactivation in pregnancy is associated with disturbed cell-mediated immunity, since pregnant women with virus reactivation had lower neutrophil counts and lymphocyte responses to PHA than did nonactivators, whereas antibody titers to the virus were comparable in the two groups and high antibody levels were often found in virus activators.[95]

*2.2.3.d. Immune Complexes.* Immune complexes containing BKV antigens have been detected by immunofluorescence in renal glomeruli from cases of lupus nephropathy, IgA nephropathy, acute poststreptococcal glomerulonephritis, membranoproliferative glomerulonephritis, and other nephropathies.[96] However, abolishment of fluorescence after absorption of anti-BKV serum with human Ig and the capability of anti-BKV serum to react with human Ig as antigens in control immunofluorescence tests demonstrated a cross-reaction between BKV capsid antigens and human Ig.[97] Therefore, if antibodies are produced against those antigenic determinants shared by BKV and human Ig, an immunologic response could be raised against Ig during natural BKV infection in humans. This could have pathogenetic significance, for instance, through amplification of Ig deposition by reaction of anti-Ig antibodies with Ig already deposited in renal glomeruli as immune complexes with the specific antigen.

BKV and JCV are highly oncogenic in rodents[98,99] and JCV in monkeys as well.[100] In hamsters, BKV induces three types of tumors: ependymomas, tumors of pancreatic islets, and osteosarcomas.[101-106] It would be interesting to investigate whether BKV is related to the same types of human tumors and also whether BKV and JCV are related to tumors of the urinary apparatus and to tumors arising in immunosuppressed patients.

### 2.2.4. B Lymphotropic Papovavirus

Recently a papovavirus with a specific tropism for B lymphocytes has been isolated from an African green monkey.[107] Characterization of the genome of this B lymphotropic papovavirus (LPV) by restriction endonuclease mapping and sequence analysis[108] has shown that it belongs to the *Polyomavirus* genus. LPV has the structural and biologic properties of polyomaviruses, including the ability to transform hamster cells *in vitro*.[109] The molecular basis of LPV-specific tropism for B lymphocytes is not clear at present. Recent experimental evidence indicates that the narrow host range of LPV may depend either on the peculiar organization of the viral transcriptional enhancer elements[110] or on the structure of the major capsid protein VP1,[111] perhaps permitting adsorption to membrane receptors present only on B cells.

The presence of antibodies to LPV in about 30% of human sera[112,113] suggests the existence of a related human virus. Isolation of the human variant could be of great interest for the possible involvement of LPV in immunosuppressive processes and in diseases of the lymphoid tissue. However, a preliminary serologic survey[113] has shown no difference in antibody frequency to LPV between normal subjects and patients with various types of tumors or with inflammatory and autoimmune diseases.

## 3. CONCLUSION

The main features of all papovaviruses are that they are ubiquitous, produce a latent infection, and are oncogenic for their natural hosts or experimental animals. The available results on the immunologic aspects of papovavirus infections emphasize the critical role of cell-mediated immunity in controlling virus latency and virus-induced tumor formation. The humoral immune response is protective in primary infections, when virus-neutralizing antibodies are produced. During reactivation, humoral immunity is stimulated by freshly produced viral antigen but seems devoid of any protective value.

We must note, however, that a discussion on the immunology of papovaviruses is far from being complete, owing to the lack of knowledge in many aspects of this field. While papovaviruses have been intensively investigated in regard to the molecular biology of their genome and to the mechanisms of neoplastic transformation, little experimental work has been dedicated to the study of the pathogenesis of the diseases, to host immunologic responses or immunopathologic mechanisms.

ACKNOWLEDGMENTS. The author's work described in this chapter was supported by Consiglio Nazionale delle Ricerche, Progetto Finalizzato Controllo della Crescita Neoplastica and by the Associazione Italiana per la Ricerca sul Cancro (A.I.R.C.).

## REFERENCES

1. Melnick, J. L., A. C. Allison, J. S. Butel, W. Eckhart, B. E. Eddy, S. Kit, A. J. Levine, J. A. R. Miles, J. S. Pagano, L. Sachs, and V. Vonka, Papovaviridae, *Intervirology*, **3**:106–120 (1974).
2. Melnick, J. L., Taxonomy and nomenclature of viruses, *Prog. Med. Virol.* **28**:208–221 (1982).
3. Tegtmeyer, P. Genetics of SV40 and polyoma virus, in: *Molecular Biology of Tumor Viruses*, Vol. 2: *DNA Tumor Viruses*, J. Tooze (ed.), P. 297–337. Cold Spring Harbor Laboratory, Cold Spring Harbor, New York (1980).
4. Pfister, H., Biology and biochemistry of papillomaviruses, *Rev. Physiol. Biochem. Pharmacol.* **99**:111–181 (1984).
5. Smith, K. T., and M. S. Campo, The biology of papillomaviruses and their role in oncogenesis, *Anticancer Res.* **5**:31–48 (1985).
6. Kidd, J. G., J. W. Beard, and P. Rous, Serological reaction with a virus causing a rabbit papilloma which becomes cancerous. II. Tests of the blood of animals carrying various tumors, *J. Exp. Mod.* **64**:63–78 (1936).
7. Seto A., K. Notake, M. Kawanishi, and Y. Ito, Development and regression of Shope papillomas induced in newborn domestic rabbits, *Proc. Soc. Exp. Biol. Med.* **156**:64–67 (1977)
8. Kreider, J. W., Neoplastic progression of the Shope rabbit papilloma, *Cold Spring Harbor Conf. Cell Prolif.* **7**:283–300 (1980).
9. Hellström, I., C. A. Evans, and K. E. Hellström, Cellular immunity and its serum-mediated inhibition in Shope-virus-induced rabbit papillomatosis, *Int. J. Cancer* **4**:601–607 (1969).
10. McMichael, H., Inhibition by methylprednisolone of regression of the Shope rabbit papilloma, *J. Natl. Cancer Inst.* **39**:55–65 (1967).
11. Jarrett, W. F. H., P. E. McNeil, W. T. R. Grimshaw, I. E. Selman, and W. I. M. McIntyre, High incidence area of cattle cancer with a possible interaction between an environmental carcinogen and a papilloma virus, *Nature (Lond.)* **274**:215–217 (1978).

12. Campo, M. S., M. H. Moar, W. F. H. Jarrett, and H. M. Laird, A new papillomavirus associated with alimentary tract cancer in cattle, *Nature (Lond.)* **286**:180–182 (1980).
13. Almeida, J. D., and A. P. Goffe, Antibody to wart virus in human sera demonstrated by electron microscopy and precipitin tests, *Lancet* **2**:1205–1207 (1965).
14. Almeida, J. D., J. D. Oriel, and L. M. Stannard, Characterization of the virus found in human genital warts, *Microbios* **3**:225–232 (1969).
15. Cubie H. A., Serological studies in a student population prone to infection with human papilloma virus, *J. Hyg.* **70**:677–690 (1972).
16. Pfister, H., and H. zur Hausen, Seroepidemiological studies of human papilloma virus (HPL-1) infections, *Int. J. Cancer* **21**:161–165 (1978).
17. Matthews, R. S. and P. V. Shirodaria, Study of regressing warts by immunofluorescence, *Lancet* **1**:689–691 (1973)
18. Shirodaria, P. V., and R. S. Matthews, An immunofluorescence study of warts, *Clin. Exp. Immunol.* **21**:329–338 (1975).
19. Ogilvie, M. M., Serological studies with human papova (wart) virus, *J. Hyg.* **68**:479–483 (1970).
20. Genner, J., Verruca vulgaris. II. Demonstration of a complement fixation reaction, *Acta Derm. Venereol. (Stockh)* **51**:365–373 (1971).
21. Pyrhönen, S., and K. Penttinen, Wart virus antibodies and the prognosis of wart disease, *Lancet* **2**:1330–1332 (1972).
22. Pyrhönen, S., and E. Johansson, Regression of warts. An immunological study, *Lancet* **1**:592–596 (1975).
23. Brodersen, I., and J. Genner, Histological and immunological observations on common warts in regression, *Acta Derm. Venereol.* **53**:461–464 (1973).
24. Pfister, H., G. Gross, and M. Hagedorn, Characterization of human papillomavirus 3 in warts of a renal allograft patient, *J. Invest. Dermatol.* **73**:349–353 (1979).
25. Pass, F., and J. V. Maizel, Wart-associated antigens. II. Human immunity to viral structural proteins, *J. Invest. Dermatol.* **60**:397–311 (1973).
26. Spencer, E. S., and H. K. Anderson, Clinically evident non-terminal infections with herpesviruses and the wart virus in immunosuppressed renal allograft patients, *Br. Med. J.* **111**:251–254 (1970)
27. Starzl, T. E., K. A. Porter, G. Andres, C. G. Halgrimson, R. Hurwitz, G. Giles, P. I. Terasaki, I. Penn, G. T. Schroter, J. Lilly, S. J. Starkie, and C. W. Putman, Long-term survival after renal transplantation in humans, *Ann. Surg.* **172**:437–472 (1970).
28. Koranda, F. C., E. M. Dehmel, G. Kahn, and I. Penn, Cutaneous complications in immunosuppressed renal homograft recipients, *JAMA* **229**:419–424 (1974).
29. Perry T. L., and L. Harman, Warts in diseases with immune defects, *Cutis* **13**:359–362 (1974).
30. Morison, W. L., Viral warts, herpes simplex and herpes zoster in patients with secondary immune deficiencies and neoplasms, *Br. J. Dermatol.* **92**:625–630 (1975).
31. Reid, T. M. S., N. G. Fraser, and I. R. Kernohan, Generalized warts and immune deficiency, *Br. J. Dermatol.* **95**:559–564 (1976).
32. Ward, M., A. Le Roux, W. P. Small, and W. Sircus, Malignant lymphoma and extensive viral wart formation in a patient with intestinal lymphangiectasia and lymphocyte depletion. *Postgrad. J.* **53**:753–757 (1977).
33. Takigawa, M., H. Tagami, S. Watanabe, A. Ogino, S. Imamura, and S. Ofugi, Recovery processes during regression of plane warts, *Arch. Dermatol.* **113**:1214–1218 (1977).
34. Oguchi, M., J. Komura, H. Tagami, and S. Ofuji, Ultrastructural studies of spontaneously regressing plane warts. Macrophages attack verruca-epidermal cells, *Arch. Dermatol. Res.* **270**:403–411 (1981).
35. Oguchi, M., J. Komura, H. Tagami, and S. Ofuji, Ultrastructural studies of spon-

taneously regressing plane warts. Langerhans cells show marked activation, *Arch. Dermatol Res.* **271**:55–61 (1981).

36. Jablonska, S., J. Dabrowski, and K. Jakubowicz, Epidermodysplasia verruciformis as a model in studies on the role of papova viruses in oncogenesis, *Cancer Res.* **32**:583–589 (1972).

37. Glinski, W., S. Jablonska, A. Langner, S. Obalek, M. Haftek, and M. Proniewska, Cell-mediated immunity is epidermodysplasia verruciformis, *Dermatologica* **153**:218–227 (1976).

38. Glinski, A., S. Obalek, S. Jablonska, and G. Orth, T cell defect in patients with epidermodysplasia verruciformis due to human papillomavirus type 3 and 5, *Dermatologica* **162**: 141–147 (1981).

39. Prawer, S. E., F. Pass, J. C. Vance, E. J. Greenberg, E. J. Yunis, and A. S. Zelickson, Depressed immune function in epidermodysplasia verruciformis, *Arch. Dermatol.* **113**:495–499 (1977).

40. Obalek, S., W. Glinski, M. Haftek, G. Orth and S. Jablonska, Comparative studies on cell mediated immunity in patients with different warts, *Dermatologica* **161**:75–83 (1980).

41. Jablonska, S., Orth, G., Glinski, G., Obalek, S., Jarzabek-Chorzelska, M., Croissant, O., Favre, M., and Rzesa, G., Morphology and immunology of human warts and familial warts, in: *Leukaemias, Lymphomas and Papillomas: Comparative Aspects* P. A. Bachmann, (ed.), pp. 107–131. Taylor and Francis, London (1980).

42. Morison, W. L., Cell-mediated immune responses in patients with warts, *Br. J. Dermatol.* **93**:553–556 (1975).

43. Chretien, J. H., J. G. Essweins, and V. F. Garagusi, Decreased T cell levels in patients with warts, *Arch. Dermatol.* **114**:213–215 (1978).

44. Seski, J. C., E. R. Reinhalter, and J. Silva, Jr., Abnormalities of lymphocyte transformations in women with condylomata acuminata, *Obstet. Gynecol.* **51**:188–192 (1977).

45. Sillman, F., A. Stanek, A. Sedlis, J. Rosenthal, K. W. Lanks, D. Buchhagen, A. Nicastri, and J. Boyce, The relationship between human papillomavirus and lower genital intraepithelial neoplasia in immunosuppressed women, *Am. J. Obstet. Gynecol.* **150**:300–308 (1984).

46. Jablonska, S., S. Obalek, G. Orth, M. Haftek, and M. Jarzabek-Chorzelska, Regression of the lesions of epidermodysplasia verruciformis, *Br. J. Dermatol.* **107**:109–116 (1982).

47. Freed, D. L. J., and K. E. Eyres, Persistent warts protected from immune attack by a blocking factor, *Br. J. Dermatol.* **100**:731–733 (1979).

48. Gissmann, L., Papillomaviruses and their association with cancer in animals and in man, *Cancer Surv.* **3**:161–181 (1984).

49. Gissmann, L., L. Wolnik, H. Ikenberg, U. Koldovsky, H. G. Schnurch, and H. zur Hausen, Human papillomavirus type 6 and 11 DNA sequences in genital and laryngeal papillomas and in some genital cancers, *Proc. Natl. Acad. Sci. U S A* **80**:560–563 (1983).

50. McCance, D. J., P. G. Walker, J. L. Dyson, D. V. Coleman, and A. Singer, Presence of human papillomavirus DNA sequences in cervical intraepithelial neoplasia, *Br. Med. J.* **287**:784–788 (1983).

51. Rowe, W. P., J. W. Hartley, L. W. Law, and R. J. Huebner, Studies of mouse polyoma virus infection. III. Distribution of antibodies in mouse colonies, *J. Exp. Med.* **109**:449–462 (1959).

52. Law, L. W., C. J. Dawe, W. P. Rowe, and J. W. Hartley, Antibody status of mice and response of their litters to paroid tumor virus (polyomavirus), *Nature (Lond.)* **184**:1420–1421 (1959).

53. Rowe, W. P., J. W. Hartley, and R. J. Huebner, The ecology of a mouse tumor virus, in: *Perspectives in Virology*, Vol.II (M. Pollard, (ed.), pp. 117–194, Rutgers University Press, New Brunswick, New Jersey (1961).

54. Rowe, W. P., J. W. Hartley, J. D. Estes, and R. J. Huebner, Growth curves of polyoma in mice and hamsters, in: *Symposium on Phenomena of the Tumor Viruses*, National Cancer Institute Monograph No. 4, pp. 187–209 (1960).

55. McCance, D. J., and C. A. Mims, Reactivation of polyoma virus in kidneys of persistently infected mice during pregnancy, *Infect. Immun. 25:* 998–1002 (1979).

56. McCance, D. J., Persistence of animal and human papovaviruses in renal and nervous tissues, in: *Polyomaviruses and Human Neurological Diseases*, J. L. Sever and D. L. Madden, (eds), pp. 343–357, Liss, New York (1983).

57. McCance, D. J., Growth and persistence of polyoma early region deletion mutants in mice, *J. Virol.* **39:**958–962 (1981).

58. Allison, A.C., Immune responses to polyoma virus-induced tumors, in: *Viral Oncology* (G. Klein, (ed.), pp. 481–487, Raven, New York (1980).

59. McCance, D. J., A. Sebesteny, B. E. Griffin, F. Balkwill, R. Tilly, and N. A. Gregson, A paralytic disease in nude mice associated with polyomavirus infection, *J. Gen. Virol.* **64:**57–67 (1983).

60. McCance, D. J., and C. A. Mims, Transplacental transmission of polyomavirus in mice, *Infect. Immun.* **18:**196–202 (1977).

61. Ashkenjai, A., and J. L. Melnick, Induced latent infection of monkeys with vacuolating SV40 papovavirus. Virus in kidneys and urine, *Proc. Soc. Exp. Biol. Med.* **111:**367–372 (1962).

62. Shah, K. V., Neutralizing antibodies to Simian Virus 40 (SV40) in human sera from India, *Proc. Soc. Exp. Biol. Med.* **121:**303–307 (1966).

63. Shah, K. V., Investigation of human malignant tumors in India for Simian Virus 40 etiology, *J. Natl. Cancer Inst.* **42:**139–145 (1969).

64. Gerber, P., Pattern of antibodies to SV40 in children following the last booster with inactivated poliomyelitis vaccines, *Proc. Soc. Exp. Biol. Med.* **125:**1284–1291 (1967).

65. Shah, K. V., H. L. Ozer, H. S. Pond, L. D. Palma, and G. P. Murphy, SV40 neutralizing antibodies in sera of US residents without history of polio immunization, *Nature (Lond.)* **231:**448–450 (1971).

66. Topp, W. C., D. B, Rifkin and U. J. Sleigh, SV40 mutants with an altered small-t protein are tumorigenic in newborn hamsters, *Virology*, **111:**341–350 (1981).

67. Aström, K. E., E. L. Mancall, and E. P. Richardson, Jr., Progressive multifocal leukoencephalopathy, *Brain* **81:**93–111 (1958).

68. Gribble, D. H., C. C. Haden, R. V. Henrickson, and L. W. Schwartz, Spontaneous progressive multifocal leukoencephalopathy (PML) in macaques, *Nature (Lond.)* **254:**602–604 (1975).

69. Holmberg, C. A., D. H. Gribble, K. K. Takemoto, P. M. Howley, C. Espana, and B. I. Osburn, Isolation of Simian Virus 40 from rhesus monkeys (*Macaca mulatta*) with spontaneous progressive multifocal leukoencephalopathy, *J. Infect. Dis.* **136:**593–596 (1977).

70. Padgett, B. L., and D. L. Walker, New human papovaviruses, *Prog. Med. Virol.* **22:**1–35 (1976).

71. Gardner, S. D., Prevalence in England of antibody to human polyomavirus (B.K.), *Br. Med. J.* **1:**77–78 (1973).

72. Shah, K. V., R. W. Daniel, and R. M. Warszawski, High prevalence of antibodies to BK virus, an SV-40 related papovavirus, in residents of Maryland, *J. Infect. Dis.* **128:**784–787 (1973).

73. Portolani, M., A. Marzocchi, G. Barbanti-Brodano, and M. La Placa, Prevalence in Italy of antibodies to a new human papovavirus (BK Virus), *J. Med. Microbiol.* **7:**543–546 (1974).

74. Brown, P., T. Tsai, and D. C. Gajdusek, Seroepidemiology of human papovaviruses: Discovery of virgin populations and some unusual patterns of antibody prevalence among remote peoples of the world, *Am. J. Epidemiol.* **102:**331–340 (1975).

75. Goudsmit, J., M. L. Baak, K. W. Slaterus, and J. van der Noordaa, Human papovavirus isolated from urine of a child with acute tonsillitis, *Br. Med. J.* **283:**1363–1364 (1981).
76. Goudsmit, J., P. Wertheim-van Dillen, A. van Strien, and J. van der Noordaa, The role of BK virus in acute respiratory tract disease and the presence of BKV DNA in tonsils, *J. Med. Virol.* **10:**91–99 (1982).
77. Hashida, J., P. C. Gaffney, and E. J. Yunis, Acute haemorrhagic cystitis of childhood and papovavirus-like particles, *J. Pediatr.* **89:**85–87 (1976).
78. Mininberg, D. T., C. Watson, and M. Desquitado, Viral cystitis with transient secondary vesicoureteral reflux, *J. Urol.* **127:**983–985 (1982).
79. Padgett, B. L., D. L. Walker, M. M. Descuitado, and D. V. Kim, BK virus and non-haemorrhagic cystitis in a child, *Lancet* **1:**770 (1983).
80. van der Noordaa, J., and P. Wertheim-van Dillen, Rise in antibodies to human papovavirus BK and clinical disease, *Br. Med. J.* **1:**1471 (1977).
81. Heritage, J., P. M. Chesters, and D. J. McCance, The persistence of papovavirus BK DNA sequences in normal human renal tissue, *J. Med. Virol.* **8:**143–150 (1981).
82. Chesters, P. M., J. Heritage, and D. J. McCance, Persistence of DNA sequences of BK virus and JC virus in normal human tissues and in diseased tissues, *J. Infect. Dis.* **147:**676–684 (1983).
83. Grinnell, B. W., B. L. Padgett, and D. L. Walker, Distribution of nonintegrated DNA from JC papovavirus in organs of patients with progressive multifocal leukoencephalopathy, *J. Infect. Disc.* **147:**669–675 (1983).
84. Drummond, J. E., K. V. Shah, and A. D. Donnenberg, Cell-mediated immune responses to BK virus in normal individuals, *J. Med. Virol.* **17:**237–247 (1985).
85. Lecatsas, G., B. D. Schoub, A. R. Rabson, and M. Joffe, Papovavirus in human lymphocyte cultures, *Lancet* **2:**907–908 (1976).
86. Lecatsas, G., E. Blignaut, and B. D. Schoub, Lymphocyte stimulation by urine-derived human polyoma virus (BK), *Arch. Virol.* **55:**165–167 (1977).
87. Portolani, M., M. Piani, G. Gazzanelli, M. Borgatti, A. Bartoletti, M. P. Grossi, A. Corallini, and G. Barbanti-Brodano, Restricted replication of BK virus in human lymphocytes, *Microbiologica* **8:**59–66 (1985).
88. Possati, L., C. Rubini, M. Portolani, G. Gazzanelli, M. Piani, and M. Borgatti, Receptors for the human papovavirus BK on human lymphocytes, *Arch. Virol.* **75:**131–136 (1983).
89. Coleman, D. V., Recent developments in the papovaviruses: the human polyomaviruses (BK virus and JC virus), in: *Recent Advances in Clinical Virology* (A. P. Waterson, (ed.), pp. 89–110, Churchill Livingstone, London (1980).
90. Coleman, D. V., A. M. Field, S. D. Gardner, K. A. Porter, and T. E. Starzl, Virus-induced obstruction of the ureteric and cystic duct in allograft recipients, *Transplant. Proc.* **5:**95–98 (1973).
91. Coleman, D. V., E. F. D. Mackenzie, S. D. Gardner, J. M. Poulding, B. Amer, and W. J. I. Russell, Human polyomavirus (BK) infection and ureteric stenosis in renal allograft recipients, *J. Clin. Pathol.* **31:**338–347 (1978).
92. Zu Rhein, G. M., Association of papova-virions with a human demyelinating disease (progressive multifocal leukoencephalopathy), *Prog. Med. Virol.* **11:**185–247 (1969).
93. Willoughby, E., R. W. Price, B. L. Padgett, D. L. Walker, and B. Dupont, Progressive multifocal leukoencephalopathy (PML): In vitro cell-mediated immune responses to mitogens and JC virus, *Neurology (NY)* **30:**256–262 (1980).
94. Borgatti, M., F. Costanzo, M. Portolani, C. Vullo, L. Osti, M. Masi, and G. Barbanti-Brodano, Evidence for reactivation of persistent infection during pregnancy and lack of congenital transmission of BK virus, a human papovavirus, *Microbiologica* **2:**173–178 (1979).
95. Coleman, D. V., S. D. Gardner, C. Mulholland, V. Fridiksdottir, A. A. Porter, R. Lilford,

and H. Valdimarsson, Human polyomavirus in pregnancy. A model for the study of defence mechanisms to virus reactivation, *Clin. Exp. Immunol.* **53:**289–296 (1983).

96. Panem, S., N. G. Ordonez, A. I. Katz, B. H. Spargo, and W. H. Kirsten, Viral immune complexes in systemic lupus erythematosus, *Lab. Invest.* **39:**413–420 (1978).

97. Donini, U., F. Poli, L. Rossi, F. Palanchetti, G. Altavilla, A. Corallini, P. Zucchelli, and G. Barbanti-Brodano, A search for BK virus immune complexes in human nephropathies: immunological cross-reaction between BK virus structural antigens and human immunoglobulins, *Microbiologica* **4:**87–100 (1981).

98. Howley, P. M., Molecular biology of SV40 and the human polyomaviruses BK and JC, in: *Viral Oncology* (G. Klein, ed.), pp. 489–550, Raven, New York (1980).

99. Padgett, B., Human papovaviruses, in: *Molecular Biology of Tumor Viruses*, Vol. 2: *DNA Tumor Viruses*, J. Tooze (ed.), pp. 339–370, Cold Spring Harbor Laboratory, Cold Spring Harbor, New York (1980).

100. London, W. T., S. A. Houff, D. L. Madden, D. A. Faccillo, M. Gravell, W. C. Wallen, A. E. Palmer, J. L. Sever, B. L. Padgett, D. L. Walker, and G. M. Zu Rhein, Brain tumors in owl monkeys inoculated with a human polyomavirus (JC virus), *Science* **210:**1246–1249 (1978).

101. Uchida, S., S. Watanabe, T. Aizawa, K. Kato, A. Furuno, and T. Muto, Induction of papillary ependymomas and insulinomas in the syrian golden hamster by BK virus, a human papovavirus, *Gann* **67:**857–865 (1976).

102. Costa, T., C. Yee, T. S. Trakla, and A. S. Rabson, Hamster ependymomas produced by intracerebral inoculation of human papovavirus (MMV), *J. Natl. Cancer Inst.* **56:**863–864 (1976).

103. Corallini, A., G. Barbanti-Brodano, W. Bortoloni, I. Nenci, E. Cassai, M. Tampieri, M. Portolani, and M. Borgatti, High incidence of ependymomas induced by BK virus, a human papovavirus, *J. Natl. Cancer Inst.* **59:**1561–1563 (1977).

104. Corallini A., G. Altavilla, M. G. Cecchetti, G. Fabris, M. P. Grossi, P. G. Balboni, G. Lanza, and G. Barbanti-Brodano, Ependymomas, malignant tumors of pancreatic islets and osteosarcomas induced in hamsters by BK virus, a human papovavirus, *J. Natl. Cancer Inst.* **61:**875–883 (1978).

105. Uchida, S., S. Watanabe, T. Aizawa, A. Furuno and T. Muto, Polyoncogenicity and insulinoma-inducing ability of BK virus, a human papovavirus, in syrian golden hamsters, *J. Natl. Cancer Inst.* **63:**119–126 (1979).

106. Corallini, A., G. Altavilla, L. Carrà, M. P. Grossi, G. Federspil, A. Caputo, M. Negrini, and G. Barbanti-Brodano, Oncogenicity of BK virus for immunosuppressed hamsters, *Arch. Virol.* **73:**243–253 (1982).

107. zur Hausen, H., and L. Gissmann, Lymphotropic papovavirus isolated from African green monkey and human cells, *Med. Microbiol. Immunol.* **167:**137–153 (1979).

108. Pawlita, M., A. Clad, and H. zur Hausen, Complete DNA sequence of lymphotropic papovavirus: Prototype of a new species of the polyomavirus genus, *Virology* **143:**196–211 (1985).

109. Takemoto, K. K., and T. Kanda, Lymphotropic papovavirus transformation of hamster embryo cells, *J. Virol.* **50:**100–105 (1984).

110. Mosthaf, L., M. Pawlita, and P. Gruss, A viral enhancer element specifically active in human haematopoietic cells, *Nature (Lond.)* **315:**597–600 (1985).

111. Kanda, T., and K. K. Takemoto, Monkey B-lymphotropic papovavirus mutant capable of replicating in T-lymphoblastoid cells, *J. Virol.* **55:**96–100 (1985).

112. Brade, L., N. Müller-Lantzsch, and H. zur Hausen, B-lymphotropic papovavirus and possibility of infection in humans, *J. Med. Virol.* **6:**301–308 (1980).

113. zur Hausen, H., Gissmann, L. Mincheva, A., and Böcher, J. F. Characterization of a lymphotropic papovavirus, in *Viruses in Naturally Occurring Cancers* (M. Essex, G. Todaro,

and H. zur Hausen, (eds). pp. 365–372, Cold Spring Harbor Laboratory, Cold Spring Harbor, New York (1980).

114. Bates, M. P., S. R. Jennings, Y. Tanaka, M. J. Tevethia, and S. S. Tevethia, Recognition of simian virus 40 T antigen synthesized during viral lytic cycle in monkey kidney cells expressing mouse H-2K$^b$- and H-2D$^b$-transfected genes by SV40-specific cytotoxic T lymphocytes leads to the abrogation of virus lytic cycle, *Virology* **162**:197–205 (1988).

# 4

# Adenoviruses

## WARREN A. ANDIMAN and MARIE F. ROBERT

## 1. ADENOVIRUSES: BASIC PROPERTIES

The adenoviruses of humans, of which there are now 41 serotypes, are naked icosahedrons, 70–90 nm in diameter, and contain a genome composed of linear double-stranded DNA. They are associated with a wide spectrum of diseases (Table I) and they have been isolated from virtually all organs, but they are primarily regarded as common pathogens of the respiratory tract and eye. Most individuals become infected early in life with at least several serotypes. To a certain extent, the kinds of adenovirus-associated diseases to which people become susceptible change as they grow older; each of these illnesses is caused by a limited number of serotypes. With rare exceptions, adenovirus infections are short-lived and self-limited.

The adenoviruses are so named because they were originally discovered in explants of human adenoid tissues that spontaneously degenerated *in vitro* producing cytopathic effects that are now known to be characteristic of the group. The first human strains to be associated with a discrete clinical illness were isolated from young army recruits suffering from an acute influenza-like syndrome, later known as acute respiratory disease syndrome (ARDS). Subsequently the adenoviruses were shown to be etiologically related to episodes of acute febrile pharyngitis in infants and young children, to pharyngoconjunctival fever, to some cases of pneumonia and pertussis, as well as to non-respiratory tract disease, e.g., epidemic keratoconjunctivitis, acute hemorrhagic cystitis, and intussusception. Most recently, it has been learned that some of the higher numbered adenoviruses that are not cultivable in standard tissue culture cells are frequent causes of infantile diarrhea (types 40 and 41) and that other hybrid adenoviruses can be isolated regularly from the urine of

WARREN A. ANDIMAN and MARIE F. ROBERT • Departments of Pediatrics and Epidemiology and Public Health, Yale University School of Medicine, New Haven, Connecticut 06510.

**TABLE I**
**Common Clinical Syndromes Associated with Adenoviruses**

| Syndrome | Common serotypes |
|---|---|
| Respiratory | |
| Infants and children | |
| Coryza, pharyngitis (endemic) | 1, 2, 3, 5, 6, 7 |
| Pharyngoconjunctival fever (epidemic) | 2, 3, 4, 7, 14 |
| Pertussis syndrome | 1, 2, 3, 5, 12, 19 |
| Adults | |
| Acute respiratory disease (ARD) of recruits (epidemic) | 4, 7, 3 |
| Pneumonia (epidemic and sporadic) | 3, 7, 7a, 21 |
| Ophthalmologic | |
| Follicular conjunctivitis | 1, 2, 3, 4, 6, 7, 9, 10, etc. |
| Epidemic keratoconjunctivitis | 8 |
| Infection in the immunosuppressed host | |
| Renal and pulmonary infection in recipients of renal allografts | 34, 35 |
| Urinary infection in patients with AIDS | 34, 35 and 34/35 hybrids |
| Miscellaneous | |
| Acute hemorrhagic cystitis | 11, 21 |
| Gastroenteritis | 2, 3, 4, 40[a], 41[a] |
| Intussusception | 1, 2, 3, 5, 6, 7 |
| Encephalitis and meningoencephalitis | 7, 1, 6, 12 |

[a]Enteric or uncultivatable.

patients with acquired immune deficiency syndrome (AIDS, types 34 and 35).

Three qualitatively different kinds of biologic behaviors have been observed in the interaction of adenoviruses with mammalian cells: lytic infection, chronic persistent infection, and cell transformation. Cell lysis, with a concomitant release of $10^4$–$10^6$ progeny virus particles per cell (most particles are noninfectious), occurs when most of the pathogenic strains of humans infect continuous epithelial cell lines. Chronic persistent infection results from the interaction of some adenoviruses with lymphoid cells. This aspect of adenovirus behavior was suspected in early studies when surgically extirpated adenoid and tonsillar tissues, generated after weeks *in vitro* in the absence of neutralizing antibody, eventually released infectious virus and manifested a characteristic cytopathic effect. More recently, some biologic aspects of the interaction of adenoviruses with mononuclear cells of primates have been studied experimentally. These observations are discussed in greater detail in Section 4. Several adenovirus serotypes, when injected subcutaneously, produce tumors in rodents after a long incubation period. Tumor induction is associated with failure of viral replication to go beyond the stage when approximately one half the early transcripts appear, probable integration of viral DNA into the host cell genome, and production of large amounts of the so-called T antigen. There is no evidence of an association between adenoviruses and any human neoplasm. It is not known whether adenoviruses can truly establish latency i.e., whether in a small number of cells the entire viral genome can move between periods of complete inexpression to periods of complete replication.

### Structural and Soluble Antigens and the Immune Response

The adenovirus capsid is composed of 240 hexons, 12 pentons, and fibers—rodlike structures with terminal knobs that project from the penton base capsomers. These three proteins become incorporated into the structure of new infectious virions and are also produced in great excess during the viral replicative cycle.

The penton bases are composed principally of group reactive antigens that are common to most members of the family. Rabbit antiserum prepared against purified penton protein has only a low titer of neutralizing activity. The hexons and fibers represent the major antigenic sites on the viral surface. The fibers are responsible for attachment of the virus to erythrocytes in hemagglutination reactions and the principal antigenic sites on the fiber seem to be primarily type specific, with some subgroup specificity. It has been suggested that the terminal knob on the fiber contains the type-specific antigenic determinant and that the subgroup determinant is carried by the shaft. Antifiber antibody can cause disruption of the penton capsomer into its two entities, penton base and fiber, and the effect is most pronounced with subgroup-specific antibody.[1] In cross-neutralization tests employing fiber from a variety of strains, numerous minor antigenic cross reactions have been observed. The hexon has at least two major antigenic sites and elicits a heterogeneous population of antibodies that produce family cross-reactivity in the complement fixation (CF) test and marked type specificity in the neutralization (NT) test. In general, antibody produced against highly purified hexon antigen neutralizes the infectivity of homologous virus only.

Historically, the adenoviruses have been divided into subgroups based on their ability to hemagglutinate monkey or rat red blood cells (RBC). In general, members of each subgroup also share other biologic properties, i.e., high, moderate, or low oncogenic potential in rodent species; ability to transform cells in tissue culture; and the percentage of guanine and cytosine residues in DNA. Although the hexons and fibers carry specificities that can induce heterotypic responses in hemagglutination inhibition (HI) and NT tests, these determinants are shared primarily by immunotypes within the same subgroup. Antigenic cross-reactivity outside the subgroup can rarely be demonstrated.

## 2. IMMUNE RESPONSES TO ADENOVIRUS

### 2.1. Antibody Responses to Natural Infection in the Normal Host

Following natural infection, antibodies begin to appear in blood by 8–10 days, reaching maximal levels 14–21 days later. Both NT and CF antibodies rise simultaneously but, whereas the CF antibodies begin to decline at 2–4 months, the NT antibodies persist for many years with only a two- to threefold decrease in titer. Minor rises in heterotypic NT antibodies occur following infection, especially if they are already present as a result of prior infection with another strain. Because natural secondary reinfections with the same serotype are rare, the magnitude of a booster response and the time of its appearance are unknown.

Antibodies to the early antigens (EA) of adenovirus appear about 5 days after the onset of clinical symptoms, reach a peak 15–30 days later, and begin to decline after a few months.[2] EA responses to proteins of adenovirus DNA homology groups A–D do not show intergroup cross-reactions, but intragroup cross-reactivity among serotypes is common. Group E adenoviruses appear to share early antigens with all the other groups.

The level of IgA in nasal secretions is inversely correlated with the severity of disease following natural infection.[3] At the time of virus isolation, U.S. Marine recruits requiring hospitalization for ARD due to adenovirus type 7 had less IgA in their nasal secretions than soldiers less clinically affected. Some of the IgA measured in the nose comes from the blood, as a result of transudation with other serum proteins. However, the major portion of the IgA probably comes from glandular epithelial cells in the nasal mucosa or sinus epithelium or from IgA-synthesizing plasma cells in the submucosa.

During adenovirus conjunctivitis, a significant rise in the tear IgG level has been observed and is accompanied by a decline in serum IgG.[4] It is not known whether IgG in the eye reflects local production, active transport from the blood, or transudation from serum.

## 2.2. Humoral and Local Antibody Responses to Adenovirus Vaccines

Soon after the high incidence and morbidity associated with ARDS in military recruits were recognized, vaccine development programs were undertaken.[5] The early trials with inactivated vaccines given by the parenteral route were successful in inducing an immune response and in reducing the number of hospitalizations associated with ARDS. However, the program using these vaccines was curtailed when it was recognized that some adenoviruses were oncogenic in lower animals and that there was recombination between the adenovirus genome and SV40 DNA present in the simian cultures used to propagate the vaccine virus. In addition, there was great variability in the antigenicity associated with each lot of vaccine. Subsequent trials have involved the use of intranasal vaccine, monovalent and bivalent vaccines given orally in the form of enteric-coated capsules, and subunit vaccines composed of soluble antigens normally expressed on the viral surface.

Enteric-coated capsules containing adenovirus types 4 and 7 have been studied extensively in the military.[6] It had been known that adenoviruses infect both the respiratory and gastrointestinal (GI) tracts but that GI symptoms rarely occur following infection. Trials of live vaccines contained in enteric capsules were predicated on the hope that live virus would induce an enteric infection and would cross-protect against type-specific respiratory disease. Initial hopes invested in this concept of immunization were soon justified when these vaccines were shown to reduce dramatically the incidence and severity of ARDS in new recruits. NT antibodies could be detected in the blood of most vaccinees by 18 days; by the third or fourth week the geometric mean titers of neutralizing antibody ranged from 1 : 38 to 1 : 100. The titers induced by dual immunization are lower than those induced by either type alone, but they are sufficient to prevent natural infection by the homologous strains. The intestinal infection remains silent and noncommunicable in barrack-mates, but transmission does occur by the fecal–oral route in the confines of the family.

Serum and intestinal antibodies are induced following ingestion of the oral adenovirus vaccines but no increase in titer of respiratory tract immunoglobulin A (IgA) antibody has been observed.[7,8] Nevertheless, the vaccine does protect against respiratory tract disease. Although some vaccinees became infected subsequently with the homologous strain by the natural route, hospitalization was rarely required. This finding suggests that serum antibody has a marked ameliorating effect on the manifestations of illness associated with infection. There is similar evidence suggesting that serum antibody to respiratory syncytial virus may protect against the severe lower airway disease associated with infection in the first 6 months of life and that some influenza vaccines induce a higher titer of nasal antibody when given by the parenteral route than by the nasal route. Whether serum antibody actually finds its way into the respiratory tract secretions or whether some unknown protective factors, such as interferon (IFN) or complement, are induced in the respiratory tract following immunization at another site is unknown. It has been observed in one study that oral administration of the lower numbered adenoviruses leads to viral shedding in the pharynx of a few subjects. The mechanisms responsible are unknown.

Adenovirus fiber and hexon antigens have been prepared as immunogens that can be given intramuscularly.[9] Both antigens induce NT antibodies that persist for months. Some vaccinees subsequently challenged by the ocular route developed mild conjunctivitis and shed small amounts of virus from the eye or rectum, but none developed disease of the respiratory tract. When heterotypic responses occurred, they were almost exclusively limited to serotypes belonging to the same immunologic group. Production problems with these components of the virus have limited the more widespread use of these types of vaccines.

## 2.3. Natural Infection in the Immunosuppressed Host: Clinical Aspects

Numerous clinical reports indicate that adenovirus infections tend to be more severe in children who are malnourished (reflected by poor weight gain or iron deficiency anemia) or who develop their infections secondary to measles.[10–12] In such persons, the infection in the lung progresses, causing severe necrotizing bronchitis, bronchiolitis obliterans, and interstitial alveolitis. In addition, the virus may disseminate to multiple organs, causing hepatitis or encephalitis. Whether adenovirus infection tends to be more severe in children with measles because of the altered T-cell function and anergy that accompany measles or because of the increased pathogenicity of adenoviruses when they supervene on an already damaged respiratory epithelium is not known.

That cell-mediated immune (CMI) mechanisms may play an important role in maintaining the latent or chronic persistent phases of adenovirus infection in the normal host has been supported in recent years by the discovery of new adenovirus serotypes in the urine of immunosuppressed patients. During the late 1970s, adenovirus types 34 and 35 were isolated for the first time from the urine and/or lungs of renal allograft recipients.[13,14] In one case, serologic studies indicated that the infection in the recipient was primary, suggesting that the new virus strain was acquired via the grafted kidney. In other instances, it could not be determined whether the grafted kidney contained the virus or whether the virus was activated in host lymphocytes during graft rejection or immunosuppression. Evidence supporting the idea that the virus is introduced with the renal allograft comes from experimental data showing that adenoviruses can persist in monkey kidney tissue for long periods and clinical data indicating that the usual source of the virus is the urine. In normal humans, the high-numbered serotypes might be responsible for large numbers of asymptomatic urinary tract infections and the establishment of persistent or latent infection. Such infections would only become apparent with the development of immunosuppression. Why serotypes 34 and 35 reactivate when the host is immunosuppressed, to the exclusion of at least six other serotypes that have been isolated from urine, is not known. It is possible that these two serotypes, 34 and 35, are only pathogenic for the compromised host. Concern that these serotypes might be partly responsible for the increased incidence of neoplasms during the post-transplant period would be heightened if it could be demonstrated definitively that these viruses truly establish latency.

More recently adenoviruses have been isolated with increasing frequency from the urine of patients with acquired immune deficiency syndrome (AIDS).[15] These strains are either type 34/35, as determined by HI or pure types 34 or 35, as determined by restriction enzyme analysis. Some of the isolates are genomic hybrids formed by the recombination of type 35 DNA with a small portion of the fiber-coding region of adenovirus 7. There is no evidence that these agents represent anything other than opportunistic pathogens; infection may be primary or due to reactivation of persistent virus.

## 3. IMMUNOSUPPRESSIVE EFFECTS OF ADENOVIRUSES

### 3.1. Modulation of T-Cell Number and Function

The natural history of the CMI response to adenovirus has not been studied well in normal subjects. However, it has been observed clinically that children with thymic alymphoplasia (DiGeorge syndrome) and Swiss-type agammaglobulinemia (severe combined immunodeficiency disease, autosomal recessive form) may develop disseminated and sometimes fatal disease following adenovirus infection.[16] In other instances, previously healthy persons have developed fulminant disease in association with an acquired lymphopenia that is both severe (absolute count 1500/mm) and prolonged.[17,18] The humoral responses in these patients has been normal. In some of these patients, leukopenia, anergy, and lymphopenia are reversed in convalescence. In other instances, clinical improvement and a return to normal of T-cell function has been attributed to the use of a humoral factor from calf thymus.[19] Both decreased E-rosette-forming activity and poor response to phytohemagglutinin (PHA) have also been observed in the course of *in vitro* infection of peripheral blood lymphocytes with adenoviruses.[20]

Several investigators have attributed the lymphopenia that accompanies adenovirus infection, as well as other viral and mycoplasmal infections, to the production of cytotoxic antibodies directed against autologous and allogeneic lymphocytes.[21,22] These antibodies are of the IgM class and their activity is complement dependent. It has been hypothesized that severe clinical disease might occur if the patient develops lymphopenia during a first infection and is then superinfected with a new agent, as would occur with adenovirus infection following measles.

## 3.2. Modulation of Macrophage Function

Intraperitoneal injection of mice with adenovirus type 6 results in a decreased antibody response to sheep red blood cells (SRBC) given 3–11 days postinfection. This indicates some degree of adenovirus-induced macrophage dysfunction.[23] Heat or ultraviolet (UV) light treatment abolishes the immunosuppressive effects of the virus. When the SRBC were given by a different route, or 1 day before or 2 weeks after infection with adenovirus type 6, the immunosuppressive effects were not observed.[24] The effect of adenovirus type 6 was thought to be selective for macrophages because silica, which is specifically toxic for macrophages, when inoculated into the peritoneum 2 hrs before the virus infection, resulted in ablation of the immunosuppressive effect.

## 3.3. Modulation of NK Cells and/or ADCC

Cook and Lewis[25] observed a difference in natural killer (NK) cell and activated macrophage responses to cells infected with nononcogenic versus oncogenic adenoviruses. *In vitro*, oncogenic and nononcogenic adenoviruses can equally transform hamster and rat cells which are both highly immunogenic.[26] Cells infected with the nononcogenic adenovirus type 2 are more readily lysed in target cell assays using hamster NK cells and bacillus Calmette–Guérin (BCG)-activated macrophages compared with similar cells infected with the oncogenic adenovirus type 12. The increased lysis of adenovirus type 12-infected cells was not affected by the presence of cytosine arabinoside (Ara C), which inhibits DNA synthesis. Therefore, investigators have speculated that an early gene product, such as T antigen, might be responsible for the variable response of some effector cells to adenovirus-infected cells; the longer survival of the adenovirus type 12 infected cells may permit viral infection of cells to progress to transformation and tumor formation.

Adenoviruses appear to enhance antibody-dependent cellular cytotoxicity (ADCC). In a study using adenovirus type 6, investigators found that cytotoxicity of chicken mononuclear cells against chicken anti-SRBC-coated SRBC was enhanced following intravenous injection of adenovirus type 6 into chickens.[27] ADCC was enhanced 14 to 24 hr after virus infection but then decreased; the preinjection level was reached after 36 hr. The apparent effect on ADCC by adenovirus type 6 appeared to involve nonphagocytic mononuclear cells, since removal of phagocytic cells by the use of carbonyl iron (thereby reducing the percentage of mononuclear cells from 5% to 1%) did not affect the results of the ADCC assay.

## 3.4. Modulation of Soluble Factors

There are conflicting data concerning the production of IFN by human cells following infection with adenoviruses.[28–30] However, human adenoviruses can cause IFN production in chickens and have been studied in chick embryo cell systems.[31,32]

Production of IFN in chick cells following infection with adenovirus type 5 requires interaction of the virus with the cell, although infectious virus is not produced. Heat or UV irradiation of the virus decreases its ability to stimulate IFN production.[33] In addition, chemical digestion of adenovirus with trypsin causes reduction of IFN production.[31,32] Different adenovirus serotypes differ in their ability to induce IFN in chick cells.[34] UV irradiation decreased the IFN-inducing ability of the more effective inducers, indicating that transcription of viral DNA might be important for IFN induction by these types. Other cells found not to produce IFN after adenovirus infection include those from mouse, monkey, and hamster. However, hamsters do produce IFN following intravenous inoculation.[35]

Although it is not clear whether adenoviruses induce IFN during human infection, there have been reports of IFN limiting infection by adenoviruses. Romano et al. used human fibroblast-derived $IFN_\beta$ to treat epidemic keratoconjunctivitis and observed a reduction in the length of the illness when compared with controls treated with corticosteroids or placebo.[36,37] It has also been shown that $2–5 \times 10^5$ reference unit daily doses of $IFN_\beta$ begun early in the course of adenovirus type 8 epidemic keratoconjunctivitis, almost totally prevented the appearance of subepithelial keratitis, which occurred in 57% of the control group. Langford et al. noted a synergistic effect of antibody and $IFN_\beta$ or $IFN_\gamma$ or inhibition of adenovirus type 3 infection of Chang human conjunctival cells or WISH cells.[30] Interestingly, antibody to $IFN_\beta$ did not affect the decrease in virus yield due to antibody to adenovirus type 3, suggesting that adenovirus type 3 did not induce IFN.

In conclusion, it is not clear whether human infection with adenoviruses leads to IFN production, although IFN of at least two types have the capacity to limit adenovirus infection. In the chick cell system, IFN production does occur after interaction of infectious virus with cells.

### 3.5. Suppression of Antibody Production

Infection of chickens with adenovirus types 6, 8, and 12 and of guinea pigs with adenovirus type 16 induces transient suppression of the antibody response to unrelated antigens.[38,39] In chickens, the primary response to a nonviral antigen (i.e., sheep red blood cells) was suppressed in that the production of serum hemagglutinins and 19S hemolytic plaque producing antibodies was reduced most markedly when the fowl were challenged with sheep cells 4–8 days after adenovirus infection. Hamsters inoculated intraperitoneally with adenovirus 16 produced less antibody to Sendai virus when challenged nine days later with Sendai, but the anti-Sendai hemagglutinins approached control levels seen in animals uninfected with adenovirus 16 within 8–10 days. These transient alterations in the humoral response to antigenic challenge have been attributed to antigenic competition, induction of interferon, or reduction in antibody production by cells of the immune system that are infected with adenoviruses.

Unlike Moloney, Rauscher, or Friend disease viruses, infection of hamsters with oncogenic viruses such as SV40 and adenovirus 12 produce little or no prolonged suppression of the immune response to sheep red cells. Transient suppression of the number of antibody-forming cells occurs during the first 2 weeks of life, but it is believed to be unlikely that such transient suppression could affect significantly the subsequent development of neoplasia.[40]

## 4. VIRAL REPLICATION WITHIN THE IMMUNE SYSTEM

The adenoviruses were originally discovered in explants of human adenoid and tonsillar tissue, but it was not until a decade later that investigations were initiated to define more precisely the relationship between these viruses and the lymphoid tissues with which they are so commonly associated. During the mid-1960s, it was shown that both epithelial and fibroblastic cells in tonsils and adenoids were susceptible to the growth of adenoviruses *in vitro*, and it was suggested that persistent infection in these cultures was maintained by the continued presence of susceptible cells in the culture and the slow release of infectious virions from such cells into the surrounding medium.[41] These early observations were extended a decade later when Lambriex and van der Veen demonstrated that adenovirus type 2 was capable of replicating to a limited extent in purified lymphocyte cultures derived from human adenoid.[42] Although as many as $10^3$–$10^4$ $TCID_{50}$ of virus could be measured in culture fluids between days 4 and 8, only 1–3 cells per million were found to produce virus. The life span of these cultures was not reduced, also reflecting the fact that very few cells were lytically infected. The growth of virus was enhanced by the presence of PHA. Interestingly, adenovirus type 4 would not replicate in these adenoid-derived lymphocyte cultures, suggesting that the tropism of adenovirus for mononuclear cells is to some extent a biologic property of some serotypes but not others.

Subsequent studies explored the possibility that lymphoid cells other than those in adenoid and tonsil could support the growth of adenoviruses. For example, it has been shown that some adenoviruses can replicate in PHA-stimulated leukocytes from Burkitt lymphoma cell lines or from human umbilical cord blood, and that lymphoblastoid cell lines derived from human umbilical cord blood by immortalization with Epstein–Barr virus (EBV) could be persistently infected with adenovirus type 5.[43,44] Continuous production of infectious virus in these cultures could be interrupted by the addition of specific NT antibody. These data indicated that persistent infection in lymphoid cells might be maintained by the continual infection of cells by complete virus. That an adenovirus-associated leukoviremia might occur naturally was later shown by the isolation of adenovirus type 2 from the peripheral blood mononuclear cells of a 5-month-old infant with documented pneumonia attributable to adenovirus and an associated atypical lymphocytosis.[45] It was hypothesized that bloodstream invasion by virus-infected white blood cells (WBC) might distribute adenoviruses to lymphoid tissues throughout the body.

A final group of studies further explored the mechanisms by which adenoviruses might persist in lymphoid cells *in vivo*.[46] After demonstrating that adenovirus types 5 and 6 could be recovered from cultures of primary umbilical cord leukocytes and from EBV-transformed lymphocytes for up to 3 months, the effects of adding homologous antibody to the cultures were examined. Infection could be obliterated from cultures of EBV-transformed simian cells following exposure to antibody. However, adenovirus could readily be recovered for long periods of time from immortalized human umbilical cord lymphocytes even though all the virus in the supernatant fluids of the same cultures had been neutralized by the antibody. It was estimated that at the peak of infection, 1–8% of the human cells released infectious virus and that each cell produced 2–8 $TCID_{50}$ of virus. These experiments indicated that chronic infection of lymphoid cells is maintained by two mechanisms: intracellular persistence of virus in a small number of cells in the presence of antibody, and cell–cell spread of small amounts of virus in the absence of antibody.

In summary, there is clear evidence that adenoviruses can replicate in immunocompetent cells derived from cultures of human tissue and in human and simian EBV-transformed lymphoblastoid cell lines. However, there are many unanswered questions regarding the nature of this infection. It is not known precisely which cell types are involved. *In vitro* experiments demonstrate that EBV-transformed cells are involved; these are B cells. However, it has also been hypothesized that macrophages derived from umbilical cord blood in culture might also be important for maintaining infection.[46] Whichever cells are involved, it is likely that adenoviruses can persist indefinitely in the human host, successfully eluding the immune system's defenses. But there is no evidence that adenoviruses can establish true latency, accompanied by intermittent activation, of the kind that is characteristic of the herpesviruses.

## 5. SIGNIFICANCE

In the vast majority of persons, adenovirus infections are short-lived, self-limited, and without consequences. NT antibodies directed against antigenic moieties on the surface of the virus appear following infection, are long-lived, and greatly limit the capacity of the virus to propagate. Antibodies and other antiviral substances produced as a result of vaccination by the oral route appear to protect against respiratory disease, thereby suggesting the existence of a general mechanism whereby induction of infection at one site prevents disease at another by way of transudation of antibody and/or IFN. In addition, CMI mechanisms contribute to limiting infection. Patients with defects in T-cell function, either genetic or acquired, as in the wake of measles, may suffer the consequences of disseminated or progressive infection. The effects of adenovirus infection on macrophage function, NK cells, ADCC, and IFN production have not been well studied in humans, but there is no significant body of data to indicate that any of these arms of the immune response are greatly affected by adenoviruses or that any play a critical role in limiting infection.

There is no evidence that adenoviruses can establish latency in humans, nor has a connection been found between any of the adenoviruses and human neoplasm. However, chronic persistent infection of mononuclear leukocytes does occur and appears to be maintained by the intracellular persistence of virus in a relatively small number of cells, the slow release of small numbers of infectious virions from such cells, and the continued presence of susceptible cells in the lymphoid tissues of the host. There is also clinical evidence to suggest that certain adenovirus serotypes may persist in renal epithelium, only to be reactivated in the recipient following transplantation and concomitant immunosuppression. Other pure adenovirus serotypes or viral hybrids may behave as opportunistic infectious agents in patients with AIDS. In these patients, infection also appears to originate in, and be limited to, the kidney.

## REFERENCES

1. Boudin, M.-L., and P. Boulanger, Antibody triggered dissociation of adenovirus penton capsomer, *Virology* 113:781–786 (1981).
2. Gerna, G., E. Cattaneo, M. Grazia Revello, M. Battaglia, and G. Achilli, Antibody to human adenovirus early antigens during acute adenovirus infections, *Infect. Immun.* 32:778–787 (1981).
3. McCormick, D. P., R. P. Wenzel, J. A. Davies, and W. E. Beam, Nasal secretion protein responses in patients with wild-type adenovirus disease, *Infect. Immun.* 6:282–288 (1972).
4. Gupta, A. K., and G. S. Sarin, Serum and tear immunoglobulin levels in acute adenovirus conjunctivitis, *Br. J. Ophthalmol.* 67:195–198 (1983).

5. Top, F. H., Jr., Control of adenovirus acute respiratory disease in U.S. Army trainees, *Yale J. Biol. Med.* **48:**185–195 (1975).

6. Top, F. H., R. A. Grossman, P. J. Bartelloni, H. E. Segal, B. A. Dudding, P. K. Russell, and E. L. Buescher, Immunization with live types 7 and 4 adenovirus vaccine. I. Safety, infectivity, antigenicity and potency of adenovirus type 7 vaccine in humans, *J. Infect. Dis.* **124:**148–160 (1971).

7. Smith, T. J., E. L. Buescher, F. H. Top, W. A. Altemeier, and J. M. McCown, Experimental respiratory infection with the type 4 adenovirus vaccine in volunteers: Clinical and immunological responses, *J. Infect. Dis.* **122:**239–248 (1970).

8. Schwartz, A. R., Y. Togo, and R. B. Hornick, Clinical evaluation of live oral types 1,2,5 adenovirus vaccines, *Am. Rev. Respir. Dis.* **109:**233–238 (1974).

9. Kasel, J. A., R. H. Alford, J. R. Lehrich, P. A. Banks, M. Huber, and V. Knight, Adenovirus soluble antigens for human immunization: A progress report, *Am. Rev. Respir. Dis.* **94:**170–174 (1966).

10. Jen, K.-F., Y. Tai, Y.-C. Lin, and H.-Y. Wang, The role of adenovirus in the etiology of infantile pneumonia and pneumonia complicating measles, *Chinese Med. J.* **81:**141–146 (1962).

11. Schonland, M., M. L. Strong, and A. Wesley, Fatal adenovirus pneumonia: Clinical and pathological features, *S. Afr. Med. J.* **50:**1748–1751 (1976).

12. Lang, W. R., C. W. Howden, J. Laws, and J. F. Burton, Bronchopneumonia with serious sequelae in children with evidence of adenovirus type 21 infection, *Br. Med. J.* **1:**73–79 (1969).

13. Keller, E. W., R. H. Rubin, P. H. Black, and M. Hirsch, Isolation of adenovirus type 34 from a renal transplant recipient with interstitial pneumonia, *Transplantation* **23:**188–190 (1977).

14. Stalder, H., J. C. Hierholzer, and M. Oxmar, New human adenovirus (candidate adenovirus type 35) causing fatal disseminated infection in a renal transplant patient, *J. Clin. Microbiol.* **6:**257–265 (1977).

15. deJong, P. J., G. Valderrama, I. Spigland, and M. S. Horwitz, Adenovirus isolates from urine of patients with acquired immunodeficiency syndrome, *Lancet* **1:**1293–1296 (1983).

16. Wigger, H. J. and W. A. Blanc, Fatal hepatitis and necrosis in adenovirus infection with thymic alymphoplasia, *N. Engl. J. Med.* **275:**870–874 (1966).

17. Dudding, B. A., S. C. Wagner, J. A. Zeller, J. T. Gmelich, E. R. French, and F. H. Top, Fatal pneumonia associated with adenovirus type 7 in three military trainees, *N. Engl. J. Med.* **286:**1289–1292 (1972).

18. Strieder, D. J., and G. Nash, Weekly clinicopathological exercises. Case 12-1975, *N. Engl. J. Med.* **292:**634–640 (1975).

19. Varsano, I., T. M. Schonfeld, Y. Matoth, B. Shohat, T. Englander, V. Rotter, and N. Trainin, Severe disseminated adenovirus infection successfully treated with a thymic humoral factor, *Acta Paediatr. Scand.* **66:**329–331 (1977).

20. Horvath, J., G. Kolesar, J. P. Ugryumov, P. Das, I. Nasz, Z. F. Barinsky, G. Simon, and J. Ongradi, Effect of adenovirus infection on human peripheral lymphocytes, *Acta Microbiol. Hung.* **30:**203–209 (1983).

21. Huang, S.-W., D. B. Lattos, D. B. Nelson, K. Reeb, and R. Hong, Antibody-associated lymphotoxin in acute infection, *J. Clin. Invest.* **52:**1033–1040 (1973).

22. Huang, S.-W. and R. Hong, Immunologic deficiences during viral infection. (Letter.), *N. Engl. J. Med.* **292:**1296 (1975).

23. Berencsi, K., M. Bakay, and I. Beladi, The role of macrophages in adenovirus-induced immunosuppression in mice, *Acta Virol.* **29:**61–65 (1985).

24. Berencsi, K., M. Bakay, and P. Kovacs, Effect of human adenovirus type 6 on the primary immune response in mice, *Acta Virol.* **26:**340–345 (1982).

25. Cook, S. L., and A. M. Lewis, Differential NK and macrophage killing of hamster cells infected with nononcogenic or oncogenic adenovirus, *Science* **224**:612–615 (1984).
26. Gallimore, P. H. and C. Paraskeva, A study to determine the reasons for differences in the tumorigenicity of rat cell lines transformed by adenovirus 2 and adenovirus 12, *Cold Spring Harbor Symp. Quant. Biol.* **44**:703–713 (1979).
27. Mandi, Y., M. Bakay, and I. Beladi, Effect of human adenoviruses on antibody-dependent cellular cytotoxicity (ADCC) in chickens, *Cell. Immunol.* **69**:395–400 (1982).
28. Lysov, V. V., O. A. Aksenon, V. I. Rudenko, S. A. Moshkin, T. I. Huyrlova, A. A. Smorodintsev, and A. A. Selivanon, Interference activity and sensitivity to interferon of original and cold strains of adenovirus types 1 and 2, *Acta Virol. (Praha)* **15**:387–392 (1971).
29. Pusztai, R., K. Berendsi, I. Beladi, and E. Szabo, Relationship between interferon production and transformation of chick cells infected with human adenovirus type 12, in: *Interferon and Interferon Inducers* (I. Foldes and M. Talas, eds.), pp. 100–108, Research Group of the Hungarian Academy of Sciences, Budapest (1976).
30. Langford, M. P., A. L. Villarreal, and G. J. Stanton, Antibody and interferon act synergistically to inhibit enterovirus, adenovirus, and herpes simplex virus infection, *Infect. Immun.* **41**:214–218 (1983).
31. Beladi, I., and R. Pusztai, Interferon-like substance produced in chick fibroblast cells inoculated with human adenovirus, *Z. Naturforsch.* **226**:165–169 (1967).
32. Ho, M., and K. Kohler, Studies on human adenoviruses as inducers of interferon in chick cells, *Arch. Gesamte Virusforsch.* **22**:69–78 (1967).
33. Pusztai, R., I. Beladi, M. Bakay, and I. Mucsi, Effect of ultraviolet irradiation and heating on the interferon inducing capacity of human adenoviruses, *J. Gen. Virol* **4**:169–176 (1969).
34. Toth, M., M. Bakay, B. Torodi, S. Toth, R. Pusztai, and I. Beladi, Different interferon-inducing ability of human adenovirus types in chick embryo cells, *Acta Virol. (Praha)* **27**:337–345 (1983).
35. Ho, M., Animal viruses and interferon formation, in: *Interferon and Interferon Inducers* (N. B. Finter ed.), pp. 29–44, North-Holland, Amsterdam (1973).
36. Romano, A., M. Revel, D. Guarari-Rotman M. Blumenthal, and R. Stein, Use of human fibroblast-derived (beta) interferon in the treatment of epidemic adenovirus keratoconjunctivitis, *J. Interferon Res.* **1**:95–100 (1980).
37. Romano, A., E. Ladizensky, D. Guarari-Rothman, and M. Revel, Clinical effect of human fibroblast-derived (beta) interferon in treatment of adenovirus epidemic keratoconjunctivitis and its complications, *Tex. Rep. Biol. Med.* **41**:559–565 (1981–1982).
38. Beladi, I., R. Pusztai, I. Mucsi, M. Bakay, and G. Bajszar, Effect of human adenoviruses on the response of chickens to sheep erythrocytes, *Infect. Immun.* **7**:22–28 (1973).
39. Hamburg, V., O. Scherbakova, and G. Svet-Moldavsky, Suppression of antibody formation against Sendai Virus in SV40 and adenovirus 16 infected hamsters, *Experientia* **26**:532–534 (1970).
40. Friedman, H., and H. Goldner, Relationship between immunologic maturation and viral oncogenesis in hamsters, *J. Natl. Cancer Inst.* **44**:809–817 (1970).
41. Strohl, W. A. and R. W. Schlesinger, Quantitative studies of natural and experimental adenovirus infections of human cells, *Virology* **26**:208–220 (1965).
42. Lambriex, M., and J. van der Veen, Comparison of replication of adenovirus type 2 and type 4 in human lymphocyte cultures, *Infect Immun.* **14**:618–622 (1976).
43. Faucon, N., and C. Desgranges, Persistence of human adenovirus 5 in human cord blood lymphoblastoid cell lines transformed by Epstein–Barr virus, *Infect. Immun.* **29**:1180–1184 (1980).
44. Faucon, N., Y. Chardonnet, M. C. Perrinet, and R. Sohier, Superinfection with adenovirus of Burkitt's lymphoma cell lines, *J. Natl. Cancer Inst.* **53**:305–307 (1974).

45. Andiman, W. A., R. I. Jacobson, and G. Tucker, Leukocyte-associated viremia with adenovirus type 2 in an infant with lower-respiratory tract disease, *N. Engl. J. Med.* **297:**100–101 (1977).
46. Andiman, W. A., and G. Miller, Persistent infection with adenovirus types 5 and 6 in lymphoid cells from humans and woolly monkeys, *J. Infect. Dis.* **145:**83–88 (1982).

# 5

# Herpes Simplex

## LAURE AURELIAN

## 1. INTRODUCTION

The term herpes has been in the medical vocabulary for at least 25 centuries. In the Hippocratic corpus, it was used to describe an assortment of cutaneous lesions, including clinical descriptions compatible with herpes simplex and herpes zoster lesions. During the early nineteenth century, six clinical entities, including facial and genital herpes, were delineated. However, they were not considered communicable, possibly because of the idiosyncratic appearance of symptoms in conjunction with disparate well-defined febrile illnesses.[1] In the years since then it has been established that two serotypes of herpes simplex virus (HSV) infect humans: type 1 (HSV-1), which primarily causes oropharyngeal lesions, and type 2 (HSV-2), which primarily causes genital disease. Characteristic of the pathogenesis of the disease, is the ability of the virus to persist in the host indefinitely, becoming periodically reactivated to cause recurrent cutaneous disease. This chapter reviews available information on HSV-induced immunity and considers the premise that immunomodulation plays a critical role in disease pathogenesis.

## 2. BASIC PROPERTIES

### 2.1. Virion Structure

Herpes simplex virus is a large (150 to 200-nm) enveloped DNA virus consisting of four distinct morphologic elements: (1) an electron-opaque core containing primarily DNA, (2) an icosahedral capsid enclosing the core, (3)

LAURE AURELIAN • Department of Pharmacology and Experimental Therapeutics and Microbiology, University of Maryland School of Medicine, Baltimore, Maryland 21201; and Divisions of Comparative Medicine and Biophysics, The Johns Hopkins Medical Institutions, Baltimore, Maryland 21205.

electron-dense amorphous material (tegument) surrounding the capsid, and (4) an envelope or membrane the outer surface of which exhibits numerous small spikes.

The diameter of the HSV capsid is approximately 100 nm, and the number of capsomers is 162. The hexameric capsomers are 9.5 × 12.5 nm in longitudinal cross section, with a channel 4 nm in diameter running part way along the length of the prism. The tegument is present in most virions, but its amount varies from virion to virion. The envelope, a triple-layered membraneous structure, is acquired by budding of the capsids from an infected cell membrane, reviewed by Spear and Roizman.[2]

HSV-1 and HSV-2 DNAs are linear and double-stranded, approximately $100 \times 10^6$ daltons and have a base composition of 67 and 69 G + C moles %, respectively.[3] They contain two sets of reiterated sequences at the terminals and are in inverted form internally.[4] These sequences subdivide the viral DNA into two parts (L and S) that structurally resemble prokaryotic DNA sequences capable of excision and insertion into the same or different DNAs. Viral DNA bears some homology to eukaryotic cell DNA.[5,6] Conceivably, both the inverted positions of the L and S segments relative to each other, and the capacity of the virus to cause nonproductive infection may hinge on the ability of the L and S components to insert and excise into either host or viral DNA.

Herpes simplex virus virions have been reported to contain 15–33 proteins.[2,7,8] The precise number has been difficult to determine because of (1) resolution problems, (2) contamination with species difficult to separate from virions, and (3) the presence in the virions of both precursor and product forms of the same protein. Only six proteins are located in the capsids. Some proteins are glycosylated, and they are located in the envelope. Two of these, glycoprotein B (gB) and gC, are encoded by the L region of HSV-1 DNA and two others, gD and gE, by the S region. HSV-2 likewise contains gB, gC, gD, and gE (reviewed by Spear).[9] In addition, it contains gG which is also encoded by the S region of the genome.[10] The probable HSV-1 equivalent of gG was recently identified.[11] The remainder of the virion proteins are probably constituents of the tegument.

Very little is known about the function of individual virion proteins. Nucleocapsid proteins are presumably essential for morphogenesis, tegument proteins for envelopment. Glycoproteins probably mediate adsorption to, and penetration into, the host cell and cell to cell fusion, such as suggested for gB.[12,13] They also appear to be the main determinants of virus-specific protective immunity (Section 3).

Antigenic and biologic markers differentiate between HSV-1 and HSV-2, although the viruses are antigenically related and their DNAs share 47–50% of their nucleotide sequences under stringent hybridization conditions.[14] Restriction endonuclease analyses revealed that epidemiologically unrelated isolates of the same HSV serotype are not identical. The major differences seen were occasional deletions and the presence or absence of restriction endonuclease cleavage sites.[15] Consistent with these differences in viral DNA within (intratypic) and between (intertypic) serotypes, few of the HSV-1 proteins have electrophoretically identical counterparts in HSV-2, and variations have been reported in the electrophoretic mobilities of proteins from various isolates of one serotype.[16]

Recent studies using monoclonal antibodies (MAbs) and synthetic peptides have identified distinct and overlapping antigenic domains (epitopes) on one viral glycoprotein, gD. The dominant epitope was a type-common continuous domain located within N-terminal amino acid residues 8–23 of the mature gD of HSV-1 (gD-1). Another epitope was located within the first 16 residues of the HSV-2 encoded gD (gD-2). Both were shown to induce antibody capable of neutralizing the infectivity of HSV-1 and HSV-2,[17,18] and one (or both) were also able to stimulate T cells to proliferate *in vitro*[19] (Section 3). Other identified epitopes include two continuous ones located, respectively, at residues 268–287 (type-common) and 340–356 (HSV-1 specific), and four discontinuous ones located within the first 260 amino acids of the mature protein.[18]

## 2.2. Replicative Cycle

The widely accepted interpretation is that adsorption of the virus to the host cells is followed by fusion of the virion envelope with the cell-surface membrane, thereby liberating the nucleocapsid into the cytoplasm.[20] It can be inferred that viral DNA is released and transported to the nucleus, where early transcription is mediated by cellular polymerase. At least three phases have been identified in the transcriptional program of HSV-1, in which both the extent of transcription of viral DNA and the accumulation of viral RNA in the cytoplasm are tightly regulated. Central to the understanding of the mechanism of replication of HSV DNA is the inversion of the L and S components relative to each other. Thus, it remains to be determined whether all four DNA isomers are functionally equivalent (for review see Spear and Roizman[2]).

The HSV-1 and HSV-2 genomes each contain $1.6 \times 10^5$ bp, 10% of which are reiterated twice such that the maximum asymmetric coding capacity, excluding possible multiple reading frames, is approximately 44,000 amino acids. Estimates of the virus-specific infected cell proteins (molecular weight 20,000 to >250,000) range between 44–50,[2,8 21,22] requiring a coding capacity for approximately 41,000 amino acids. They form at least three groups ($\alpha$, $\beta$, $\gamma$), the synthesis of which is coordinately regulated and sequentially ordered in cascade fashion.[23] The $\alpha$- or immediate early (IE) proteins are made before any other viral protein. Six IE proteins have been identified and of these, at least four have regulatory functions.[24–26] The $\beta$- or early (E) proteins include the enzymes involved in viral DNA replication. Their synthesis peaks at 5–7 hr postinfection (p.i.). The $\gamma$- or late (L) proteins are primarily structural virion components. Their synthesis requires the presence of functional IE and E proteins and they are made at increasing rates until 15–18 hr p.i. Both the $\beta$- and $\gamma$-groups can be subdivided into two subgroups in terms of their kinetics of synthesis and some of their properties. The precise role of the various proteins in immunity is still unclear (see Section 3).

Of major significance from the standpoint of immunity are the HSV-induced alterations in infected cell plasma membranes. These alterations are manifested, at least in part, by changes in the behavior of infected cells relative to each other, by the acquisition of a receptor for the Fc domain(s) of immunoglobulin (Ig) (for review see Spear and Roizman[2] and Spear[9]), and for HSV-1, by the acquisition of a receptor for complement component fragment C3b.[27] They are presumably acquired as a consequence of the incorporation of virus-specified glycoproteins into the plasma membrane. gE is at least part of the Fc receptor,[28] and the C3b receptor function has been assigned to gC.[27]

What is the role of these receptors in HSV infection? It has been suggested that the Fc receptor (1) is a protective mechanism by the virus (backward binding blocks specific antibody), (2) facilitates cellular penetration by antibody-sensitized virus, and (3) is involved in superinfection by facilitating nonspecific (Fc mediated) binding of an unrelated pathogen. Since C3b receptor purified from human erythrocytes converts C3b to C3bi, thereby inactivating the complement system, it has also been suggested that the HSV-1-induced C3b receptor protects the infected cell from complement-mediated injury.[27]

## 3. MODULATION OF IMMUNE RESPONSES BY HERPESVIRUS

The immune response induced by HSV infection includes (1) an early nonspecific containment phase characterized primarily by local host defense factors, and (2) a later specific effector phase directed toward virus eradication and consisting of humoral and cell-mediated immune (CMI) functions. Since these responses generally do not prevent establishment of regional (ganglionic) latency and recurrent disease, consideration of the role of immunity in HSV disease must address two problems: (1) the definition of the function, specificity, and regulation of virus-induced immunity; and (2) the role that it plays in the reduction of acute peripheral virus replication and its relationship to the establishment and maintenance of an asymptomatic (latent) state.

### 3.1. Antibody Production

Humoral immunity to HSV antigens is entirely dependent on T-helper functions. Animals with impaired T-cell functions, such as congenitally athymic nude mice (nu/nu), neonatally thymectomized mice, or mice treated with antithymocyte serum, have a severely decreased ability to produce anti-HSV antibody.[29] Seroconversion in the IgM class is first detected at 5–15 days p.i. and is followed by the appearance of IgG (Fig. 1). The latter persists indefinitely, presumably reflecting the propensity of the latent virus to periodically reactivate, such that infectious virus[30] and viral antigen[31] are shed by asymptomatic hosts. However, kinetic and quantitative details of serum antibody detection depend on the nature of the antigen (virus type and protein specificity), the method of antigen presentation (with adjuvant, in liposomes, route of immunization), and the animal species used in the particular study.

In one rather comprehensive study,[32] the kinetics of antibody ap-

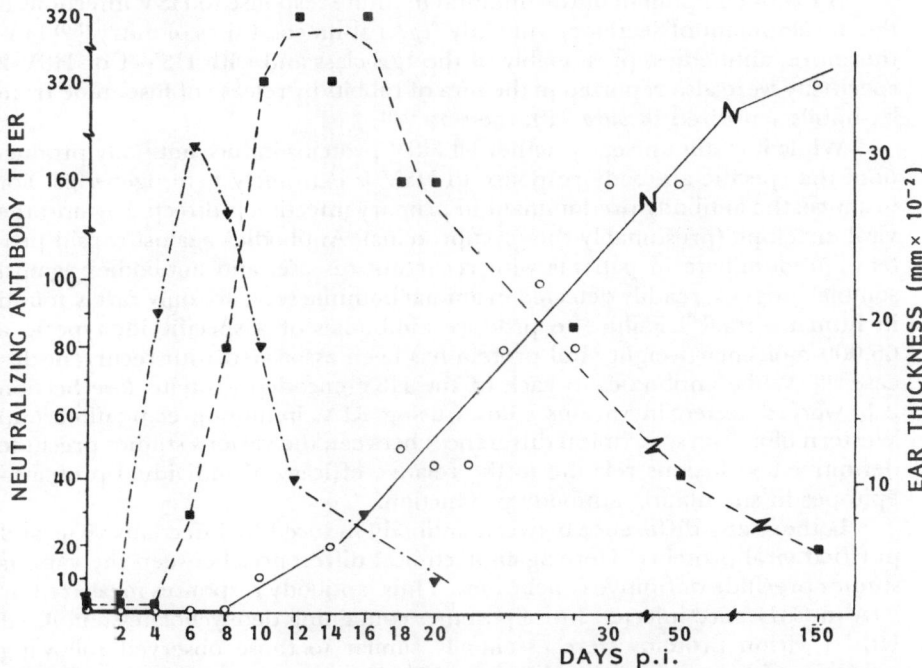

**FIGURE 1.** Immune response in mice after i.d. infection with HSV-2. (○) IgG anti-HSV neutralizing antibody. (■) IgM anti-HSV neutralizing antibody. Each point represents the mean of at least seven mice. (▲) Ear-thickness responses represented as the arithmetic mean.

pearance after labial infection of BALB/c mice with HSV-1 were studied by neutralization (NT), complement fixation (CF), enzyme-linked immunosorbent assay (ELISA), radioimmunoassay (RIA), and antibody-dependent cellular cytotoxicity (ADCC). Antibody detected by ELISA, RIA, or ADCC was present in most mice at 5 days p.i., when lip lesions were first observed. Antibody detectable by NT or CF was not observed until day 10 p.i., when the lip lesions were already healing. NT antibodies were not observed in intravaginally infected BALB/c mice, while the same mouse strain infected intravenously (i.v.) with the same HSV-2,[33] and guinea pigs infected intradermally (i.d.) or subcutaneously (s.c.)[34,35] were NT antibody positive.

In humans, almost all subjects with primary genital infections develop IgG1 subclass antibodies with HSV specificity. IgG2, IgG3, and IgG4 antibodies are detected in acute-phase sera, most often in patients with recurrent genital HSV, but in none of those with primary infections. IgG4 antibodies are significantly more frequent in sera from men than in those from women with recurrent genital infections.[36] The titers of NT antibodies are higher in subjects with a history of recurrent disease than in those without such history (seropositive controls).[34,37]

A major component of the humoral immune response to HSV infections is the development of secretory antibody (IgA) at mucosal sites of entry.[30] Furthermore, antibodies, presumably of the IgE class and with HSV-1 or HSV-2 specificity were also reported in the sera of rabbits by release of histamine from basophils sensitized *in vitro* with the sera.[38]

While it is still unclear whether all HSV proteins induce antibody production, the specific antibody response to HSV is extremely heterogeneous. For instance, the antibody predominant in primary infection is directed against the viral envelope (presumably the glycoproteins). Antibodies against capsid proteins predominate in patients with recurrent disease, and antibodies against soluble antigens, readily detected in animal immune sera, are only rarely found in human sera.[39] Failure to produce antibodies of a specific idiotype to a 66,000-molecular weight viral protein has been associated with recurrent disease.[40] While antibodies to each of the HSV-encoded proteins (see Section 2.1) were detected in various studies using RIA, immunoprecipitation, and Western blot assays, technical differences between the various studies preclude definitive conclusions relating to the relative efficacy of individual proteins–epitopes in stimulating antibody production.

Is there any difference between antibody induced by infectious virus and purified viral proteins? Here again, technical differences between the various studies preclude definitive conclusions. Thus, antibody responses measured by RIA in C3H mice injected with liposomes containing deoxycholate-solubilized HSV-1 virion proteins were essentially similar to those observed following infection with live virus.[41] However, HRS/J mice immunized with a lectin-purified fraction primarily containing gC-1 produced anti-gC antibodies, but they were non-NT.[42] Glycoproteins isolated by other investigators, including gB-1, gC-1, gE-1,[28,43] and gC-2,[44] induced NT antibodies in rabbits.

The most reasonable interpretation of the findings is that individual epitopes induce functionally different antibody. In support of this conclusion is the finding that MAbs specific for gB, gC, gE, and gD exhibit NT activity, albeit of different titers.[17,18,45–49] Furthermore, of 33 MAbs selected for their ability to bind to HSV-1 and HSV-2 virions and displaying specificity for various glycoproteins, only 6 had NT activity, and all were specific for gD.[49] While these findings suggest that gD represents the most critical target or elicitor of NT antibody, it should be pointed out that two MAbs to gD had little NT activity,[49] consistent with the interpretation that epitopes differ in their potential to induce NT antibody.[17,18]

## 3.2. T Lymphocytes

It is commonly believed that at least as relates to recovery from infection, immunity to HSV primarily involves the T-cell system. This is based on the findings that HSV lesions are abnormally severe in immunosuppressed persons whose humoral responses to HSV are apparently normal,[50] and on adoptive transfer studies that have established a major role for T lymphocytes in recovery from lethal HSV-1 infection in mice.[51,52]

Effector mechanisms recognized in cellular immunity to HSV include cytotoxic T lymphocytes (CTL),[53-55] delayed-type hypersensitivity (DTH),[56] and effector lymphokines such as immune interferon (IFN$_\gamma$),[57] interleukin 2 (IL-2), and other soluble factors. T-helper cells and/or the lymphocytes responsible for DTH and lymphokine release (which can be provisionally included within the T-helper subset) show I-region compatibility requirements. Conversely, CTL recognize antigen in association with products of the K and D regions of the H-2 complex in the mouse T-cell recognition of both I and K/D-coded self-antigens seems important in immune protection against HSV-1 in mice, with long lasting immunity being conferred by I-region compatibility.[52]

Delayed-type hypersensitivity is an early response, first detected in mice at 4–7 days after i.d. or s.c. infection with HSV-1 (see Fig. 1), concomitant with the onset of clinical (labial) symptoms and the development of virus-specific antibody as measured by ELISA/RIA or ADCC, but preceding the appearance of NT antibody.[32,58] The DTH response is HSV type common[59] and remains inducible for at least 2 years p.i.[58] However, the ability to transfer DTH adoptively with draining lymph node cells is only observed at 6–10 days p.i., suggesting that DTH producing cells are absent at late times p.i. or that they are under the control of suppressor lymphocytes.[58]

At 5 days p.i., the frequency of HSV-specific IL-2 producing helper T cells in lymph node suspensions was estimated at 1 : 2,470–1 : 5,800. The detection of this Lyt1 + cell population required that cells from mice primed with infectious (but not noninfectious) virus be stimulated *in vitro* by exposure to viral antigen for 9 days, suggesting that helper T cells are strictly antigen dependent. Consistent with the interpretation that suppressor cells are present in the lymph nodes, the frequency of helper T-cell precursors was increased two- to threefold by deletion of the Lyt2 + cells prior to limiting dilution assays.[60]

Herpes simplex virus-mediated blastogenesis coincides with DTH[32] and shows a gradient of antigen exposure first observed in lymphoid cells from the lymph node draining the site of infection, followed by spleen cells (SC) and then by peripheral blood lymphocytes (PBL).[61] While the response appears to have an HSV type-specific component,[61] it is primarily type common.[62] To obtain further information on the development of HSV specific T cells responses as they relate to the generation of *in vitro* secondary responses, we followed the development of proliferative responses in HSV-2 infected guinea pigs.[62] Virus-specific blastogenesis was observed in cultures of SC obtained at 7 days p.i. The response reached maximal levels at 3 days in culture and decreased thereafter. At 10 days p.i., the response continued to increase in magnitude through 6 days in culture. The trend of the kinetic curves to become linear and increase in magnitude as a function of both time p.i. and antigen concentration continued through 14–28 days p.i., approaching patterns characteristic of those seen at 150 days p.i. By analogy to the progression toward higher-affinity antibody, the progressive increase in proliferation was shown to reflect an increase in the relative number of cells that respond to low antigen concentrations (immune maturation).

The antigenic determinants recognized in DTH and proliferation are not yet totally identified. Recent studies indicate that gB, although not possessing epitopes that induce NT antibodies, contains antigenic determinants that activate helper T cells capable of adoptively protecting mice from lethal HSV challenge.[63] Similarly, a synthetic peptide representing the type-common epitope(s) located at residues 1–23 on the mature gD shown to induce NT antibodies[17,18] activates human helper T cells to proliferate *in vitro*.[19]

In general, it has been difficult to induce CTL to HSV. Activity was only detected following cyclophosphamide pretreatment of mice prior to HSV immunization and/or extended (3 days) *in vitro* culture of the effector cells before assay.[53,54,64,65] Furthermore, in contrast to other viruses, CTL to HSV were demonstrated only in draining lymph nodes.[64] These CTL were first seen at 4 days p.i., reached maximal activity at 6–9 days, and were no longer detectable at 12 days p.i.[64] However, like other CTL, they were also K/D restricted.[53] The frequency of HSV-specific CTL precursors was similar to that of helper T-cell precursors. However, unlike the latter, they were not antigen-dependent.[66]

Target cell recognition by the CTL involves HSV glycoproteins[67] and is associated with the H-2K$^b$ self-antigen.[55] Type-common and type-specific epitopes were implicated by different studies.[67] Inactivated HSV and viral proteins failed to stimulate secondary CTL responses *in vitro*. However, irradiated SC from mice primed with infectious virus or IL-2 were able to supply helper activity for deficient *in vitro* cultures,[68,69] suggesting that (1) helper T cells fail to recognize inactivated virus, or (2) inactivated virus preferentially stimulates suppressor cell activity.

CTL clones were recently established from HSV-1- or HSV-2-stimulated human PBL. Some recognized type-common, others type-specific antigenic determinants. In contrast to other virus systems, all the HSV-specific clones had a helper cell phenotype (OKT4+ in humans) and were restricted by HLA class II MB or DR antigens.[70] All were bifunctional in that they also proliferated in response to HSV stimulation and in the absence of exogeneous IL-2, and the HSV-type specificity and histocompatibility restriction of the proliferative response was identical to that of the cytotoxic activity. HSV-specific stimulation resulted in the production of a factor (presumably IL-2), which induced proliferation of an IL-2-dependent T-cell line, demonstrating that the CTL clones also provide helper cell activity.[71] HSV-1-specific CTL clones were directed against gB-1, gD-1, or gE-1; HSV-2-specific clones were directed against gC-2, gD-2, gE-2, or gG.[72]

## 3.3. Macrophages

The nonspecific containment phase of the immune response involves polymorphonuclear leukocytes (PMNL), mononuclear phagocytes (MP), and soluble factors generated by their activation. MP play a crucial role in resistance to HSV infection as evidenced by the finding that (1) MP transfer antiviral resistance to neonatal animals that form a model (albeit incomplete) of MP deficiency, (2) early after HSV infection peritoneal MP exhibit extrinsic antiviral resistance defined as the ability to lyse HSV-infected targets and mediate ADCC, (3) activation of MP by immunomodulators such as pyran or bacillus Calmette–Guérin (BCG) increases resistance of mice to HSV, while increased susceptibility is observed in mice treated with MP toxicants, and (4) tissue MP play a significant role in protection of mice depleted of bonemarrow derived cells by radioactive $^{89}$Sr.[73]

MP antiviral activity includes extrinsic resistance, the mechanism of which is presumably cell lysis[73] and/or inhibition of protein and DNA synthesis,[74] and intrinsic activity the mechanism of which is MP abortive infection. Extrinsic resistance is displayed by murine peripheral MP elicited with thioglycollate, activated with immunomodulators, or elicited in response to HSV-2 infection. The MP lytic activity is enhanced by IFN; in the presence of antibody, MP can mediate ADCC.[75] Generally, resident MP do not exhibit this response.[76] While the specific antigenic determinants are unknown, human monocytes activated in vitro specifically lyse virus-infected but not uninfected cells.[77]

Mononuclear phagocytes exhibit variable degrees of intrinsic resistance for HSV. Since MP from mice that are genetically resistant to HSV (C57BL/6) are more resistant in vitro to HSV infection than are MP from sensitive strains (BALB/c), it was suggested that intrinsic resistance is the sole determinant of the in vivo phenomenon. However, recent studies have failed to demonstrate a similar correlation,[77,78] and the in vivo resistance was not decreased by depletion of circulating monocytes.[79] Both intrinsic and extrinsic resistance can be modified by immune lymphocytes, antibody, or lymphokines, suggesting that in addition to MP, restriction of HSV infection in vivo involves other immune parameters.

Mononuclear phagocytes are also involved in the induction, regulation, and amplification of specific immune responses. On the positive side of the regulatory pathway, the role of MP includes antigen processing and presentation and the synthesis of interleukin 1 (IL-1) required for both T- and B-cell activation. On the negative side, it involves the synthesis of prostaglandins (PG).[80] The role of PG in HSV pathogenesis is discussed in Sections 4 and 5.2.

Tissue MP are particularly significant in HSV infections. Immunologically active elements in the skin include the keratinocytes, which produce an IL-1-like factor (ETAF),[81] and Langerhans cells (LC), which are bonemarrow derived, bear Fc and C3b receptors, express Ia antigens, and are involved in antigen processing and presentation.[82] These Ia+ epidermal cells have accessory cell function for HSV-2-induced T-cell proliferation of immune lymphoid cells.[83,84] Helper T cells primed by coculture with LC caused a marked reduction in virus titers in the skin of HSV infected nude mice and prevented the formation of zosteriform skin lesions,[84] indicating that LC are involved both in immunoregulation and in the control of cutaneous HSV infection.

## 3.4. NK Cells

Natural killer (NK) cells display spontaneous non-MHC restricted cytotoxicity against tumor, normal, and virus-infected cells. Like the MP, NK cells acting early in infection might limit HSV-1 growth and reduce the virus load. NK cells are nonphagocytic, not thymus dependent, and bear no surface Ig. However, they are a heterogeneous population.[85,86] Recent studies[87] showed that HSV-1-infected mouse cell lines are susceptible to a wide range of cytolytic activities mediated by NK cells or by cells lacking NK cell-surface markers such as Qa-5, NK1.1, and Asialo-GM1(aGM1), do not respond to IFN, and most closely resemble natural cytotoxic (NC) cells.[88]

Besides their heterogeneity relative to target selectivity, NK cells may also differ in their hemopoietic lineage. Thus, NK cells that lyse HSV-1-infected cells are negative for the mature T-cell antigen (OKT3)[89] present on several NK cell lines.[90] Clones of murine CTL were shown to express low concentrations of NK alloantigens and were induced to express NK-like lytic activity by culture with high concentrations of SC supernatants.[91,92] Grossman and Herberman[93] recently suggested that the differentiation of NK and that of T cells is intimately interrelated, with the divergence in their characteristics becoming initiated upon rearrangement of the genes for the T-cell receptor.

The antigenic determinants recognized by NK cells on HSV-infected targets have been the subject of recent investigation. Fab fragments of human HSV-1 positive (but not negative) Ig and MAbs specific for different epitopes on HSV-1 glycoproteins gB and gC blocked NK activity against HSV-1 infected cells, suggesting that NK cells interact directly with viral glycoproteins.[94]

As already discussed relative to MP, the *in vivo* resistance of certain mouse strains to HSV infection has also been correlated with genetically high NK cell activity.[95] Using the same experimental approach as that used for MP, it was shown that depletion of NK cell activity with $^{89}$Sr[79] or with antibody to aGM1[96] caused increased HSV-1 growth in the brain, liver, and spleen of infected mice and increased mortality. However, it should be stressed that animals with high NK cell activity also have high IFN levels,[97] raising some doubt about the antiviral role of NK cells relative to IFN. Some investigators[96] concluded that the decline in resistance against HSV-1 infection following treatment with anti-aGM1 is due to NK cell depletion, since repeated IFN injections did not prolong the life of NK cell-depleted mice. Likewise, NK cells (Leu 11b+) were shown to limit viral replication *in vitro*. The effect was not due to IFN, since the antiviral activity was not reduced by anti-IFN serum.[98] Finally, the observation that persons who are particularly susceptible to unusually severe disease (newborns, patients with Wiscott–Aldrich syndrome) have very low levels of NK activity against HSV-1-infected targets[99] also supports the relative importance of NK cells.

## 3.5. Other Cells

There are relatively few experimental studies about the interactions between granulocytes and HSV. Several studies have shown that granulocytes are able to kill HSV-infected targets in an ADCC-like fashion, i.e., if the targets are coated with specific antibodies against HSV.[100] Human PMNL and lymphocytes have also been shown to mediate ADCC.[101,102] A new mechanism of antiviral immunity and a role for granulocytes was proposed by Grewal *et al.*,[103] who showed that highly enriched populations of bovine neutrophils were able to destroy HSV-infected cells in the presence of complement. This mechanism of cytotoxicity was termed complement-dependent cytotoxicity.

## 3.6. Soluble Factors

The mechanism underlying involvement of the I-region-restricted helper T cells in protection against, and recovery from, HSV infections may involve the production of lymphokines. In order to determine the relationship between blastogenesis and the production of lymphokines, we chose to assay for leukocyte migration inhibition factor (LIF) in supernatants from cultures of HSV-stimulated immune SC.[62] We found an early (1 day in culture) component at 3 days p.i. This component (early LIF) was presumably made by cells that were antigen-driven *in vivo* and were therefore fully differentiated at the onset of culture. A second component (late LIF) that required at least 3 days of *in vitro* exposure to HSV antigen and was presumably made by cells that must become antigen differentiated in culture was first observed at 14 days p.i. Both components were present in cultures of SC obtained at 28 and 150 days p.i. The reappearance of early LIF coincided with our ability to isolate HSV-2 from the site of infection. These LIF components, also observed in cultures of human PBL,[104] are made by helper T cells (unpublished data).

What, if any, is the role of lymphokines in recovery from HSV infection? In one study,[105] SC from HSV-2 immunized animals were adoptively transferred to mice infected with HSV-2 i.p. or i.v., and virus content in the liver was measured at 48 hr post-transfer. Under these conditions of infection, untreated animals developed focal necrotizing hepatitis. However, there was a significant reduction in liver virus content 48 hr after the transfer of SC obtained as early as 4 days after HSV-2 immunization. The activity coincided with the production of high levels of macrophage migration inhibition factor (MIF) and with the accumulation of mononuclear cells in the liver. It was suggested that MIF is instrumental in macrophage recruitment into the liver, thereby playing a critical role in recovery from infection.

Interferon, another lymphokine made early in infection, 3–20 days,[106] has also been associated with resistance of certain mouse strains to HSV infection. In one such study,[107] administration of anti-IFN serum was shown to enhance HSV-1 growth and disease severity. Pure cultures of peritoneal or splenic MP from mouse strains susceptible to HSV *in vivo* were shown to produce high IFN titers when infected with HSV *in vitro* and vice versa.[108] SJL mice, which are low in NK cell activity but which produce high levels of IFN in response to HSV-1 infection, are relatively resistant to HSV-1 infection. Antibody to aGM1 did not influence HSV-1-induced mortality in this system.[109] Also supporting the significance of IFN, rather than NK cells, neonatal mice were protected from fatal HSV-1 challenge by adoptive transfer of adult mouse leukocytes depleted of NK cells with anti-aGM1 antibody. However when the mice were treated with anti-IFN antibody, HSV-1 growth was enhanced and the protective effect of the leukocytes was abrogated.[110]

The peritoneal exudate cells (PEC) that produce IFN in response to HSV challenge are MP. Pure cultures of HSV stimulated mouse T cells did not produce IFN when treated with infectious HSV, whereas IFN was produced by pure cultures of mouse bone marrow derived MP. Noninfectious HSV was incapable of eliciting IFN production in PEC, or *in vivo*, in the peritoneal cavity of mice.[108,111]

Lymphokines may be involved in recovery and/or protection from HSV infection by virtue of their role in the differentiation of HSV specific CTL. The induction of active CTL requires the participation of at least three cell types: (1) Ia+ accessory cells,[112] helper T cells (Lyt 1+ in mice),[113] and CTL precursors. The helper cells are stimulated by antigen and IL-1 to proliferate and secrete soluble factors, including the signals required for the differentiation of the antigen-driven CTL precursors.[114] Soluble factors required for CTL induction include IL-2 and $IFN_\gamma$.[115] IL-2 is necessary for the production of $IFN_\gamma$, and $IFN_\gamma$ seems to induce IL-2 receptors on T lymphocytes.[116] $IFN_\gamma$ enhances DR expression on murine and human monocytes[117] and may play a role in enhancing antigen presentation. Furthermore, $IFN_\gamma$ is at least one of several macrophage activating factors (MAF)[118]; locally, it may trigger induction of CTL or enhancement of NK activity (a function that it shares with IL-2), thereby leading to eradication of HSV infected cells.

In humans, lymphokine production (including IFN) has been used as a major indicator of the involvement of the T-cell system in HSV infections.[119–122] Most of these studies have concentrated on the patient with recurrent disease. However, in their totality, they support the conclusion that lymphokine production is an early response that may be involved in curtailing the severity of the systemic HSV disease.

## 4. MECHANISM(S) OF IMMUNOSUPPRESSION

Herpes simplex virus replicates in cultures of adult human lymphocytes, provided they are stimulated to blastogenesis with PHA, Con A, pokeweed mitogen anti-lymphocyte Ig, or IL-2. Several human studies have shown that only T lymphoblasts are permissive for HSV replication with growth confined to 2–3% of the cells.[123,124] Others indicated that both T and B cells are permissive,[125,126] and in the mouse, only activated B cells support HSV growth.[127] Furthermore, T cells stimulated by exposure to different activators differ with respect to their permissiveness, as indicated by the higher virus titers observed in PHA/IL-2-stimulated human T cells.[128] It is not clear why mitogenic stimulation is required in order to render adult lymphocytes permissive for HSV, particularly since this treatment also induces IFN production.[129] Possibly permissiveness requires increased cellular DNA synthesis or the acquisition of cell-surface receptors for HSV. However, newborn lymphocytes are permissive without any prestimulation.[130] To our knowledge, there is no published evidence of HSV replication in lymphocytes activated by specific antigens or by HSV antigen.

Cultured T cells can be persistently infected with HSV. High virus titers were obtained for 500 days p.i. of one such line, although only 3% of the cells were infected and cell viability was not decreased. Within the first 120 days of this infection, there was a 24-day interval when virus production spontaneously diminished to undetectable levels. Virus production could then be reactivated by treating the cultures with PHA.[126]

HSV-1 suppresses the induction of antibody to other antigens, such as diphtheria toxoid, in cultures of human tonsils,[124] a suppression that seems to result from nonproductive infection of a small population of helper T cells.[131] By contrast, inactivated HSV-1 or HSV-2 have been shown to inhibit mitogen or antigen induced *in vitro* proliferation,[125,132] suggesting that inhibition of the proliferative response does not require viral replication in the lymphocytes. Recent studies indicate that HSV complexes directly to IL-2, making it unavailable to the lymphocytes. Indeed, the addition of IL-2 to these cultures restored cellular proliferation.[132]

HSV-infected adult murine MP undergo abortive infection (intrinsic resistance), the precise mechanism of which is unclear (see Section 3.3). However, MP can be rendered permissive for HSV growth by several days of *in vitro* growth prior to infection,[77] by growth in conditioned medium or by SV40 transformation.[133] While cytolytic infection would functionally remove the MP system, causing a decrease in subsequent resistance to infection, it is not clear whether this relates to the finding that HSV infection inhibits ADCC activity.[134]

*In vitro* studies have focused on changes in MP receptors, phagocytosis, oxidative metabolism, phagocyte–lysosome fusion, and microbicidal killing.[73] Few studies, however, have addressed the effects of the virus on immunoregulatory functions of the MP. Crucial questions remaining are: When are viral functions completely inactivated? When do viral genes persist, if at all, and if so can they be reactivated? How do viral genes alter MP functions? and, What is the precise mechanism of this abortive infection, and how is it related to the pathogenesis of HSV disease?

Another aspect of HSV-induced immunosuppression appears to involve prostaglandin (PG) synthesis. Thus, indomethacin, an inhibitor of PG synthetase, was shown to increase human mononuclear and NK cell cytotoxicity of HSV-infected targets. The indomethacin-treated mononuclear cells protected mice from lethal HSV challenge.[135] PG, phosphodiesterase inhibitors, or dibutyryl-cAMP have no effect on complement-mediated lysis. However, at physiologic concentrations, $PGE_1$ or $PGE_2$, $PGF_{2\alpha}$, or Al PGB1 and other reagents that increase intracellular levels of cAMP inhibit ADCC, apparently by blocking attachment of the mononuclear cells to the antibody coated target cells.[136] Incubation of MP or lymphocytes with reagents that elevate cAMP was also shown to cause suppression of Fc-receptor expression,[137] and PGE caused a significant (100–1000%) increase in HSV-1 titers *in vitro;* inhibitory effects were mediated by $PGF_{2\alpha}$.[138]

Herpes simplex virus infection has also been associated with an increase in the suppressor T-cell subpopulation. However, the route of virus inoculation plays a critical role within this context. As opposed to the s.c. route, which leads to the induction of T- and B-cell immunity, at least in the draining lymph node, mice infected with HSV-1 by the i.v. route develop T cells that specifically suppress the induction of DTH. These cells are Lyt1 + at early times p.i. (7 days), but they are both Lyt1 + and Lyt2 + at 4 weeks p.i.[139] It remains to be determined whether the Lyt1 + cells are suppressor-inducer cells or an unusual suppressor cell subpopulation uniquely identified in HSV infection. HSV-1 inoculation in the anterior chamber of the eye also produces specific T-cell nonresponsiveness in DTH, similar to that seen upon i.v. infection.[140]

Immunosuppression was also observed in immune guinea pigs exposed to a secondary antigen dose by reinfection.[141] In these studies, SC obtained at 3 days postreinfection, had markedly depressed HSV-induced proliferative responses and were negative for LIF production. Cell-mixing experiments indicated that these SC preparations had suppressor activity, but expression of the suppressor signal required a prior antigen-dependent differentiation step *in vitro*. Significantly, at this time, the reinfecting virus strain was able to replicate at the site of infection and was also isolated from the local sensory ganglia. Essentially similar findings were reported by Nick *et al.*,[144] who found that i.p. infection of mice and rats by HSV-2 results in suppression of antibody formation on subsequent challenge with HSV-1 or HSV-2. This suppression was not observed in HSV-1-infected animals. Furthermore, it was significantly reduced by injection of silica on the day of the secondary HSV-2 challenge, suggesting that macrophages play a central role in the induction of this immunosuppression. The precise function of the macrophage in immunosuppression is unclear, but it may involve PG synthesis (see Section 5.2). The proportion of suppressor T cells was also significantly increased in HSV-2-infected guinea pigs[142] and humans[104,122,143] re-exposed to HSV antigen by virtue of virus reactivation during recrudescent disease.

## 5. IMMUNOMODULATION AND PATHOGENESIS

While it is well established that the immune system is involved in protection from fatal HSV infection (see Sections 3 and 4), the respective contributions of the humoral versus CMI responses is still somewhat controversial. The widely accepted interpretation is that antibody is not important in promoting recovery. However, there is also evidence in support of the role of antibody: (1) patients with disseminated primary HSV-1 infections have a decreased ability to form serum antibody to HSV-1,[145] (2) passively transferred NT antibody prevents the spread and multiplication of HSV-1 in the PNS and CNS and protects mice from fatal HSV disease,[146–148] and (3) NT MAbs protect mice from zosteriform cutaneous lesions provided that they are given before the virus has completed one round trip along the nerves (within 60 hr p.i.).[149] In any case, it should be recalled that antibody cannot be effective during primary infection, since HSV reaches the ganglia within 24–48 hr p.i., long before a detectable antibody response has occurred.

## 5.1. Immunity and the Establishment/Maintenance of Latency

It has been suggested that the development of virus-specific immunity predisposes to the establishment of a latent infection.[150,151] Indeed, passively administered antiviral antibody given 1 day after virus challenge reduced productive ganglionic infection but potentiated latency. Likewise, active immunity induced by i.p. virus injection between days 1–3 before challenge reduced acute viral replication but not latency. However, this conclusion is not supported by other studies,[152] and the role of immunity in the establishment of ganglionic latency remains a matter of controversy.

Hypothetical mechanisms proposed[153] for the role of the immune response in establishing ganglionic latency include the following: (1) ganglionic cells are permissive for HSV; the immune response modulates infection by converting a potentially lytic infection into a nonlytic (latent) one; (2) there are two populations of ganglionic cells, only one of which is permissive; latent infection occurs in the nonpermissive cells and does not involve the immune system; and (3) ganglionic cells are generally nonpermissive for HSV replication; they are converted to a permissive state by various triggers including the immune response. While presently available data do not differentiate among these interpretations, acceptance of the premise that virus-specific immunity potentiates ganglionic infection implicitly presumes that the immune response plays a role in the establishment and/or maintenance of latency, arguing against the second interpretation.

The role of the immune response in the maintenance of latency is even less well understood. Stevens and Cook[154] found that virus reactivation could be prevented in latently infected ganglia transplanted into noninfected mice by passive administration of antiviral IgG and suggested that antiviral antibody is involved in the maintenance of the latent state. However, opposite conclusions were reached in another study[153] in which BALB/c mice passively immunized with anti-HSV-1 serum were virtually free of antibody at 4 months p.i. while remaining positive for latent ganglionic virus.

Human studies have failed to demonstrate a role for humoral immunity in maintaining the asymptomatic (quiescent) latent state. Patients with a history of recurrent disease have significantly higher titers of NT antibodies than do seropositive controls,[37,104,122] a finding confirmed in latently infected and vaccinated guinea pigs.[34,155] Recurrent disease occurs in the face of high levels of circulating NT and binding antibody,[30,156] an observation confirmed in the guinea pig[34,35,155] and mouse ear[157] models. Despite rare claims to the contrary,[40] there seems to be no qualitative or quantitative defect in the production of antibodies to particular viral antigens in patients with frequent recurrent disease.[158] Finally, although it had been suggested that recurrent disease is associated with low levels of serum IgA,[159] this was not confirmed by a more recent study.[160]

The role of CMI, if any, in the maintenance of a latent state is also unclear. It is generally accepted that patients with recurrent disease have an intact HSV immune memory.[37,104,122,156] However, there is disagreement pertaining to the magnitude of the response in the different patient groups. According to one study,[156] patients with quicker and more intense virus-specific proliferative responses had shorter duration of virus shedding and of clinical symptoms, quicker healing lesions, and less new lesion formation than did those whose *in vitro* proliferative responses to HSV antigens were low. Likewise, Lopez and O'Reilly[37] concluded that the HSV type-specific response was suppressed during recurrent disease. However, differences in the magnitude of the blastogenic response of patients with recurrent disease as compared with seropositive controls were not seen in other studies,[104,122] possibly reflecting technical differences. Contradictory findings were also reported using other *in vitro* assays of CMI, with some studies[161,162] reporting no defects in NK cell activity of patients with recurrent disease and others[163] describing decreased responses.

By contrast, there is general agreement that, as compared with seropositive controls, patients with recurrent disease have reduced lymphokine responses. Mononuclear cells from patients with frequent recurrences have a slight decrease in HSV-induced IFN production,[164] and there is a strong correlation between peak IFN levels and the time to the next recurrence, such that the lower the IFN level, the more frequent the recurrent episodes. We[104,122], and others[119] found that patients with recurrent disease have decreased LIF production when studied at recrudescence, and T lymphocytes from patients with frequent recurrences failed to produce MIF in response to HSV.[120,165,166] How do these defects arise? Do they reflect the failure of the immune system to contain the virus in a latent state? Are they the result or the cause of the recurrent episode? Do they arise from the specific immunosuppression that can occur during primary infection?[139,140] Answers to these questions remain a major goal for future investigation.

## 5.2. Immunity and Recurrent Disease

Studies of the mouse ear model of HSV infection have led to the hypothesis that a peripheral stimulus produces local changes in the skin that lead both to virus reactivation in the ganglia and the development of recurrent disease (for review see Hill[167]). However, this hypothesis does not accommodate the observation that often virus reactivation is not accompanied by recurrent symptoms. Accordingly, we proposed that the development of recurrent disease hinges on the ability of the reactivated virus to replicate in the ganglia and/or the epidermis, causing visible cytopathology. According to this interpretation, factors (such as an effective immune response) that interfere with viral replication will prevent recurrent disease even though the latent virus was reactivated.[62,104,123,141]

Although cause and effect remains to be established, available data support this interpretation. In humans, the risk of recurrent disease was correlated with abnormal lymphokine production.[104,119–122,141,164,168,174] In a prospective study of 30 patients first seen during primary HSV-2 infection, we showed that the risk of developing subsequent recurrent disease was strongly correlated with the loss (at 20–30 days p i.) of the ability to produce lympho-kine in response to HSV stimulation.[122,141] The lymphokine unrespon-siveness acquired at 20–30 days p.i. is associated with an increase in the propor-tion of OKM2+ (MP) and OKT8+ (cytotoxic/suppressor) cells.[174]

Connecting between the containment and effector phases of the HSV immune response, we have recently shown that supernatants from cultures of PBL obtained immediately before and during recrudescence are negative for both LIF activity and for lymphokines (IFN and IL-2) that enhance NK ac-tivity.[104,122,168] This was associated with an increase in the proportion of OKT8+, OKIa+, and OKM2+ cells. Depletion of these cell subpopulations by complement-mediated lysis restored both activities. Likewise, PBL collected at recrudescence inhibited virus-specific lymphocyte proliferation,[104] unless previously depleted of the OKT8+ and/or OKM2+ cells. These findings do not reflect a unique cellular compartmentalization, since similar results were obtained in cultures of SC from latently infected guinea pigs.[62,142]

The suppression involved both MP-produced PG and suppressor factor(s) presumably made by the suppressor T cells. Indeed, recrudescent superna-tants failed to augment NK activity, although the levels of IFN and IL-2 were similar to those observed in supernatants from seropositive controls. Dialysis as well as addition of indomethacin to the cultures restored the NK-enhancing activity. However, PGE levels were similar in recrudescent and seropositive control supernatants.[168] More recent studies indicate that besides PG, sup-pression requires a 8200-dalton suppressor factor that appears to block the binding of the lymphokine to its specific target cell.[174] In the guinea pig model, we found that HSV-2-activated suppressor cells, observed at recrudes-cence, elaborate both virus-specific and -nonspecific soluble suppressor fac-tor(s) that respectively dampen the HSV-2 or mitogen-stimulated proliferation of immune or nonimmune cells.[142]

Confirming our conclusions about the potentially significant role played by suppressor cells in HSV disease, Horohov et al.[169] found that murine Lyt1+ and Lyt2+ HSV immune splenocytes proliferate in response to HSV antigen in vitro. The suppressor cell is Lyt+ IJ+. However, suppression also requires the contribution of Lyt1+ T cells and IJ+ antigen-presenting cells. More recently,[170] these investigators found that suppression is mediated by a soluble factor (molecular weight 90,000–100,000) that contains an IJ+ anti-idiotypic protein. Consistent with our findings in the guinea pig model,[142] suppression mediated by this factor was antigen specific.[170] However, as pre-viously reported by us in the guinea pig model,[142] nonspecific suppressor activity could also be generated, for instance, by incubating the IJ+ suppressor factor with Lyt1+ cells from HSV-immune mice.[170]

Taken in toto, these findings support the interpretation that immu-nomodulation plays a significant role in the pathogenesis of HSV cutaneous disease and provide a plausible mechanism for unifying the two phases of the HSV immune response and local stimuli that are presumably significant in cutaneous disease. These stimuli include the PGs which are produced by epi-dermal cells (EC)[83,168,117] in response to triggering factors (e.g., UV light) that induce HSV recurrent disease. Skin irritants (e.g., cellophane stripping) that induce recurrent disease in the mouse ear model cause a significant increase in the levels of $PGE_2$ in the skin, while a similar increase is not seen after DMSO application, a treatment that reactivates latent virus but does not cause recurrent symptoms.[167]

Local stimuli also involve suppressor factors. Thus, EC-mediated HSV-induced proliferation of immune lymphoid cells is inhibited by UVB irradiation.[83,84] PGE does not mediate the inhibition, since equal levels of PGE are secreted by UVB-irradiated and -nonirradiated EC. The inhibition appears to be mediated by soluble suppressor factor(s) including one (molecular weight: 30,000–40,000) that appears to suppress HSV-induced DNA synthesis of immune LNC specifically and another one (molecular weight: 80,000–100,000) that suppresses both HSV and mitogen-induced responses.[83] Whether the various suppressor factors represent structurally similar or different molecular entities is unknown. Furthermore, the mechanism of action of the suppressor factor(s) is unclear, thereby precluding final conclusions pertaining to the suppressive effects of these apparently different moieties, as well as their respective role in the pathogenesis of herpetic disease. Possibly, UVB-irradiated EC preferentially activate virus-specific suppressor cells that secrete soluble suppressor factors. Alternatively, UVB directly stimulates the production by EC (or a subpopulation thereof) of substances that inhibit T-cell proliferation.[172] Recent findings from our laboratory indicate that the production of the soluble factors detected under these conditions require $Lyt1^+$ $L3T4^+$ T cells consistent with suppressor inducers and $Lyt2^+$ suppressor T cells, consistent with antigen-induced suppressor effectors. Suppressor factors are antigen-specific and nonspecific. The antigen-specific suppressor factor (HSV-SF) is a 115-kilodalton protein consisting of two disulfide-bound components with molecular sizes of 70 and 52 kilodaltons. By analogy with hapten systems, it seems reasonable to assume that HSV-SF is part of the regulatory Ts-cell circuit that modulates HSV-induced immunity.[173,175]

The concept that immunosuppression is a major component of the pathogenesis of recurrent HSV disease raises a number of questions of both basic and practical (for vaccine development) significance. Are all HSV proteins capable of inducing the suppression network? Is suppression induced by unique epitopes on certain viral proteins? If suppression is induced by unique antigenic domains, how do they differ from those that induce immunity? Is viral replication in the epidermis required for the induction of suppression? Is antigen presentation by tissue MP (e.g., LC) the determining factor in the induction of the suppression network? Recent progress in molecular biology, including the development of vaccinia virus recombinants that contain specific HSV genes, or fragments thereof, may provide the necessary tools to address these significant problems.

ACKNOWLEDGMENTS.   Work done in this laboratory and quoted in this manuscript was supported by U.S. Public Health Service grant AI-22192 from the National Institute of Allergy and Infectious Diseases.

# REFERENCES

1. T. S. L. Beswick, The origin and the use of the word herpes, *Med. Hist.* **6:**214–232, (1962).
2. Spear, P. G., and B. Roizman, Herpes simplex viruses. In DNA Tumor Viruses J. Tooze (Ed.) p 615–747 Cold Spring Harbor Laboratory, Cold Spring Harbor, New York (1980).
3. Kieff, E. D., S. L. Bachenheimer, and B. Roizman, Size composition and structure of the deoxyribonucleic acid of herpes simplex virus subtype 1 and 2, *J. Virol.* **8:**125–132 (1971).
4. Sheldrick, P., and N. Berthelot, Inverted repetitions in the chromosome of herpes simplex virus, *Cold Spring Harbor Symp. Quant. Biol.* **39:**667–678 (1975).
5. Peden, K., P. Mounts, and G. S. Hayward, Homology between mammalian cell DNA sequences and human herpesvirus genomes detected by a hybridization procedure with high complexity probe, *Cell* **31:**71–80 (1982).
6. Wu, J. R., C. W. Diffenbach, D. M. Torres, L. Aurelian, and P. O. P. Ts'o, DNA sequences homologous to the HSV-2 transforming DNA fragment in normal and transformed cells, *J. Cell. Biol.* **97:**135a (1983).
7. Spear, P. G., and B. Roizman, Proteins specified by herpes simplex virus. V. Purification of structural proteins of the herpesvirion, *J. Virol.* **9:**143–159 (1972).
8. Strnad, B., and L. Aurelian, Proteins of herpesvirus type 2. I. Virion, nonvirion, and antigenic polypeptides in infected cells, *Virology* **69:**438–452 (1976).
9. P. G. Spear, Herpesviruses, in: *Cell Membranes and Viral Envelopes*, Vol. 2 H. A. Blough and J. M. Tiffany, eds., pp. 709–750, Academic, London (1980).
10. Roizman, B., B. Norrild, C. Chan, and L. Pereira, Identification and preliminary mapping with monoclonal antibodies of a herpes simplex virus 2 glycoprotein lacking a known type 1 counterpart, *Virology* **133:**242–247 (1984).
11. Richman, D. D., A. Buckmaster, S. Bell, C. Hodgman, and A. C. Minson, Identification of a new glycoprotein of herpes simplex virus type 1 and genetic mapping of the gene that codes it, *J. Virol.* **57:**647–655 (1986).
12. Sarmiento, M., M. Haffey, and P. G. Spear, Membrane proteins specified by herpes simplex viruses. III. Role of glycoprotein VP7 (B2) in virion infectivity, *J. Virol.* **29:**1149–1160 (1979).
13. Manservigi, R., Spear, P. G., and A. Buchan, Cell fusion induced by herpes simplex virus is promoted and suppressed by different viral glycoproteins, *Proc. Natl. Acad. Sci. USA* **74:**3913–3917 (1977).
14. Kieff, E. D., B. Hoyer, S. L. Bachenheimer, and B. Roizman, Genetic relatedness of type 1 and type 2 herpes simplex viruses, *J. Virol.* **9:**738–745 (1972).
15. Buchman, T. G., B. Roizman, G. Adams, and H. Stover, Restriction endonuclease fingerprinting of herpes simplex DNA: A novel epidemiology tool applied to a nosocomial outbreak, *J. Infect. Dis* **138:**488–498 (1978).
16. Cassai, E., D. DiLuca, R. Manservigi, M. Tognon, and A. Rotola, Comparative analysis of the virion polypeptides specified by herpes simplex virus type 2 strains, *Arch. Virol.* **64:**35–45 (1980).
17. Cohen, G. H., B. Dietzschold, M. Ponce de Leon, D. Long, E. Golub, A. Varrichio, L. Pereira, and R. J. Eisenberg, Localization and synthesis of an antigenic determinant of HSV glycoprotein D that stimulates production of neutralizing antibody, *J. Virol.* **49:**102–108 (1984).
18. Eisenberg, R. J., D. Long, M. Ponce de Leon, J. T. Matthews, P. G. Spear, M. G. Gibson, L. A. Lasky, P. Berman, E. Golub, and G. H. Cohen, Localization of epitopes of herpes simplex virus type 1 glycoprotein D, *J. Virol.* **53:**634–644 (1985).
19. DeFreitas, E. C., B. Dietzschold, and H. Koprowski, Human T-lymphocyte response *in*

*vitro* to synthetic peptides of herpes simplex virus glycoprotein D, *Proc. Natl. Acad. Sci. USA* **82:**3425–3429 (1985).

20. Morgan, C. H., M. Rose, and B. Mednis, Electron microscopy of herpes simplex virus, I. Entry, *J. Virol.* **2:**507–516 (1968).

21. Morse, L. S., L. Pereira, B. Roizman, and P. A. Schaffer, Anatomy of herpes simplex virus (HSV) DNA. X. Mapping of viral genes by analysis of polypeptides and functions specified by HSV-1 and HSV-2 recombinants, *J. Virol.* **26:**389–410 (1978).

22. Honess, R. W., and B. Roizman, Proteins specified by herpes simplex virus. XI. Identification and relative molar rates of synthesis of structural and nonstructural herpes virus polypeptides in the infected cells, *J. Virol.* **12:**1347–1365 (1973).

23. Honess, R., and B. Roizman, Regulation of herpesvirus macromolecular synthesis. I. Cascade regulation of the synthesis of 3 groups of viral proteins, *J. Virol.* **14:**8–19 (1974).

24. O'Hare, P., and G. S. Hayward, Evidence for a direct role for both 175,000- and 110,000-molecular weight immediate early proteins of herpes simplex virus in transactivation of delayed early promoters, *J. Virol.* **53:**751–760 (1985).

25. Sacks, W. R., S. C. Greene, D. P. Aschman, and P. A. Schaffer, Herpes simplex virus type 1 ICP27 is an essential regulatory protein, *J. Virol.* **55:**796–803 (1985).

26. Sears, A. E., I. W. Halliburton, B. Meignier, S. Silver, and B. Roizman, Herpes simplex virus 1 mutant deleted in the α22 gene: Growth and gene expression in permissive and restrictive cells and establishment of latency in mice, *J. Virol.* **55:**338–346 (1985).

27. Smiley, M. L., J. A. Hoxie, and H. M. Friedman, Herpes simplex virus type 1 infection of endothelial, epithelial, and fibroblast cells induces a receptor for C3b, *J. Immunol.* **134:**2673–2678 (1985).

28. Para, M. F., R. B. Bauck, and P. G. Spear, Glycoprotein gE of HSV-1: Effects of anti-gE on virion infectivity and on virus-induced Fc binding receptors, *J. Virol.* **41:**129–136 (1982).

29. Burns, W. H., L. C. Billups, and A. L. Notkins, Thymus dependence of viral antigens, *Nature (Lond.)* **256:**654–656 (1975).

30. Douglas, G. R., and R. B. Couch, A prospective study of chronic herpes simplex virus infection and recurrent labialis in humans, *J. Immunol.* **104:**289–295 (1970).

31. Aurelian, L., and I. I. Kessler, Subclinical herpesvirus infections of the genital tract are commonly associated with viral shedding, *Cervix* **3:**235–248 (1985).

32. Morahan, P. S., T. A. Thomson, S. Kohl, and B. K. Murray, Immune responses to labial infection of BALB/c mice with herpes simplex virus type 1, *Infect. Immun.* **32:**180–187 (1981).

33. Morahan, P. S., M. C. Breinig, and M. B. McGeorge, Immune responses to vaginal or systemic infection of BALB/c mice with herpes simplex virus type 2, *J. Immunol.* **119:**2030–2036 (1977).

34. Donnenberg, A. D., E. Chaikoff, and L. Aurelian, Immunity to herpes simplex virus type 2: Cell mediated immunity in latently infected guinea pigs, *Infect. Immun.* **30:**99–109 (1980).

35. Scriba, M., and F. Tatzber, Pathogenesis of herpes simplex virus infections in guinea pigs, *Infect. Immun.* **34:**655–661 (1981).

36. Coleman, R. M., A. J. Nahmias, S. C. Williams, D. J. Phillips, C. M. Black, and C. B. Reimer, IgG subclass antibodies to herpes simplex virus, *J. Infect. Dis.* **151:**929–936 (1985).

37. Lopez, C., and R. J. O'Reilley, Cell mediated immune responses in recurrent herpesvirus infections. I. Lymphocyte proliferation assay, *J. Immunol.* **118:**895–902 (1977).

38. Day, R. P., J. Bienenstock, and W. E. Rawls, Basophil-sensitizing antibody response to herpes simplex viruses in rabbits, *J. Immunol.* **117:**73–78 (1976).

39. Kalino, K. O. K., R. J. Marttila, K. Granfors, and M. K. Viljanen, Solid-phase radioimmunoassay of human immunoglobulin M and immunoglobulin G antibodies against

herpes simplex virus type 1 capsid, envelope and excreted antigens, *Infect. Immun.* **15:**883–889 (1977).

40. Ashley, R. L., and L. Corey, Effect of acyclovir treatment of primary genital herpes on the antibody response to herpes simplex virus, *J. Clin. Invest.* **73:**681–688 (1984).

41. Naylor, P. T., H. S. Larsen, L. Huang, and B. T. Rouse, In vivo induction of anti-herpes simplex virus immune response by type 1 antigens and lipid A incorporated into liposomes, *Infect. Immun.* **36:**1209–1216 (1982).

42. Zweering, H. J., D. Martinez, R. J. Lynch, and L. W. Stanton, Immune responses in mice against herpes simplex virus: Mechanisms of protection against facial and ganglionic infections, *Infect. Immun.* **31:**267–275 (1981).

43. Eberle, R., and R. J. Courtney, Preparation and characterization of specific antisera to individual glycoprotein antigens comprising the major glycoprotein region of HSV-1, *J. Virol.* **35:**902–917 (1980).

44. Zezulak, K. M., and P. G. Spear, Characterization of a herpes simplex virus type 2 75,000 molecular weight glycoprotein antigenically related to herpes simplex virus type 1 glycoprotein, *J. Virol.* **47:**553–562 (1983).

45. Pereira, L., D. V. Dondero, D. Gallo, V. Devlin, and J. D. Woodie, Serological analysis of herpes simplex virus types 1 and 2 with monoclonal antibodies, *Infect. Immun.* **35:**363–367 (1982).

46. Showalter, S. D., M. Zweig, and B. Hampar, Monoclonal antibodies to HSV-1 proteins including the immediate early protein ICP4, *Infect. Immun.* **34:**684–692 (1981).

47. Balachandran, N. D., W. E. Harnish, W. E. Rawls, and S. Bacchetti, Glycoproteins of herpes simplex virus type 2 as defined by monoclonal antibodies, *J. Virol.* **44:**344–355 (1982).

48. Holland, T. C., S. D. Marlin, M. Levine, and J. Glorioso, Antigenic variants of herpes simplex virus selected with glycoprotein specific monoclonal antibodies, *J. Virol.* **45:**672–682 (1983).

49. Para, M. F., M. L. Parish, G. Noble, and P. G. Spear, Potent neutralizing activity associated with anti-glycoprotein D specificity among monoclonal antibodies selected for binding to herpes simplex virions, *J. Virol.* **55:**483–488 (1985).

50. Pass, R. F., R. J. Whitley, J. D. Whelchel, A. G. Diethelm, D. W. Reynolds, and C. A. Alford, Identification of patients with increased risk of infection with herpes simplex virus after renal transplantation, *J. Infect. Dis.* **140:**487–492 (1979).

51. Nagafuchi, S., H. Oda, R. Mori, and T. Taniguchi, Mechanisms of acquired resistance to herpes simplex virus as studied in nude mice, *J. Gen. Virol.* **44:**715–723 (1979).

52. Howes, E. L., W. Taylor, N. A. Mitchison, and E. Simpson, MHC matching shows that at least two T cell subsets determine resistance to HSV, *Nature (Lond.)* **277:**67–68 (1979).

53. Pfizenmaier, K., H. Jung, A. Starzinski-Powitz, M. Rollinghoff, and H. Wagner, The role of T cells in anti-herpes simplex virus immunity. I. Induction of antigen-specific cytotoxic T lymphocytes, *J. Immunol.* **119:**939–944 (1977).

54. Lawman, M. J. P., B. T. Rouse, R. J. Courtney, and R. D. Walker, Cell mediated immunity against herpes simplex. Induction of cytotoxic T lymphocytes, *Infect. Immun.* **27:**133–139 (1980).

55. Jennings, S. R., P. L. Rice, S. Pan, B. B. Knowles, and S. S. Tevethia, Recognition of herpes simplex virus antigens on the surface of mouse cells of the H-2b haplotype by virus-specific cytotoxic T lymphocytes, *J. Immunol.* **132:**475–481 (1984).

56. Nash, A. A., J. Phelan, and P. Wildy, Cell mediated immunity in herpes simplex virus infected mice: H-2 mapping of the delayed type hypersensitivity response and the antiviral T cell response, *J. Immunol.* **126:**1260–1261 (1981).

57. Green, J. A., T. J. Yeh, and J. C. Overall, Jr., Sequential production of IFN α and immune specific IFNα by human mononuclear leukocytes exposed to herpes simplex virus, *J. Immunol.* **127:**1192–1196 (1981).

58. Nash, A. A., H. J. Field, and R. Quartey-Papafio, Cell mediated immunity in herpes simplex virus infected mice: Induction, characterization and antiviral effects of delayed type hypersensitivity, *J. Gen. Virol.* **48:**351–357 (1980).

59. Schreir, R. D., L. I. Pizer, and J. W. Moorhead, Delayed hypersensitivity to herpes simplex virus: Murine model, *Infect. Immun.* **35:**566–571 (1982).

60. Prymowicz, D., R. N. Moore, and B. T. Rouse, Frequency of herpes simplex virus specific helper T lymphocyte precursors in the lymph node cells of infected mice, *J. Immunol.* **134:**2683–2688 (1985).

61. Jacobs, R. P., L. Aurelian, and G. H. Cole, Cell mediated immune response to herpes simplex virus: Type specific lymphoproliferative responses in lymph nodes draining the site of primary infection, *J. Immunol.* **116:**1520–1525 (1976).

62. Donnenberg, A. D., R. B. Bell, and L. Aurelian, Immunity to herpes simplex virus type 2 (HSV-2). I. Development of virus-specific lymphoproliferative and leukocyte migration inhibition factor responses in HSV-2-infected guinea pigs, *Cell Immunol.* **56:**526–539 (1980).

63. Chan, W. L., M. L. Lukig, and F. Y. Liew, Helper T cells induced by an immunopurified herpes simplex virus type 1 (HSV-1) 115 kilodalton glycoprotein (gB) protect mice against HSV-1 infection, *J. Exp. Med.* **162:**1304–1318 (1985).

64. Nash, A. A., R. Quartey-Papafio, and P. Wildy, Cell mediated immunity in herpes simplex virus-infected mice: Functional analysis of lymph node cells during periods of acute and latent infection, with reference to cytotoxic and memory cell, *J. Gen. Virol.* **49:**309–317 (1980).

65. Pfizenmaier, K., A. Strazinski-Powitz, M. Rollinghoff, D. Falke, and H. Wagner, T cell mediated cytotoxicity against herpes simplex virus-infected target cells, *Nature (Lond.)* **265:**630–632 (1977).

66. Rouse, B. T., H. S. Larsen, and H. Wagner, Frequency of cytotoxic T lymphocyte precursors to herpes simplex virus type 1 as determined by limiting dilution analysis, *Infect. Immun.* **39:**785–792 (1983).

67. Carter, V. C., P. L. Rice, and S. S. Tevethia, Intratypic and intertypic specificity of lymphocytes involved in the recognition of herpes simplex virus glycoproteins, *Infect. Immun.* **37:**116–126 (1982).

68. Schmid, D. S., H. S. Larsen, and B. T. Rouse, Role of Ia antigen expression and secretory function of accessory cells in the induction of cytotoxic T lymphocyte responses against herpes simplex virus, *Infect. Immun.* **37:**1138–1147 (1982).

69. Schmid, D. S., and B. T. Rouse, Cellular interaction in the cytotoxic T lymphocyte response to HSV antigens: Differential antigen activation requirements for the helper T lymphocyte and cytotoxic T lymphocyte precursors, *J. Immunol.* **131:**479–484 (1983).

70. Yasukawa, M., and J. M. Zarling, Human cytotoxic T cell clones directed against herpes simplex virus infected cells. I. Lysis restricted by HLA class II MB and DR antigens, *J. Immunol.* **133:**422–427 (1984).

71. Yasukawa, M., and J. M. Zarling, Human cytotoxic T cell clones directed against HSV infected cells. II. Bifunctional clones with cytotoxic and virus induced proliferative activities exhibit HSV type 1 and 2 specific or type common reactivities, *J. Immunol.* **133:**2736–2742 (1984).

72. Yasukawa, M., and J. M. Zarling, Human cytotoxic T cell clones directed against herpes simplex virus infected cells. III. Analysis of viral glycoproteins recognized by CTL clones by using recombinant herpes simplex viruses, *J. Immunol.* **134:**2679–2682 (1985).

73. Morahan, P. S., J. R. Conner, and K. R. Leary, Viruses and the versatile macrophage, *Br. Med Bull.* **41:**51–21 (1985).

74. Morse, S. S., and P. S. Morahan, Activated macrophages mediate interferon-independent inhibition of herpes simplex virus, *Cell Immunol.* **58:**72–84 (1981).

75. Kohl, S., S. E. Starr, J. M. Olske, S. L. Shore, R. B. Ashman, and A. J. Nahmias, Human monocyte–macrophage mediated antibody dependent cytotoxicity to herpes simplex virus infected cells, *J. Immunol.* **118:**729–735 (1977).

76. Morahan, P. S., L. A. Glasgow, J. L. Crane, Jr., and E. Kern, Comparison of antiviral and antitumor activity of activated macrophages, *Cell Immunol.* **28**:404–415 (1977).

77. Lopez, C., and G. Dudas, Replication of herpes simplex virus type 1 in macrophages from resistant and susceptible mice, *Infect. Immun.* **23**:432–437 (1979).

78. Armerding, D., P. Mayer, M. Scriba, A. Hien, and H. Rossiter, *In vivo* modulation of macrophage functions by herpes simplex virus type 2 in resistant and sensitive inbred mouse strains, *Immunobiology* **106**:217–227 (1981).

79. Lopez, C., R. Ryshke, and M. Bennett, Marrow dependent cells depleted by [89]Sr mediate genetic resistance to herpes simplex virus type 1 infection in mice, *Infect. Immun.* **28**:1028–1032 (1980).

80. Oppenheim, J. J., and I. Gery, Interleukin 1 is more than an interleukin, *Immunol. Today* **3**:113–119 (1982).

81. Luger, T. A., B. M. Stadler, S. I. Katz, and J. J. Oppenheim, Epidermal cell-derived thymocyte activating factor (ETAF), *J. Immunol.* **127**:1493–1498 (1981).

82. Stingl, G., K. Tamaki, and S. I. Katz, Origin and function of epidermal Langerhans cells, *Immunol. Rev.* **53**:149–174 (1980).

83. Hayashi, Y., and L. Aurelian, Immunity to herpes simplex virus type 2: Viral antigen presenting capacity of epidermal cells and its impairment by ultraviolet irradiation, *J. Immunol.* **136**:1087–1092 (1986).

84. Yasumoto, S., N. Okabe, and R. Mori, Role of epidermal langerhans cells in resistance to herpes simplex virus infection, *Arch. Virol.* **90**:261–271 (1986).

85. Herberman, R. B. (ed.), *Natural Cell Mediated Immunity Against Tumors*, Academic, New York (1980).

86. Ortaldo, J. R., and R. B. Herberman, Heterogeneity of natural killer cells, *Annu. Rev. Immunol.* **2**:359–394 (1984).

87. Colmenares, C., and C. Lopez, Enhanced lysis of herpes simplex virus type 1 infected mouse cell lines by NC and NK effectors, *J. Immunol.* **136**:3473–3480 (1986).

88. Lattime, E. C., G. A. Pecoraro, and O. Stutman, Natural cytotoxic cells against solid tumors in mice. III. A comparison of effector cell antigenic phenotype and target cell recognition structures with those of NK cells, *J. Immunol.* **126**:2011–2014 (1981).

89. Bishop, G. A., J. C. Glorioso, and S. A. Schwartz, Relationship between expression of HSV glycoproteins and susceptibility of target cells to human natural killer activity, *J. Exp. Med.* **157**:1544–1561 (1983).

90. Hercend, T., E. L. Reinherz, S. Meuer, S. F. Scholssman, and J. Ritz, Phenotypic and functional heterogeneity of human cloned natural killer cell lines, *Nature (Lond.)* **301**:158–160 (1983).

91. Brooks, C. G., R. C. Burton, S. B. Pollack, and C. S. Henney, The presence of NK alloantigens on cloned cytotoxic T lymphocytes, *J. Immunol.* **131**:1391–1395 (1983).

92. C. G. Brooks, Reversible induction of natural killer cell activity in cloned murine cytotoxic T lymphocytes, *Nature (Lond.)* **305**:155–158 (1983).

93. Grossman, Z., and R. B. Herberman, Natural killer cells and their relationship to T-cells. Hypothesis on the role of T cell receptor gene rearrangement on the course of adoptive differentiation, *Cancer Res.* **46**:2651–2658 (1986).

94. Bishop, G. A., S. D. Marlin, S. A. Schwartz, and J. C. Glorioso, Human natural killer cell recognition of herpes simplex virus type 1 glycoproteins: Specificity analysis with the use of monoclonal antibodies and antigenic variants, *J. Immunol.* **133**:2206–2213 (1984).

95. C. Lopez, Resistance to herpes simplex virus type 1 (HSV-1), *Curr. Top. Microbiol. Immunol.* **92**:15–24 (1981).

96. Habu, S., K. I. Akametsu, N. Tamaoki, and K. Okumura, *In vivo* significance of NK cell on resistance against virus (HSV-1) infections in mice, *J. Immunol.* **133**:2743–2747 (1984).

97. Engler, H., R. Zawatzky, H. Kirchner, and D. Armerding, Experimental infection of inbred mice with herpes simplex virus. VI. Comparison of interferon production and

natural killer cell activity in susceptible and resistant adult mice, *Arch. Virol.* **74:**239–247 (1982).

98. Fitzgerald, P. A., M. Mendelsohn, and C. Lopez, Human natural killer cells limit replication of herpes simplex virus type 1 *in vitro, J. Immunol.* **134:**2666–2672 (1985).

99. Lopez, C., D. Kirkpatrick, S. E. Read, P. A. Fitzgerald, J. Pitt, S. Pahwa, C. Y. Ching, and E. M. Smithwick, Correlation between low natural kill of fibroblasts infected with herpes simplex virus type 1 and susceptibility to herpesvirus infections, *J. Infect. Dis.* **147:**1030–1035 (1983).

100. Siebens, H., S. S. Tevethia, and B. M. Babior, Neutrophil mediated antibody-dependent killing of herpes simplex virus infected cells, *Blood* **54:**88–94 (1979).

101. Melewicz, F. M., S. L. Shore, E. W. Ades, and D. J. Phillips, The mononuclear cells in human blood which mediate antibody-dependent cellular cytotoxicity to virus infected target cells. II. Identification as a K cell, *J. Immunol.* **118:**567–573 (1977).

102. Gale, R. P., and J. Zighelboim, Polymorphonuclear leukocytes in antibody dependent cytotoxicity, *J. Immunol.* **114:**1047–1051 (1975).

103. Grewal, A. S., B. T. Rouse, and L. A. Babiuck, Mechanisms of recovery from viral infections: Destruction of infected cells by neutrophils and complement, *J. Immunol.* **124:**312–319 (1980).

104. Sheridan, J. F., A. D. Donnenberg, and L. Aurelian, Immunity to herpes simplex virus type 2. IV. Impaired lymphokine production correlates with a perturbation in the balance of T lymphocyte subsets, *J. Immunol.* **129:**326–331 (1982).

105. S. C. Mogensen, Macrophage migration inhibition as a correlate of cell mediated immunity to herpes simplex virus type 2 in mice, *Immunobiology* **162:**28–38 (1982).

106. Kirchner, H., R. Zawatsky, and H. M. Hirt, *In vitro* production of immune interferon by spleen cells of mice immunized with herpes simplex virus, *Cell Immunol.* **40:**204–210 (1978).

107. Zawatzky, R., I. Gresser, E. Demayer, and H. Kirchner, The role of interferon in the resistance of C57BL/6 mice to various doses of herpes simplex virus type 1, *J. Infect. Dis.* **146:**405–410 (1982).

108. Brucher, J., T. Domke, C. H. Schroder, and H. Kirchner, Experimental infection of inbred mice with HSV. VI. Effect of IFN on *in vitro* virus replication in macrophages, *Arch. Virol.* **82:**83–93 (1984).

109. Chmielarczyk, W., I. Domke, and H. Kirchner, Role of interferon in the resistance of C3H/HeJ mice to infection with herpes simplex virus, *Antiviral Res.* **5:**55–59 (1985).

110. Bukowski, J. F., and R. M. Welsh, The role of natural killer cells and interferon in resistance to acute infection of mice with herpes simplex virus type 1, *J. Immunol.* **136:**3481–3485 (1986).

111. Kirchner, H., H. Engler, C. H. Schroder, R. Zawatzky, and E. Starch, Herpes simplex virus type 1 induced interferon production and activation of natural killer cells in mice, *J. Gen. Virol.* **64:**437–441 (1983).

112. Kreeb, F., and R. M. Zinkernagel, Role of the H-2i region in the generation of an antiviral cytotoxic T cell response *in vitro, Cell Immunol.* **53:**285–297 (1980).

113. Okado, M., and C. S. Henney, The differentiation of cytotoxic T cells *in vitro*. III. The role of helper T cells and their products in the differentiation of cytotoxic cells from "memory" cell populations, *J. Immunol.* **125:**850–857 (1980).

114. Wagner, H., C. Hardt, B. T. Rouse, M. Rollinghoff, P. Schaurich, and K. Pfizenmaier, Dissection of the proliferative and differentiative signals controlling murine cytotoxic T lymphocyte responses, *J. Exp. Med.* **155:**1876–1881 (1982).

115. Farrar, W. L., H. M. Johnson, and J. J. Farrar, Regulation of the production of immune interferon and cytotoxic T lymphocytes by interleukin 2, *J. Immunol.* **126:**1120–1125 (1981).

116. Johnson, H. M., and W. L. Farrar, The role of α-interferon like lymphokine in the

activation of T cells for expression of interleukin 2 receptors, *Cell. Immunol.* **75**:154–159 (1983).

117. Basham, T. J., and T. C. Merigan, Recombinant IFNα increases HLA-DR synthesis and expression, *J. Immunol.* **130**:1492–1494 (1983).

118. Roberts, W. K., and A. Vasil, Evidence for identity of murine α-interferon and macrophage activating factor, *J. Interferon Res.* **2**:519–532 (1982).

119. O'Reilly, R., A. Chibbaro, E. Anger, and C. Lopez, Cell mediated immune responses in patients with recurrent herpes simplex infections. II. Infection associated deficiency of lymphokine production in patients with recurrent herpes labialis or herpes progenitalis, *J. Immunol.* **118**:1095–1102 (1977).

120. Shillitoe, E. J., J. M. A. Wilton, and T. Lehner, Sequential changes in cell-mediated immune responses to herpes simplex virus after recurrent herpetic infection in humans, *Infect. Immun.* **18**:130–137 (1977).

121. Rattray, M. C., G. M. Peterman, L. C. Altman, L. Corey, and K. K. Holmes, Lymphocyte derived chemotactic factor synthesis in initial genital herpesvirus infection: Correlation with lymphocyte transformation, *Infect. Immun.* **30**:110–116 (1980).

122. Sheridan, J. F., and L. Aurelian, Immunity to herpes simplex virus type 2. V. Risk of recurrent disease following primary infection: Modulation of T cell subsets and lymphokine (LIF) production, *Diagno. Immunol.* **1**:245–256 (1983).

123. Bouroncle, B. A., K. D. Clausen, and E. M. Dorner, Replication of HSV in cultures of phytohemagglutinin stimulated human lymphocytes, *J. Natl. Cancer Inst.* **44**:1065–1078 (1970).

124. Pelton, B. K., R. C. Imrie, and A. M. Denman, Susceptibility of human lymphocyte populations to infection by herpes simplex virus, *Immunology* **32**:803–810 (1977).

125. Plaeger-Marshall, S., and J. W. Smith, Inhibition of mitogen and antigen induced lymphocyte blastogenesis by herpes simplex virus, *J. Infect. Dis.* **138**:506–511 (1978).

126. Rinaldo, C. R., B. S. Richter, P. H. Black, R. Callery, L. Chess, and M. S. Hirsch, Replication of herpes simplex virus and cytomegalovirus in human leukocytes, *J. Immunol.* **120**:130–136 (1978).

127. Kirchner, H., H. M. Hirt, C. Kleinicke, and K. Munk, Replication of herpes simplex virus in mouse spleen cell cultures stimulated by lipopolysaccharide, *J. Immunol.* **117**:1753–1756 (1976).

128. Hammer, S. M., and J. M. Gillis, Herpes simplex virus replication in interleukin-2 stimulated human T cells, *J. Infect. Dis.* **151**:544–548 (1985).

129. Friedman, R. M., and H. L. Cooper, Stimulation of IFN production in human lymphocytes by mitogen, *Proc. Soc. Exp. Biol. Med.* **125**:901–905 (1967).

130. D. Westmoreland, Herpes simplex virus type-1 and human lymphocytes: Virus expression and the response to infection of adult and foetal cells, *J. Gen. Virol.* **40**:559–575 (1978).

131. Pelton, B. K., I. B. Duncan, and A. M. Denman, Herpes simplex virus depresses antibody production by affecting T cell function, *Nature (Lond.)* **284**:176–177 (1980).

132. Wainberg, M. A., J. D. Portnoy, B. Clencer, S. Hubschman, J. Legace-Simard, N. Rabinovitch, Z. Remer, and J. Mendelson, Viral inhibition of lymphocyte proliferative responsiveness in patients suffering from recurrent lesions caused by herpes simplex virus, *J. Infect. Dis.* **152**:441–448 (1985).

133. Sethi, K. K., and H. Brandis, In vitro acquisition of resistance against herpes simplex virus by permissive murine macrophages, *Arch. Virol.* **59**:157–172 (1979).

134. Plaeger-Marshall, S., L. A. Wilson, and J. W. Smith, Alteration of rabbit alveolar and peritoneal macrophage function by herpes simplex virus, *Infect. Immun.* **41**:1376–1379 (1983).

135. Kohl, S., D. M. Jansen, and L. S. Loo, Indomethacin enhancement of human natural killer cytotoxicity to herpes simplex virus infected cells *in vitro* and *in vivo*, *Prostaglandins Leukotrienes Med.* **9**:159–166 (1982).

136. Trofatter, K. F. Jr., and C. A. Daniels, Interaction of human cells with prostaglandins and cyclic AMP modulators. I. Effects on complement-mediated lysis and antibody-dependent cell-mediated cytolysis of herpes simplex virus-infected human fibroblasts, *J. Immunol.* **122:**1363–1370 (1979).

137. J. Rhodes, Modulation of macrophage Fc receptor expression *in vitro* by insulin and cyclic nucleotides, *Nature (Lond.)* **257:**597–599 (1985).

138. A. A. Newton, Effect of cyclic nucleotides on the response of cells to infection by various herpesviruses, in: *Oncogenesis and Herpes Viruses.* Vol. III. (G. de The, W. Henle, and F. Rapp, eds.), pp. 381–387, IARC, Lyon (1978).

139. Nash, A. A., and P. G. H. Gell, Membrane phenotype of murine effector and suppressor T cells involved in delayed hypersensitivity and protective immunity to herpes simplex virus, *Cell. Immunol.* **75:**348–355 (1983).

140. Whittum, J. A., J. Y. Niederhorn, J. P. McCulley, and J. W. Streilein, Role of suppressor T cells in herpes simplex virus-induced immune deviation, *J. Virol.* **51:**556–558 (1984).

141. L. Aurelian, Mechanism of recurrent herpes infections and prospects for vaccination, in: *Infections in Reproductive Health,* Vol. I: *Common Infections* (L. G. Keith, G. S. Berger, D. A. Edelman, eds.), pp. 115–136, MTP Press, Lancaster, Pennsylvania (1985).

142. Iwasaka, T., J. F. Sheridan, and L. Aurelian, Immunity to herpes simplex virus type 2: Recurrent lesions are associated with the induction of suppressor cells and soluble suppressor factors, *Infect. Immun.* **42:**955–964 (1983).

143. Schooley, R. T., M. S. Hirsch, R. B. Colvin, A. B. Cosimi, N. E. Tolkoff-Rubin, R. T. McCluskey, R. C. Burton, P. S. Russell, J. T. Herrin, F. L. Delmonico, J. V. Giorgi, W. Henle, and R. H. Rubin, Association of herpesvirus infections with T-lymphocyte-subset alterations, glomerulopathy and opportunistic infections after renal transplantation, *N. Engl. J. Med.* **308:**307–313 (1983).

144. Nick, S., P. Kampe, A. Knoblich, B. Metzger, and D. Falke, Suppression and enhancement of humoral antibody formation by herpes simplex virus types 1 and 2, *J. Gen. Virol.* **67:**1015–1024 (1986).

145. Hirsch, M. S., S. H. Cheeseman, and S. M. Hammer, Human herpesvirus infections: Pathogenesis and clinical implications, in: *Seminars in Infectious Disease,* Vol. II (L. Weinstein and B. N. Fields, eds.), pp. 217–264, Stratton Intercontinental Medical Book, New York (1979).

146. McKendall, R. R., T. Klassen, and J. R. Baringer, Host defenses in herpes simplex infections of the nervous system: Effect of antibody on disease and viral spread, *Infect. Immun.* **23:**305–311 (1979).

147. Worthington, M., McG. A. Conliffe, and A. Baron, Mechanism of recovery from systemic herpes simplex virus infection. I. Comparative effectiveness of antibody and reconstitution of immune spleen cells in immunosuppressed mice, *J. Infect. Dis.* **142:**163–174 (1980).

148. Kino, J., Y. Hayashi, I. Hayashida, and R. Mori, Dissemination of herpes simplex virus in nude mice after intracutaneous inoculation and the effect of antibody in the course of infection, *J. Gen. Virol.* **63:**475–479 (1982).

149. Simmons, A., and A. A. Nash, Role of antibody in primary and recurrent herpes simplex virus infection, *J. Virol.* **53:**944–948 (1985).

150. Openshaw, H., L. V. S. Asher, C. Wohlenberg, T. Sekizawa, and A. L. Notkins, Acute and latent herpes simplex virus ganglionic infection: Immune control and viral reactivation, *J. Gen. Virol.* **44:**205–215 (1979).

151. Price, R. W. and J. Schmitz, Route of infection, systemic host resistance, and integrity of ganglionic axons influence acute and latent herpes simplex virus infection of the superior cervical ganglion, *Infect. Immun.* **23:**373–383 (1979).

152. Waltz, M. A., H. Yamamoto, and A. L. Notkins, Immunological response restricts numbers of cells in sensory ganglia infected with herpes simplex virus, *Nature (Lond.)* **264:**554–556 (1976).

153. Openshaw, H., T. Sekizawa, C. Wohlenberg, and A. L. Notkins, The role of immunity in latency and reactivation of herpes simplex viruses, in: *The Human Herpesviruses* (A. J. Nahmias, W. R. Dowdle, and R. F. Schinazi, eds.), pp. 289–296, Elsevier, New York (1980).

154. Stevens, J. G., and M. L. Cook, Maintenance of latent herpetic infection: An apparent role for antiviral IgG, *J. Immunol.* **113**:1685–1693 (1974).

155. M. Scriba, Vaccination against herpes simplex virus: Animal studies on the efficacy against acute, latent and recurrent infections, in: *Herpetic Ocular Diseases* (R. Sundmacher, ed.), pp. 67–72, Springer-Verlag (Bergmann), Berlin (1981).

156. Corey, L., W. C. Reeves, and K. K. Holmes, Cellular immune response in genital herpes simplex virus infection, *N. Engl. J. Med.* **299**:986–991 (1978).

157. Darville, J. M., and W. A. Blyth, Neutralizing antibody in mice with primary and recurrent herpes simplex virus infection, *Arch. Virol.* **71**:303–310 (1982).

158. Zweerink, H. J., and L. W. Stanton, Immune response to HSV infections: Virus specific antibodies in sera from patients with recurrent facial infections, *Infect. Immun.* **31**:624–630 (1981).

159. Tokumaru, T. A possible role for αA-immunoglobulin in herpes simplex virus infection in man, *J Immunol.* **97**:248–259 (1966).

160. Friedman, M. G., and N. Kimmel, Herpes simplex virus-specific serum immunoglobulin A: Detection in patients with primary or recurrent herpes infections and in healthy adults, *Infect. Immun.* **37**:374–377 (1982).

161. Fujimiya, J., L. A. Babiuk, and B. T. Rouse, Direct lymphocytotoxicity against herpes simplex virus infected cells, *Can. J. Microbiol* **24**:1076–1081 (1978).

162. Russell, A. S., J. Percy, and T. Kovithavongs, Cell mediated immunity to herpes simplex in humans: Lymphocyte cytotoxicity measured by ⁸⁹Sr release from infected cells, *Infect. Immun.* **11**:355–359 (1975).

163. Steele, R. W., M. M. Vincent, S. A. Hensen, D. A. Fucillo, I. A. Chapas, and L. Canales, Cellular immune responses to herpes simplex virus type 1 in recurrent herpes labialis. *In vitro* blastogenesis and cytotoxicity to infected cell lines, *J. Infect. Dis.* **131**:528–534 (1975).

164. Cunningham, A. L., and T. C. Merigan, Gamma interferon production appears to predict time of recurrence of herpes labialis, *J. Immuno.* **130**:2397–2400 (1983).

165. Wilton, J. M. A., L. Ivanyi, and T. Lehner, Cell mediated immunity in herpesvirus hominis infections, *Br. Med. J.* **1**:723–726 (1972).

166. Gange, R. W., A. D. Bats, J. R. Park, C M. Bradstreet, and E. L. Rhodes, Cellular immunity and circulating antibody to herpes simplex virus in subjects with recurrent herpes simplex lesions and controls as measured by the mixed leucocyte migration inhibition test and complement fixation, *Brt. J. Dermatol.* **97**:539–544 (1975).

167. Hill, T. J., Herpes simplex virus latency, in: *The Viruses*, Vol. 3: *The Herpesviruses* (B. Roizman, ed.), pp. 175–240, Plenum, New York (1985).

168. Sheridan, J. F., M. Beck, L. Aurelian, and M. Radowsky, Immunity to herpes simplex virus: Virus reactivation modulates lymphokine activity, *J. Infect. Dis.* **152**:339–456 (1985).

169. Horohov, D. W., R. N. Moore, and B. T. Rouse, Regulation of herpes simplex virus-specific lymphoproliferation by suppressor cells, *J. Virol.* **56**:1–6 (1985).

170. Horohov, D. W., J. H. Wyckoff, III, R. N. Moore, and B. T. Rouse, Regulation of herpes simplex virus specific cell-mediated immunity by a specific suppressor factor, *J. Virol.* **58**:331–338 (1986).

171. Goldyne, M. E., Prostaglandins and cutaneous inflammation, *J. Invest. Dermatol.* **64**:377–385 (1975).

172. Granstein, R. D., A. Lowy, and M. I. Greene, Epidermal antigen-presenting cells in activation of suppression: Identification of a new functional type of ultraviolet radiation-resistant epidermal cell, *J. Immunol.* **132**:563–565 (1984).

173. Yasumoto, S., Y. Hayashi, and L. Aurelian, Immunity to herpes simplex virus type 2: Suppression of virus-induced immune responses in UVB irradiated mice, *J. Immunol.* **139**:2788–2793 (1987).

174. Sheridan, J. F., Beck, M., Smith, C. C., and L. Aurelian, Reactivation of herpes simplex virus is associated with production of a low molecular weight factor that inhibits lymphokine activity *in vitro. J. Immunol.* **138**:1234–1239 (1987).

175. Aurelian, L., Yasumoto, S., and Smith, C. C., Antigen-specific immune-suppressor factor in herpes simplex virus type 2 infections of UVB-irradiated mice. *J. Virol.* **63**:2520–2524 (1988).

# 6

# Human Cytomegalovirus

JOSEPH L. WANER

## 1. INTRODUCTION

### 1.1. Type Strains of Human Cytomegalovirus

The modern era of investigation of human cytomegalovirus (HCMV) began with the independent isolation of viruses from three different laboratories.[1-3] An insightful historic account of those reports was given by Weller.[4] The AD169 strain isolated by Rowe et al.[3] from cultures of adenoids and the Davis strain isolated by Weller et al.[2] from a liver biopsy became prototype strains for study. More recently, the Towne strain isolated from urine and studied by Plotkin et al.[5] as a vaccine candidate joined AD169 and Davis as the three most studied strains of HCMV.

### 1.2. Properties of the Virus

Cytomegaloviruses (CMV) are members of the family Herpesviridae (herpesviruses) and the subfamily β-herpesvirinae. Generally recognized criteria for classification of HCMV are a molecular weight of $150 \times 10^6$ for the DNA genome, a slow replicative cycle, recovery of virus only from humans, a restricted host range in cell culture, and no animal model of infection. The virus becomes latent in the host following primary infection and retains the potential to reactivate and replicate under conditions that promote immunosuppression.

Cytomegaloviruses of other species exist, of which the murine cytomegalovirus (MCMV) is the most extensively studied. The close evolutionary relationship of a CMV with its host species, however, should always caution against extensive generalizations regarding properties of CMV.

JOSEPH L. WANER • Department of Pediatrics, University of Oklahoma, Health Sciences Center, Oklahoma City, Oklahoma 73190; and Virology Laboratory, Oklahoma Children's Memorial Hospital, Oklahoma City, Oklahoma 73104.

The absence of an animal model for HCMV necessitates that knowledge regarding the immune response must derive from observations of human infection and disease. The various manifestations of active symptomatic or asymptomatic HCMV infections as they occur in different age groups complicate attempts to dissect the immune response. In addition, the various components of the immune response should be viewed as mechanisms of recognition, protection, and/or pathogenesis in relationship to HCMV infection and disease. This chapter discusses immunity to HCMV relative to the principal circumstances of transmission to healthy hosts and the resulting infection with or without manifestation of disease. References to studies of MCMV are made where insights relative to human infection are provided.

## 2. INFECTION

### 2.1. Congenital

The multiple studies reviewed by Ho[6] indicate that the frequency of congenital infection ranges from 0 to 3.4% of all births, with an average of 1% commonly cited. Approximately 10% of congenitally infected infants are clinically symptomatic at birth and are usually associated with primary maternal infections that occurred during pregnancy;[7–9] 5–15% of the asymptomatic infected newborns, nevertheless, may develop symptoms within the first 2 years of life.[10–13]

Most primary maternal infections are asymptomatic.[7–9,14–17] The incidence of congenital infection is higher among infants born to women who are seropositive before pregnancy, reflecting reactivated infection in the mother as the source of fetal infection.[9,18,19] These newborns are far less likely, however, to have clinical sequelae.[9,18–21] Congenitally infected infants routinely produce antibody but show certain deficiencies in cell-mediated immunity (CMI).[22–25]

### 2.2. Perinatal

The number of infants infected during birth by passage through an infected uterine cervix, or within the first few months of life, exceeds the number of congenitally infected infants. Infants born at term and infected perinatally are more likely to be asymptomatic; those who show symptoms rarely have serious, long-term sequelae.[26–29] Passively transferred maternal antibody,[20,26,29] including neutralizing antibody,[30] does not prevent infection but may modulate disease. Premature infants are at higher risk of serious clinical consequences of perinatal infection than are full-term infants,[31,32] perhaps reflecting a more immature immune system.

Although the presence of immunity in the mother and of antibody in the infant is associated with minimal or no clinical sequelae, the relative roles of maternal immunity, passively transferred antibody in the fetus or newborn, and age of the fetus or newborn in determining symptomatic infection are not clear. If a threshold of inoculum exists over which natural fetal immunity is overcome and fetal infection becomes disseminated, maternal immunity may function to keep the viral inoculum low enough to prevent severe disease, but not infection. Asymptomatic infants, in fact, excrete significantly less virus than do symptomatic infants.[9,20]

## 2.3. HCMV Mononucleosis

Acquired HCMV infection in older children and adults is usually asymptomatic but may cause a heterophile-negative mononucleosis syndrome.[33] Major contributions to the understanding of the immune response to HCMV infection have derived from studies of these patients and are referred to in the following sections.

## 2.4. Immunocompromised Hosts

Latent HCMV may be reactivated by immunosuppression, indicating the equilibrium that exists between the latent virus and the healthy immunocompetent host. Excretion of HCMV in transplant and cancer patients due to the immunosuppressive nature of either the disease or treatment, or both, is a common occurrence that may have severe clinical consequences. Like the congenitally infected infant, most immunocompromised patients retain intact humoral immune functions against HCMV but suffer dysfunctions of CMI associated with reactivated infection. Immunocompromised patients who acquire a primary infection are much more likely to have severe disease than those who have reactivated infections [reviewed by Ho[34]]. In a revealing study, rheumatology patients being treated with corticosteroids or cylophosphamide were observed for excretion of HCMV. Only patients immunosuppressed by cyclophosphamide became HCMV excretors.[35] In addition, the use of antithymocyte serum in renal transplant patients was associated with greater HCMV disease, as exemplified by increased viremia.[36] Thus, efforts that compromise CMI are associated with greater HCMV activity.

Cytomegalovirus is also an important opportunistic pathogen in patients with the acquired immunodeficiency syndrome (AIDS).[137–139] In addition to being a major cause of morbidity and mortality in AIDS patients, HCMV may also be an additive or complementary immunosuppressive agent.[138–140]

## 3. ANTIGENS OF HCMV

Immunologists are interested in the proteins of HCMV as they interact with the immune system. Definitions of the proteins in terms of their immunogenic and antigenic characteristics are important to the current and future understanding of the immune response.

Antigens on the surface of the virion are likely to be targets of circulating antibodies, whereas viral antigens on the surface of infected cells would serve as targets for either antibodies or mechanisms of CMI, or both. The salient questions of any analysis of virus–immune interactions refer to delineation of which viral antigens are targets of a protective immune response. Dissection of the HCMV antigenic mosaic has only begun. An evaluation of the antigenic composition of HCMV should discern between the investigational approaches used experimentally to ascertain antigens of HCMV and the host's response to the antigens relative to the different conditions of natural infection.

## 3.1. Physical and Biochemical Analysis

Cytomegalovirus virions possess a minimum of 35 structural proteins.[37–39] Eleven glycosylated proteins were identified, eight of which were routinely detected, and three of higher-molecular weights identified as precursors.[40] Virions contain at least five glycoproteins in their surface membranes, and each of the carbohydrate moieties appears to have structural differences.[41] In a comprehensive study, only 17 of the 35 polypeptides identified in virions were also seen in infected cells.[39] All eight of the glycoproteins identified in virions, however, were also seen in infected cells; one additional glycoprotein (100 KDa) was present in infected cells but was not identified in virions. Discrepancies exist between laboratories in characterizing HCMV proteins, due in part to the different methods employed.

Synthesis of HCMV proteins is sequential, being effectively divided between proteins synthesized prior to, and independent of viral DNA synthesis and those produced following and dependent on DNA synthesis;[42,43] the former are, for the most part, important in governing viral replication and the latter are structural proteins. More than 50 polypeptides are synthesized in infected cells, with 10 of these synthesized within 6 hr of infection.[43]

Stinski et al. [44] showed that HCMV glycoproteins appear on the plasma membrane of infected cells within 24 hr of infection and as long as 2 days before the release of virus progeny. Preparations of viral glycoproteins were immunogenic in rabbits, with the resulting antisera successfully precipitating the respective glycoprotein antigens;[40,43] this work supported the original observation that human antisera may contain antibodies against HCMV antigens expressed on the surface of infected cells.[45] Rabbit antisera against viral glycoproteins also neutralized infectivity of HCMV.[40]

## 3.2. Immunologic Analysis

### 3.2.1. Analysis Using Monoclonal Antibodies

A different perspective on the antigenic composition of HCMV was provided with monoclonal antibodies. Rather than identifying viral proteins through their physicochemical properties, HCMV polypeptides were characterized through immunogenic and antigenic properties of specific epitopes. Kim et al.[46] produced monoclonal antibodies against purified virions of the

AD169 strain of HCMV. Five of nine antibodies produced reacted with a single glycopeptide of 66 KDa that can be detected within 2 hr of infection, was produced in large quantities, and was common to many strains of HCMV.[46,47]

By contrast, Pereira et al.[48] prepared monoclonal antibodies against cells infected with the AD169 strain of HCMV. Twelve different glycoproteins were identified with 14 monoclonal antibodies. The glycoproteins were present on the surface of infected cells, although only 9 of the 14 antibodies reacted by immunofluorescence with the surface membrane of infected cells; one of the three antibodies that possessed neutralizing activity did not react with the surface of infected cells.[48] Thus, divergent expression of HCMV epitopes occurs in the membranes of infected cells. This may be due in part to the existence of polymorphic forms of the glycoproteins.[49] Virion glycoproteins may also vary in molecular weight between strains of HCMV.[50]

*Neutralizing Epitopes.* Rasmussen and co-workers produced a monoclonal antibody against an 86-kDa protein of strain AD169 that neutralized several strains of HCMV[51]; the protein was localized in the cytoplasm of infected cells. The same laboratory also reported two monoclonal antibodies that recognized the same 55- and 130-kDa polypeptides but possessed different immune properties; one antibody neutralized only with complement, while the second showed no neutralizing activity.[52] Similarly, Kari et al.[141] described monoclonal antibodies that reacted by immunofluorescence with several strains of HCMV but showed divergent neutralizing properties.

Britt[53] identified three glycoproteins of 160, 116, and 55kDa with monoclonal antibodies that also had neutralizing activity. The glycoproteins were believed to be structured within virions as covalently linked disulfide-bonded protein complexes;[53] only the 116- and 55-kDa glycoproteins, however, seem to exist as final products in the virion envelope.[54] Rasmussen et al. [52] detected similar complex structures in cell-free virus and in extracts of infected cells. A gene expressing two glycosylated forms (145 and 55 kDa) of a CMV glycoprotein was expressed in a recombinant vaccinia virus;[142] antibody produced in rabbits following infection with the recombinant neutralized HCMV *in vitro.*

### 3.2.2. Fc Receptors

In addition to specific HCMV antigens, receptors for Fc fragments are induced in the cytoplasm and on the plasma membrane of infected cells.[55,56] The purpose of the receptors has not been elucidated but may bear on protecting the infected cell from lysis by antibody and complement or by cytolytic effector cells.

## 4. HOST ANTIBODY RESPONSE AGAINST HCMV

The antibody response to primary or reactivated HCMV infections does not appear to be impaired by the infectious process. It should be noted that, in contrast, the host antibody response to MCMV is markedly suppressed.[57,58]

Common serologic tests have been applied to the detection of HCMV antibody. References to serologic tests will be made only within the context of discussing the antibody response to HCMV as a component of the immune response.

## 4.1. Complement-Fixing Antibodies

The complement-fixation (CF) test has been the most extensively used and studied serologic assay. Glycine-buffer extraction of infected cells results in a more effective antigen preparation. The principal viral component of a glycine-extracted antigen preparation appears to be a 66-kDa glycoprotein[47,59,60] that is probably held in common by all strains of CMV; a 140-kDa polypeptide[47] and a 50-kDa glycoprotein,[60] however, were also reported; Kim et al.[38] had previously identified the 66-kDa glycoprotein as the most prominent structural protein of the virion. The 66- and 50-kDa glycoproteins, in particular, were well precipitated by human sera with CF antibody. Thus, conventional CF tests primarily detect antibody production against major structural proteins of the virus.

The CF antigens react primarily with IgG antibodies and are not reliable in detecting specific IgM antibodies against HCMV.[61] Studies and discussions by Pereira et al.[48,60] suggest that CF antibody fluctuations and indications of strain differences observed by CF[62] may be due to underproduction of particular CF antigens by some strains of HCMV and/or a manifestation of varying mixtures of type-common and type-different determinants in antigen preparations.

## 4.2. IgG Subclasses

Immunoglobulin G (IgG) antibodies against HCMV in the sera from healthy donors and patients as measured by enzyme-linked immunosorbent assays (ELISA) were primarily of IgG1 and IgG3 subclasses, the most efficient IgG subclasses in CF reactions. IgG3 antibodies were detected first in primary disease. IgA1 and IgA2 antibodies against HCMV were also detected.[63]

## 4.3. Immunoprecipitation–Immunoblot Analysis

Antibodies in human convalescent sera also recognize glycoproteins of 52, 67, 95, 130, and 250 kDa in the envelopes of virions.[46] Six glycoproteins (130, 110, 96, 66, 50, and 25kDa) and four nonglycosylated proteins were immune precipitated from infected cells by IgG or IgM antibodies of human antisera.[60] Evidence was also presented that antibodies to the various glycoproteins occur at different concentrations and have a temporal appearance after congenital or perinatally acquired infections.[64] The sera from symptomatic children continued to precipitate greater amounts of HCMV antigen over a longer period of time than did sera from asymptomatic patients.

Early (1 month) and late (6–8 months) patterns of antibody development to HCMV antigens were noted in adult patients following primary infection. Immunoblotting detected antibodies to proteins of 66, 55, 40, 35, and 30 kDa within one month of seroconversion; antibodies to 100, 60, and 22 kDa proteins were detected six to eight months after seroconversion. A third pattern of antibody development was detected but varied from patient to patient. A similar study indicated that bone marrow transplant recipients make antibodies to HCMV proteins in a similar pattern to that of immunocompetent persons.[144] The number of bands and the molecular weights of the HCMV proteins reported varies between laboratories, and a concensus regarding the development of antibodies has yet to be established.

## 4.4. Secretory IgA

Secretory IgA antibody against HCMV was demonstrated in cervical secretions of adult patients,[65] and in saliva from seropositive virus-excreting and -nonexcreting infants.[66] Low neutralizing activity was associated with the IgA antibody in saliva but was not looked for in the cervical secretions. The absence of virus excretion did not correlate with the presence of secretory IgA antibody in either study.

## 4.5. IgE Antibody

Specific IgE antibody against HCMV was reported to be a good indicator of primary infection. Only 9% of patients with recurrent infection had detectable IgE antibody, while 96% of patients with primary infections were positive.[67]

## 4.6. Neutralizing Antibody

Weller et al.[68] showed neutralizing activity in sera from infants with active HCMV infections. These workers further demonstrated antigenic differences between the virus isolates and the AD169 strain in neutralization tests using sera obtained from these infants. Many reports have since confirmed the heterogeneity of HCMV isolates, but the clinical implications of those reports and that original study are yet to be resolved.

Neutralizing antibodies would be expected to afford some degree of immune protection. The formation of IgG antibody–HCMV complexes, however, does not necessarily prevent productive infection.[69] Indeed, HCMV–antibody complexes were absorbed to the host cell as efficiently as noncomplexed virus.[70] Monoclonal antibodies prepared against the AD169 strain of HCMV and possessing neutralizing activity precipitated glycoproteins of 160, 116, and 55 kDa; human immune sera with neutralizing activity also precipitated these glycoproteins in addition to proteins of 200, 145, 100, 66, and 34 kDa.[50]

The evidence indicates that glycoproteins of HCMV are strongly immunogenic in the host and serve as major targets for neutralizing antibodies on the surface of virions. Glycoproteins are also present prominently on the surface of infected cells and would be expected to serve as targets of cytolytic immune mechanisms. The expression of glycoproteins on the surface of infected cells may, however, be different from that on virions.[71–74]

### 4.7. Immune Complexes

The enhanced and extended production of antibody in symptomatic pediatric patients draws attention to the report of immune complex formation in congenitally infected infants.[75] Renal glomeruli of three symptomatic infants studied who subsequently died had a heavy deposition of immune complexes on the basement membranes. Increased quantities of circulating immune complexes were also found in symptomatic infants in comparison with asymptomatic HCMV excretors. The elevated quantities of antibody produced in clinically apparent infections, therefore, may play a role in the pathogenesis of HCMV disease.

### 4.8. Cytolytic Antibody

The strongest experimental data for the protective role of antibody derive from studies showing cytolytic activity of IgM and IgG antibodies.[72,73] The original report associated complement-dependent cytolytic antibody activity with symptomatic infection;[72] the responsible immunoglobulin was IgM. Accordingly, cytolytic activity disappeared 1–2 months after the appearance of symptoms. Middeldorp et al.,[73] however, demonstrated convincingly that IgM and IgG antibodies from symptomatic patients may be cytolytic by way of the classic pathway of complement activation. The target antigens were on the surface of infected cells and, although not specifically identified, were primarily late antigens; attempts to lyse infected cells showing only early antigens with homologous reactive antisera were unsuccessful. The early and late HCMV antigens on the surface of infected cells appear to exist as individual sets of antigens that may also differ in part from antigens on the surface of virions.[73,74] The best clinical indication for a protective role of specific antibody comes from trials using immune globulin to prevent or modulate HCMV disease in high-risk patients.[76–79]

## 5. LYMPHOCYTE BLASTOGENIC RESPONSE

Lymphocytes of immunocompetent, HCMV-seropositive individuals undergo blastogenesis upon exposure to HCMV antigen.[80–82] Lymphocyte transformation occurs in response to purified virions or antigen extracts of infected cells. The presence of antibody in the assay does not affect lymphocyte blastogenesis.[82]

Antigens of HCMV stimulate memory T cells to undergo blastogenesis.[83] Lymphocytes appear to recognize different antigens on virions and infected cells.[84] The retention of a blastogenic response against HCMV infected cells was associated with suppression of HCMV in renal transplant recipients; all patients showed depression of the lymphocyte response against virions.[85]

Strain-specific responses of lymphocytes from tonsils and peripheral blood were reported using three strains of HCMV, including AD169 and Davis.[86] The findings concerning strain-specific responses of lymphocytes from peripheral blood, however, were not confirmed using purified antigen preparations of AD169, Davis, and Towne strains.[87]

Healthy persons seropositive for both late and early antigens showed significantly higher blastogenic responses against HCMV antigens than did those with only antibody against late antigen.[88] In a later study, Waner et al.[89] used preparations of early antigens and antigens from productively infected (PI) cells to evaluate the blastogenic response in seropositive healthy donors and virus-excreting patients. Seropositive healthy persons showed blastogenic responses to early antigens whether or not antibodies to early antigens were detectable. Four of 9 patients, however, with reactive lymphocyte responses to PI antigen did not have detectable responses to early antigens. The early antigen preparation contained viral polypeptides of 40kd, 56kd, 62kd, and 83kd as identified by SDS-PAGE.

Lymphocytes from immunocompetent persons experiencing a primary HCMV infection with the clinical syndrome of mononucleosis showed a diminished response to the mitogen concanavalin A (Con A) but a normal response to phytohemagglutinin (PHA).[90] Concurrently, such patients showed a reversal in the ratio of helper (OKT4) to suppressor-cytotoxic (OKT8) lymphocytes.[91] Convalescence was characterized by a return of normal Con A responses and normal OKT4/OKT8 ratios. The blastogenic response to HCMV antigen in HCMV mononucleosis patients was also depressed. Characteristically, development of specific antibody was detectable early in the illness, but the blastogenic response of lymphocytes was delayed and did not reach normal levels for months.[92] Patients with previously acquired immunity against herpes simplex virus (HSV) and varicella–zoster virus (VZV) also had diminished proliferative responses of lymphocytes against the homologous antigens indicating a generalized suppressive effect of the active infection.

The immunosuppressive effect of HCMV in healthy adults may bear on the occurrence of congenital infection. Congenitally infected infants and their mothers showed varying degrees of suppression of the blastogenic response.[22,23,93] The mothers of infants with symptomatic infections showed a higher degree of suppression than mothers of infants with asymptomatic infections. The lymphocyte response of symptomatic infants was significantly more depressed than that of asymptomatic infants.[93] Stimulation of blastogenesis by PHA or herpes simplex antigen was normal in the mothers, indicating a specific defect against HCMV. Gehrz et al. [94] reported that seropositive non-virus-excreting pregnant subjects show specific suppression of lymphocyte blastogenesis against HCMV that is greatest in the third trimester. Maternal immunity as assessed by blastogenic assays may not, however, be a determinant of viral excretion during gestation. Faix *et al.*[95] studied a population of adolescents longitudinally during pregnancy and reported that a blastogenic response was not detectable on 40% of the test occasions in seropositive subjects; furthermore, significant differences were not observed between mothers excreting virus (asymptomatic) and seropositive mothers not excreting virus.

Most congenitally infected infants do not develop a blastogenic response for years;[24,25] the acquisition of a response, however, is associated with termination of viral excretion.[24] The immune defect is specific for HCMV, as patients with antibody against HSV exhibit blastogenic responses.[24] Nor is significant suppression of PHA-stimulated blastogenesis observed in congenital or postnatally infected patients.[25]

# 6. NATURAL KILLER CELL ACTIVITY

## 6.1. *In Vitro* Assay

Diamond *et al.*[96] reported the "unexpected lysis" of HCMV-infected fibroblasts by peripheral blood mononuclear cells from healthy seronegative donors; uninfected fibroblasts were minimally lysed under the same conditions. Natural killer (NK) cell activity against infected fibroblasts was eventually characterized as being mediated by non-B, non-T lymphocytes with Fc receptors; lymphocytes from seronegative or seropositive donors effected lysis that was not HLA restricted.[97,98] Lytic activity was also associated with effector cell populations enriched for large granular lymphocytes.[98,99] Preincubation of lymphocytes with interferon (IFN) enhanced the lytic activity.[98] NK-cell activity, however, was shown to be independent of IFN production in the assay.[99,100]

The appearance of HCMV early antigens on the surface of infected cells was sufficient to induce susceptibility to lysis by NK cells.[100] A principal target of NK activity may, however, be a determinant(s) present on uninfected fibroblasts whose expression is enhanced[99,100] or diminished,[99] depending on the strain of infecting virus.[99] These latter considerations may be important in determining the course of primary infection before the advent of acquired immune mechanisms.

## 6.2. Clinical Studies

The role of NK activity in CMV infection and disease has not been elucidated. A study of lymphocyte-mediated cytolysis of eight virus-excreting congenitally infected infants and six of their mothers reported greater impairment of lytic activity in the mothers than in the infants.[101] Harrison and Waner[102] reported on NK cell activity in 39 infants and children excreting HCMV by assessing lytic activity against uninfected fibroblasts, HCMV-infected fibroblasts, and K562 cells. Differential responses against the three targets were observed and correlated with age and clinical status of the patients. Patients with interstitial pneumonitis (IP) or with IP and hepatitis constituted an acutely ill population; those under 6 months of age had significantly higher levels of NK activity than did healthy, nonexcreting controls, while patients over 6 months of age had slightly depressed activity (insignificant) against HCMV infected targets and significantly depressed activity against K562 cells. Interestingly, both age groups showed significantly higher levels of lytic activity against uninfected fibroblasts. Patients without IP constituted a chronically ill study population that did not display different levels of NK activity against uninfected fibroblasts than did comparable controls. The chronically ill patients 6 months of age or older, however, showed significantly depressed NK activity against HCMV infected targets and K562 cells; patients under 6 months of age showed normal levels of activity. All viral excretors showed pronounced NK activity regardless of relative suppression or enhancement relative to controls. The significantly elevated levels of NK activity against uninfected fibroblasts was associated with the most acute disease. Starr *et al.*[103] also reported increased levels of lytic activity against uninfected fibroblasts in patients with HCMV disease following renal transplantation. Thus, NK activity against uninfected cells may have a role in the pathogenesis of HCMV disease.

The NK response of patients with HCMV mononucleosis did not differ from that of normal donors when measured only against K562 cells.[104] Non-HLA-restricted cytolytic activity (NK activity) was also reported as an important component in the recovery of bone-marrow-transplant recipients with HCMV infection;[105] patients with fatal infections had depressed levels of activity before and during the infection.

## 6.3. Genetic Influence on the Murine NK-Cell Response

Natural killer cell responses are detected in MCMV infected mice within 3 days of inoculation,[106] suggesting an important role in early defense. Mice depleted of NK activity by treatment with a specific monoclonal antibody (anti-asialo GM1) against NK cells suffer a more severe infection of longer duration; the treatment with antibody did not interfere with other immune mechanisms.[107] The effects of depletion of NK activity were most profound in the first 5 days of infection. In addition, high titers of MCMV in salivary glands 1 month after inoculation in antibody-treated mice suggested a role for NK cells in regulating persistent infection.

Strains of mice differ in their susceptibility to infection,[108,109] which correlates with NK activity.[110] Beige mice are genetically deficient in NK activity and were more susceptible to lethal infection by MCMV than heterozygous littermates with NK activity;[111] infected animals also developed 33- to 43-fold greater titers of virus in liver, spleen, and kidney than did the heterozygotes that experienced sublethal infection.

In the mouse model of infection, therefore, the abrogation of the NK response early in the course of infection is associated with a poor prognosis. Strains of mice that are particularly susceptible to MCMV infection have genetic dispositions to low NK activity. The inability of the host, therefore, to manage the early infectious process successfully before the advent of acquired immune responses may be a major determining factor in the clinical outcome for mice and perhaps humans.

## 7. ANTIBODY-DEPENDENT CELL-MEDIATED CYTOTOXICITY

Enhancement of cytotoxicity of HCMV-infected fibroblasts following preincubation of the target cells with human antisera to HCMV was reported.[97] Lymphocytes from seropositive and seronegative donors mediated the reaction and were found in the same lymphocyte fraction as NK cells, confirming other reports that the effector cells of NK and antibody-dependent cell-mediated cytotoxicity (ADCC) activity may be identical.[112] The parameters of ADCC in humans using HCMV-infected targets have not been elucidated, nor has the assay been applied to ascertain the role of ADCC in HCMV disease. This would appear to be a fertile area of investigation, given the recent findings discussed above concerning the antigenic composition of the HCMV infected cell.

## 8. HLA-RESTRICTED CYTOTOXIC T-CELL ACTIVITY

Three bone marrow transplant recipients with primary HCMV infections developed an HLA-restricted cytolytic response against HCMV-infected targets that was mediated by T cells;[113] the responses were induced specifically by the CMV infection. Borysiewicz et al.[114] subsequently showed that precursors of cytotoxic T lymphocytes (CTL) are present in the peripheral blood of normal, seropositive donors. Clonal expansion of cytotoxic T-cell subsets occurs upon incubation of peripheral blood mononuclear cells with fibroblasts infected with HCMV; similar incubations with cell-free virus, however, resulted in helper lines. Cytolysis mediated by the cytotoxic lymphocyte lines (Leu 2a+) was HLA restricted. HCMV-infected cells displaying only early antigens were lysed, in addition to productively infected cells.

Cytolytic activity of mononuclear cells from seropositive adults and infants with active infection was enhanced by incubation of the lymphocytes with cell-free HCMV antigens for 6 days prior to assay.[115] Depletion experiments using monoclonal antibodies indicated that the effector cells were composed of CTL (OKT8) and an NK cell population.

The CTL response appears to be an important factor in the recovery of bone marrow transplant recipients from active HCMV infections;[116] patient survival was always associated with a response. The response occurs prior to the median onset of viral shedding or rises in serum antibody levels. A similar conclusion was drawn from a study of renal transplant recipients.[117] In addition, the specific cytotoxic response may have protected against HCMV injury to the graft.

## 9. INTERACTION OF HCMV WITH LEUKOCYTES

The original observations on the effect of HCMV on the cellular elements of peripheral blood derived from clinical investigations. Recent studies have begun to clarify the earlier reports.

### 9.1. T-Cell Subset Ratios

Concurrent with diminished blastogenic responses, lymphocyte populations from patients with acute HCMV mononucleosis show a reversal of the normal OKT4:OKT8 cell ratio.[91] The reversal is characterized by relative and absolute decreases in helper cells with concomitant increases in suppressor cells. The normal ratio of the T-cell subpopulations returns during convalescence. The atypical lymphocytes seen during the HCMV mononucleosis syndrome are predominantly in the OKT8 subpopulation. [118] The negligible effect of HCMV infection on antibody production is supported by the report that T-cell ratio changes usually precede rises in antibody production in cardiac transplant recipients.[119]

Symptomatic congenitally infected infants under 1 year of age also may show increases in the percentage of OKT8 and a decreased ratio of OKT4:OKT8 cells. The decreased ratio was primarily due to reduced numbers of helper cells. The ratio of T-cell subpopulations did not differ from controls, however, in asymptomatic children or in symptomatic children older than 1 year.[120] This is reminiscent of Harrison and Waner's observations of an age-related association with NK activity.[102] Thus, the overall effect of active HCMV infection on the OKT4:OKT8 ratio was less pronounced in infants than in adults with HCMV mononucleosis. This difference may reflect the chronic nature of the infection in congenitally infected infants versus the acute nature of the disease in adults.

### 9.2. HCMV Infection of Leukocytes

Cytomegalovirus was isolated predominantly from the polymorphonuclear leukocyte fractions and to a lesser degree from lymphocyte fractions of peripheral blood of patients with HCMV mononucleosis.[90,121] The threat of acquiring HCMV through transfusion of blood from healthy seropositive donors is associated with transfer of leukocytes, indicating carriage of virus in the cells;[122] isolation of virus from healthy seropositive donors was reported once.[123] Using DNA probes that code for four immediate early regions of the HCMV genome, Schrier et al.[124] detected HCMV messenger RNA (mRNA) in peripheral blood mononuclear cells of healthy seropositive donors. In eight seropositive subjects, the level of hybridization ranged between 0.03 and 2% of the PBM examined. Most lymphocytes purified by a fluorescence-activated cell sorter and reacting with the probe were OKT4 (2.4%) subpopulation of T cells, with 0.8% of the reacting lymphocytes being OKT8.

Initial laboratory investigations of infection of human leukocytes and lymphoblastoid cell lines by HCMV were inconclusive and not very enlightening when analyzed within the context of clinical studies. Infection of peripheral leukocyte cultures or lymphoblastoid cell lines with the laboratory-passaged AD169 or Towne strains of HCMV produced persistently infected cultures.[125–127] It was not resolved, however, whether the persistence of detectable virus was due to retention of inoculum or slow replication with an equilibrium established with dividing uninfected cells. CMV-induced cell DNA synthesis was shown in two cell lines with a constant number of viral DNA copies per cell detected in the cultures for approximately 2 weeks.[128]

Isolates of HCMV, passaged fewer than 10 times, induced early antigen production in approximately 2% of inoculated mononuclear leukocytes. By contrast, inoculation with the multipassaged AD169 strain resulted in early antigen production in 0.01–0.02% of leukocytes; there was no indication of productive viral replication. Neither fresh isolates of HCMV nor the AD169 strain was effective in infecting polymorphonuclear leukocytes.[129] Depletion experiments and morphologic studies of separated cell fractions further indicated that monocytes were the principal targets of infection.

Rice et al.[130] contributed important additional findings. Seven of eight low-passaged isolates of HCMV induced the 72-kDa immediate–early protein in less than 1–15% of peripheral blood mononuclear cells; in only 1 of 10 instances was the 72-kDa protein induced by the AD169 strain and then, only in less than 1% of the cells. Late antigens, reflective of productive infection, were not detected in any experiments. Monocytes comprised the greatest percentage of 72-kDa reactive cells; smaller percentages of OKT4- and OKT8-positive lymphocytes, B lymphocytes, and NK cells were also abortively infected. In assessments of the effect of infection on functional characteristics, low-passaged isolates of HCMV suppressed PHA- and antigen-induced proliferation as well as NK cell activity; the AD169 strain only evoked minimal suppression of antigen-induced proliferation, while no HCMV strain affected ADCC. Low-passaged clinical isolates were subsequently shown to suppress cytotoxic T-lymphocyte activity specific for HCMV.[145] Sing and Garnett[131] also reported depression of the PHA response of T cells when infected with AD169 HCMV; in addition, a reversal of OKT4 : OKT8 cell ratios was seen 6 hr after infection of the cultures. Viable virus was required to effect these observations, which were made with lymphocytes from seronegative donors. Braun and Reiser[146] reported replication of HCMV in a small subset of T cells consisting primarily of the T3+ and T8+ phenotype, although T4+ cells may also be susceptible.

## Activation of B Cells

The direct activation of B cells in vitro by HCMV was reported.[132] Virus apparently functioned as a nonspecific polyclonal activator. The B-cell response to HCMV occurred in cells from seronegative donors, was independent of virus replication, and did not require T cells. By contrast, Yachie et al.[133] showed activation of T cells and B cells by the AD169, Davis, and Towne strains of HCMV with resulting Ig production only in lymphocytes cultured from seropositive donors and only in the presence of T cells. B cells failed to produce Ig when OKT4 cells were removed from the cultures, while the removal of OKT8 cells had no effect.

### 9.3. Immune Suppression in HCMV Mononucleosis

The depression of the blastogenic response of lymphocytes to Con A from patients with HCMV mononucleosis[90] was subsequently shown to be associated with the adherent cell fraction of mononuclear cells.[134] Suppression was diminished by culturing mononuclear leukocytes for 1–7 days prior to addition of Con A and to a greater degree by depletion of the adherent cells after the culture period; blastogenesis of recultured nonadherent cells was also significantly suppressed by the addition of fresh adherent cells from the same patient. Suppressor activity was not found in the sera of patients.

Cytomegalovirus was isolated from the monocytes of some patients with HCMV mononucleosis.[135] Infected monocytes from patients and monocytes from uninfected donors subsequently infected in culture exerted a suppressive action on autologous lymphocyte responses to Con A. Suppressed Con A responses could only be reversed in lymphocyte cultures of patients showing clinical symptoms for 3 weeks or more; lymphocytes from patients with symptoms for 2 weeks or less were refractory to reversal of suppression by culturing of lymphocytes or removal of monocytes.

The immunosuppressive effect exerted by HCMV infected monocytes may in part be due to an inhibitory protein of interleukin 1 (IL-1) activity. The protein is approximately 95kDa and was produced following infection by HCMV of monocytes from seronegative donors. The inhibition of IL-1 production is HCMV-specific but does not require viral replication: in fact, no expression of HCMV activity was detected.[136] These experiments were performed with the AD169 strain of HCMV. Dudding and Garnett[147] reported expression of HCMV immediate-early antigens in monocytes infected with a low-passaged clinical isolate; antigens were rarely detected in monocytes infected with AD169. Infection with the clinical isolate also resulted in enhanced suppression of lymphocyte proliferation over that seen with AD169; addition of IL-1 to the infected monocyte cultures restored the proliferative response of lymphocytes.

Differentiation of cells of the immune system may be important for the ultimate outcome of HCMV infection. Following treatment with a phorbol ester, a monocyte cell line was infected with HCMV and produced a productive infection characterized by expression of late antigens and infectious virions.[148]

## 10. PERSPECTIVE

A picture is emerging of HCMV infection directly exerting an immunosuppressive effect in the host through infection of immune cells of different lineages. The immunosuppression may result from an effect of HCMV on effector cells, such as T cells and NK cells or through the dysfunction of normal regulatory mechanisms, such as suppression of IL-1 and IFN production.[23,92] Viral replication in the infected immune cells may not be required, indicating a possible selected evolutionary advantage of HCMV. Defective particles produced during the course of infection and possibly in greater quantity than infectious particles might be capable of infecting immune cells and exerting the described effects on the immune system through the expression of early functions.

Infection with only expression of early functions may persist longer and continually exert an immunosuppressive effect in contrast to productive infection, which would destroy the cell. The clinical studies documenting increased virus excretion in symptomatic patients may be reflective of the immune system succumbing to a greater load of both infectious and noninfectious particles. Pathogenesis would ensue from disseminated infection to multiple organs with subsequent cell destruction due to viral replication, the formation of immune complexes, and possibly autoimmune phenomena.

## REFERENCES

1. Smith, M. G., Propagation in tissue cultures of a cytopathogenic virus from human salivary gland virus (SGV) disease, *Proc. Soc. Exp. Biol. Med.* **92**:424–430 (1956).
2. Weller, T. H., J. C. Macauley, J. M. Craig, and P. Wirth, Isolation of intranuclear inclusion producing agents from infants with illnesses resembling cytomegalic inclusion disease, *Proc. Soc. Exp. Biol. Med.* **94**:4–12 (1957).
3. Rowe, W. P., J. W. Hartley, S. Waterman, H. C. Turner, and R. J. Huebner, Cytopathogenic agent resembling human salivary gland virus recovered from tissue cultures of human adenoids, *Proc. Soc. Exp. Biol. Med.* **94**:418–424 (1956).
4. Weller, T. H., Cytomegalovirus: The difficult years, *J. Infect. Dis.* **122**:532–539 (1970).
5. Plotkin, S. A., T. Furukawa, N. Zygraich, and C. Huygelen, Candidate cytomegalovirus strain for human vaccination, *Infect. Immun.* **12**:521–527 (1975).
6. Ho, M., Congenital and perinatal human cytomegalovirus infections, in: *Cytomegalovirus, Biology and Infection: Current Topics in Infectious Diseases* (W. B. Greenough III and T. C. Merigan, eds.), pp. 131–149, Plenum, New York (1982).
7. Griffiths, P. D., A. Campbell-Benzie, and R. B. Heath, A prospective study of cytomegalovirus infection in pregnant women, *Br. J. Obstet. Gynecol.* **87**:308–314 (1980).
8. Grant, S., E. Edmond, and J. Syme, A prospective study of cytomegalovirus infection in pregnancy. I. Laboratory evidence of congenital infection following maternal primary and reactivated infection, *J. Infect.* **3**:24–31 (1981).
9. Stagno, S., R. F. Pass, M. E. Dworsky, R. E. Henderson, E. G. Moore, P. D. Walton, and C. A. Alford, Congenital cytomegalovirus infection. The relative importance of primary and recurrent maternal infection, *N. Engl. J. Med.* **306**:945–949 (1982).
10. Melish, M. E., and J. B. Hanshaw, Congenital cytomegalovirus infection: Developmental progress of infants detected by routine screening, *Am. J. Dis. Child.* **126**:190–194 (1973).
11. Saigal, S., O. Lunyk, R. P. B. Larke, and M. A. Chernesky, The outcome in children with congenital cytomegalovirus infection: A longitudinal follow-up study, *Am. J. Dis. Child.* **136**:896–901 (1982).
12. Kumar, M. L., G. A. Nankervis, I. B. Jacobs, C. B. Ernhart, C. E. Glasson, P. M. McMillan, and E. Gold, Congenital and postnatally acquired cytomegalovirus infections: Long-term follow-up, *J. Pediatr.* **104**:674–679 (1984).
13. Reynolds, D. W., S. Stagno, K. G. Stubbs, A. J. Dahle, M. M. Livingston, S. S. Saxon, and

C. A. Alford, Jr., Inapparent congenital cytomegalovirus infection with elevated IgM levels: Causal relationship with auditory and mental deficiency, *N. Engl. J. Med.* **290**:291–296 (1974).

14. Nankervis, G. A., M. L. Kumar, F. E. Cox, and E. Gold, A prospective study of maternal cytomegalovirus infection and its effect on the fetus, *Am. J. Obstet. Gynecol.* **149**:435–440 (1984).

15. Ahlfors, K., M.Forsgren, S.-A. Ivarsson, S. Harris, and L. Svanberg, Congenital cytomegalovirus infection on the relation between type and time of maternal infection and infant's symptoms, *Scand. J. Infect. Dis.* **15**:129–138 (1983).

16. Monif, G. R. G., E. A. Egan, II, B. Held, and D. V. Eitzman, The correlation of maternal cytomegalovirus infection during varying stages in gestation with neonatal involvement, *J. Pediatr.* **80**:17–20 (1972).

17. Stern, H., and S. M. Tucker, Prospective study of cytomegalovirus infection in pregnancy, *Br. Med. J.* **2**:268–270 (1973).

18. Stagno, S., D. W. Reynolds, E. S. Huang, S. D. Thames, R. J. Smith, and C. A. Alford, Jr., Congenital cytomegalovirus infection: Occurrence in an immune population, *N. Engl. J. Med.* **296**:1254–1258 (1977).

19. Schopfer, K., E. Laube, and U. Krech, Congenital cytomegalovirus infection in newborn infants of mothers infected before pregnancy, *Arch. Dis. Child.* **53**:536–539 (1978).

20. Stagno, S., D. W. Reynolds, A. Tsiantos, D. A. Fuccillo, W. Long, and C. A. Alford, Jr., Comparative serial virologic and serologic studies of symptomatic and subclinical congenitally and naturally acquired cytomegalovirus infections, *J. Infect. Dis.* **132**:568–577 (1975).

21. Stagno, S., M. E. Dworsky, J. Tores, T. Mesa, and T. Hirsh, Prevalence and importance of congenital cytomegalovirus infection in three different populations, *J. Pediatr.* **101**:897–900 (1982).

22. Reynolds, D. W., P. H. Dean, R. F. Pass, and C. A. Alford, Jr., Specific cell-mediated immunity in children with congenital and neonatal cytomegalovirus infection and their mothers, *J. Infect. Dis.* **140**:493–499 (1979).

23. Starr, S. E., M. D. Tolpin, H. M. Friedman, K. Pauker, and S. A. Plotkin, Impaired cellular immunity to cytomegalovirus in congenitally infected children and their mothers, *J. Infect. Dis.* **140**:500–505 (1979).

24. Pass, R. F., S. Stagno, W. J. Britt, and C. A. Alford, Specific cell-mediated immunity and the natural history of congenital infection with cytomegalovirus, *J. Infect. Dis.* **148**:953–961 (1983).

25. Okabe, M., S. Chiba, T. Tamura, Y. Chiba, and T. Nakao, Longitudinal studies of cytomegalovirus-specific cell-mediated immunity in congenitally infected infants, *Infect. Immun.* **41**:128–131 (1983).

26. Reynolds, D. W., S. Stagno, T. S. Hosty, M. Tiller, and C. A. Alford, Jr., Maternal cytomegalovirus excretion and perinatal infection, *N. Engl. J. Med.* **289**:1–5 (1973).

27. Kumar, M. L., E. Gold, and G. A. Nankervis, Risk of acquired cytomegalovirus infection in infants of maternal CMV excretors, *Pediatr. Res.* **9**:512, 1975.

28. Stagno, S., D. W. Reynolds, R. F. Pass, and C. A. Alford, Jr., Breast milk and the risk of cytomegalovirus infection, *N. Engl. J. Med.* **302**:1073–1074 (1980).

29. Numazaki, Y., N. Yano, T. Morizuka, S. Takai, and N. Ishida, Primary infection with human cytomegalovirus: Virus isolation from healthy infants and pregnant women, *Am. J. Epidemiol.* **91**:410–417 (1970).

30. Reynolds, D. W., S. Stagno, R. Reynolds, and C. A. Alford, Jr., Perinatal cytomegalovirus infection: influence of placentally transferred maternal antibody, *J. Infect. Dis.* **137**:564–567 (1978).

31. Ballard, R. B., W. L. Drew, K. G. Hufnagle, and P. A. Riedel, Acquired cytomegalovirus infection in pre-term infants, *Am. J. Dis. Child.* **133**:482–485 (1979).

32. Yeager, A. S., P. E. Palumbo, N. Malachowski, R. L. Ariagno, and D. K. Stevenson,

Sequelae of maternally derived cytomegalovirus infections in premature infants, *J. Pediatr.* **102**:918–922 (1983).

33. Klemola, E., R. von Essen, G. Henle, and W. Henle, Infectious-mononucleosis-like disease with negative heterophile agglutination test. Clinical features in relation to Epstein–Barr virus and cytomegalovirus antibodies, *J. Infect. Dis.* **121**:608–614 (1970).

34. Ho, M., Human cytomegalovirus infections in immunosuppressed patients, in: *Cytomegalovirus, Biology and Infection: Current Topics in Infectious Diseases* (W. B. Greenough III and T. C. Merigan, eds.), pp. 171–201, Plenum, New York (1982).

35. Dowling, J. N., A. R. Saslow, J. A. Armstrong, and M. Ho, Cytomegalovirus infection in patients receiving immunosuppressive therapy for rheumatologic disorders, *J. Infect. Dis.* **133**:399–408 (1976).

36. Pass, R. F., D. W. Reynolds, J. D. Whelchel, A. G. Diethelm, and C. A. Alford, Impaired lymphocyte transformation response to cytomegalovirus and phytohemagglutinin in recipients of renal transplants: Association with antithymocyte globulin, *J. Infect. Dis.* **143**:259–265 (1981).

37. Sarov, I., and I. Abody, The morphogenesis of human cytomegalovirus. Isolation and polypeptide characterization of cytomegalovirions and dense bodies, *Virology* **66**:464–473 (1975).

38. Kim, K. S., V. Sapienza, R. I. Carp, and H. M. Moon, Analysis of structural polypeptides of purified human cytomegalovirus, *J. Virol.* **20**:604–611 (1976).

39. Stinski, M. F., Synthesis of proteins and glycoproteins in cells infected with human cytomegalovirus, *J. Virol.* **23**:751–767 (1977).

40. Stinski, M. F., Human cytomegalovirus: Glycoproteins associated with virions and dense bodies, *J. Virol.* **19**:594–609 (1976).

41. Farrar, G. H., and J. D. Oram, Characterization of human cytomegalovirus envelope glycoproteins, *J. Gen. Virol.* **65**:1991–2001 (1984).

42. Furukawa, T., A. Fioretti, and S. A. Plotkin, Growth characteristics of cytomegalovirus in human fibroblasts with demonstration of protein synthesis early in viral replication, *J. Virol.* **11**:991–997 (1973).

43. Stinski, M. F., Sequence of protein synthesis in cells infected by human cytomegalovirus: Early and late virus-induced polypeptides, *J. Virol.* **26**:686–701 (1978).

44. Stinski, M. F., E. S. Mocarski, D. R. Thomsen, and M. L. Urbanowski, Membrane glycoproteins and antigens induced by human cytomegalovirus, *J. Gen. Virol.* **43**:119–129 (1979).

45. The, T. H., and M. M. A. C. Langenhuysen, Antibodies against membrane antigens of cytomegalovirus infected cells in sera of patients with a cytomegalovirus infection, *Clin. Exp. Immunol.* **11**:475–482 (1972).

46. Kim, K. S., V. J. Sapienza, G.-M. J. Chen, and K. Wisniewski, Production and characterization of monoclonal antibodies specific for a glycosylated polypeptide of human cytomegalovirus, *J. Clin. Microbiol.* **18**:331–343 (1983).

47. Kim, K. S., H. M. Moon, V. Sapienza, and R. I. Carp, Complement-fixing antigen of human cytomegaloviruses, *J. Infect. Dis.* **135**:281–288 (1977).

48. Pereira, L., M., Hoffman, D. Gallo, and N. Cremer, Monoclonal antibodies to human cytomegalovirus: Three surface membrane proteins with unique immunological and electrophoretic properties specify cross-reactive determinants, *Infect. Immun.* **36**:924–932 (1982).

49. Pereira, L.,T.. Hoffman, M. Tatsuno, and D. Dondero, Polymorphism of human cytomegalovirus glycoproteins characterized by monoclonal antibodies, *Virolgy* **139**:73–86 (1984).

50. Nowak, B., C. Sullivan, P. Sarnow, R. Thomas, F. Bricout, J. C. Nicolas, B. Fleckenstein, and A. J. Levine, Characterization of monoclonal antibodies and polyclonal immune sera directed against human cytomegalovirus virion proteins, *Virology* **132**:325–338 (1984).

51. Rasmussen, L. E., R. M. Nelson, D. C. Kelsall, and T. C. Merigan, Murine monoclonal antibody to a single protein neutralizes the infectivity of human cytomegalovirus, *Proc. Natl. Acad. Sci. U.S.A.* **81:**876–880 (1984).

52. Rasmussen, L., J. Mullenax, R. Nelson, and T. C. Merigan, Viral polypeptides detected by a complement-dependent neutralizing murine monoclonal antibody to human cytomegalovirus, *J. Virol.* **55:**274–280 (1985).

53. Britt, W. J., Neutralizing antibodies detect a disulfide-linked glycoprotein complex within the envelope of human cytomegalovirus, *Virology*, **135:**369–378 (1984).

54. Britt, W. J., and D. Auger, Synthesis and processing of the envelope gp 55-116 complex of human cytomegalovirus, *J. Virol.* **58:**185–191 (1986).

55. Furukawa, T., E., Hornberger, S. Sakuma and S. A. Plotkin, Demonstration of immunoglobulin G receptors induced by human cytomegalovirus, *J. Clin. Microbiol.* **2:**332–336 (1975).

56. Westmoreland, D., S. St. Jeor, and F. Rapp, The development of cytomegalovirus infected cells binding affinity for normal human immunoglobulin, *J. Immunol.* **116:**1566–1570 (1976).

57. Osborn, J. E., and D. N. Medearis, Jr., Suppression of interferon and antibody and multiplication of Newcastle disease virus in cytomegalovirus infected mice, *Proc. Soc. Exp. Biol. Med.* **124:**347–353 (1967).

58. Osborn, J. E., A. A. Blazkovec, and D. L Walker, Immunosuppression during acute murine cytomegalovirus infection, *J. Immunol.* **100:**835–844 (1968).

59. Waner, J. L., Partial characterization of a soluble antigen preparation from cells infected with human cytomegalovirus: Properties of antisera prepared to the antigen, *J. Immunol.* **114:**1454–1457 (1975).

60. Pereira, L., M., Hoffman, and N. Cremer, Electrophoretic analysis of polypeptides immune precipitated from cytomegalovirus-infected cell extracts by human sera, *Infect. Immun.* **36:**933–942 (1982).

61. Cremer, N. E., M. Hoffman, and E. H. Lennette, Analysis of antibody assay methods and classes of viral antibodies in serodiagnosis of cytomegalovirus infection, *J. Clin. Microbiol.* **8:**153–159 (1978).

62. Waner, J. L., T. H. Weller, and S. V. Kevy, Patterns of cytomegalovirus complement-fixing antibody activity: A longitudinal study of blood donors, *J. Infect. Dis.* **127:**538–543 (1973).

63. Linde, G. A., L. Hammarstrom, M. A. A. Persson, C. I. E. Smith, V.-A. Sundqvist, and B. Wahren, Virus-specific antibody activity of different subclasses of immunoglobulins G and A in cytomegalovirus infections, *Infect. Immun.* **42:**237–244 (1983).

64. Pereira, L., S. Stagno, M. Hoffman, and J. E. Volanakis, Cytomegalovirus-infected cell polypeptides immuneprecipitated by sera from children with congenital and perinatal infections, *Infect. Immun.* **39:**100–108 (1983).

65. Waner, J. L., D. R. Hopkins, T. H. Weller, and E. N. Allred, Cervical excretion of cytomegalovirus: Correlation with secretory and humoral antibody. *J. Infect. Dis.* **136:**805–809 (1977).

66. Tamura, T., S. Chiba, Y. Chiba, and T. Nakao, Virus excretion and neutralizing antibody response in saliva in human cytomegalovirus infection. *Infect. Immun.* **29:**842–845 (1980a).

67. van Loon, A. M., J. T. M. van der Logt, F. W. A. Heessen, and J. van der Veen, Quantitation of immunoglobulin E antibody to cytomegalovirus by antibody capture enzyme-linked immunosorbent assay, *J. Clin. Microbiol.* **21:**558–561 (1985).

68. Weller, T. H., J. B. Hanshaw, and D. E. Scott, Serologic differentiation of viruses responsible for cytomegalic inclusion disease, *Virology* **12:**130–132 (1960).

69. Rundell, B. B., and R. F. Betts, Interaction of cytomegalovirus immune complexes with host cells, *Infect Immun.* **3:**658–665 (1981).

70. Betts, R. F., and B. B. Rundell, Physical properties of cytomegalovirus immune com-

plexes prepared with IgG neutralizing antibody anti-IgG, and complement, *J. Immunol.* **124:**337–342 (1980).

71. Middledorp, J. M., J. Jongsma, and T. H. The, Immunity to human cytomegalovirus (HCMV). II. Latently-infected donors have high levels ot T cell memory to HCMV-specific membrane antigens (CMV-MA) in the absence of detectable antibodies to CMV-MA, in: *CMV: Pathogenesis and Prevention of Human Infection* (S. A. Plotkin, S. Michelson, J. S. Pagano, and F. Rapp, eds.), pp. 445–449, Liss, New York (1984).

72. Betts, R. F., and S. G. Schmidt, Cytolytic IgM antibody to cytomegalovirus to primary cytomegalovirus infection in humans, *J. Infect. Dis.* **143:**821–826 (1981).

73. Middeldorp, J. M., J. Jongsma, and T. H. The, Killing of human cytomegalovirus-infected fibroblasts by antiviral antibody and complement, *J. Infect. Dis.* **153:**48–55 (1986).

74. Middeldorp, J. M., J. Jongsma, and T. H. The, Cytomegalovirus early and late membrane antigens detected by antibodies in human convalescent sera, *J. Virol.* **54:**240–244 (1985).

75. Stagno, S., J. E. Volanakis, D. W. Reynolds, R. Stroud, and C. A. Alford, Jr., Immune complexes in congenital and natal cytomegalovirus infections of man, *J. Clin. Invest.* **60:**838–845 (1977).

76. Condie, R. M., B. L. Hall, and R. J. Howard, Treatment of life threatening infections in renal transplant recipients with high dose intravenous human IgG, *Transplant Proc.* **11:**66–72 (1979).

77. Winston, D. J., R. B. Pollard, W. G. Ho, J. G. Gallagher, L. E. Rasmussan, S. N.-Y. Huang, C.-H. Lin, T. G. Gossett, T. C. Merigan, and R. P. Gala, Cytomegalovirus immune plasma in bone marrow transplant recipients, *Ann. Intern. Med.* **97:**11–18 (1982).

78. Myers, J. D., J. Lesczynski, J. A. Zaia, N. Flournoy, B. Newton, D. R. Szydman, G. G. Wright, M. J. Levin, and E. D. Thomas, Prevention of cytomegalovirus infection by cytomegalovirus immune globulin after marrow transplantation, *Ann. Intern. Med.* **98:**442–446 (1983).

79. Condie, R. M., and R. J. O'Reilly, Prevention of cytomegalovirus infection by prophylaxis with an intravenous, hyperimmune, native, unmodified cytomegalovirus globulin. Randomized trial in bone marrow transplant recipients, *Am. J. Med.* **76:**(3A):124–141 (1984).

80. Møller-Larsen, A., H. K. Anderson, I. Heron, and I. Sarov, *In vitro* stimulation of human lymphocytes by purified CMV, *Intervirology* **6:**249–257 (1975/76).

81. Ten Napel, C. H. H., T. H. The, J. Bijker, G. C. DeGast, and M. M. A. C. Langenhuysen, Cytomegalovirus-directed lymphocyte reactivity in healthy adults tested by a cytomegalovirus-induced lymphocyte transformation test, *Clin. Exp. Immuno.* **29:**52–60 (1977).

82. Waner, J. L., and J. E. Budnick, Blastogenic response of human lymphocytes to human cytomegalovirus, *Clin. Exp. Immuno.* **30:**44–49 (1977).

83. Wahren, B., K.-H. Robert, and S. Nordlund, Conditions for cytomegalovirus stimulation of lymphocytes, *Scand. J. Immunol.* **13:**581–586 (1981).

84. Schirm, J., H. W. Roenhorst, and T. H. The, Comparison of *in vitro* lymphocyte proliferations induced by cytomegalovirus-infected human fibroblasts and cell-free cytomegalovirus, *Infect. Immun.* **30:**621–627 (1980).

85. Roenhorst, H. W., J. M. Middledorp, J. M. Beelen, J. Schirm, A. M. Tegzess, and T. H. The, Maintenance of cytomegalovirus (CMV) latency and host immune responses of long-term renal allograft survivors. I. Prolonged suppression of *in vitro* lymphocyte responses against CMV infected fibroblasts related to previous secondary CMV infection, *Clin. Exp. Immunol.* **59:**709–715 (1985).

86. Beutner, K. R., A. Morag, R. Deibel, B. Morag, D. Raiken, and P. L. Ogra, Strain-specific local and systemic cell-mediated immune responses to cytomegalovirus in humans, *Infect. Immun.* **20:**82–87 (1978).

87. Starr, S. E., B. Dalton, T. Garrabrant, K. Paucker, and S. A. Plotkin, Lymphocyte blastogenesis and interferon production in adult human leukocyte cultures stimulated with cytomegalovirus antigens, *Infect. Immun.* **30**:17–22 (1980).
88. Ten Napel, C. H. H., and T. H. The, Acute cytomegalovirus infection and the host immune response. I. Development and maintenance of cytomegalovirus induced *in vitro* lymphocyte reactivity and its relationship to production of cytomegalovirus antibodies, *Clin. Exp. Immunol.* **39**:263–271 (1980).
89. Waner, J. L., N. Kong, and S. Biano, Blastogenic response of human lymphocytes to early antigen(s) of human cytomegalovirus, *Infect. Immun.* **41**:1084–1088 (1983).
90. Rinaldo, C. R., Jr., P. H. Black, and M. S. Hirsch, Interactions of cytomegalovirus with leukocytes from patients with mononucleosis due to cytomegalovirus, *J. Infect. Dis.* **136**:667–678 (1977).
91. Carney, W. P., R. H. Rubin, R. A. Hoffman, W. P. Hansen, K. Healey, and M. S. Hirsch, Analysis of T lymphocyte subsets in cytomegalovirus mononucleosis, *J. Immunol.* **126**:2114–2116 (1981).
92. Levin, M. J., C. R. Rinaldo, Jr., P. L. Leary, J. A. Zaia, and M. S. Hirsch, Immune response to herpesvirus antigens in adults with acute cytomegaloviral mononucleosis, *J. Infect. Dis.* **140**:851–857 (1979).
93. Gehrz, R. C., S. C. Marker, S. O. Knorr, J. M. Kalis, and H. H. Balfour, Jr., Specific cell-mediated immune defect in active cytomegalovirus infection of young children and their mothers, *Lancet* **2**:844–847 (1977).
94. Gehrz, R. C., W. R. Christianson, K. M. Linner, M. M. Conroy, S. A. McCue, and H. H. Balfour, Jr., Cytomegalovirus-specific humoral and cellular immune responses in human pregnancy, *J. Infect. Dis* **143**:391–395 (1981).
95. Faix, R. G., S. E. Zweig, J. F. Kummer, D. Moore, and D. J. Lang, Cytomegalovirus-specific cell-mediated immunity during pregnancy in lower socioeconomic class adolescents, *J. Infect. Dis.* **148**:621–629 (1983).
96. Diamond, R. D., R. Keller, G. Lee, and D. Finkel, Lysis of cytomegalovirus-infected human fibroblasts and transformed human cells by peripheral blood-lymphoid cells from normal human donors, *Proc. Soc. Exp. Biol. Med.* **154**:259–364 (1977).
97. Kirmani, N., R. K. Ginn, K. K. Mittal, J. F. Manischewitz, and G. V. Quinnan, Jr., Cytomegalovirus-specific cytotoxicity mediated by non-T lymphocytes from peripheral blood of normal volunteers, *Infect. Immun.* **34**:441–447 (1981).
98. Starr, S. E., and T. Garrabrant, Natural killing of cytomegalovirus-infected fibroblasts by human mononuclear leukocytes, *Clin. Exp. Immunol.* **46**:484–492 (1981).
99. Waner, J. L., and J. A. Nierenberg, Natural killing (NK) of cytomegalovirus (CMV)-infected fibroblasts: A comparison between two strains of CMV, uninfected fibroblasts and K562 cells, *J. Med. Virol.* **16**:233–244 (1985).
100. Borysiewicz, L. K., B. Rodgers, S. Morris, S. Graham, and J. G. P. Sissons, Lysis of human cytomegalovirus infected fibroblasts by natural killer cells: Demonstration of an interferon-independent component requiring expression of early viral proteins and characterization of effector cells, *J. Immunol.* **134**:2695–2701 (1985).
101. Rola-Pleszczynski, M., L. D. Frenkel, D. A. Fuccillo, S. A. Hensen, M. M. Vincent, D. W. Reynolds, S. Stagno, and J. A. Bellanti, Specific impairment of cell-mediated immunity in mothers of infants with congenital infection due to cytomegalovirus, *J Infect Dis.* **135**:386–391 (1977).
102. Harrison, C. J., and J. L. Waner, Natural killer cell activity in infants and children excreting cytomegalovirus, *J. Infect. Dis.* **151**:301–307 (1985).
103. Starr, S. E., L. Smiley, C. Wlodaver, H. M. Friedman, S. A. Plotkin, and C. Barker, Natural killing of cytomegalovirus-infected targets in renal transplant recipients, *Transplantation* **37**:161–164 (1984).
104. Rinaldo, C. R., Jr., M. Ho, W. H. Hamoudi, X.-E. Gui, and R. L. DeBiasio, Lymphocyte

subsets and natural killer cell responses during cytomegalovirus mononucleosis, *Infect. Immun.* **40**:472–477 (1983).

105. Quinnan, G. V., Jr., N. Kirmani, A. H. Rook, J. F. Manischewitz, L. Jackson, G. Moreschi, G. W. Santos, R. Saral, and W. B. Burns, HLA-restricted T-lymphocyte cytotoxic responses correlate with recovery from cytomegalovirus infection in bone-marrow-transplant recipients, *N. Engl. J. Med.* **307**:7–13 (1982).

106. Quinnan, G. V., and J. E. Manischewitz, The role of natural killer cells and antibody dependent cell-mediated cytotoxicity during murine cytomegalovirus infection, *J. Exp. Med.* **150**:1549–1554 (1979).

107. Bukowski, J. F., B. A. Woda, and R. M. Welsh, Pathogenesis of murine cytomegalovirus infection in natural killer cell-depleted mice, *J. Virol.* **52**:119–128 (1984).

108. Selgrade, M. K., and J. E. Osborn, Role of macrophages in resistance to murine cytomegalovirus, *Infect. Immun.* **10**:1383–1390 (1974).

109. Chalmer, J. E., J. S. Mackenzie, and N. F. Stanley, Resistance to murine cytomegalovirus linked to the major histocompatibility complex of the mouse, *J. Gen. Virol.* **37**:107–114 (1977).

110. Bancroft, G. J., G. R. Shellam, and J. E. Chalmer, Genetic influences on the augmentation of natural killer (NK) cells during murine cytomegalovirus infection: Correlation with patterns of resistance, *J. Immunol.* **126**:988–994 (1981).

111. Shellam, G. R., J. E. Allan, J. M. Papadimitriou, and G. J. Bancroft, Increased susceptibility to cytomegalovirus infection in beige mutant mice, *Proc. Natl. Acad. Sci. USA* **78**:5104–5108 (1981).

112. Kay, H. D., G. D. Bonnard, W. H. West, and R. B. Herberman, A functional comparison of human Fc-receptor bearing lymphocytes active in natural cytotoxicity and antibody-dependent cellular cytotoxicity, *J. Immunol.* **118**:2058–2066 (1977).

113. Quinnan, G. V., N. Kirmani, E. Esber, R. Saral, J. E. Manischewitz, J. L. Rogers, A. H. Rook, G. W. Santos, and W. H. Burns, HLA-restricted cytotoxic T lymphocyte and non-thymic cytotoxic lymphocyte responses to cytomegalovirus infections of bone marrow transplant recipients, *J. Immunol.* **126**:2036–2041 (1981).

114. Borysiewicz, L. K., S. Morris, J. D. Page, and J. G. P. Sissons, Human cytomegalovirus-specific cytotoxic T lymphocytes: Requirements for *in vitro* generation and specificity, *Eur. J. Immunol.* **13**:804–809 (1983).

115. Gehrz, R. C., and S. R. Rutzick, Cytomegalovirus (CMV)-specific lysis of CMV-infected target cells can be mediated by both NK-like and virus-specific cytotoxic T lymphocytes, *Clin. Exp. Immunol.* **61**:80–89 (1985).

116. Quinnan, G. V., Jr., W. H. Burns, N. Kirmani, A. H. Rook, J. Manischewitz, L. Jackson, G. W. Santos, and R. Saral, HLA-restricted cytotoxic T lymphocytes are an early immune response and important defense mechanism in cytomegalovirus infections, *Rev. Infect. Dis.* **6**:156–163 (1984).

117. Rook, A. H., G. V. Quinnan, Jr., W. J. R. Frederick, J. F. Manischewitz, and N. Kirmani, Importance of cytotoxic lymphocytes during cytomegalovirus infection in renal transplant recipients, *Am. J. Med.* **76**:385–392 (1984).

118. Felsenstein, D., W. P. Carney, V. R. Iacoviello, and M. S. Hirsch, Phenotypic properties of atypical lymphocytes in cytomegalovirus-induced mononucleosis, *J. Infect. Dis.* **152**:198–203 (1985).

119. Maher, P., C. M. O'Toole, T. G. Wreghitt, D. J. Spiegelhalter, and T. A. H. English, Cytomegalovirus infection in cardiac transplant recipients associated with chronic T cell subset ratio inversion with expansion of Leu-7$^+$T$_{s-c}^+$ subset, *Clin. Exp. Immunol.* **62**:615–624 (1985).

120. Pass, R. F., M. A. Roper, and A. M. August, T lymphocyte subpopulations in congenital cytomegalovirus infection, *Infect. Immun.* **41**:1380–1382 (1983).

121. Fiala, M., J. E. Payne, T. V. Berne, T. C. Moore, W. Henle, J. Z. Montgomerie, S. N.

Chatterjee, and L. B. Guze, Epidemiology of cytomegalovirus infection after transplantation and immunosuppression, *J. Infect. Dis.* **132**:421–433 (1975).

122. Winston, D. J., W. G. Ho, C. L. Howell, M. J. Miller, R. Mickey, W. J. Martin, C. Lin, and R. P. Gale, Cytomegalovirus infections associated with leukocyte transfusions. *Ann. Intern. Med.* **93**:671–675 (1980).

123. Diosi, P., E. Moldovan, and N. Tomescu, Latent cytomegalovirus infection in blood donors, *Br. Med. J.* **4**:660–662 (1969).

124. Schrier, R. D., J. A. Nelson, and M. B. A. Oldstone, Detection of human cytomegalovirus in peripheral blood lymphocytes in a natural infection, *Science* **230**:1048–1051 (1985).

125. St. Jeor, S., and A. Weisser, Persistence of cytomegalovirus in human lymphoblasts and peripheral leukocyte cultures, *Infect. Immun.* **15**:402–409 (1977).

126. Furukawa, T., N. Yoshimura, J. H. Jean, and S. A. Plotkin, Chronically persistent infection with human cytomegalovirus in human lymphoblasts, *J. Infect. Dis.* **139**:211–214 (1979).

127. Tocci, M. J., and S. C. St. Jeor, Susceptibility of lymphoblastoid cells to infection with human cytomegalovirus, *Infect. Immun* **23**:418–423 (1979).

128. Tocci, M. J., and S. C. St. Jeor, Persistence and replication of the human cytomegalovirus genome in lymphoblastoid cells of B and T origin, *Virology* **96**:664–668 (1979).

129. Einhorn, L., and A. Ost, Cytomegalovirus infection of human blood cells, *J. Infect Dis.* **149**:207–214 (1984).

130. Rice, G. P. A., R. D. Schrier, and M. B. A. Oldstone, Cytomegalovirus infects human lymphocytes and monocytes: Virus expression is restricted to immediate–early gene products, *Proc. Natl. Acad. Sci. USA* **81**:6134–6138 (1984).

131. Sing, G. K., and H. M. Garnett, The effects of human cytomegalovirus challenge *in vitro* on subpopulations of T cells from seronegative donors, *J. Med. Virol* **14**:363–371 (1984).

132. Hutt-Fletcher, L. M., N. Balachandran, and M. H. Elkins, B cell activation by cytomegalovirus, *J. Exp. Med.* **158**:2171–2176 (1985).

133. Yachie, A., G. Tosato, S. E. Straus, and R. M. Blaese, Immunostimulation by cytomegalovirus (CMV): Helper T cell-dependent activation of immunoglobulin production *in vitro* by lymphocytes from CMV-immune donors, *J. Immunol.* **135**:1395–1400 (1985).

134. Rinaldo, C. R., Jr., W. P. Carney, B. S. Richter, P. H. Black, and M. S. Hirsch, Mechanisms of immunosuppression in cytomegaloviral mononucleosis, *J. Infect. Dis.* **141**:488–495 (1980).

135. Carney, W. P., and M. S. Hirsch, Mechanisms of immunosuppression in cytomegalovirus mononucleosis. II. Virus–monocyte interactions, *J Infect. Dis.* **144**: 47–54 (1981).

136. Rodgers, B. C., D. M. Scott, J. Mundin, and J. G. P. Sissons, Monocyte-derived inhibitor of interleukin 1 induced by human cytomegalovirus, *J. Virol.* **55**: 527–532 (1985).

137. Drew, W. L., M. A. Conant, R. C. Miner, E. S. Huang, J. L. Ziegler, J. R. Groundwater, J. H. Gullett, P. Volberding, D. I. Abrams, and L. Mintz, Cytomegalovirus and Kaposi's sarcoma in young homosexual men, *Lancet* **2**:125–127 (1982).

138. Macher, A. M., C. M. Reichert, S. E. Straus, D. L. Longo, J. Parrillo, H. C. Lane, and A. S. Fauci, Death in the AIDS patient: Role of cytomegalovirus, (Letter.) *N. Engl. J. Med.* **309**:1454 (1983).

139. Quinnan, G. V. Jr., H. Masur, A. H. Rook, G. Armstrong, W. R. Frederick, J. Epstein, J. F. Manischewitz, A. M. Macher, L. Jackson, J. Ames, H. A. Smith, M. Parker, G. R. Pearson, J. Parrillo, C. Mitchell, and S. E. Straus, Herpesvirus infections in the acquired immune deficiency syndrome, *JAMA* **252**:72–77 (1984).

140. Hirsch, M. S., R. T. Schooley, D. D. Ho, and J. C. Kaplan, Possible viral interactions in the acquired immunodeficiency syndrome (AIDS), *Rev. Infect. Dis.* **6**:726–731 (1984).

141. Kari, B., N. Lussenhop, R. Goertz, M. Wabuke-Bunoti, R. Radeke, and R. Gehrz, Char-

acterization of monoclonal antibodies reactive to several biochemically distinct human cytomegalovirus glycoprotein complexes, *J. Virol.* **60**:345–352 (1986).

142. Cranage, M. P., T. Kouzarides, A. T. Bankier, S. Satchwell, K. Weston, P. Tomlinson, B. Barrell, H. Hart, S. E. Bell, A. C. Minson, and G. L. Smith, Identification of the human cytomegalovirus glycoprotein B gene and induction of neutralizing antibodies via its expression in recombinant vaccinia virus, *EMBO.* **5**:3057–3063 (1986).

143. Gold, D., R. Ashley, H. H. Handsfield, M. Verdon, L. Leach, J. Mills, L. Drew, and L. Corey, Immunoblot analysis of the humoral immune response in primary cytomegalovirus infection, *J. Infect. Dis.* **157**:319–326 (1988).

144. Zaia, J., S. J. Forman, Y-P. Ting, E. Vanderwal-Urbina, and K. G. Blume. Polypeptide-specific antibody response to human cytomegalovirus after infection in bone marrow transplant recipients, *J. Infect Dis.* **153**:780–787 (1986).

145. Schrier, R. D., and M. B. A. Oldstone, Recent clinical isolates of cytomegalovirus suppress human cytomegalovirus-specific human leukocyte antigen-restricted cytotoxic T-lymphocyte activity, *J. Virol.* **59**:127–131 (1986).

146. Braun, R. W., and H. C. Reiser, Replication of human cytomegalovirus in human peripheral blood T cells, *J. Virol.* **60**:29–36 (1986).

147. Dudding, L. R., and H. M. Garnett, Interaction of strain AD169 and a clinical isolate of cytomegalovirus with peripheral monocytes: The effect of lipopolysaccharide stimulation, *J. Infect. Dis.* **155**:891–896 (1987).

148. Weinshenker, B. G., W. Sharon, and G. P. A. Rice, Phorbol ester-induced differentiation permits productive human cytomegalovirus infection in a monocyte cell line, *J. Immunol.* **140**:1625–1631 (1988).

# Epstein–Barr Virus-Induced Immune Deficiency

DAVID T. PURTILO and HELEN L. GRIERSON

## 1. INTRODUCTION

Epstein–Barr virus (EBV) is an ubiquitous herpesvirus infecting human beings and primates, exclusively. Lymphocytic proliferation associated with EBV infection, the normal immune responses controlling the virus, the transient anergy occurring during acute infectious mononucleosis (IM), and the acquired progressive immune defects occurring in males with the X-linked lymphoproliferative syndrome (XLP) are described in this chapter. The results of a decade of investigation of the Duncan family are summarized to illustrate mechanisms of virus-induced immune suppression.

## 2. PROPERTIES OF EPSTEIN–BARR VIRUS

Epstein–Barr virus was initially observed in 1964 in an electron micrograph of cultured Burkitt lymphoma (BL).[1] The C3d receptor on B cells is the attachment site for EBV. Whether receptors for EBV are present on oropharyngeal epithelial cells is debated; however, EBV genome is found in virtually all undifferentiated nasopharyngeal carcinomas (NPC)[2] and is present in salivary glands.[3]

Epstein–Barr virus is shed in saliva and on infecting B cells; two pathways of viral expression can transpire. First, following uncoating of virus, Epstein–

---

DAVID T. PURTILO and HELEN L. GRIERSON • Departments of Pathology and Microbiology, and Pediatrics, and the Eppley Institute for Research in Cancer and Allied Diseases, University of Nebraska Medical Center, Omaha, Nebraska 68105-1065.

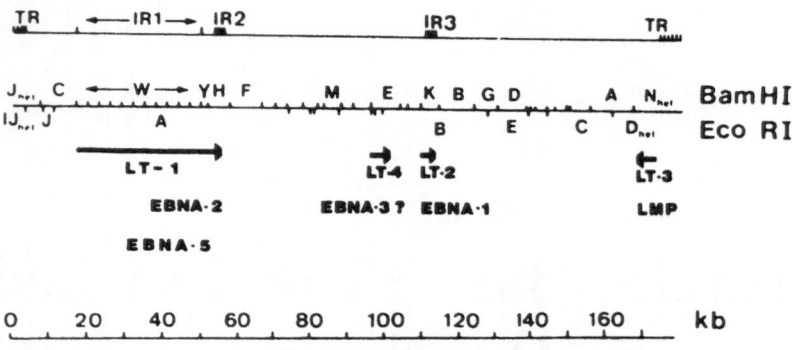

**FIGURE 1.** Epstein–Barr virus genome map. (Courtesy of Ingemar Ernberg.)

Barr nuclear antigen (EBNA) is demonstrable by anticomplement immunofluorescence staining within 6–8 hour postinfection.[4] The virus endows an
unlimited proliferative capacity to the transformed B cells, and viral latency is
rapidly established. Infected cells express EBNA. Virus is maintained integrated in host DNA and nonintegrated in plasmids (circular) in the infected
cells. Second, the productive pathway of the virus results in viral particle formation and cellular death.[5] Following expression of EBNA, early antigen
(EA) and viral capsid antigen (VCA) are sequentially expressed. Virions can be
observed by electron microscopy in the cells that stain for VCA.

Although the EBV genome has been sequenced,[6] the gene products are
incompletely characterized. The synthesis of polypeptides of EBV genome and
the development of monoclonal antibodies to them have led to identification of
at least five different EBNA.[7] In addition, latent membrane protein (LMP) is
expressed in the surface membrane of the infected B cells. LMP may be the
LYDMA (lymphocyte-defined membrane antigen) that likely serves as the target for cytotoxic T cells (CTL) (Fig. 1). A differential expression of EBNA 5
and LMP occurs in polyclonal diploid lymphoblastoid cell lines (LCL) versus
BL lines; expression of LMP (8) and EBNA 5 are decreased in BL lines. This
may account for increased resistance of BL versus LCL to CTL.[9]

## 3. MODULATION OF IMMUNITY BY EBV

Depending on the time in the life cycle during which EBV infects an
individual and the presence or absence of immunodeficiency, various outcomes occur. Silent seroconversion almost always occurs in young children,
whereas approximately two thirds of adolescents who become infected manifest infectious mononucleosis (IM). The clinical signs and symptoms,[10]
atypical lymphocytosis and hyperimmunoglobulinemia of the patients are expressions of the immunologic struggle ongoing between the CTL, natural killer
(NK) cells, macrophages, and humoral effectors responding to the trans-

formed proliferating B cells carrying EBV. During acute IM, approximately one B lymphocyte per $10^4$ circulating B cells contains EBV genome, whereas during latency only one B cell per $10^6$ circulating B cells studied by limiting dilution analysis harbors EBV genome.[11] B cells infected with EBV are held in check by memory cytotoxic T cells.[12] Delayed hypersensitivity skin-test responses often become negative,[13,14] and *in vitro* lymphocyte responses to mitogens,[14–16] soluble antigens,[14,15] and allogeneic cells[15,17] are usually depressed in acute IM.

## 3.1. Antibody Production during Acute EBV Infection

Bahna *et al.*[18] demonstrated polyclonal B-cell activation characterized by markedly increased total quantitative immunoglobulin E (IgE) (i.e., 175% above normal), IgM (140% increased), IgA (160% increased), IgG (135% increased), and IgD (130% increased) in acute IM. Also well known is the appearance of autoantibodies in patients with acute IM.[19] These autoantibodies may be deleterious. For example, autoantibodies against neutrophils occur concurrently with the polyclonal activation of B cells in the vast majority of cases of acute IM.[20] Neutropenia often occurs in IM.

To study the impact of primary EBV infection on the capacity to produce antigen-specific antibody to bacteriophage $\Phi X174$, Junker *et al.*[21] immunized 17 college students who had acute IM. During the early phase of the disease, the proportion of peripheral blood lymphocytes displaying Ia and T8 (CD8) phenotypes was increased and the T helper/suppressor (T4:T8) ratio was decreased to <1. These abnormalities disappeared during convalescence. The investigators demonstrated a depressed humoral immune response to bacteriophage $\Phi X174$ both *in vivo* and *in vitro* in most patients. *In vitro* coculture experiments demonstrated that Ia-positive suppressor T (Ts) cells inhibited antibody production and isotype switching from IgM to IgG antibody. Removal of the T8-positive lymphocytes from cultures normalized *in vitro* antibody synthesis. These studies demonstrate that EBV causes a broad-based transient immune deficiency in patients with uncomplicated IM.

## 3.2. T-Lymphocyte Function during Acute Infectious Mononucleosis

During the first week following onset of symptoms of acute IM, the number of T cells greatly expands. Junker *et al.*[21] found no significant changes in the proportion of cells expressing markers for monocytes, NK cells, E-rosette receptors, mature T cells, or helper T cells. By contrast, the number of CD8-bearing cells expressing Ia antigens increased markedly.[22] The T4:T8 ratio generally decreased below 1.0 during the early phase of IM. Within 4–10 weeks following acute IM, the numbers of cells returned to normal proportions. The functions of T lymphocytes were impaired in response to cutaneous challenge with antigens. However, CTL showed increased killing, especially of autologous LCL, and non-HLA-restricted killing was enhanced.[23] Anomalous killing may be responsible for the subtle damage occurring in the liver, bone marrow, and other organs in immune-competent persons with acute IM. In immunoincompetent males with XLP, the lesions are extensive.

Immune responses to EBV occur independent of macrophages; however, these cells are activated in hematopoietic tissues of acutely infected persons. The number of macrophages and NK cells is not increased in the peripheral blood of patients with acute IM[21]; however, small granulomas can be found in the bone marrow during acute IM.[24] Moreover, if a person is immune incompetent, virus-associated hemophagocytic syndrome (VAHS) characterized by a rash, fever, hepatosplenomegaly, and pancytopenia associated with erythrophagocytosis in bone marrow and lymph nodes can occur.[25,26] Also important is the elaboration of prostaglandins by macrophages activated by EBV. Prostaglandins may suppress T-cell function. Elaboration of lymphokines such as interferon (IFN) and interleukin 2 (IL-2) has been imcompletely studied in acute EBV infection. Both NK cells[27] and IFN suppress the transformation of B cells by EBV in vitro.[28] Perhaps the production of IL-2 and IFN might be reduced transiently by the marked increase in Ts cell activity that occurs during acute IM.

Epstein—Barr virus replicates within B cells and oropharyngeal and salivary epithelial tissues. The virus persists in cells in the plasmid form, and cytotoxic T memory cells maintain latency.[12] Supporting this view is the finding that cyclosporin A suppression of RNA polymerase II in T lymphocytes[29] or depletion of T cells by monoclonal antibodies in bone marrow of non-HLA-matched donors can lead to life-threatening lymphoproliferative diseases.[30]

## 4. OTHER MECHANISMS OF EBV-INDUCED IMMUNOSUPPRESSION

Alternative mechanisms of immune suppression have been suggested by Menezes et al.[31] In addition to viral activation of Ts cells and macrophages abrogating host defenses, shedding of viral structural antigens by EBV-infected B cells could lead to production of blocking antibodies, formation of immune complexes, and activation of suppressor cells that can block host immunologic defenses. Also, the transformed cells may elaborate plasminogen activating factor and prostaglandins that impede host immune defenses.

## 5. CLINICAL SIGNIFICANCE OF EBV-INDUCED IMMUNOMODULATION IN VIRAL PATHOGENESIS OF DISEASES IN MALES WITH XLP

In the immune-competent person with acute IM, the resulting immune suppression may potentially be responsible for the frequent development of streptococcal and pneumococcal oropharyngeal infections. Also, reactivation of other herpesviruses can be observed on occasion. Perhaps a more serious complication of the immune responses to EBV-transformed B cells is the anomalous killing of hepatocytes,[32,33] bone marrow,[25] and other cells.[34] The suppression of cytotoxic T cells in males with XLP normally induced during acute IM to control EBV-transformed B cells may allow a genetically altered cell destined to become a malignant clone to sneak by the defective immune surveillance. These foregoing immunopathologic features in IM can be exaggerated in immune-suppressed patients.

In our initial report of the Duncan kindred, we described the inheritance and phenotypic expressions of the EBV-induced diseases observed as resulting from the X-linked recessive progressive combined variable immune-deficiency disease.[35] To illustrate EBV-induced immune suppression in genetically defective persons, the immunopathologic manifestations and mechanisms responsible for diverse diseases occurring in the Duncan kindred during the past decade are summarized. XLP serves as a model for studying and demonstrating exaggerated pathophysiologic mechanisms of viral-induced immune suppression and the resulting diseases. In 1975, fatal infectious mononucleosis had led to the demise of three of seven young brothers in the Duncan kindred, two maternally related males died with malignant lymphoma, and another male cousin succumbed with acquired hypogammaglobulinemia following IM.[35]

Our survey of the family in 1974 revealed a 10-year-old male (Table I and II No.V-026) with partial IgA deficiency who had experienced neonatal thrush and sequentially *Neisseria meningitides* meningitis and vaccinia following smallpox vaccination during childhood. In November 1974, he became infected with EBV from his sister who had recently experienced IM. His serum IgM level increased from 230 to 656 mg/dl. Although we could culture EBV in throat washings, we could not detect EBV antibodies in his serum.[36] In 1981, the patient developed anemia associated with erythroblastopenia.[37] One month later, delayed onset of acute IM occurred (Table I). This illness was characterized by the usual clinical signs and symptoms of IM, appearance of IgM VCA antibodies, and activation of Ts cells, with an inverted T4 : T8 ratio in peripheral blood. Next, he acquired hypogammaglobulinemia. B cells could not be detected in his peripheral blood with monoclonal antibodies. He acquired a defect in NK cell activity. An HLA-matched bone marrow transplant resulted in temporary restoration of his T- and B-cell functions and NK cell activity. Regrettably, the patient succumbed to adenovirus-induced hepatitis.[38,39]

We have continued to investigate additional members of the Duncan family (Fig. 2, Table II). Males with XLP show variously decreased IgM, IgA, or IgG levels in serum and fail to switch from IgM to IgG antibody responses on secondary challenge with bacteriophage ΦX174.[40] Even before EBV infection, affected males manifest defective antibody response to ΦX174 comparable to that of patients with acute IM.[21] The immune defect detected in response to challenge with ΦX174 before EBV infection may explain why patients are so vulnerable to EBV infection. The mortality among 184 patients in the XLP registry was 85% by 10 years of age and 100% by 40 years of age[33,41] (Fig. 3).

Prospective studies by Seeley *et al.*[42] of patients with XLP pre-EBV infec-

**TABLE I**

**Reactivation of Epstein–Barr Virus Infection Associated with Partial Red Blood Cell Aplasia and Delayed Onset of Infectious Mononucleosis Followed by Agammaglobulinemia[a,b]**

| Clinical events and date | Age (years) | Immunoglobulin levels (mg/dl) | | | Reciprocal titers | | | | | | OKT4/T8 |
| | | | | | Antiviral capsid antigen | | | Anti-early antigen | | Anti-Epstein–Barr nuclear-associated antigen | |
| | | IgG | IgA | IgM | IgG | IgA | IgM | DR | D | | |
| Epstein–Barr virus infection (virus isolated from blood and saliva) | | | | | | | | | | | |
| 8/10/74 | 11 | 940 | 20 | 230 | <5 | NT | <2 | <10 | <10 | <2 | |
| 12/16/74 | | 1120 | 15 | 656 | <5 | NT | <2 | <10 | <10 | <2 | |
| 4/22/75 | | 1430 | 24 | 393 | 80 | NT | <2 | <10 | <10 | 5 | |
| Encephalitis | | | | | | | | | | | |
| 7/19/77 | 14 | 1285 | 5 | 438 | 40 | NT | <2 | <10 | <10 | 2 | |
| Asymptomatic | | | | | | | | | | | |
| 3/28/78 | | 1150 | 4 | 456 | <5 | NT | <2 | <10 | <10 | <2 | |
| 2/26/81 | | 1285 | 3 | 496 | <5 | <2 | <2 | <10 | <10 | <2 | |
| Red blood cell aplasia (treated with packed erythrocytes and gammaglobulin) | | | | | | | | | | | |

| Patient | Date | | | | | | | | | | | |
|---|---|---|---|---|---|---|---|---|---|---|---|---|
| 18 | 12/5/81 | 1230 | 3 | 580 | <5 | <2 | 10 | <5 | <5 | <5 | | 2.7 |
| | 12/29/81 | 1230 | 5 | 650 | ≤5 | <2 | <2 | <5 | <5 | <5 | | |
| | 1/5/82 | 1160 | 4 | 720 | ≤5 | <2 | <2 | <5 | <5 | <5 | | |
| Infectious mononucleosis (virus isolated from lymph node) | | | | | | | | | | | | |
| | 1/14/82 | 980 | 4 | 548 | 5 | <2 | <2 | <5 | <5 | <5 | <5 | |
| | 1/25/82 | 3 | 612 | 5 | <2 | <5 | <5 | <5 | <5 | ≤5 | <5 | |
| | 2/9/82 | 3 | 824 | 80 | <2 | <5 | <5 | <5 | <5 | ≤5 | <5 | |
| | 2/16/82 | 3 | 1088 | 40 | 20 | <5 | <5 | <5 | <5 | ≤5 | <5 | 0.2 |
| | 2/15/82 | 2 | 904 | 80 | 40 | <5 | <5 | <5 | <5 | ≤5 | <5 | |
| Partial recovery of red cell aplasia | | | | | | | | | | | | |
| | 3/10/82 | 2 | 596 | 80 | <2 | ≤5 | <5 | <5 | <5 | <5 | <5 | 0.5 |
| | 4/22/82 | NT | 23 | 5 | <2 | <2 | <5 | <5 | <5 | | | |
| Agammaglobulinemia (no detectable B cells) | | | | | | | | | | | | |
| 19 | 6/11/82 | 564 | <4 | 9 | ≤5 | <2 | <2 | <5 | <5 | <5 | <5 | |
| | 6/23/82 | 450 | NT | 0.1 | <2 | 2 | <2 | <5 | <5 | <5 | <5 | 0.4 |
| | 7/1/82 | 520 | 0.9 | 4.6 | 10 | <2 | <2 | <5 | <5 | <5 | <5 | 1.1 |
| B cells detected | | | | | | | | | | | | |
| | 9/2/82 | 515 | <4.5 | 6.0 | 20 | <2 | <2 | <5 | <5 | <5 | <5 | 1.3 |
| | 9/22/82 | 419 | <4.5 | <6.0 | 80 | <2 | <2 | <5 | <5 | <5 | <5 | 2.4± |
| | | 650–1700 | 50–250 | 30–350 | 20–160 | | | | | | >10–160 | 0.8 |

[a]From Purtilo et al.[14]

[b]NT, not tested.

## TABLE II
### Males in the Duncan Kindred Affected with X-Linked Lymphoproliferative Syndrome[a]

| Pedigree number | Age (years) at death/ present age | Phenotypes | EBV serology | Serum immunoglobulin concentration | Response to bacteriophage ΦX174 challenge |
|---|---|---|---|---|---|
| V-005 | 9.5 | IM→hypogamma | NT | Hypo G, A, M after IM | NT |
| V-009 | 18.0 | IM→lymphoma | NT | Hypo A | NT |
| V-010 | 3.5 | Lymphoma | Defective anti-EBNA | | NT |
| V-017 | 3.0 | Fatal IM | NT | | NT |
| V-020 | 8.5 | Fatal IM | NT | Hyper M and G after IM | NT |
| V-023 | 2.0 | Fatal IM | NT | Hyper G, A, M after IM | NT |
| V-026 | 18.5 | Hypo IgA, hyper M RCA, IM, agamma | Defective response tp EBV | Hypo A→ Hyper M→ Hypo G, A, M | Defective |
| V-047 | 15.0 alive | At risk | Seronegative | Hypo A, M | Borderline |
| V-048 | 13.5 alive | At risk | Seronegative | Hypo A | Normal |
| VI-004 | 13.0 alive | Hypogamma | Seronegative | Hypo A, hyper M | Defective |
| VI-005 | 8.0 alive | Hypogamma | Seronegative | Hypo G, A, hyper M | Defective |
| VI-010 | 6.0 alive | Hypogamma | Seronegative | Hypo A, M | Defective |
| VI-011 | 9.0 alive | Hypogamma | Defective | Hypo G, A, M | Borderline |
| VI-012 | 6.0 alive | Hypogamma | Defective response to anti-EBNA | Hypo G, A, M | Borderline |

[a]EBV, Epstein–Barr virus; IM, infectious mononucleosis; NT, not tested; G, IgG; A, IgA; M, IgM; EBNA, Epstein–Barr nuclear-associated antigen; RCA, red cell aplasia.

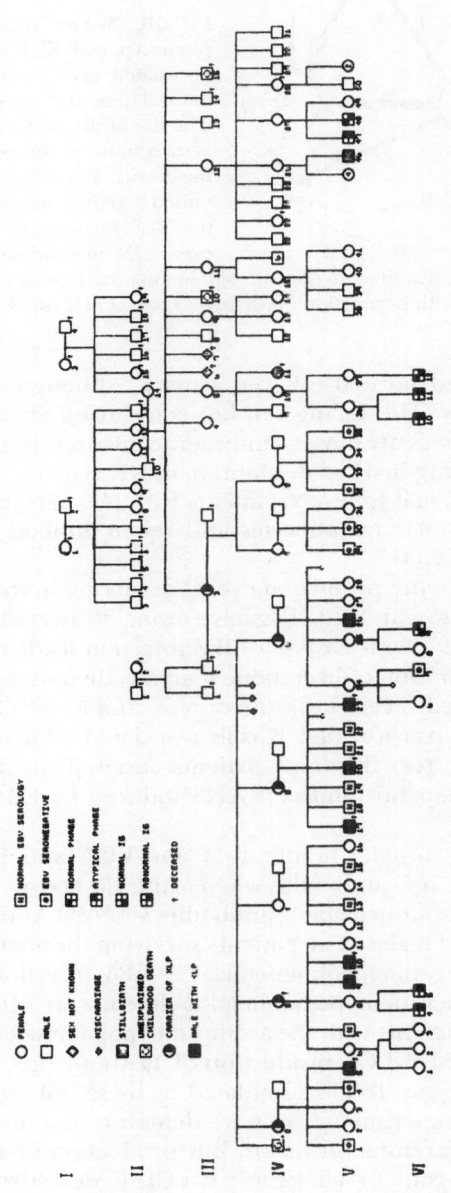

FIGURE 2. Pedigree of the Duncan kindred in 1987.

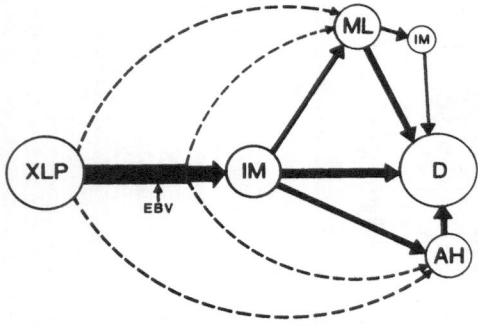

**FIGURE 3.** The frequency of the various phenotypes of XLP is depicted with the postulated events occurring during the natural history of patients with the defect. The size of the circles and thickness of the arrows indicate the relative frequency of the events. These events have been determined based on analysis of 161 patients in the XLP Registry. (- - -) Postulated rare events. IM, infectious mononucleosis; ML, malignant lymphoma, AH, acquired hypogammaglobulinemia; D = death. (From Grierson and Purtilo[33]; published with permission of *Annals of Internal Medicine.*)

tion have demonstrated normal NK cell activity. Although exaggerated NK activity and anomalous CTL killing can be seen during acute IM in patients with XLP,[43] several patients have silently seroconverted in response to EBV infection without showing marked diminution in NK activity. Notably, many of these patients show normal Ig levels. Patients with NK defects possess normal numbers of large granular lymphocytes and retain antibody-dependent cellular cytotoxicity (ADCC).[44]

Shown in Fig. 4 are the postulated events occurring in the natural history of EBV-induced diseases in XLP. Possibly, owing to decreased numbers of circulating CTL in males with XLP,[45] EBV infection leads to B- and T- cell proliferation with infiltration of liver, bone marrow, thymus, and other organs. Numerous EBV-infected B cells infiltrate organs, and fewer CD8-positive cells are admixed with the transformed B cells associated with necrosis of these organs.[26,32,34] Nearly two thirds of patients succumb to infectious mononucleosis. Fulminant hepatitis and/or VAHS induced by EBV are the major causes of death.

Patients with XLP usually mount IgM anti-VCA and EA antibody responses as well as VCA IgG antibodies when acute IM occurs. However, silent infection can occur without detectable antibodies.[36,37,46] Antibodies to EBNA are defective in affected males. The patients surviving the acute primary infection often acquire hypogammaglobulinemia. Pre-EBV infection, partial IgM or IgA deficiency, or borderline hypogammaglobulinemia are often found (Table II). *In vitro* studies of patients with the acquired hypogammaglobulinemia phenotype show suppressed B-LCL production of IgA and IgG, but not IgM, in the presence of autologous T cells stimulated by these cell lines.[47] In similar challenge and mixing experiments, we have demonstrated normal production of IL-2 and B-cell differentiation factor, but production of IFN-$_\gamma$ by T cells stimulated with autologous or allogeneic B-LCL is defective.[48] These and other immune defects identified in the Duncan kindred and others in the XLP Registry afflicted with XLP are summarized in Table III.

We have postulated (Fig. 4) that the malignant lymphomas occur due to defective CTL responses to EBV and genetic alterations in an infected cell.

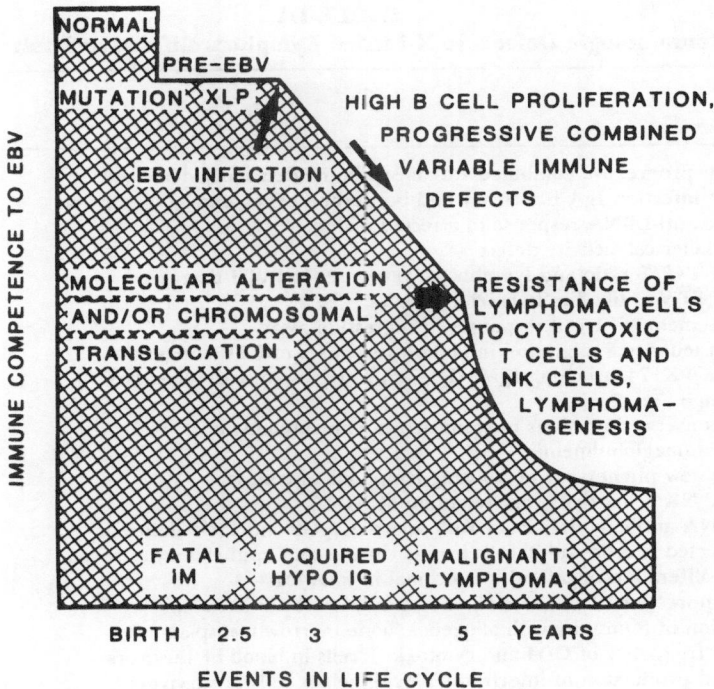

**FIGURE 4.** Postulated events and mechanisms responsible for Epstein–Barr virus (EBV)-induced fatal infectious mononucleosis (IM), acquired hypogammaglobulinemia (hypogamma), and malignant lymphoma in males with the X-linked lymphoproliferative syndrome.

This is likely related to sustained Ts cell activity, which impairs CTL. Invariably, the malignant lymphomas occur as diffuse intermediate to high-grade lesions chiefly involving the gastrointestinal (GI) tract. Surviving patients almost always show acquired hypogammaglobulinemia. On the basis of the hypothesis regarding mechanisms of African Burkitt lymphomagenesis put forth by Klein and Klein,[49] we have postulated that a proliferating B cell may undergo a molecular or chromosomal translocation involving c-*myc* in chromosome 8 and the heavy-chain locus in chromosome 14 or light-chain loci in chromosomes 22 and 2, respectively.[50] The juxtaposition of c-*myc* with the active Ig locus located at the breakpoints is thought to activate the c-*myc* proto-oncogene and freeze the cell in a growth phase of the cycle. Also, the cell may become more sensitive to growth signals, and the expression of latent membrane protein may be downregulated.[8] Thus, the CTL may be deprived of a viral target on the surface of the malignant lymphoma cell.[9]

Our hypothesis that the polyclonal proliferation of B cells following EBV infection of immune incompetent persons occasionally (about 20% of cases) progresses to a monoclonal B-cell lymphoma is supported by our recent finding of oligoclonal and monoclonal EBV-infected B cells in tissues from patients

## TABLE III
### Immunologic Defects in X-Linked Lymphoproliferative Syndrome

| Findings | Year reported |
|---|---|
| X-linked progressive combined variable immune deficiency described | 1975 |
| Pre-EBV infection IgA deficiency and borderline hypogammaglobulinemia | 1978 |
| Defective anti-EBNA response in affected males | 1979 |
| Natural killer cell activity defect | 1980 |
| Inverted T4/T8 ratio and lymphocyte responses to PWM decreased | 1982 |
| Elevated EBV antibodies in carrier females | 1982 |
| Defective memory T-cell responses in regression assay | 1982 |
| Deficient leukocyte migration inhibition responses to EBV antigens | 1982 |
| Defective $\Phi$X174 switching IgM$\rightarrow$IgG antibody response to secondary challenge | 1983 |
| Delayed onset of infectious mononucleosis with subsequent hypogammaglobulinemia | 1984 |
| Evolving new phenotype: necrotizing lymphoid vasculitis | 1985 |
| Defective NK cell activity, but retention of ADCC | 1986 |
| Anti-EBNA antibody deficiency with selective IgG2 and IgG3 deficiency | 1986 |
| EBV-infected B cells and T cells invade thymus and epithelium destroyed | 1986 |
| B-cell proliferation oligo and monoclonal in fatal IM | 1986 |
| T-cell suppression of production of IgA and IgG by LCL in vitro | 1986 |
| Restoration of immunity with allogeneic bone marrow transplantation | 1986 |
| Reduced frequency of CD4 and cytotoxic T cells in blood of survivors | 1987 |
| Decreased production of interferon-Y by T helper cells of survivors | 1987 |

with fatal IM. Immunoglobulin gene rearrangements were detected within a few weeks of onset of symptoms of IM using Southern blot analysis of JH gene probes hybridized with DNA extracted from lesions of patients.[51]

## 6. SUMMARY

In normal immune-competent persons, transient immune deficiency occurs with acute IM. This anergy is corrected within a few weeks to months following the onset of symptoms. EBV produces explosive polyclonal B- and T-cell proliferation. Normal immune-competent persons carefully orchestrate a barrage of humoral and cellular effectors to control EBV-transformed B cells. These effectors are carefully modulated in part by Ts cells that transiently depress normal immune responses.

By contrast, certain immune-incompetent persons, especially males with XLP who have a defective lymphoproliferative control locus in the X chromosome and children with Chediak–Higashi syndrome,[52] are highly vulnerable to the progressive immune defects acquired following EBV infection and fatal lymphoproliferation occurs. Many patients with XLP who survive acute IM show sustained Ts cell activity leading to acquired hypogammaglobulinemia.

Patients with severe hypogammaglobulinemia require lifelong Ig therapy. Ts cell activity in patients with malignant lymphoma may account for their survival.[47,49] Finally, evidence is growing that molecular or cytogenetic alterations occurring in proliferating B cells may allow tumor cells to evade host-immune surveillance, resulting in an aggressive monoclonal B-cell lymphoma.

ACKNOWLEDGMENTS. This work was supported in part by U.S. Public Health Service Department of Health and Human Services, National Cancer Institute CA30196 grant awarded by the National Cancer Institute, Laboratory Research Center support grant CA36727, the State of Nebraska Department of Health LB506, and the Lymphoproliferative Research Fund.

# REFERENCES

1. Epstein, M. A., and Y. M. Barr, Cultivation in vitro of human lymphoblasts from Burkitt's malignant lymphoma, *Lancet* **1**:251–252 (1964).
2. Wolf H., H. zur Hausen, and V. Becker, EB Viral genomes in epithelial nasopharyngeal carcinoma cells, *Nature New Biol.* **224**:245–248 (1973).
3. Wolf, H. Biology of Epstein–Barr virus, in: *Immune Deficiency and Cancer: Epstein–Barr Virus and Lymphoproliferative Malignancies* (D. T. Purtilo, (ed.) pp. 233–242 Plenum, New York (1985).
4. Menezes, J., M. Jondal, W. Leibold, and G. Dorval, Epstein–Barr virus interactions with human lymphocyte subpopulations: Virus adsorption, kinetics of expression of Epstein–Barr virus-associated nuclear antigen, and lymphocyte transformation, *Infect. Immun.* **13**:303–310 (1976).
5. Henle, W., G. Henle, and E. T. Lennette, The Epstein–Barr virus, *Sci. Am.* **241**:48–59 (1979).
6. Baer, R., A. T. Bankier, M. D. Biggin, P. L. Deininger, P. J. Farrell, T. J. Gibson, C. Hatfull, G. S. Hudson, S. C. Satchwell, C. Seguin, P. S. Tuffnell, and B. G. Barrell, DNA sequence and expression of the B95-8 Epstein–Barr virus genome, *Nature (Lond.)* **310**:207–211 (1984).
7. Dillner, J., B. Kallin, H. Alexander, I. Ernberg M. Uno, Y. Ohno, G. Klein, and R. A. Lerner, A novel Epstein–Barr virus determined nuclear antigen (EBNA-5) partly encoded by the transformation-associated Bam WYH region of EBV DNA: Preferential expression in lymphoblastoid cell lines, *Proc. Natl. Acad. Sci. USA* **83**:6641–6645 (1986).
8. Modrow, S., and H. Wolf, Characterization of two related Epstein–Barr virus-encoded membrane proteins that are differentially expressed in Burkitt lymphoma and in vitro-transformed cell lines, *Proc. Natl. Acad. Sci. USA* **83**:5703–5707 (1986).
9. Rooney, C. M., C. F. Edwards, D. Lenwar, H. Repone, and A. B. Rickinson, Differential activation of cytotoxic responses by Burkitt's lymphoma (BL)-cell lines: Relationship to the BL-cell surface phenotypes. *Cell. Immunol.* **102**:99–112 (1986).
10. Finch, S. C., Clinical symptoms and signs of infectious mononucleosis, in: *Infectious Mononucleosis*, R. L. Carter, and H. G. Penman, (eds.), p. 19–25 Blackwell, Oxford (1969).
11. Tosato, G., and R. M. Blaese, Epstein–Barr virus infection and immunoregulation in man, *Adv. Immunol.* **37**:99–149 (1985).
12. Rickinson, A. B., D. J. Moss, L. E. Wallace, M. Rowe, I. S. Misko, M. A. Epstein, and J. H. Pope, Long-term T-cell-mediated immunity to Epstein–Barr virus, *Cancer Res.* **41**:4216–4221 (1981).
13. Haider, S., M. de L. Coutinho, R. T. D. Emond, and R. N. P. Sutton, Tuberculin anergy and infectious mononucleosis, *Lancet* **2**:74 (1973).

14. Mangi, R. J., J. C. Niederman, J. E. Kelleher, J. M. Dwyer, A. S. Evans, and F. S. Kantor, Depression of cell-mediated immunity during acute infectious mononucleosis, *N. Engl. J. Med.* **291**:1149–1153 (1974).

15. Reinherz, E. L., C. O'Brien, P. Rosenthal, and S. F. Schlossman, The cellular basis for viral-induced immunodeficiency: Analysis by monoclonal antibodies, *J. Immunol.* **125**:1269–1274 (1980).

16. Johnsen, H. E., M. Madsen, and T. Kristensen, Lymphocyte subpopulations in man: Suppression of PWM-induced B-cell proliferation by infectious mononucleosis T cells, *Scand. J. Immunol.* **10**:251–255 (1979).

17. Twomey, J. J. Abnormalities in the mixed leukocyte reaction during infectious mononucleosis, *J. Immunol.* **112**:2278–2281 (1974).

18. Bahna, S. L., D. C. Heiner, and C. A. Horwitz, Sequential changes of the five immunoglobulin classes and other responses in infectious mononucleosis, *Int. Arch. Allergy Appl. Immunol.* **74**:1–8 (1984).

19. Purtilo, D. T., Immunopathology of infectious mononucleosis and other complications of Epstein-Barr virus infections, in: *Pathology Annual.* Part 1, S. C. Sommers and P. P. Rosen, (eds.), pp. 253–229 Appleton-Lange, East Norwalk, Connecticut (1980).

20. Schooley, R. T., H. D. Densen, D. Felsenstein, M. S. Hirsch, W. Henle, and S. Weitzman, Antineutrophil antibodies in infectious mononucleosis, *Am. J. Med.* **76**:85–90 (1984).

21. Junker, A., H. D. Ochs, E. A. Clark, M. L. Puterman, and R. J. Wedgwood, Transient immune deficiency in patients with acute Epstein–Barr virus infection, *Clin. Immunol. Immunopathol.* **40**:436–446 (1986).

22. Tosato, G., I. Magrath, I. Koski, N. Dolley, and M. Blaese, Activation of suppressor T-cells during Epstein–Barr virus-induced infectious mononucleosis, *N. Engl. J. Med.* **301**:1133–1137 (1979).

23. Seeley, J., E. Svedmyr, O. Weiland, G. Klein, E. Moffer, E. Eriksson, K. Anderson, and L. Van Der Waal, Epstein–Barr virus selective T cells in infectious mononucleosis are not restricted to HLA-A and B antigens, *J. Immunol.* **127**:293–298 (1981).

24. Rothwell, D. J., Bone marrow granulomas and infectious mononucleosis, *Arch. Pathol. Lab Med.* **99**:508–512 (1975).

25. Risdall, R. J., R. W. McKenna, M. E. Nesbit, W. Krivit, H. H. Balfour, R. L. Simmons, and R. D. Brunning, Virus-associated hemophagocytic syndrome, *Cancer* **44**:993–1002 (1979).

26. Mroczek, E., D. D. Weisenburger, H. L. Grierson, R. Markin, and D. T. Purtilo, Fatal infectious mononucleosis and virus-associated hemophagocytic syndrome, *Arch Pathol. Lab. Med.* **111**:530–535 (1987).

27. Kaplan, J., and T. C. Shope, Natural killer cells inhibit outgrowth of autologous Epstein–Barr virus-infected B lymphocytes, *Natl. Immunol. Cell Growth Regul.* **4**:40–47 (1985).

28. Thorley-Lawson, D. A., The transformation of adult but not newborn human lymphocytes by Epstein–Barr virus and phytohemagglutinin is inhibited by interferon, *J. Immunol.* **126**:829–833 (1981).

29. Neitzel, H., A routine method for the establishment of permanent growing lymphoblastoid cell lines, *Hum. Genet.* **73**:320–326 (1986).

30. Purtilo, D. T., Clonality of EBV-induced lymphoproliferative diseases in immune deficient patients, *N. Engl. J. Med.* **311**:191 (1984).

31. Menezes, J., S. K. Sundar, and C. A. Ahronheim, Immunosuppressive effects of Epstein–Barr virus infection. in: *Viral Mechanisms of Immunosuppression,* N. Gilmore and M. A. Wainberg, (eds.), pp. 115–134, Liss, New York (1985).

32. Markin, R., J. Linder, K. Zuerlein, E. Mroczek, H. L. Grierson, B. Brichacek, and D. T. Purtilo, Hepatitis in infectious mononucleosis, *Gastroenterology* **93**:1210–1217 (1987).

33. Grierson, H., and D. T. Purtilo, Epstein–Barr virus infections in males with the X-linked lymphoproliferative syndrome, *Ann. Intern. Med.* **106**:538–545, 1987.

34. Mroczek, E., T. Seemayer, H. L. Grierson, R. Markin, J. Linder, B. Brichacek, and D. T. Purtilo, Thymic lesions in fatal infectious mononucleosis, *Clin. Immunol. Immunopathol.* **43:**243–255 (1987).

35. Purtilo, D. T., J. P. S. Yang, C. K. Cassel, P. Harper, S. R. Stephenson, B. H. Landing, and G. F. Vawter, X-Linked recessive progressive combined variable immunodeficiency (Duncan's disease), *Lancet* **1:**935–950 (1975).

36. Purtilo, D. T., L. M. Hutt, S. Allegra, C. Cassel, J. P. S. Yang, and F. S. Rosen, Immunodeficiency to the Epstein–Barr virus in the X-linked recessive lymphoproliferative syndrome, *Clin. Immunol. Immunpathol.* **9:**174–156 (1978).

37. Purtilo, D. T., L. Zelkowitz, S. Harada, C. D. Brooks, T. Bechtold, A. K. Saemundsen, H. L. Lipscomb, J. Yetz, and G. Rogers, Delayed onset of infectious mononucleosis associated with acquired agammaglobulinemia and red cell aplasia, *Ann. Intern. Med.* **101:**180–186 (1984).

38. Purtilo, D. T., R. White, L. Filipovich, J. Kersey, and L. Zelkowitz, Fulminant liver failure induced by adenovirus post bone marrow transplantation, *N. Engl. J. Med.* **312:**1707–1708 (1985).

39. Filipovich, A. H., B. R. Blazar, N. K. C. Ramsey, J. H. Kersey, L. Zelkowitz, S. Harada, and D. T. Purtilo, Allogeneic bone marrow transplantation for X-linked lymphoproliferative syndrome, *Transplantation* **42:**222–224 (1986).

40. Ochs, H. D., J. L. Sullivan, R. J. Wedgwood, J. K. Seeley, K. Sakamoto, and D. T. Purtilo, X-linked lymphoproliferative syndrome: Abnormal antibody responses to bacteriophage ΦX174, in: *Primary Immunodeficiency Disease* (R. Wedgwood and F. Rosen (eds.), pp. 321–323 Liss, New York (1983).

41. Purtilo, D. T., K. Sakamoto, V. Barnabei, J. Seeley, T. Bechtold, G. Rogers, J. Yetz, and S. Harada, Epstein–Barr virus-induced diseases in males with the X-linked lymphoproliferative syndrome (XLP), *Am. J. Med.* **73:**49–56 (1982).

42. Seeley, J. K., T. Bechtold, T. Lindsten, and D. T. Purtilo, NK deficiency in X-linked lymphoproliferative syndrome, in: *NK Cells and Other Natural Effector Cells*, Vol. 2 pp. 1211–1218 Academic New York (1982).

43. Sullivan, J. L., K. S. Byron, F. E. Brewster, S. M. Baker, and H. D. Ochs, X-linked lymphoproliferation syndrome. Natural history of the immunodeficiency, *J. Clin. Invest.* **71:**1765–1778 (1983).

44. Argov, S. D. Johnson, M. Collins, H. S. Koren, H. Lipscomb, and D. T. Purtilo, Defective natural killing (NK) activity but retention of lymphocyte mediated antibody-dependent cellular cytotoxicity (ADCC) in patients with the X-linked lymphoproliferative syndrome (XLP), *Cell. Immunol.* **100:**1–9 (1986).

45. Lai, P. K., and D. T. Purtilo, Frequencies of reactive cells in patients with XLP induced by autologous B-LCL. In: *Epstein–Barr Virus and Human Diseases* (P. H. Levine, D. V. Ablashi, M. Monoyama, G. R. Pearson, and R. Glaser, (eds.), pp. 397–398, Humana Press, Clifton, New Jersey (1987).

46. Sakamoto, K., H. Freed, and D. T. Purtilo, Antibody responses to Epstein–Barr virus in families with the X-linked lymphoproliferative syndrome, *J. Immunol.* **125:**921–925 (1980).

47. Yasuda, N., P. K. Lai, and D. T. Purtilo, Immunoglobulin production by B cells in patients with the X-linked lymphoproliferative syndrome (XLP), *Fed. Proc.* **45:**959 (1986) (abstr.).

48. Yasuda, N., P. K. Lai, J. R. Davis, and D. T. Purtilo, Production of interferon-gamma, interleukin-2 and B cell differentiation factor in T cells from males with X-linked lymphoproliferative syndrome, *Fed. Proc.* **46:**1028 (1987) (abstr.).

49. Klein, G., and E. Klein, Conditioned tumorigenicity of activated oncogenes, *Cancer Res.* **46:**3211–3224 (1986).

50. Harrington, D. S., D. D. Weisenburger, and D. T. Purtilo, Malignant lymphomas in the X-linked lymphoproliferative syndrome, *Cancer* **59:**1419–1429 (1987).

51. Brichacek, B., J. Davis, and D. T. Purtilo, Presence of monoclonal and oligoclonal B-cell proliferation in fatal infectious mononucleosis, in: Epstein–Barr Virus and Human Diseases (P. H. Levine, D. V. Ablashi, M. Nonoyama, G. R. Pearson, and R. Glaser eds.), pp. 53–54, Clifton, New Jersey (1987).
52. Merino, F., M. Collins, and D. T. Purtilo, Chediak-Higashi syndrome derived T cell lines manifest giant lysosomal granules, normal natural killer cell and lectin-mediated cytotoxicity, *J. Clin. Lab. Anal.* **1:**72–76 (1987).

# 8

# Immunosuppression by Bovine Herpesvirus 1 and Other Selected Herpesviruses

LORNE A. BABIUK, M. J. P. LAWMAN, and P. GRIEBEL

## 1. INTRODUCTION

Viruses can cause immunosuppression by a variety of mechanisms. Immunosuppression can occur as a result of direct or indirect effects of the virus on various leukocyte populations.[1-5] In the case of direct effects, viruses may infect and destroy the specific leukocytes involved in the development and expression of immunity. In addition, viral components that can be released into the extracellular environment may interact directly with specific cells and affect either accessory or effector cell functions.[6] Indirect effects can be produced by the release of mediators, such as hormones, complement, or prostaglandins.[7-9] Inhibition of mediator release following viral infection can also reduce cellular reactivity and subsequent development of immunity.[10-12] In some instances, immunosuppression is confined to the specific antigen(s) causing the suppression, whereas in other instances there is generalized immunosuppression to a wide variety of antigens.[3,13] Thus, it is clearly evident that the phenomenon of virus-induced immunosuppression can occur via a wide variety of different pathways, and in many cases a combination of factors appears to act in concert.

The first part of this chapter describes what is known concerning immunosuppression in cattle caused by bovine herpesvirus (BHV-1). The second

LORNE A. BABIUK, M. J. P. LAWMAN, and P. GRIEBEL • Department of Veterinary Microbiology and Veterinary Infectious Disease Organization, University of Saskatchewan, Saskatoon, Saskatchewan, Canada S7N 0W0.

part summarizes a few selected herpesviruses of domestic animals (Table I), birds (Table II), and rodents (Table III). In this latter part, we try to demonstrate how different viruses interact with the host to cause disease or viral persistence and how the immune response can alter these interactions. In the BHV-1 model, functional defects of macrophages, polymorphonuclear neutrophils (PMNs), and lymphocytes occur between 4 and 7 days postinfection. Although this coincides with peak viral replication in nasal passages, immunosuppression is presumed to occur as a result of both direct and indirect mechanisms. This chapter describes the kinetics and the degree of immunosuppression associated with various leukocyte functions following viral infection. Initially, a description of the sequence of events relating to immunosuppression is presented in an attempt to explain how immunosuppression may occur, as well as the consequences of immunosuppression. Second, attempts are made to propose a model whereby all the direct and indirect effects of immunosuppression interact to make the animal much more susceptible to secondary bacterial infections and how these may influence clearance of the virus.

## 2. BOVINE HERPESVIRUS

### 2.1. Virology

Cattle, like most other species, are blessed with a number of different herpesviruses that cause a variety of diseases ranging from respiratory, genital, and localized skin infections to systemic infections. These viruses are classified as bovid herpesvirus 1 to 6 (BHV-1 to BHV-6). Bovid HV-1, -2, and -6 are classified within the subfamily of the α-herpesvirinae, whereas BHV-3 is a γ-herpesvirinae, with BHV-4 belonging to the β-herpesvirinae. Although there have been reports regarding the pathology and proposed mechanisms of pathogenesis for all these herpesviruses, very little information has been presented regarding the immune responses to these herpesviruses or their ability to cause immunosuppression with the exception of studies on BHV-1.[14] Therefore, the major emphasis of this review is on the BHV-1 model. BHV-1 also called infectious bovine rhinotracheitis (IBR) virus, is an important pathogen of cattle that can cause a variety of syndromes ranging from severe respiratory infections, vulvovaginitis, abortions, conjunctivitis, and meningoencephalitis to generalized systemic infections in young animals.

Infectious bovine rhinotracheitis is the most common form of BHV-1 observed in situations in which large numbers of animals are intensively reared, such as in feedlots. This disease is associated with a rapid rise in fever, which occurs within 2 days of infection. As a result of infection, there is inflammation of the mucosal surfaces of the upper respiratory tract with distinct ulceration or plaques, where viral replication and infection occurs. This initial infection with BHV-1 of the respiratory tract alters the respiratory mucosal surfaces sufficiently to allow large quantities of *Pasteurella hemolytica* to repli-

**TABLE I**

**Herpesvirus of Domestic Animals**

| Species | Virus designation | Common name | Natural host | Subfamily | Comments regarding immunosuppression |
|---------|-------------------|-------------|--------------|-----------|--------------------------------------|
| Bovine | BHV-1 (Bovid herpesvirus-1) | Infectious bovine rhinotracheitis Infectious vulvovaginitis | Cattle | α | See text discussion |
| | BHV-2 (Bovid Herpesvirus-2) | Bovine mammilitis | Cattle | α | |
| | BHV-3 (Acelaphine herpes 1,2) | Malignant catarrhal fever | Cattle Wildebeest | γ | |
| | BHV-4 | Various strain designation cytomegalovirus | Cattle | β | |
| Caprine[a] | CHV-1 (BHV-5) | Sheep herpesvirus | Sheep | ? | |
| | CHV-2 (BHV-6) | Goat herpesvirus | Goat | α | |
| Equine | EHV-1 | Equine rhinopneumitis Equine abortion | Horses | α | Infects macrophages |
| | EHV-2 | Equine cytomegalovirus-like | Horses | β | Infects mononuclear cells Weakly pathogenic |
| | EHV-3 | Equine coital exanthama | Horses | α | |
| Feline | FHV-1 | Infectious rhinotracheitis | Cats | α | Isolated from spleen but not from leukocytes? |
| | FHV-2 | Cat cytomegalovirus | Cats | β | |
| Canine | CHV-1 | Canine herpesvirus | Dog | α | |
| Porcine | SHV-1 | Pseudorabies, Aujesky disease | Pig | α | Replicates in lymphoid tissue, macrophage, and lymphocyte |
| | SHV-1 | Inclusion-body rhinitis Pig cytomegalovirus | Pig | β | Limited information |

[a]Caprine herpesvirus 1 and 2 are also classified as BHV-5 and BHV-6, respectively.

## TABLE II
## Herpesviruses of Birds

| Virus designation | Common name | Subfamily | Comments |
|---|---|---|---|
| Gallid herpesvirus 1 | Infectious laryngotracheitis virus (ILT) | α | Virus infects macrophages and can cause lymphopenia |
| Gallid herpesvirus 2 | Marek disease herpesvirus | γ | Immunosuppression and immunostimulation/ tumor production—generally nonproductive infection of lymphocytes |
| Meleagrid herpesvirus 1 | Turkey herpesvirus | γ | Weakly pathogenic but can cause reduced lymphocytes function 3–7 days postinfection |
| Gruid herpesvirus 1 | Crane herpesvirus 1 | ?. | |
| Anatid herpesvirus 1 | Duck plaque herpesvirus | ?. | |
| Ciconiid herpesvirus 1 | Black stork herpesvirus | ?. | Very limited information is available regarding immunosuppression and infection of lymphoid tissue |
| Peredicid herpesvirus 1 | Bobwhite quail herpesvirus | ?. | |
| Phalacrocoracid herpesvirus 1 | Cormorant herpesvirus | ?. | |
| Psittacid herpesvirus 1 | Parrot herpesvirus, recently rediscovered Pacheco disease virus | ?. | |
| Strigid herpesvirus 1 | Virus of hepatosplenitis of owls | ?. | |

## TABLE III
### Some Selected Herpesviruses of Rodents That Infect Leukocytes

| Species | Virus designation | Name | Comments |
|---|---|---|---|
| Mouse | Murid herpesvirus 1 | Mouse cytomegalovirus | Infects predominantly monocytes |
| | Murid herpesvirus 3 | Mouse thymic virus | Infects T cells |
| Guinea pig | Caviid herpesvirus 1 | Guinea pig herpesvirus | Infects T and B cells |
| Rabbit | Leporid herpesvirus 1 | Herpesvirus sylvilagus | Infects T and B cells |

cate, migrate down into the lower respiratory tract and cause severe pneumonia. Reflecting the frequent occurrence of this disease following movement of cattle to feedlots, it is often referred to as shipping fever. Since BHV-1 is not the only virus involved in this particular disease syndrome, this disease is also termed bovine respiratory disease complex. However, BHV-1 appears to be the major initiator of this bovine respiratory disease complex. The immunosuppression caused by BHV-1 following respiratory infection appears to be the major reason for enhanced colonization of the lower respiratory tract and furthermore for the inability of the host to clear the bacteria from the lower lung.

Keratoconjunctivitis is often observed in animals suffering from BHV-1-induced respiratory infections, indicating that the virus has spread systemically to infect other mucosal surfaces. Keratoconjunctivitis can also occur in the absence of respiratory infections or following mild respiratory infections that are often undetected. Conjunctivitis, in the absence of severe respiratory infections, generally occurs more frequently in adult dairy cattle. In many cases, this is the only evidence of acute viral infection. Following conjunctivitis, however, pregnant adult cattle may abort, indicating that even though the disease may appear to be a relatively mild localized infection, the virus does spread systemically to infect the fetus. Depending on the stage of gestation, the fetus can either be aborted or, if infection occurs late in gestation, be born prematurely and die shortly after birth.

Like most herpesviruses, BHV-1 has a predilection for cells of the central nervous system (CNS). Thus, latency is maintained in cells of neurologic origin;[15] however, encephalitis is rare in cattle following infection with BHV-1. Although it has been reported to occur in adult cattle, young calves are much more likely to suffer from encephalitis than are adult animals.[16,17]

In sexually mature animals, BHV-1 infection can spread by the genital route. The virus can be disseminated through a herd by asymptomatic bulls during natural breeding or through artificial insemination wherein virus is present in the semen. As is the case with respiratory infections, the virus causes pustules on the vagina or penal mucosal surfaces. These pustules can progress to necrotic erosions. Although these lesions heal within approximately 1 week following primary infection, animals can remain carriers of the virus.

As a result of these generalized infections and the economic importance of the specific disease, extensive studies have been undertaken in attempts to identify (1) the specific cells in which the virus replicates, (2) the immunologic events that occur as a result of viral infection, and (3) how the virus itself can cause alteration of immune responses.

*In vitro*, the virus replicates to very high titers in bovine epithelial fibroblast or monocyte cultures. Although there have been reports of the virus replicating in cells from other species, only rabbit and porcine cells appear to replicate virus to sufficient levels for investigational work. Considering the locations of the pathologic lesions *in vivo*, it appears that the virus can infect a wide variety of epithelial, fibroblast, or monocytic cells and actually spreads as a result of infection of a wide variety of cells within the animal. Although it is believed that virus spreads by the hematogenous route, possibly by infection of monocytes, very little *in vivo* evidence for this method of spread has been presented.

## 2.2. Immunology

Investigations have been conducted into the kinetics of specific antiviral defense mechanisms *in vivo* following infection with BHV-1.[18-21] A correlation of the kinetics of killing between these various effector mechanisms and viral replication at the individual cell level and expression of antigens on the infected cell surface has also been attempted. The mechanisms by which virus-infected cells could be killed include T-cell-mediated cytotoxicity,[21,22] antibody-complement-mediated cytotoxicity,[18] antibody-dependent-cell-mediated cytotoxicity (ADCC),[23] complement-facilitated antibody-dependent-cell-mediated cytotoxocity (ADCC-C),[24,25] natural killer (NK) cell activity, and complement-dependent neutrophil-mediated cytotoxicity (CDNC).[26,27] Although these mechanisms have been shown to be able to kill virus-infected cells *in vitro*, this does not mean that each of these effector mechanisms function *in vivo* to help curtail viral replication and spread. It is therefore crucial to determine whether these specific defense mechanisms can kill virus-infected cells before completion of viral replication and spread to adjacent uninfected cells. In most *in vitro* studies, killing by a single mechanism has been investigated. However, it is probable that *in vivo,* all these effector mechanisms interact sufficiently to be able to curtail viral replication and spread. In this regard, it has been shown that surface antigen expression is present on virus-infected cells at approximately 3–4 hr postinfection (p.i.). However, there does not appear to be sufficient antigen for the effector mechanisms to recognize and lyse the virus-infected cell until approximately 9–10 hr p.i.[18] At this time, the virus is already beginning to spread and infect adjacent uninfected cells by the intracellular route. Thus, these mechanisms alone do not appear to be very effective in curtailing viral replication and spread.

It has been demonstrated that in the presence of interferon (IFN), the killing of virus-infected cells occurs earlier and at a faster rate.[28] In the host, where IFN synthesis occurs very rapidly following viral infection,[29,30] it is possible that these effector mechanisms are important in limiting virus spread between cells. Investigations into the specific effector mechanisms that may be important in primary infections have clearly demonstrated that IFN and cytotoxic T cells (CTL) and possibly NK cells are important in clearing BHV-1 in the initial stages.[22,30,31] Virus clearance begins approximately 8–10 days p.i., a time when IFN and CTL are evident with very little antibody being detected.[22,31] Since there have been extensive reports of the various effector mechanisms in recovery from BHV-1 infections and a recent review has been published in this regard,[32] this aspect of immunity to bovine herpesviruses is not discussed further here.

## 2.3. Immunosuppression

Immunosuppression caused by BHV-1 is considered one of the major reasons for enhanced colonization of the lower lung with *Pasteurella hemolytica* and subsequent development of pneumonia. Aerosol infection of healthy animals with *Pasteurella hemolytica* results in rapid infiltration of neutrophils into the lung and clearance of the bacteria within 4 hr of exposure.[11] However, aerosol challenge with *Pasteurella hemolytica* 3–7 days post-BHV-1 infection results in a reduced rate of cellular infiltration into the lungs and increased colonization of the lungs with bacteria such that more than 50% of infected animals will develop severe pneumonia and die. In an attempt to explain this increased susceptibility to bacterial superinfection, numerous investigations have been directed at correlating changes in susceptibility to alteration in leukocyte numbers and functions. In most cases in which such studies have been performed, there have been indications of leukopenia and altered leukocyte functions, with reduction of leukocyte function being more dramatic than depletion of any specific leukocyte type. These studies also indicate that the depression of leukocyte number and reduced function is different for the various cell types.

Although the depletion or reduced cell function of one cell type may have dramatic repercussions on functional activities of other leukocytes, for the sake of clarity, in the initial description of immunosuppression, the different effects observed with each specific cell type are described. Subsequently, attempts are made to correlate the interactions generating both specific and nonspecific immunosuppression, in order to provide a unified hypothesis of how BHV-1 may induce this in cattle. It is hoped that this approach will provide a clearer picture of the events that occur during viral infection and eventually aid in developing methods for reducing the level of immunosuppression.

Despite a slight decrease in lymphocyte numbers following viral infection, there is almost total paralysis of lymphocyte proliferation *in vitro*. The precise mechanism(s) of suppression of lymphocyte function are unknown, but it has been postulated that both specific and nonspecific suppression can occur. In Sections 2.3.1 and 2.3.2, we discuss what has been shown to occur following BHV-1 infection and then try to provide a plausible explanation for the observed phenomena. It is our contention that BHV-1 causes immunosuppression by more than one process.

### 2.3.1. Specific Suppression

The classic work of Gershon during the early 1970s led to the universally accepted phenomenon of a suppressor regulatory network.[33] The discovery of suppressor T lymphocytes (Ts) is important, as these cells may play a significant role in the regulation of the immune response. These regulatory T cells have been shown to be both nonspecific in function[34] and antigen specific.[35] Most studies investigating the mechanism of action by Ts have been carried out using the murine and human cells. This has been greatly helped by the availability of phenotypic markers in the characterization of these cells. In these systems, Ts appear early during infection or generation of the immune response and seem to act at the level of T-cell proliferation preventing clonal expansion.[36] The target cells appear to be the T-helper (Th) lymphocyte with some activity on B lymphocytes and macrophages being recorded. Ts appear to be antigen specific and therefore capable of antigen recognition. In the case of nonspecific activation of Ts by lectins, there is some thought that the activation, by nonspecific means, may also nonspecifically induce antigen-specific clones.[36] It has been proposed that Ts inhibit the clonal expansion of antigen activated T lymphocytes by preventing the induction of interleukin 2 (IL-2) responsiveness.[37] This nonresponsiveness to IL-2 is thought to be induced by the production of a suppressor lymphocyte-associated molecule (SAM). The SAM has been proposed to bind the erythrocyte (E) receptor (on human T cells, or E-equivalent receptor on T cells of other mammalian species) and thereby produce a state of suppression.

If Ts are to be a factor in suppression they, like other T-cell subsets, must expand clonally, and this has been shown to be dependent on IL-2.[38] Ts must therefore be able to suppress antigen-specific Th responsiveness to IL-2 as well as induce IL-2 production by Th.[37] The observation that in cattle infected with BHV-1, IL-2 production peaks at the time of maximal immunosuppression make the existence of BHV-1-specific Ts highly probable. It should be noted that factor(s) other than IL-2 have also been isolated and shown to cause activation and proliferation of Ts.[39] One factor, suppressor cell induction factor (SIF), has been shown to be distinct from other lymphokines, in particular IL-2.[39] Whether this factor is produced during herpesvirus infections or in cattle is unknown.

Herpesviruses are known to induce suppressor cell activity.[40] In the case of bovine herpesvirus, however, it has not been possible to demonstrate suppressor cell activity conclusively following BHV-1 infection, even though they can be easily induced by concanavalin A (Con A).[41] Although this does not exclude the possibility that suppressor cells do exist or are generated by BHV-1 antigens, they do not appear to be important in the reduction of lymphocyte functions following BHV-1 infections. It must be emphasized, however, that these studies are just beginning, and it is possible that at a later date antigen-specific suppressor cells may be identified.

### 2.3.2. Nonspecific Suppression

*2.3.2.a. Lymphocytes.* Within 1–3 days post-BHV-1 infection, lymphocyte responsiveness to phytohemagglutinin (PHA) or Con A becomes significantly depressed and remains depressed for up to 9 days postchallenge.[30,42] These decreased lymphocyte responses cannot be restored by the addition of exogenous IL-2. Furthermore, exogenous IL-2 cannot act as a nonspecific activator of resting cells obtained from virus-infected animals, even though it can activate normal bovine lymphocytes. In an attempt to explain the depressed mitogen response, Th cell function was investigated. Although the results appear to be somewhat variable, in most cases the generation of IL-2 is enhanced following infection of animals with BHV-1. This enhancement appears to be correlated with the initial rise in production of IFN in the nasal passages 2 days after BHV-1 infection. Administration of exogenous IFN by the nasal route also increases IL-2 production by lymphocytes from normal animals in a transient manner. Although animals appear to produce higher levels of IL-2 following infection with a virus, lymphocytes from virus-infected animals are not responsive to IL-2. This lack of responsiveness may be attributable to (1) downregulation of IL-2 receptors on cells, (2) viral interference with their expression, or (3) margination of activated T cells to local drainage lymph node or sites of BHV-1 replication. Unpublished experimental evidence from flow cytometric analysis demonstrates a decline of E-rosette-positive cells in animals infected with BHV-1. This evidence still does not discriminate between the loss of these cells or downregulation of E-rosette receptors. The downregulation of T-cell proliferation by high antigen dose has been shown for other systems.[43] That study showed that T lymphocytes produced IL-2 but were unable to respond because of the loss of IL-2 receptor expression. BHV-1 has been shown to be a potent inducer of IL-2.[44] In this context, it is worth reiterating that during BHV-1 infection, IL-2 production peaks at the time of maximal virus shedding in nasal passages. The peak of viral antigen and IL-2 production also correlates with the time that T cells become unresponsive to IL-2. This finding appears to parallel what Ceredig and Corradin[43] reported in terms of IL-2 production and downregulation of IL-2 receptors in the face of high antigen doses.

In a recent article, it was shown that, under certain conditions, human peripheral blood mononuclear cells (PBMC) could be induced to release IL-2 receptors.[45] These conditions involve activation by lectins and toxoids (tetanus). HTLV-I-positive T-cell clones were also shown to release large amounts of IL-2 receptors. Whether these observations have any relevance for cattle is unknown. It is tempting to speculate that bovine T cells may lose responsiveness to IL-2 due to the loss of IL-2 receptors following binding of virus or viral glycoproteins to lymphocytes. It was recently demonstrated that BHV-1 can bind to T-cell clones[46] (P. Griebel and L. A. Babiuk, unpublished observations). Thus, if HTLV-1 binding and infection of lymphocytes can cause the release of IL-2 receptors, could the same mechanism also operate in the BHV infection? Unfortunately, monoclonal antibodies that recognize human IL-2 receptors do not recognize bovine IL-2 receptors. Thus, until specific antibody to the bovine IL-2 receptor is developed, this question cannot be fully explored. Alternatively, as has been observed in the murine system, antigen-specific suppressor cells may induce a state of Th cell unresponsiveness.[37]

It is known that exposure of lymphocytes to some viral antigens can lead to nonresponsiveness of these lymphocytes in culture.[47] Since BHV-1 is a systemic virus, such exposure *in vivo* could account for the lower responses to mitogens at the peak of viral replication and during clinical disease. Exposure of lymphocytes from normal animals to BHV-1 antigens *in vitro* does result in decreased blastogenic responses.[48] Regardless of whether the antigen was inactivated or live virus when used in the assays, it was found that high doses of antigen dramatically suppressed mitogen induced lymphocyte proliferation *in vitro*.[48] Specific antigen stimulation of lymphocytes from immune animals was also suppressed at high antigen doses. Since suppression occurs with live or inactivated virus and since there is no evidence of even an abortive infection of bovine lymphocytes with BHV-1, it appears that viral replication is not required to induce immunosuppression in lymphocytes. It is possible that the mere association of virus with lymphocytes can alter their function. Although lymphocytes cannot be infected with BHV-1, the virus can bind to lymphocytes[46] (P. Griebel, M. Lawman, and L. A. Babiuk, unpublished observations). This interaction appears to occur especially if T lymphocytes are activated.

Using both primary bovine T-cell cultures and T-cell clones, it has been observed that infectious BHV-1 decreases the viability of these T cells without detectable viral replication. Furthermore, both infectious and inactivated BHV-1 decreases IL-2 responsiveness and activation of these T-cell clones by phorbol esters (P. Griebel, M. Lawman, and L. A. Babiuk, unpublished observations). This finding supports the hypothesis that BHV-1 directly affects T-cell function and responsiveness without the induction of antigen-specific Ts. It was recently reported that herpes simplex virus (HSV) will replicate in activated human T lymphocytes.[49] Our contention is that BHV-1 may infect/replicate in a small subpopulation of activated T cells (making its detection by standard techniques, such as infectious center assays, difficult) and that these infected T cells would be eliminated. This could reflect a major defect in the CMI response without being an obvious defect in total T-cell populations. This does not explain the observation that the general population of cells are nonresponsive to mitogens.

No reports have described the specific protein or glycoprotein involved in the suppression of lymphocyte function, but it is highly likely that one of the viral glycoproteins is involved. Since these glycoproteins are released into the extracellular environment during viral replication[50-53] and virus—cell interactions are initiated by glycoprotein, it is highly likely that interaction of the glycoprotein with the lymphocyte host cell membrane alters their function. The availability of purified BHV glycoproteins makes this hypothesis feasible to test.

Interferon production in nasal passages is evident within 2–3 days p.i. Although very little free IFN is detectable in plasma from infected calves, it is possible that it is rapidly cell associated and thereby alters lymphocyte reactivity. Studies involving the pharmacokinetics of clearance of IFN clearly demonstrates that exogenous IFN is rapidly cleared (half-life approximately 2 hr). Coupled with the observation that IFN can alter lymphocyte function dramatically, this suggests that rapid cell association is highly likely.[30] IFN has been shown to inhibit proliferation of bovine T cells *in vitro* (P. Griebel and L. A. Babiuk, unpublished observations) and reduce their responsiveness to IL-2 and phorbol ester (P. Griebel and L. A. Babiuk, unpublished observations). The observations that bovine IFN activity is enhanced at elevated temperatures up to 40°C,[54] maximal immunosuppression correlates with peak temperature responses and that IFN is present in nasal secretions suggest that even low levels of serum IFN may be important in reducing lymphocyte reactivity.

A further observation that may or may not have relevance to BHV-1 is the production of a suppressor lymphokine by Ts. This lymphokine has been included with nonspecific mechanisms because despite its production by Ts, the lymphokine is activated nonspecifically by phagocytic cells. The substance in question is the soluble immune response suppressor (SIRS), first described by Aune and Pierce[55] in the mouse and subsequently described in humans.[56] SIRS is produced by Lyt 2+ cells (murine Ts) in an inactive nonoxidized form. It is activated (by oxidation) by $H_2O_2$ produced by macrophages and presumably polymorphonuclear leukocytes (PMNs). Once the substance is oxidized, it is inhibitory not only to neoplastic cell replication but to immune cell functions as well. If a bovine counterpart to the murine and human SIRS exists, infection with BHV-1, which causes elevated production of toxic oxygen species, may well indirectly activate this suppressor molecule and cause immune suppression.

*2.3.2.b. Monocytes.* Within the lung, alveolar macrophages are vital as the main protectors of the alveolar airspaces against microorganisms and foreign substances. Like other members of the mononuclear phagocytic system, the alveolar macrophage may also participate in immunologic reactions by virtue of their accessory cell capacity.[57,58] In addition to these functions, however, which are generally considered beneficial to the host, the alveolar macrophage also possess the capacity to synthesize and secrete a wide armamentarium of mediators, including enzymes, monokines, complement components, and prostaglandins, which have the potential to modulate the immune response or induce lung injury either directly or through stimulation of other inflammatory and immune cells.[59] Since the expression of various surface antigens on macrophages can be correlated with functional activity, initial attempts were made at identifying surface markers on alveolar macrophages and peripheral blood macrophages following infection both *in vivo* and *in vitro* with BHV-1.

Following *in vitro* exposure of bovine alveolar macrophages to BHV-1, it was found that although macrophages were more resistant to infection with BHV-1 than were fibroblasts or epithelial cells, they could be infected. Once infected, the number of Fc receptors on macrophages is dramatically reduced, as is the ability of these infected macrophages to phagocytize erythrocytes or *Candida*.[5] By contrast, there is an increase in complement receptors for the first 3–12 hr p.i. At latter times postinfection, there is a decrease in complement receptors as the cells begin to suffer the effects of virus replication. In these studies it did not matter whether the macrophages were lavaged from IBR-seropositive or -seronegative animals, suggesting that arming of the macrophages with antibody did not have any effect on the susceptibility to BHV-1 infection or their ability to participate in immune responses following BHV-1 infection.

Since alveolar macrophages could potentially be involved in the elimination of viral infections by a variety of mechanisms, the ability of infected macrophages to function as effector cells in ADCC was investigated.[5] Within 2–4 hr post-BHV-1 infection *in vitro,* alevolar macrophages could not mediate ADCC killing of virus-infected cells. Although it was possible to infect alveolar macrophages *in vitro,* retrieval of alveolar macrophages from virus-infected animals demonstrated that a very small percentage of the macrophages were actually infected. However, infection of cells per se cannot totally explain the reduced functional capacity of bovine macrophages following BHV-1 infection, as many macrophages retrieved from lungs of infected animals are not infected but are functionally inactive. Similarly, blood monocytes collected from infected animals, which have never been shown to be infected, are also paralyzed with respect to their ability to produce various mediators.[60] These results suggest that indirect effects of the virus on leukocyte function may be as important as the direct effects of viral infection on a specific subpopulation of macrophages.

Studies with human and rat blood macrophages and monocytes and alveolar macrophages have shown them to have both enhancing and suppressive effects on immunoglobulin secretion and mitogen-induced proliferation of peripheral blood lymphocytes.[61–67] Detailed studies in the human have shown that functional heterogeneity of alveolar macrophages exists and that the expression of one function in preference to the other is dependent on cell numbers.[68] This dependence on cell density is important with respect not only to the macrophage but to the PBL as well. In the above study, it was shown that low numbers of PBMC and of alveolar macrophages enhanced lymphocyte proliferation. However, in the presence of elevated numbers of monocytes or macrophages inhibition of lymphocyte proliferation occurs. These investigators also stated that alveolar macrophages were 10-fold more suppressive than peripheral blood monocytes.

In cattle, alveolar macrophages have been reported to suppress mitogen-induced proliferation of PBMC.[69] This suppression was evident at levels at which the macrophage comprised only 1% of the cells. These investigators clearly demonstrated that this suppression was mediated via a soluble mediator released by the macrophages. Similar suppression of mitogen induced lymphocyte proliferation can be demonstrated *in vitro* by increasing the ratio of peripheral blood monocytes to lymphocyte.[67]

Using flow cytometric analysis of Ia-positive monocytes and nonspecific esterase staining of monocytes, it has been shown that there is a significant rise in the number of monocytic cells in the PBMC, 3 and 8 days post-BHV-1 infection (P. Griebel, M. Lawman, and L. A. Babiuk, unpublished observations). This rise in monocyte numbers can increase to 10–30% of the PBMC population. On the basis of the studies reported above, this increase in the number of macrophages could easily be responsible for at least some of the suppression observed 4–8 days p.i. It has also been shown by flow cytometry that in addition to an increase in the number of Ia-positive cells in peripheral blood 4 days after BHV-1 infection, there is also a shift (increase) in size. Whether increased size is also correlated with the production of immunomodulating factors (i.e., prostaglandins) complement and IFN-$_\alpha$ need to be investigated. Since all these mediators have been reported to have suppressive effects on the immune response, it is highly likely that the macrophage may be pivital in BHV-1-induced immunosuppression.

Natural cytotoxicity is another function exhibited by a subpopulation of mononuclear leukocytes. This activity, expressed against xenogeneic cell lines, decreases following infection with BHV-1. As was the case for blastogenesis, depression was seen 3–5 days p.i. but returned to normal more quickly and actually rose above preinfection levels by 7 days p.i.[30] This increase in natural cytotoxicity appears to be correlated with an increase in the number of monocytes present in the peripheral blood.

*2.3.2.c. Macrophage-Soluble Mediators of Suppression.* Macrophages are a major source of $IFN_\alpha$. Bovine $IFN_\alpha$ (recombinant) has been shown to downregulate the production of IL-2 and the number of circulating lymphocytes.[30] Since IFN is produced very early in the infection cycle, these observations suggested that IFN may be important in early immunosuppression. $IFN_\alpha$ was also shown to have direct antiproliferative activity on bovine T-cell cultures and to have inhibited the cytotoxic activity of cloned bovine CTL (M. Lawman and L. A. Babiuk, unpublished observations). Cultured bovine T cells were rendered IL-2 unresponsive in the presence of $IFN_\alpha$; furthermore, activation of bovine T cells by phorbol esters was suppressed (P. Griebel, M. Lawman, and L. A. Babiuk, unpublished observations). Even though IFN levels in the peripheral blood are not elevated,[30] this may not preclude the possibility that IFN was removed by binding to target cells (*in vivo*) and therefore unavailable for *in vitro* assays. Letchworth *et al.*[54] have shown that the antiviral effect of bovine IFN is increased at 40°C. During BHV-1 infection, immunosuppression is greatest between 3 and 5 days; it is at this time that the pyrexic response is greatest (>41°C). On the basis of these observations, it would be interesting to determine whether elevated temperatures also alter the immunomodulating role of IFN.

One of the major immunosuppressive factors produced by macrophages is prostaglandin (PG). PGs have been shown to downregulate various parameters of the immune response. PGE-2, in particular, may inhibit T-cell function blocking both proliferation[70] and IFN production.[71] PGs have also been shown to be responsible for alteration in macrophage function and receptor expression.[72] Production of IL-1 is inhibited by PGE-2.[73] Thus, PGE-2, which may be produced in response to IL-1,[74] may serve in the regulation of IL-1 production by macrophage/monocytes. PGs have also been shown to be responsible for alteration of macrophage function and receptor expression.[72]

Unpublished data from our laboratory indicate that serum levels of PGE-2 increase significantly 6–10 days post-BHV-1 infection. These levels of PGE-2 correlate with the concentration required to suppress mitogen-induced proliferation *in vitro*. Plasma from BHV-1 infected cattle (4–10 days p.i.) is capable of suppressing (>80%) lymphocyte responses significantly in blastogenic assays at dilutions as high as 1 : 100. $PGE_2$ can also suppress IL-2 production.[75] Furthermore, Chouaib and co-workers also have shown that $PGE_2$ induces suppressor T lymphocytes. PG appears to be significant in modulating the immune response to BHV-1 and this modulation may well be amplified by its multiple effects.

The monokine IL-1 is an important immunoregulatory molecule, with the characteristics of upregulating the expression of cell-surface receptors, effector cell functions, and proliferation of lymphocyte subsets.[76] One interesting function of IL-1 that may have relevance to BHV-1 infection is that it has been shown to interact with vascular endothelial cells and to cause the concomitant release of $PGE_2$.

Testing of alveolar macrophages from BHV-1-infected animals collected from day 1–6 p.i. demonstrated a progressive decrease in IL-1 production or activity. It was recently demonstrated that human macrophages, when exposed to influenza or respiratory syncytial virus, are a source of an IL-1 inhibitory substance.[77] These two observations suggest that a similar situation may occur in BHV-1 infections. Alternatively, the release from the liver of acute phase proteins during early BHV-1 infection could be responsible for suppressing IL-1 production.[73]

During the inflammatory reaction observed in BHV-1 infection, activated phagocytic cells release toxic oxygen species. These oxygen radicals have been shown to cause damage to DNA of innocent bystander cells.[78] Further work by Carson et al.[79] has shown that DNA strand breakage caused by toxic oxygen radicals can cause significant lymphocyte dysfunction of resting PBL. These lymphocytes show marked suppression of proliferation in blastogenic assays. This may also be of significance in BHV-1 infection. We have shown that BHV-1 infection in cattle causes PMN and macrophages to produce elevated levels of $H_2O_2$ and superoxide at the time of maximum immunosuppression. The role of superoxide/$H_2O_2$ production in lymphocyte suppression in cattle by means of a mechanism as described by Carson et al.[79] needs critical evaluation.

*2.3.2.d. Polymorphonuclear Leukocytes.* Polymorphonuclear leukocytes (PMN) are important components of the host defenses against bacterial infections. Their function in viral infections is somewhat unclear. They have been shown to participate in ADCC in both humans and animals. In humans, they have also been shown to enhance lymphocyte stimulation through the release of soluble mediators.[80,81] More importantly, in terms of this review, PMN have also been shown to be inhibitory to lymphocyte proliferation in humans[82,83] and horses.[84] Most of these studies have centered around depression of lymphocyte proliferation by increasing the PMN content of the PBL culture. In cattle affected by shipping fever, an increase in the number of PMN in PBMC populations isolated by ficoll-hypaque has been noted (M. Lawman, unpublished observations). In some instances, this increase in the number of PMN has been as high as 40% (range of 10–40%). If PMN are added to PBL within the range seen *in vivo*, mitogen-induced proliferation is significantly depressed (M. Lawman, unpublished observations). PMN, like macrophages, are a potent source of toxic oxygen radicals. BHV-1 infection causes elevated production of these toxic oxygen species at the time of maximum suppression. Therefore, the activation of PMN, in terms of their respiratory burst, could be a source of toxic oxygen that could potentially suppress lymphocyte function by causing DNA damage. In our laboratory, the activation of bovine PMN by recombinant $IFN_\alpha$ and/or adjuvants was shown to enhance their ability to suppress blastogenesis (M. Lawman, unpublished data). PMN also are a source of a soluble mediator capable of suppressing IL-1 production.[85] This inhibitor is specific for IL-1, hence does not modulate other lymphokines. Since PMN become activated in BHV-1 infection, they could modulate both lymphocyte and monocyte/macrophage function(s).

2.3.2.e. *Stress.* Bovine respiratory disease is often associated with the movement and mixing of animals. These factors are all considered to stress the animals sufficiently to make them more susceptible to viral and bacterial infection. One explanation for this increased susceptibility has been the elevated cortisol often found in stressed animals. It has recently been demonstrated that cortisol can decrease IL-2 production by bovine PBL.[86] Through reduction of Th function, cortisol may be important in limiting the amplitude of the immune response. Although not directly demonstrated, cortisol may also inhibit production of IFN, which is closely linked with IL-2 production.[87] Elevated adrenocorticotropic hormone (ACTH), which is responsible for increased cortisol production, may in itself have direct immunosuppressive effects on T cells.[88] In addition, stress may be important in enhancing the replication of BHV-1.[89] BHV-1 replication *in vitro* has been shown to be enhanced by inclusion of serum from stressed calves. Thus, it is possible that some factors present in the serum do allow higher levels of BHV-1 virus and antigens to be present in the serum and thereby further cause immunosuppression by the mechanisms previously described. In addition, the state of nervous excitement that occurs during shipping may be of importance. Recent evidence suggests that immune responses may be modulated by both the peripheral nervous system and the CNS. In this regard, IL-1 has been implicated as an important link between CNS and T-cell activation. There is also evidence to suggest that the peripheral nervous system can modulate immunologic responses on mucosal surfaces.[90] Such substances as somatostatin and vasoactive intestinal peptides (VIP) act at the level of T lymphocytes decreasing proliferation and decreasing colony-stimulating activity.[90] There is insufficient evidence to determine whether stress increases or decreases these specific peptides or what their precise function is in immune modulation of animals infected with BHV-1. It should also be noted that following BHV-1 infection, animals become anorexic and may completely go off feed for 3–4 days. Peak anorexia occurs approximately 3–7 days p.i. It is possible that this anorexia has some impact on various CNS-released peptides, which could then further cause immunosuppression.

Substance P, a sensory neuropeptide found in a subpopulation of different neurons of the nervous system that supplies skin and mucous membranes,[91] has been shown to interact with human PMN and to activate their basic phagocytic functions and increase $H_2O_2$ production.[90] $H_2O_2$ production can suppress lymphocyte proliferative responses by causing DNA breaks and by oxidizing SIRS to its active suppressor molecule.

Lymphocytes have also been shown to be a source of endorphins.[92] It has also been shown that both $\gamma$- and $\alpha$-endorphins are immunosuppressive, with $\alpha$-endorphins being more highly active.[93] Enkephalins have also been shown to be suppressive for antibody production in *in vitro* assays.[93] Thus it seems possible that local inflammatory responses caused by BHV-1 infection induce the production of various neuropeptides or neuroendocrine hormones that could shut down or suppress immune responses. These possibilities need further investigation with respect to viral infections in general and bovine herpesviruses in particular.

   *2.3.2.f. Other Nonspecific Mechanisms of Immunosuppression.* Recent studies have shown the importance of iron-binding proteins, iron itself, and storage proteins (transferrin, ferritin, and lactoferrin) in the immune response. A recent symposium and articles suggest that the iron-binding property of certain proteins may in fact be a major factor in immune surveillance.[94] Transferrin has been shown to be important in lymphocyte proliferation, and the amount of iron bound to transferrin is critical for proper immune regulation. Thus, too much or too little can cause the suppression of Con A responsiveness. It has been suggested that differences in iron binding to transferrin may be important in Th or Ts activity. This suggestion has been supported in part by reports showing that activated $T_4{}^+$(Th) can synthesize transferrin, but $T_8{}^+$(Ts) cannot.[95] Ferritin, the iron storage protein, has also been shown to have an inhibitory effect on mitogen-induced T-cell proliferation.[96] Since BHV-1 does cause extensive cell death and hemorrhaging, it is highly likely that BHV modulates iron availability by (1) releasing extracellular iron into the circulation, (2) altering the efficiency of iron binding to transport proteins, and (3) regulating the expression of transferrin receptors.

## 2.4. Summary

   In the preceding section of this review, we have identified a number of effects that BHV-1 infection exerts on the bovine immune system. In many cases, the discussion centered on a specific direct or indirect interaction of the virus with a specific cell type. However, we would like to emphasize that *in vivo* during the development of an immune response, individual cells or their products interact with other cell types that can have dramatic repercussions on the development of the immune response. In this final section, we would like to summarize the various ways BHV-1 infection can either enhance or suppress immune responses. In some cases, enhanced activity of one specific cell type may have a negative effect on another cell. Thus, it is clearly evident that the consequences of any modulation of the immune response by any pathogen, including BHV-1, may be very complex. Figure 1 attempts, at least partially, to explain the multitude of effects BHV-1 infection has on immune responses in cattle. The events appearing in the upper portion of Fig. 1 are considered enhancing, whereas those in the lower portion are clearly immunosuppressive. It should be noted, however, that the components situated on the positive side can have or trigger negative effects as well. Following BHV-1 infection, the virus can interact with macrophages, which are good producers of $IFN_\alpha$ and

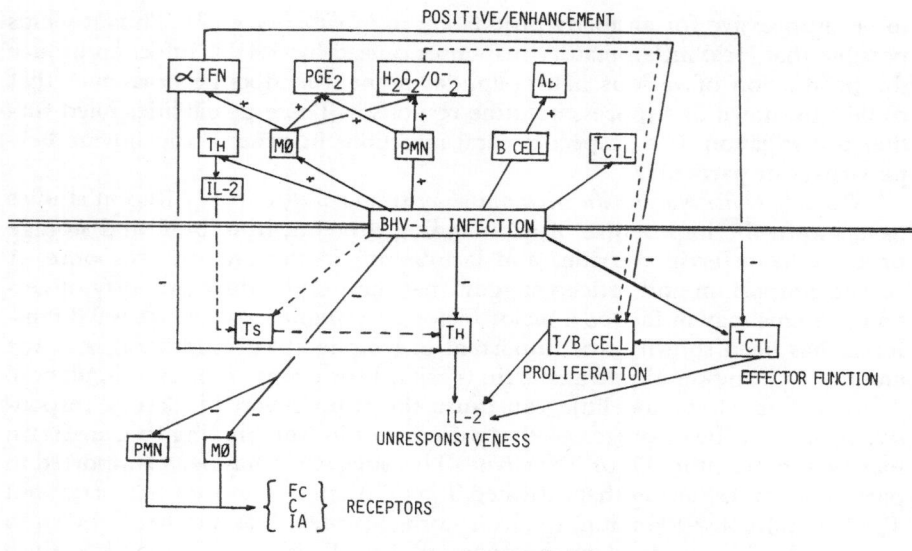

**FIGURE 1.** Summary of the effects BHV-1 has on various leukocyte functions. (—) Data are available to support the interactions; (- - -) areas in which concrete evidence is lacking.

$PGE_2$. These two components are known to be able to depress the proliferation of T and B lymphocytes, as well as have a negative impact on the effector function of these cells. Thus, even though BHV-1 infection does stimulate CTL and Th cells that produce IL-2, the negative effect of the components secreted by macrophages reduces their activity. Similarly, BHV-1 infection activates PMN to secrete leukotrienes and reactive oxygen molecules. The precise effect of the oxygen-free radicals on T- and B-cell proliferation and effector cell function requires further investigation. In addition, BHV-1 infection alters various cell-surface markers of macrophages and PMN, such as Fc receptors, complement receptors, and Ia antigens. These receptors are extremely important during phagocytosis and antigen presentation. Any decrease in these particular receptors may have devastating effects on antigen presentation by macrophages, thereby inhibiting the development of an effective immune response. Little is known about how BHV-1 infection affects Ts and Th or about the production and responsiveness of these cells to IL-2 or other lymphokines. This also is an area in need of further investigation.

From this brief summary, it is evident that the interaction of BHV-1 with the various immune cells of the host is extremely complex and that in many areas further investigation is required before the therapeutic enhancement of a specific function can be predicted to have a beneficial or negative effect on the animal's ability to deal with the infection.

## 3. OTHER HERPESVIRUSES

This section briefly reviews some of the other herpesviruses that infect leukocytes. No attempt is made to discuss all the potential mechanisms of specific and nonspecific suppression, except to identify whether the viruses infect specific leukocytes and whether the infection leads to a lytic infection or transformation of leukocytes, which may lead to disruption of function.

### 3.1. Equine Herpesviruses

Infection of horses by various equine herpesviruses (EHV) can range from very mild localized or systemic infection, as seen with EHV-2, to more severe disease, as observed with EHV-1. In most instances, EHV appear to become associated with leukocytes either in a latent nonproductive state or a lytic state.[97] EHV-2 can be isolated from apparently healthy horses and can persist in the presence of high levels of antibody. In these instances, no apparent effect on immune responses is evident, and it has not been possible to establish any direct association among EHV-2 infection, disease, and immunosuppression, even though the virus can be isolated from leukocytes.[97,98] By contrast, infection with EHV-1 does result in functional and ultrastructural changes in neutrophils and lymphocytes.[99,100] These alterations appear to be age related, as does the severity of the disease. Thus, young animals are more severely affected both with respect to disease and immunosuppression. In the case of young animals, motility of PMN is increased 2 days p.i., whereas their ability to ingest *Staphylococcus aureus* is decreased. Although there is an increase in PMN migration, there is a decrease in their ability to act as effector cells in ADCC.[99] In both instances, migration and ADCC effects are transient and return to normal within 5 days p.i. Suppression of lymphocyte function as measured by responsiveness to mitogens occurs approximately 10 days p.i.[100] Infection of T and B lymphocytes and monocytes could be detected as early as 2 days p.i. and peaked 4 days p.i., using infectious center assays.[101] These studies also demonstrated that the T cell was the predominant leukocyte cell type harboring the virus, indicating that EHV-1 is a lymphotropic virus.

Since maximal T-cell infection occurred 4 days p.i. and maximum inhibition of mitogen responses occurred 10 days p.i., these results suggest infection of lymphocytes may not be the sole reason for reduced lymphocyte reactivity. These investigators suggested that a serum factor was not responsible for inhibition of mitogen responses 10 days p.i., since serum from infected animals could not consistently suppress immune responses of lymphocytes from healthy animals. It is interesting to note that maximal suppression of lymphocyte responses occurred at the time when blastogenic responses to EHV-1 antigen was maximal.[100] This may suggest that the lymphocytes were responding to EHV-1 antigen and releasing some soluble factors such as $IFN_\gamma$ that could then suppress mitogen responses of lymphocytes.

### 3.2. Pseudorabies Virus

Pseudorabies virus (PRV), also commonly referred to as Aujesky disease, can cause acute infection of pigs as well as a wide variety of other domestic species. In many cases, most of the signs are directly related to CNS involvement with death occurring within 1 week p.i. In addition to infecting the CNS and inducing latent infection,[102] virus can be demonstrated in many other organs. One of the primary sites of initial virus multiplication appears to be in the tonsillar area.[103] These studies, however, did not clearly demonstrate whether the virus was replicating in the epithelial cells of the lymphoid nodules of the tonsils or actually in lymphoid tissue. More recent studies indicate that the virus can replicate *in vitro* in both macrophages and lymphocytes obtained from swine[104] (T. Molitor, personal communication). In addition to being able to replicate in pig lung macrophage cultures, the virus appears to be able to infect macrophages *in vivo* and result in immunosuppression to the extent that animals are more susceptible to secondary bacterial infection with *Pasteurella multocida* similar to that observed with BHV-1 (M. Fuentes and C. Pijoin, personal communication).

*In vitro* infection of alveolar macrophages indicates that their function is suppressed as early as 2 hr p.i., well before cytopathology or virus assembly is evident. Furthermore, virulent field strains of the virus are extremely cytotoxic to both macrophages and lymphocytes but do not produce large quantities of infectious virus. By contrast, some mutants, e.g., the Norden vaccine strain, are not very cytotoxic for macrophages or lymphocytes; however, they do produce large quantities of infectious virus ($10^6$–$10^7$ PFU) and eventually affect lymphocyte replication and function. In an attempt to determine whether specific genes of pseudorabies virus are correlated with their ability to infect macrophages and lymphocytes, Molitor (personal communication) has produced deletion mutants of pseudorabies virus and tested their ability to infect macrophages and lymphocytes in culture. Mutants, especially with deletions in the unique short region, do not replicate at all in leukocytes. These results clearly indicate that the leukocytes are susceptible to pseudorabies virus and that specific genes within the virus determine the extent of viral replication and cytotoxicity. These results also indicate that complete viral replication is not required for suppression of function and cytotoxicity of leukocytes.

### 3.3. Avian Herpesviruses

Viruses capable of infecting avian species are listed in Table II. Although a large number of avian herpesviruses have been isolated and investigated with respect to their ability to cause disease in a variety of birds, there is very limited information regarding their interaction with lymphoid tissue. The exceptions are Marek disease virus (MDV), turkey herpesvirus (HVT), and infectious laryngotracheitis (ILT). MDV, like many other herpesviruses, can cause both productive infections and nonproductive infections, depending on the specific cell type involved.[105,106] Furthermore, the final outcome of the disease is related to the strain of virus as well as the genetic susceptibility of the animals. Thus, it is possible to select chickens with specific disease-resistance genes. These genetic resistance genes are associated with the major histocompatibility loci of chickens.[107,108] In addition to the strain of chicken, the age at the time of infection and the strain of virus also determine the final outcome of the disease. HVT does not result in tumors but can protect chickens from virulent Marek disease.[109,110] The strain of virus and chicken will determine the extent of disease, as well as the extent of immunosuppression observed following infection with MDV or HVT.[111,112] In both cases, suppression of mitogen-induced lymphocyte proliferation occurs at 7 days p.i.[112] This initial transient suppression of mitogen-induced lymphoproliferation coincides with various stages of virus lymphoid–tissue interactions. Acute cytolytic infection of lymphoid tissues, notably bursal, thymus, and spleen, becomes apparent at 3 days p.i. and reaches a peak at 5–7 days. This cytolytic infection is so extensive that it results in the dramatic loss of bursal and thymus weight. Virus can infect both T and B cells.[110] Certain strains of MDV can cause mortality during this phase of the disease. Shortly after the initial cytolytic infection, immunologic responses begin returning to normal and can even be enhanced[113] before undergoing a second depression when birds begin developing lymphomas. During the secondary depression of immune responses, deficiencies in humoral and cell-mediated responses to various antigens are depressed,[114] allograph rejection is delayed, and tuberculin hypersensitivity responses are impaired.[114] In birds infected with nononcogenic MDV or HVT, following the initial immunosuppression, the immune responses return to normal or are elevated approximately 2 weeks p.i. These birds are then resistant to virulent MDV.

Although lymphoid destruction occurs during the acute cytolytic stage of infection by MDV, it is unlikely to account entirely for the early immunosuppression, since similar immunosuppression is observed with nononcogenic MDV or HVT.[113] This suppression may be partially caused by one or more viral glycoproteins present on the envelope of the viruses. Support for this suggestion is forthcoming from studies by Wainberg et al.,[115] who demonstrated that in vitro mitogenic responses can be inhibited by either infectious or noninfectious MDV. Thus, viral replication is not a prerequisite for suppression. A second explanation for immunosuppression has been suggested to be attributed to suppressor macrophages present in infected spleens.[116] It is also possible that as a result of cytolytic infection, depletion of the responsive cells results in immunosuppression. Suppressor cells have also been identified in chickens bearing transplantable lymphomas.[117,118] Therefore, it is possible that in the early stages of infection, immunosuppression may occur as a result of direct virus or glycoprotein interaction with lymphoid cells, with subsequent immunosuppression resulting from lymphoma formation and induction of suppressor cells.

Infectious laryngotracheitis is an acute infection of chickens, resulting in severe respiratory distress. Following infection, maximal febrile responses occur approximately 5 days p.i. Peak febrile responses and respiratory distress correlate with maximal lymphopenia.[119–121] In an attempt to explain the possible cause for lymphopenia, a number of investigators have attempted to determine whether leukocytes could be infected by virus. *In vitro* studies clearly indicate that macrophages are susceptible to infection with ILT.[119] However, they also demonstrate that both the cell genotype and the virus genotype are important in determining the extent of viral replication. Thus, in some strains, viral replication is restricted.

An even more interesting observation is that attenuated vaccine strains of ILT replicate more efficiently in macrophages than do virulent virus strains.[119] It must be emphasized that although infection of leukocytes with ILT does occur *in vitro*, there is no evidence for leukocyte infection *in vivo*. Whether this is an indication of the inability of virus to infect cells *in vivo* or whether the infected cells are lost during the purification process remains to be determined. A similar observation has been seen with BHV-1 wherein macrophages can easily be infected *in vitro*, but it is extremely difficult to demonstrate infection *in vivo*.[5] Since many herpesviruses are lymphotrophic, it would not be surprising if some leukocytes could be infected *in vivo* and result in a viremia. Whether leukopenia occurs as a result of direct infection of circulating leukocytes or preventing production of macrophages (bone marrow-derived chicken macrophages are susceptible to infection with ILT virus) remains to be determined. Although infection of macrophages occurs, replication appears to be restricted. Thus, large quantities of virus-specific macromolecules are produced, but virus assembly appears to be blocked in some way, similar to that observed with herpes simplex virus infection of mouse macrophages.[122] Spleen lymphocytes, peripheral blood lymphocytes, thymocytes, bursal lymphocytes, and activated T cells are refractory to infection with ILT virus.[119]

## 3.4. Rodent Herpesviruses

Mouse thymic virus (MTV) is a herpesvirus that causes extensive necrosis of the thymic cortex and medulla 10–14 days p.i.[123] As a result of infection of the thymus, animals have severe temporary impairment of T cell functions. Lymphocytes obtained from these mice fail to undergo stimulation by T-cell mitogens or to participate *in vitro* in antibody responses to T-dependent antigens.[124] This virus is restricted to T cells, since B-cell numbers and their functional ability to make antibody and proliferate when stimulated with B-cell mitogens is unimpaired.[124] Although MTV can infect mice of all ages, the neonatal mouse is the only one that suffers from thymic atrophy, suggesting a specific tropism of the virus for thymocytes of newborn mice. Alternatively, it is possible that mature mice can mount a very effective and rapid immune response against the virus, thereby restricting viral replication to the salivary glands.[123] Infection of mice with MTV also results in a marked decrease in the capacity of thymocytes to respond to alloantigens in mixed lymphocyte reactions and to generate CTL while sparing other T-cell functions. These studies suggest that MTV selectively affects subpopulations of T cells. The virus has no effect on the evolution of the capacity of B cells to respond to T-cell-independent antigens. Whether persistence of virus in salivary glands has any relationship to the initial infection or alteration of T-cell immunity remains to be determined. Although cellular damage produced by MTV infection affects normal T-cell functions during the acute phase, these functions return to normal 6–8 weeks later. However, these animals failed to produce a humoral antibody response to MTV. Whether these animals became tolerant to the virus or whether suppressor T-cell populations, which are elevated following MTV infection,[123] resulted in depression of antibody production remains to be determined. One of the most interesting aspects of MTV infection is that, despite dramatic thymic destruction with suppressed immunologic functions, many of these functions are transient, whereas some may become permanent. This is generally a rare occurrence in most herpesviruses.

Guinea pig herpes-like virus (GPHV), like other rodent viruses, has a predilection for lymphocytic tissue present in the spleen, with virus infecting both T and B cells as well as macrophages during acute infection.[125–128] In addition to infection of most leukocyte cell types, the virus can also induce latency in these cells.[125,128] In contrast to many other herpesviruses in which there is lymphopenia, GPHV causes an increase in lymphocyte numbers.[126] Studies indicate that this increase is predominantly of the T-cell subpopulation.

Herpesvirus sylvilagus is a natural pathogen of cottontail rabbits. Other rabbit species cannot be infected with this virus. Following infection of cottontail rabbits, there is extensive leukocytosis, splenomegaly, lymphoadenopathy, and virus can be isolated from circulating mononuclear cells. In animals infected with the virus, there is an elevation of up to 2.5 times the normal mononuclear blood cell counts, as well as increased numbers of spleen cells (up to six-fold increase) within 6 days p.i., with spleen size increasing approximately 10-fold. Approximately 0.2% of all cells could develop infectious center assays 2 weeks p.i., indicating that lymphoid cells are the major site of replication of the virus.[130] Virus continues to persist in these cells for extended periods of time postinfection. In contrast to GPHV and mouse CMV, herpesvirus sylvilagus does not appear to infect macrophages.[129] The increase in lymphocyte numbers following infection with the virus is predominantly associated with a T-cell population.[128] Thus, in the rodent viruses, there is a spectrum of interactions; mouse CMV predominantly infects mononuclear cells, MTV predominantly infects thymic cells, and GPHV and herpesvirus sylvilagus infect both T and B lymphocytes *in vivo* and *in vitro*.[125–127,129] As a result of these different interactions, viruses can persist in leukocytes and establish latent infection or lead to transformation and result in lymphoproliferative disorders. All these rodent herpesvirus interactions can serve as excellent models to provide an understanding of herpesvirus persistence, transformation, and immunomodulation. These rodent models also clearly emphasize that although they can interact with lymphoid cells, the interactions are very specific concerning the virus and host in question. In some cases, these are extremely restrictive to a specific species or genotype of virus and host, whereas in other instances the virus can infect a large number of different cell types from a wide variety of different species. Whether this indicates a specific state of activation of lymphoid cells *in vivo* or whether there are indeed restrictions due to surface structures responsible for virus attachment or potential transport and uncoating within the cell remains to be determined.

## 3.5. Varicella Zoster

Until recently, it often has been difficult to demonstrate a viremic phase following natural infection with varicella zoster virus (VZV). However, the hematogenous dissemination of VZV clearly has been demonstrated in immunocompromised patients. More recently, hematogenous spread has been shown in healthy children. Virus has been recovered from both adherent and nonadherent cells obtained from patients suffering from acute VZV infection.[130,131] Whether this indicates that virus can infect both lymphocytes and monocytes requires further characterization using monoclonal antibodies directed against various lymphocytic subpopulations, to determine whether indeed lymphocytes can be infected with VZV. Ozaki *et al.*[131] suggested that since there is decreased lymphocyte stimulation during VZV infection, there is a lymphocyte-associated viremic stage at the onset of varicella infection. Although this is highly possible, one must clearly demonstrate that the decrease in lymphocyte activity is directly related to viral infection of these lymphocytes. Thus, infection of monocytes by the virus may lead to lymphocyte dysfunction. It has been shown that macrophages infected with VZV can lead to reduced mixed lymphocyte reactions.[132] Thus, at least in some instances, virus-related suppression of lymphocyte responses *in vitro* may be due to dysfunction of macrophages rather than of lymphocytes directly. Twomey's group also suggested that the anergy often observed during VZV infection is correlated very closely with viral infection of macrophages.[132] It is known that monocytes can have a suppressive effect on lymphocyte stimulation responses.[133] In addition, it has been shown that even removal of phagocytic cells did not remove immunosuppression following infection with varicella. Twomey *et al.* clearly demonstrated that although the percentage of T lymphocytes was unchanged during the infectious process, there was an alteration in the helper to suppressor/cytotoxic subsets with the suppressor/cytotoxic cells increasing, resulting in a lower ratio between the two subsets.[133] This alteration is similar to that observed with other herpesviruses of humans. Whether preactivation of lymphocytes *in vitro* leads to nonresponsiveness in PHA or pokeweed mitogen stimulation responses remains to be determined. It is also possible that infection of macrophages *in vitro* may be at least partially responsible for the transient anergy observed during viral infections.

## REFERENCES

1. Abramson, J. S., G. S. Gibrik, E. L. Mills, and P. Quie, Polymorphonuclear leukocyte dysfunction during influenza virus infection in cinchillas, *J. Infect. Dis.* **143**:836–846 (1981).
2. Astry, G. L. and G. J. Jakab, Influenza virus induced immune complexes suppress macrophage phagocytosis, *J. Virol.* **50**:287–292 (1984).

3. Wainberg, M. A., S. Vydelingum, and R. G. Margolese, Viral inhibition of lymphocyte mitogenesis: Interference with synthesis and functionally active T cell growth factor (TCGF) activity and reversal of inhibition by the addition of the same, *J. Immunol.* **130**:2372–2378 (1983).
4. Casali, P., G. P. Price, and M. B. A. Oldstone, Viruses disrupt functions of human lymphocytes: Effects of measles virus and influenza virus on lymphocyte-mediated killing and antibody production, *J. Exp. Med.* **159**:1322–1377 (1984).
5. Forman, A. J., and L. A. Babiuk, Effect of infectious bovine rhinotracheitis virus infection on bovine alveolar macrophage function, *Infect. Immun.* **35**:1041–1047 (1982).
6. Jakab, G. J., Viral bacterial interactions in respiratory tract infections: A review of the mechanisms of virus-induced suppression of pulmonary antibacterial defenses, in: *Bovine Respiratory Disease, A Symposium* (R. W. Loon, ed.), pp. 223–286, Texas A & M University Press, College Station, Texas (1984).
7. Crabtree, G. R., S. Gillis, K. A. Smith, and A. Munck, Mechanisms of glucocorticoid-induced immunosuppression: Inhibitory effects on expression of Fc receptors and production of T cell growth factors, *J. Steroid Biochem.* **12**:445–449 (1980).
8. Egwang, T. G., and A. D. Befus, The role of complement in the induction and regulation of the immune response, *Immunology* **51**:207–224 (1984).
9. Walker, C., F. Kristensen, F. Bettens, and A. L. deWeck, Lymphokine regulation of activated (G1) lymphocytes. I. Prostaglandin $E_2$-induced inhibition of interleukin 2 production, *J. Immunol.* **130**:1770–1773 (1983).
10. Jaffe, M. I., and A. R. Rabson, Defective helper factor (LMF) production in patients with acute measles infection, *Clin. Immunol. Immunopathol.* **20**:215–223 (1981).
11. McGuire, R. L., and L. A. Babiuk, Evidence for defective neutrophil functions in calves exposed to infectious bovine rhinotracheitis, *Vet. Immunol. Immunopathol.* **5**:259–271 (1984).
12. Schorlemmer, H. U., and A. C. Allison, Effects of activated complement components on enzyme secretion by macrophages, *Immunology* **31**:781–788 (1976).
13. Von Pirquet, C. E., Das verhalten der kautanen tuberculinreaktion wahrend der masern, *Dtsch. Med. Wopchenschi Wochenschrl.* **34**:1297–1300 (1908).
14. Gibbs, E. P. J. and M. M. Reyweyemamu, Bovine herpesvirus. I. Bovine herpesvirus-1, *Vet. Bull.* **47**:317–323 (1977).
15. Homan, E. J., and B. C. Easterday, Isolation of bovine herpesvirus-1 from trigeminal ganglia of clinically normal cattle, *Am. J. Vet. Res.* **41**:1212–1216 (1980).
16. Baxter, G. M., Neonatal meningoencephalitis associated with IBR virus, *Bovine Pract.* **19**:41–44 (1984).
17. Hill, B. D., M. W. M. Hill, Y. S. Chung, and R. J. Whittle, Meningoencephalitis in calves due to bovine herpesvirus type 1 infection, *Aust. Vet. J.* **61**:242–243 (1984).
18. Babiuk, L. A., R. C. Wardley, and B. T. Rouse, Defense mechanisms against bovine herpesviruses: Relationship of virus–host cell events to susceptibility to antibody-complement lysis, *Infect. Immun.* **12**:958–963 (1975).
19. Rouse, B. T., and L. A. Babiuk, Host defence mechanisms against infectious bovine rhinotracheitis virus. I. *In vitro* stimulation of sensitized lymphocytes by virus antigen, *Infect. Immun.* **10**:681–687 (1975).
20. Rouse, B. T. and L. A. Babiuk, Host defence mechanisms against infectious bovine rhinotracheitis virus. II. Inhibition of viral plaque formation by immune peripheral blood lymphocytes, *Cell. Immunol.* **17**:43–46 (1975).
21. Rouse, B. T., and L. A. Babiuk, Host defence mechanisms against infectious bovine rhinotracheitis virus. III. Isolation and immunological activities of bovine T lymphocytes, *J. Immunol.* **113**:1391–1398 (1975).
22. Rouse, B. T., and L. A. Babiuk, The direct antiviral cytotoxicity by bovine lymphocytes is not restricted by genetic incompatibility of lymphocytes and target cells, *J. Immunol.* **118**:618–624 (1977).
23. Wardley, R. C., B. T. Rouse, and L. A. Babiuk, Antibody dependent cytotoxicity medi-

ated by neutrophils as a possible mechanism of antiviral defense, *J. Reticuloendothel. Soc.* **19:**323–332 (1976).

24. Rouse, B. T., L. A. Babiuk, and Y. Fujimiya, Enhancement of antibody dependent cell cytotoxicity of herpesvirus infected cells by complement, *Infect. Immun.* **18:**660–665 (1977).

25. Rouse, B. T., A. S. Grewal, and L. A. Babiuk, Complement enhances antiviral antibody dependent cell cytotoxicity, *Nature (Lond.)* **226:**456–458 (1977).

26. Grewal, A. S., B. T. Rouse, and L. A. Babiuk, Mechanisms of recovery from viral infections: Destruction of infected cells by neutrophils and complement, *J. Immunol.* **124:**312–319 (1980).

27. Grewal, A. S., and L. A. Babiuk, Complement dependent polymorphonuclear neutrophil mediated cytotoxicity of herpesvirus infected cells: Possible mechanisms of cytotoxicity, *Immunology* **40:**151–161 (1980).

28. Babiuk, L. A., and B. T. Rouse, Interaction between effector cell activity and lymphokines: Implications for recovery from herpesvirus infections, *Int. Arch. Allergy App. Immunol.* **57:**62–73 (1978).

29. Babiuk, L. A., H. Bielefeldt Ohmann, G. Gifford, C. Czarniecki, V. T. Scialli, and E. B. Hamilton, Effect of bovine-1 interferon on bovine herpesvirus-1 induced respiratory disease, *J. Gen. Virol.* **66:**2838–2894 (1985).

30. Bielefeldt Ohmann, H., and L. A. Babiuk, Viral bacterial pneumonia in calves: Effect of bovine herpesvirus-1 on immunological functions, *J. Infect. Dis.* **151:**937–947 (1985).

31. Babiuk, L. A., and B. T. Rouse, Immune interferon production by lymphoid cells: Role in the inhibition of herpesvirus, *Infect. Immun.* **13:**1567–1578 (1976).

32. Splitter, G. A., L. Eskra, M. Miller-Edge, and J. L. Splitter, Bovine herpesvirus: Interactions between animal and virus, *CRC Comparative Pathobiology of Viral Diseases,* Vol. 1, pp. 57–88 (R. G. Olsen, S. Krakawka, and J. R. Blakesley, eds.), CRC Press, Boca Raton, Florida (1984).

33. Gershon, R. K., and K. Kondo, Cell interactions in the induction of tolerance: The role of thymic lymphocytes, *Immunology* **18:**723–737 (1970).

34. Rich, R. R., and C. W. Pierce, Biological expressions of lymphocyte activation: Effects of phytomytogens on antibody synthesis *in vitro, J. Exp. Med.* **137:**205–223 (1973).

35. Tada, T., and T. Takemori, Selective roles of thymus derived lymphocytes in the antibody response. I. Differential suppressive effect of carrier-primed T cells on hapten-specific IgM and IgG antibody responses, *J. Exp. Med.* **140:**239–252 (1974).

36. Klein, J., *Immunology the Science of Self-Nonself Discrimination,* Wiley, New York (1982).

37. Herbert, A. G., and J. D. Watson, T-cell ontogeny: The role of a stimulator suppressor cell, *Immunol. Today* **7:**72–76 (1986).

38. Bensussan, A., O. Acuto, R. E. Hussey, C. Milanese, and E. L. Reinherz, T3–Ti receptor triggering of T8+ suppressor T cells leads to immunoresponsiveness to interleukin-2, *Nature (Lond)* **311:**565–567 (1984).

39. Kasakura, S., M. Taguchi, Y. Watanabe, T. Okubo, T. Murachi, H. Uchino, and M. Hanaoka, Suppressor cell induction factor: A new mediator released by stimulated human lymphocytes and distinct from previously described lymphokines, *J. Immunol.* **130:**2720–2726 (1983).

40. Horohov, D. W., J. H. Wyckoff, R. N. Moore, and B. T. Rouse, Regulation of herpesvirus simplex virus-specific cell-mediated immunity by a specific suppressor factor, *J. Virol.* **58:**331–338 (1986).

41. Bielefeldt Ohmann, H., L. G. Filion, and L. A. Babiuk, Bovine monocytes and macrophages: An accessory role in suppressor-cell generation by ConA and in lectin-induced proliferation, *Immunology* **50:**189–197 (1983).

42. Filion, L. G., R. L. McGuire, and L. A. Babiuk, Non-specific suppressive effect of bovine herpesvirus-1 on bovine leukocyte functions, *Infect. Immun.* **42:**106–112 (1983).

43. Ceredig, R., and G. Corradin, High antigen concentration inhibits T cell proliferation

but not interleukin 2 production: Examination of limiting dilution microcultures and T cell clones, *Eur. J. Immunol.* **16**:30–34 (1986).

44. Miller-Edge, M. A., T-cell mediated immune responses to bovine herpesvirus 1: The role of interleukin-2, in: *Characterization of the Bovine Interferon Immune System and the Genes Regulating Expression of Immunity with Particular Reference to Their Role in Disease Resistance: A Symposium* (W. C. Davis, J. N. Shelton, and C. W. Weems, eds.), pp. 99–118, Washington State University Press, Pullman, Washington (1985).

45. Rubin, L. A., C. C. Kurman, M. E. Fritz, W. E. Biddison, B. Boutin, R. Yarchoan, and D. L. Nelson, Soluble interleukin 2 receptors are released from activated human lymphoid cells *in vitro*, *J. Immunol.* **135**:3172–3177 (1985).

46. Splitter, G. A., and L. Eskra, Bovine T lymphocyte response to bovine herpesvirus-1: Cell phenotypes, viral recognition and acid-labile interferon production, *Vet. Immunol. Immunopathol.* **11**:235–250 (1986).

47. Woodruff, J. F., and J. J. Woodruff, The effect of viral infections on the function of the immune system in: *Viral Immunology and Immunopathology* (A. L. Notkins, ed.), pp. 393–418, Academic, New York (1975).

48. Babiuk, L. A., and H. Bielefeldt Ohmann, Bovine herpesvirus-1 (BHV-1) infection in cattle as a model for viral induced immunosuppression, in: *Viral Mechanisms of Immunosuppression* (M. Kende, J. Gainer, and H. Chirigos, eds.), pp. 99–114, Liss, New York (1985).

49. Braun, R. W., H. K. Teute, H. Kirchner, and K. Munk, Replication of herpes simplex virus in human T lymphocytes: Characterization of the viral target cell, *J. Immunol.* **132**:914–919 (1984).

50. Isfort, R. J., R. A. Stringer, H. J. Kung, and L. F. Velicer, Synthesis, processing and secretion of Marek's disease herpesvirus A antigen glycoprotein, *J. Virol.* **57**:464–474 (1986).

51. Maes, R. K., S. L. Fritsch, L. L. Herr, and P. A. Rota, Immunogenic proteins of feline rhinotracheitis virus, *J. Virol.* **51**:259–262 (1984).

52. Okuno, T., K. Yamanishi, K. Shiraki, and M. Takahashi, Synthesis and processing of glycoproteins of varicella-zoster virus (VZV) as studies with monoclonal antibodies to VZV antigens, *Virology* **129**:357–368 (1983).

53. Rea, T. J., J. G. Timmins, G. W. Lony, and L. E. Post, Mapping and sequence of the gene for the pseudorabies virus glycoprotein which accumulates in the medium of infected cells, *J. Virol.* **54**:21–29 (1985).

54. Letchworth, G. J. III, and L. E. Carmichael, The effect of temperature on production and function of bovine interferons, *Arch. Virol.* **82**:211–221 (1982).

55. Aune, T. M., and C. W. Pierce, Identification and initial characterization of a nonspecific suppressor factor produced by soluble immune response suppressor-treated macrophages, *J. Immunol.* **127**:1828–1833 (1981).

56. Schnaper, H. W., Identification and initial characterization of concanavalin A- and interferon-induced human suppressor factors: Evidence for a human equivalent of murine soluble immune response suppressor (SIRS), *J. Immunol.* **132**:2429–2436 (1984).

57. Lipscomb, M. F., G. B. Toews, G. R. Lyons, and J. W. Uhr, Antigen presentation by guinea pig alveolar macrophages, *J. Immunol.* **126**:286–292 (1981).

58. Towes, G. B., W. C. Vial, and M. M. Dunn, The accessory cell function of human alveolar macrophages in specific T cell proliferation, *J. Immunol.* **132**:181–186 (1984).

59. Slauson, D. O., The mediation of pulmonary inflammatory injury, *Adv. Vet. Sci. Compar. Med.* **26**:99–120 (1982).

60. Bielefeldt Ohmann, H., and L. A. Babiuk, Alteration of some leukocyte functions following *in vivo* and *in vitro* exposure to recombinant bovine alpha- and gamma- interferon, *J. Interferon Res.* **6**:123–136 (1986).

61. Holt, P. G., Alveolar macrophages. II. Inhibition of lymphocyte proliferation by peripheral macrophages from rat lung, *Immunology* **37**:429–436 (1979).

62. Holt, P. G., Alveolar macrophages. III. Studies on the mechanism of inhibition of T-cell proliferation, *Immunology* **37**:437–445 (1979).

63. Yeager, H., Jr., J. A. Sweeney, H. B. Herscowitz, I. S. Barsaum, and E. Kagan, Modulation of mitogen induced proliferation of autologous peripheral blood lymphocytes by human alveolar macrophages, *Infect. Immun.* **38**:260–266 (1982).

64. Lawrence, E. C., G. J. Theodore, and R. R. Martin, Modulation of pokeweed mitogen induced immunoglobulin secretion by human bronchoalveolar cells, *Am. Rev. Respir. Dis.* **126**:248–252 (1982).

65. Twomey, J. J., A. Laughter, and M. F. Brown, A comparison of the regulatory effects of human monocytes, pulmonary alveolar macrophages (PAMs) and spleen macrophages upon lymphocyte responses, *Clin. Exp. Immunol.* **52**:449–545 (1983).

66. Rinehart, J. J., M. Orser, and M. E. Kaplan, Human monocyte and macrophage modulation of lymphocyte proliferation, *Cell. Immunol.* **44**:131–143 (1979).

67. Rice, L., A. H. Laughter, and J. J. Twomey, Three suppressor systems in human blood that modulate lymphoproliferation, *J. Immunol.* **122**:991–996 (1979).

68. Liu, M. C., Human lung macrophages enhance and inhibit lymphocyte proliferation, *J. Immunol.* **132**:2895–2904 (1984).

69. Bendixen, P. H., P. E. Shewen, and B. N. Wilkie, The influence of bovine alveolar macrophages on the blastogenic response of peripheral blood mononuclear cells, in: *The Ruminant Immune System* (J. E. Butler, ed.), p. 814, Plenum, New York (1981).

70. Goldyne, M. E., and J. D. Stobo, Immunoregulatory role of prostaglandins and related lipids, *CRC Crit. Rev. Immunol.* **2**:189–242 (1981).

71. Hasler, F., H. G. Bluestein, N. J. Zvaifler, and L. B. Epstein, Analysis of the defects responsible for impaired regulation of EBV-induced B cell proliferation by rheumatoid arthritis lymphocytes. II. Role of monocytes and increased sensitivity of rheumatoid arthritis lymphocytes to prostaglandin E, *Eur. J. Immun.* **131**:768–774 (1983).

72. Kunkel, S. L., D. A. Cambell, S. W. Chensue, and G. I. Higashi, Species-dependent regulation of monocyte–macrophage Ia antigen expression and antigen presentation by prostaglandin E, *Cell. Immunol.* **97**:140–152 (1986).

73. Dinarello, C. A., Human interleukin-1. The importance of its multiple activities for immunoregulation, in: *Sixth International Congress on Immunology* (B. Cinader and R. G. Miller, eds.), p. 41, Academic Press, Toronto (1986).

74. Albrightson, G. R., N. L. Baenziger, and P. Needleman, Exaggerated human vascular cell prostaglandin biosynthesis mediated by monocytes: Role of monokines and interleukin 1, *J. Immunol.* **135**:1872–1877 (1985).

75. Chouaib, S., L. Chatenoud, D. Klatzmann, and D. Fradelizi, The mechanisms of inhibition of human IL-2 production II. PGE-2 induction of suppressor T lymphocytes, *J. Immunol.* **132**:1851–1857 (1984).

76. Oppenheim, J. J., E. J. Kovacs, K. Matsushima, and S. K. Durum, There is more than one interleukin 1, *Immunol. Today* **7**:45–56 (1986).

77. Roberts, N. J., A. H. Prill, and T. N. Mann, Interleukin 1 and interleukin 1 inhibitor production by human macrophages exposed to influenza virus or respiratory syncytial virus. Respiratory syncytial virus is a potent inducer of inhibitory activity, *J. Exp. Med.* **163**:511–520 (1986).

78. Weitzman, S. A., and T. P. Stossel, Mutation caused by human phagocytes, *Science* **212**:546–547 (1981).

79. Carson, D. A., S. Seto, and D. B. Wasson, Lymphocyte dysfunction after DNA damage by toxic oxygen species: A model of immunodeficiency, *J. Exp. Med.* **163**:746–752 (1986).

80. Rodrick, M. L., I. B. Lamster, S. T. Sonis, S. G. Pender, A. B. Kolodkin, J. E. Fitzgerald, and R. E. Wilson, Effect of supernatants of polymorphonuclear neutrophils recruited by different inflammatory substances on mitogen responses of lymphocytes, *Inflammation* **6**:1–11 (1982).

81. Yoshinaga, M., K. Nishime, S. Nakamura, and F. Goto, A PMN-derived factor that enhances DNA-synthesis in PHA or antigen-stimulated lymphocytes, *J. Immunol.* **124**:94–99 (1980).

82. Hsu, C. C., M. B. Wu, and J. Rivera-Arcilla, Inhibition of lymphocyte reactivity *in vitro* by autologous polymorphonuclear cells (PMN), *Cell Immunol.* **48**:288–295 (1979).

83. Starke, I. D., Granulocyte content and titrated thymidine uptake of mononuclear cells, preparations from patients with ovarian cancer, *Clin. Oncol.* **8**:243–249 (1982).

84. Judson, D. G., and J. B. Dixon, Depression of lymphocyte reactivity by granulocytes in equine whole blood culture, *Vet. Immunol. Immunopathol.* **8**:289–295 (1985).

85. Tiku, K., M. L. Tiku, S. Liu, and J. L. Skosey, Normal human neutrophils are a source of a specific interleukin 1 inhibitor, *J. Immunol.* **136**:3686–3692 (1986).

86. Blecha, F., and P. E. Baker, Effect of cortisol *in vitro* and *in vivo* on production of bovine interleukin 2, *Am. J. Vet. Res.* **47**:841–845 (1986).

87. Emery, D. L., J. H. Duffy, and P. R. Wood, An analysis of cellular proliferation and production of lymphokines and specific antibody *in vitro* by leukocytes from immunized cattle, in: *First International Veterinary Immunology Symposium* (B. Wilkie, P. E. Shewan, K. Nielson, J. R. Duncan, and B. W. Stemshorn, eds.), p. 57, University of Guelph Press, Guelph, Ontario (1986).

88. Blalock, J. E., Production and action of lymphocyte-derived neuroendocrine peptide hormones: A summary, in: *Sixth International Congress on Immunology* (B. Cinader and R. G. Miller, eds.), p. 48, Academic Press, Toronto (1986).

89. Blecha, F., and H. C. Minocha, Suppressed lymphocyte blastogenic responses and enhanced *in vitro* growth of bovine rhinotracheitis virus in stressed feeder calves, *Am. J. Vet. Res.* **44**:2145–2148 (1983).

90. Payan, D. G., and E. J. Goetzl, Modulation of lymphocyte function by sensory neuropeptides, in: *Neuromodulation of Immunity and Hypersensitivity, J. Immunol.* **135**(suppl.):783–786 (1985).

91. Payan, D. G., J. D. Levine, and E. J. Goetzl, Modulation of immunity and hypersensitivity by sensory neuropeptides, *J. Immunol.* **132**:1601–1604 (1984).

92. Smith, E. M., D. Harbour-McMenamin, and E. J. Blalock, Lymphocyte production of endorphins and endorphin-mediated immunoregulatory activity, in: *Neuromodulation of Immunity and Hypersensitivity, J. Immunol.* **135**(suppl.):779–782 (1985).

93. Johnson, H. M., E. M. Smith, B. A. Torres, and J. E. Blalock, Neuroendocrine hormone regulation of an *in vitro* antibody production, *Proc. Natl. Acad. Sci. USA* **79**:4171–4174 (1982).

94. Brock, J. H., and M. DeSousa, Immunoregulation by iron-binding proteins, *Immunol. Today* **7**:30–31 (1986).

95. Lum, J. F., A. J. Infante, D. M. Makker, F. Yang, and B. H. Bowman, Transferrin synthesis by inducer T lymphocytes, *J. Clin. Invest.* **77**:841–850 (1986).

96. Matzner, Y., C. Hershko, A. Polliack, A. M. Konijn, and G. Izak, Suppressive effect of ferritin on *in vitro* lymphocyte function, *Br. J. Haematol.* **42**:345–353 (1979).

97. Gleeson, L. J., and L. Coggins, Equine herpesvirus type 2: Cell–virus relationship during persistent cell associated viremia, *Am. J. Vet. Res.* **46**:19–23 (1986).

98. Dutta, S. K., and A. C. Myrup, Infectious center assay of intracellular virus and infective virus titer for equine mononuclear cells infected *in vivo* and *in vitro* with equine herpesvirus, *Can. J. Compar. Med.* **47**:64–69 (1983).

99. Coignoul, F. L., T. A. Bertram, and N. F. Cheville, Functional and ultrastructural changes in neutrophils from mares and foals experimentally inoculated with a respiratory tract strain of equine herpesvirus-1, *Am. J. Vet. Res.* **45**:1972–1975 (1984).

100. Dutta, S. K., Myrup, A. and M. K. Bumgardner, Lymphocyte responses to virus and mitogen in ponies during experimental infection with equine herpesvirus-1, *Am. J. Vet. Res.* **41**:2066–2068 (1980).

101. Scott, J. C., S. K. Dutta, and A. C. Myrup, *In vivo* harboring of equine herpesvirus-1 in

leukocyte populations and subpopulations and their quantitation from experimentally infected ponies, *Am. J. Vet. Res.* **44:**1344–1348 (1983).

102. Wittman, G., V. Ohlinger, and H. J. Rziha, Occurrence and reactivation of latent Aujesky's disease virus following challenge in previously vaccinated pigs, *Arch. Virol.* **75:**29–41 (1983).

103. Narita, M., S. Inui, and Y. Shimizu, Tonsillar changes in pigs given pseudorabies (Aujesky's disease) virus, *Am. J. Vet. Res.* **45:**247–251 (1984).

104. Smid, B., L. Valick, and A. Sabol, Morphogenesis of Aujesky's disease virus in pig lung macrophages, *Acta Vet. Brno* **50:**79–87 (1981).

105. Jakowski, R. M., T. N. Fredrickson, T. W. Chomiak, and R. E. Luginbuhl, Hematopoietic destruction in Marek's disease, *Avian Dis.* **14:**374–383 (1970).

106. Witter, R. L., J. M. Sharma, and A. M. Fadley, Pathogenicity of variant Marek's disease virus isolates in vaccinated and unvaccinated chickens, *Avian Dis.* **24:**210–219 (1980).

107. Longenecker, B. M., F. Pazderka, J. S. Gavora, J. L. Spencer, E. A. Stephens, and R. L. Witter, Role of major histocompatability complex in resistance to Marek's disease: Restriction of growth of JMV-MD tumor cells in genetically resistant birds, *Adv. Exp. Med. Biol.* **88:**287–296 (1977).

108. Briles, W. E., R. W. Briles, W. H. McGibbon, and H. A. Stone, Identification of B alloalleles associated with resistance to Marek's disease, pp. 395–416, in: *Resistance and Immunity to Marek's Disease* (P. M. Biggs, ed.), CEC Publication Eur 6470, Luxembourg (1982).

109. Calnek, B. W., J. C. Carlisle, J. Fabricant, K. K. Murthy, and K. A. Schat, Comparative pathogenesis studies with oncogenic and nononcogenic Marek's disease virus and turkey herpesvirus, *Am. J. Vet. Res.* **40:**541–548 (1979).

110. Schierman, L. W.. G. A. Theis, and R. A. McBride, Preservation of a T cell mediated immune response in Marek's disease virus-infected chickens by vaccination with a related virus, *J. Immunol.* **116:**1497–1507 (1976).

111. Powell, P. C., *In vitro* stimulation of blood lymphocytes by phytohemagglutinin during the development of Marek's disease, *Avian Pathol.* **9:**471–485 (1980).

112. Shek, W. R., B. W. Calnek, K. A. Schat, and C. H. Chen, Characterization of Marek's disease virus-infected lymphocytes: Discrimination between cytolytically and latently infected cells, *J. Natl. Cancer Inst.* **70:**485–491 (1983).

113. Lee, L. F., J. M. Sharma, K. Nazerian, and R. L. Witter, Suppression and enhancement of mitogen response in chickens infected with Marek's disease virus and herpesvirus of turkeys, *Infect. Immun.* **21:**474–479 (1978).

114. Schierman, L. W., G. A. Theis, and R. A. McBride, Preservation of a T cell mediated immune response in Marek's disease virus-infected chickens by vaccination with a related virus, *J. Immunol.* **116:**1497–1499 (1976).

115. Wainberg, M. A., B. Beiss, and E. Israel, Viral-mediated abrogation of chicken lymphocyte responsiveness to mitogenic stimulus, *Avian Dis.* **24:**580–591 (1980).

116. Lee, L. F., J. M. Sharma, K. Nazerian, and R. C. Witter, Suppression of mitogen induced proliferation of normal spleen cells by macrophages from chickens inoculated with Marek's disease virus, *J. Immunol.* **120:**1554–1559 (1978).

117. Theis, G. A., Subpopulations of suppressor cells in chickens infected with cells of a transplantable lymphoblastic leukemia, *Infect. Immun.* **34:**526–534 (1981).

118. Theis, G. A., Effect of lymphocytes from Marek's disease infected chickens on mitogenic responses of syngeneic normal chicken spleen cells, *J. Immunol.* **118:**887–894 (1977).

119. Calnek, B. W., K. J. Fahey, and T. J. Bagust, *In vitro* infection studies with infectious laryngotracheitis virus, *Avian Dis.* **30:**327–336 (1986).

120. Chang, P. W., F. Sculco, and V. J. Yates, An *in vivo* and *in vitro* study of infectious laryngotracheitis virus in chicken leukocytes, *Avian Dis.* **21:**492–500 (1977).

121. Bagust, T. J., B. W. Calnek, and K. J. Fahey, Gallid-1 herpesvirus infection in the

chicken. 3. Reinvestigation of the pathogenesis of infectious laryngotracheitis in acute and early post-acute disease, *Avian Dis.* **30:**179–190 (1986).

122. Stevens, J. G., and M. L. Cook, Restriction of herpes simplex virus by macrophages: An analysis of cell-virus interaction, *J. Exp. Med.* **13:**19–38 (1971).

123. Cross, S. S., J. C. Parker, W. C. Rowe, and M. L. Robbins, Biology of mouse thymic virus, a herpesvirus of mice and the antigenic relationship to mouse cytomegalovirus, *Infect. Immun.* **26:**1186–1195 (1979).

124. Cohen, P. L., S. S. Cross, and D. E. Mosier, Immunologic effects of neonatal infection with mouse thymic virus, *J. Immunol.* **115:**706–710 (1975).

125. Griffith, B. P. and G. D. Hsiung, Persistence and expression of herpesvirus in guinea pig B and T cells, *Proc. Soc. Exp. Biol. Med.* **162:**202–206 (1978).

126. Dowler, K. W., S. McCormick, J. A. Armstrong, and G. D. Hsiung, Lymphoproliferative changes induced by infection with a lymphotropic herpesvirus of guinea pigs, *J. Infect. Dis.* **150:**105–111 (1984).

127. Tenser, R. B., and G. D. Hsiung, Infection of thymus cells *in vivo* and *in vitro* with a guinea pig herpes-like virus and the effect of antibody on virus replication in organ culture, *J. Immunol.* **110:**552–560 (1973).

128. Gonzalez-Serva, A., and G. D. Hsiung, Expression of herpesvirus in adherent cells from bone marrow of latently infected guinea pigs, *Am. J. Pathol.* **9:**483–496 (1978).

129. Kramp, W. J., P. Medveczky, C. Mulder, H. C. Hinze, and J. L. Sullivan, Herpes sylvilagus infects both B and T lymphocytes *in vivo*, *J. Virol.* **56:**60–65 (1985).

130. Arbeit, R. D., J. A. Zaia, M. A. Valerio, and M. J. Levin, Infection of human peripheral blood mononuclear cells by varicella zoster virus, *Intervirology* **18:**56–65 (1982).

131. Ozaki, T., T. Ichikawa, Y. Matsui, H. Konda, T. Nagain, Y. Asano, K. Yamanishi, and M. Takahashi, Lymphocyte-associated viremia in varicella, *J. Med. Virol.* **19:**249–253 (1986).

132. Twomey, J. J., F. Gyorkey, and S. M. Norris, The monocyte disorder with herpes zoster, *J. Exp. Clin. Med.* **83:**768–777 (1974).

133. Arneborn, P. and G. Biberfeld, T lymphocyte subpopulation in relation to immunosuppression in measles and varicella, *Infect. Immun.* **39:**29–37 (1983).

# 9

# Poxviruses

## DAVID S. STRAYER

## 1. INTRODUCTION

The relationship of poxviruses and the immune system is the oldest recorded association of its kind in medical literature. It began in 1798, when Jenner showed that immunity to one orthopoxvirus, cowpox, prevented the development of disease due to a related orthopoxvirus, variola, or smallpox. Initially, the use of cowpox inoculation to prevent smallpox was not universally successful. Other less benign vaccination techniques, including attenuated smallpox, were used. Finally, the introduction of vaccinia virus vaccination for smallpox and its widespread use resulted in the elimination of variola virus from the list of human scourges a decade ago.

Despite the historical intimacy between immunology and poxvirus virology, and despite a wealth of information on the genetics and biochemistry of poxvirus replication, little is known about the effects of poxviruses on the immune system. This chapter reviews the current understanding of the effects of poxvirus infections on immunologic function and of the means whereby poxviruses alter lymphocyte function.

## 2. PROPERTIES OF POXVIRUSES

### 2.1. Physical Characteristics

Poxviruses are the largest, and probably most complex, of all viruses. They are roughly ovoid and vary from 195–400 nm in length and 115–285 nm in width. They are double-stranded DNA viruses with total genome weights rang-

DAVID S. STRAYER • Department of Pathology and Laboratory Medicine, University of Texas Health Science Center, Houston, Texas 77030

ing up to 150 MDa. Poxvirus DNA is circular, the ends of the two complementary strands of DNA being linked at their terminals by hairpin loops.

Structurally, they are composed of an outer envelope derived from the plasma membrane of infected cells. This envelope is present only on extracellular virions, as it is formed during the budding process. An inner envelope is also present. It is topographically marked by surface tubular elements that vary in size and arrangement, depending on the virus. Internal to this are a variable number of lateral bodies of unknown function and the virus nucleic acid core. The complexity of this arrangement is still being elucidated.

## 2.2. Genetics

To date, more than 110 virion proteins have been found in vaccinia virus. The functions of these various proteins are mostly unknown. However, a great deal of work has been done on some of them. The best studied of these is the DNA-dependent RNA polymerase, which is required for viral gene transcription, as poxviruses appear to remain entirely in the cytoplasm of infected cells. Endonucleases and exonucleases have also been found associated with the virions of poxviruses. Many other enzymes, as well as nonenzymatic proteins and glycoproteins, have been identified in vaccinia and other poxviruses. This literature was well reviewed recently by Dales and Pogo[1] and by Moss.[2]

Perhaps the most intriguing of the known gene products of poxviruses is the 19-kDa protein, which acts as an agonist at the receptor level for epidermal growth factor (EGF) in EGF-responsive mammalian cells.[3] The importance of this protein for vaccinia virus function is unknown. However, the EGF receptor, a protein related to the oncogene *v-erb^b*, may function as a cellular receptor for vaccinia.[4] It has also recently been reported that vaccinia also codes for a 28-kDa protein that shows sequence homologies to the *v-erb^b* protein and the EGF receptor.[5]

## 2.3. Classification

Poxviruses are classified principally by their host range. In 1979, the International Committee on Taxonomy of Viruses proposed a classification of the family Poxviridae. Table I is adapted from their classification[6] and shows the major genera of vertebrate poxviruses and representative members of these genera.

## 2.4. Host Range

The host ranges of poxviruses may be either narrow or wide. Variola virus may cause slight lesions in rabbits, for example, without systemic illness. *Molluscum contagiosum* apparently infects only humans. Rabbit myxoma and fibroma viruses do not infect rodents. However, vaccinia has a wide range of potential targets, as do the viruses of cowpox and monkeypox. One of the insect poxviruses may even infect plants, under appropriate circumstances.

## TABLE I
### Classification of Vertebrate Poxviruses

*Orthopoxviruses:* Vaccinia, cowpox virus, variola virus, ectromelia virus, rabbitpox virus
*Avipoxviruses:* Turkeypox virus, fowlpox virus, pigeonpox virus
*Capripoxviruses:* Sheep pox virus, goatpox virus
*Leporipoxviruses:* Rabbit myxoma virus, Shope fibroma virus, malignant rabbit fibroma virus
*Parapoxviruses:* Orfvirus, bovine pustular stomatitis virus
*Suipoxviruses:* Swinepox virus
*Unclassified: Molluscum contagiosum,* Yaba monkey tumor virus

Cellular tropisms of poxviruses vary from one virus to the next. Skin is the most universally involved organ and is the principal portal of entry for many poxviruses into the natural host. For some poxviruses, the skin may be the only tissue infected. *Molluscum contagiosum* virus seems only to infect the squamous cells of the skin. For other poxviruses, cutaneous involvement may be only a part of a generalized disease, e.g., smallpox.

### 2.5. Tumor Formation

Poxvirus-induced tumors are well known. Shope fibroma virus infects fibroblasts and other nonepidermal cells in the leporine subcutis. Infected cells proliferate and produce a highly mucoid tumor rich in fibroblasts, inflammatory cells, and vasculature. Virus antigen, however, seems to be confined to the spindle cells comprising the bulk of the cellular portion of the tumor. Yaba tumor virus induces a local proliferation of histiocytes at the inoculation site in the subcutis. In both cases, the tumors behave in a benign fashion regressing over time. The nature of the regression is not entirely clear but is thought to involve cytotoxic T lymphocytes.

Rabbit myxoma virus, and its relative, malignant rabbit fibroma virus, infect fibroblasts as well as other cell types (e.g., renal tubular cells, conjunctiva, skin). Tumors produced by these viruses disseminate rapidly. Infected animals die within a few weeks of inoculation, often within a week of the onset of symptoms. Death in these cases is usually due to infection by gram-negative-bacteria. The nature of tumors induced by these viruses is unclear. Incorporation of virus DNA into host genome has not been demonstrated following *in vivo* virus inoculation. Similarly, although transformed cell lines have been established with Shope fibroma virus infected SIRC cells, and although SFV DNA has been found in the host nucleus, these cell lines lose their transformed phenotype by 50 passages.[7]

## 3. EFFECTS OF POXVIRUSES ON IMMUNE FUNCTION

Although considerable information is available regarding poxvirus genetics and structure, little is known about the immunosuppressive effects of poxvirus infection. Some data are available in the case of vaccinia and variola, but most of what we know about the effects of poxviruses on the immune system derives from the study of the oncogenic poxvirus, malignant rabbit fibroma virus. The known effects of variola and vaccinia on the immune system are therefore discussed briefly, followed by a more detailed review of the oncogenic leporipox-viruses.

## 3.1. Smallpox

The effects of variola virus infection on the human immune system are summarized by Fenner[8] but are not well understood. Following initial infection of epithelium, usually of the upper respiratory tract, the rapidly replicating virus is thought to infect local macrophages. These virus-bearing macrophages act as the conduit for dissemination of variola infection. Virus spreads to and replicates first in lymph nodes, then in bone marrow and elsewhere.

In some cases, antibody and cytotoxic lymphocyte responses are adequate to contain the infection. Infected macrophages and lymphocytes produce interferon (IFN). At other times, viremia persists in high titer and is associated with thrombocytopenia. Death ensues. In the latter case, the antibody and cytotoxic lymphocyte responses to the virus are believed to be poor. The nature of the factors determining the differences between these two outcomes is unclear. Although the lymphocytotropism of variola seems established, little is known about either variola replication in lymphocytes and macrophages or its effects on the functions of these cells.

As smallpox is no longer a clinical disease, the likelihood of our understanding the effects of variola virus on the immune system is small. Monkeypox, a simian virus similar to variola, produces comparable disease. Human cases of monkeypox are quite uncommon and have not been well studied from the immunologic perspective.

## 3.2. Vaccinia

Vaccinia virus infects the skin, underlying soft tissues, and lymphocytes. A brief viremia normally follows vaccinia virus or cowpox virus inoculation. Unless a primary immunologic deficiency is present, however, little systemic illness occurs. The observation that patients with primary cellular, but not humoral, immune deficits handled vaccinia infection poorly was one of the first indications of the importance of cell-mediated immunity for resisting viral infection.[9,10]

Systemic immunologic dysfunction has been reported in vaccinia virus infection *in vivo*. Poor mitogen responses to concanavalin A (Con A) have been noted in infected rabbits during the acute phase of the infection. These responses recover as infection clears. On reinfection, the extent of depressed Con A-induced mitogenesis is decreased both in level and in duration .[11] Vaccinia virus infects lymphocytes *in vitro* and has been recovered from them by infectious center assay. The level of immune dysfunction incurred by these cells has not been studied. However, as Bloom *et al.*[12] amply demonstrated for almost all lymphocytotropic viruses, appreciable replication of vaccinia in lymphocytes does not occur in the absence of mitogenic stimulation. Its ability to infect lymphocytes should make vaccinia a profoundly immunosuppressive agent, as it turns off most ongoing host cell functions on initiation of infection. Protein synthesis, DNA transcription, and RNA translation on the part of the host cell are all converted to virus functions. The replicative cycle of vaccinia is also quick and efficient. Although the virus replicates efficiently, vaccinia infection is benign in its course and has not typically been associated with systemic immunologic defects.

Some studies of vaccinia replication in macrophages have been performed. Naive macrophages appear to support viral replication. However, macrophages from immune or convalescent animals do not. Views differ as to the nature of the block in this condition. Buchmeier et al.[13] describe abortive replication of vaccinia in macrophages from previously infected rabbits. They observed normal adsorption, uncoating, and production of virus DNA, but very few mature virions, and conclude that a block occurs late in viral replication. Tompkins et al.,[14] however, noted that vaccinia did not even penetrate macrophages from immune rabbits.

One report indicates that mice infected with vaccine virus show depressed T lymphocyte responsiveness. These authors indicate that suppressor cells of both adherent and nonadherent populations are induced by vaccine infection in vivo.[14a]

## 3.3. Leporipoxviruses

Of all the poxviruses, leporipoxviruses have been most thoroughly characterized regarding their immunosuppressive effects. This group includes three principal oncogenic viruses: Shope fibroma virus (SFV), rabbit myxoma virus (MYX), and malignant rabbit fibroma virus (MV).

These three viruses are closely related. Antisera to each will cross-neutralize the others. All show similarities by restriction endonuclease analysis and Southern blotting. MV appears to be a recombinant between SFV and MYX. Depending on the enzymes used, restriction endonuclease analysis of MV and MYX shows that 75–85% of the restriction sites in MV are also present in MYX. The remainder of the restriction fragments are either shared with SFV or are unique. McFadden et al.[15] found that the area of recombination between SFV and MYX to form MV lies in the inverted terminal repeat region.

Clinically, SFV, MV, and MYX produce distinct syndromes. All are tumorigenic. In the case of SFV, tumors remain localized. Local inoculation of $\approx 10^7$ focus-forming units (FFU) SFV induces benign tumors within 1 week. These tumors regress by 2 weeks later, leaving rabbits immune to all three viruses. The course of SFV infection varies somewhat according to the strain of virus used to infect immunocompromised rabbits.[16] Rabbits bearing tumors induced by SFV are immunologically normal. These rabbits, their spleen cells, and SFV therefore often serve as controls for the immunologic dysfunction induced by MV.

Very small doses of MYX and MV (for MV, $LD_{100} \leq 20$ pfu) induce progressive tumors that spread rapidly. These are complicated by lethal gram negative infection. Clinical and pathologic differences between MV and MYX have been noted.[17,18] However, most of the following comments regarding the effects of MV on the immune system probably apply to MYX as well.

## 3.4. Effects on the Immune System of Malignant Rabbit Fibroma Virus

### 3.4.1. Effects on Cellular Proliferation

Malignant rabbit fibroma virus infection is associated with profound and complex immunologic deficiency, both humoral and cellular. Spleen cells from rabbits given MV proliferate poorly in response to both B- and T-lymphocyte mitogens (Fig. 1). This impaired proliferative response is first noted about 5–6 days following *in vivo* inoculation of 1000 PFU MV. Poor responsiveness is observed thereafter until about day 11.

Similarly, lymphocyte proliferative responses are inhibited following primary *in vitro* exposure to MV. Dose–response curves indicate that a multiplicity of infection (MOI) virus-to-lymphocyte ratio of at least 0.01 is needed to observe this effect. At a MOI of 0.05, we observe an *in vitro* lag time of 2 days before radionucleotide incorporation is decreased.[19] The magnitude of the

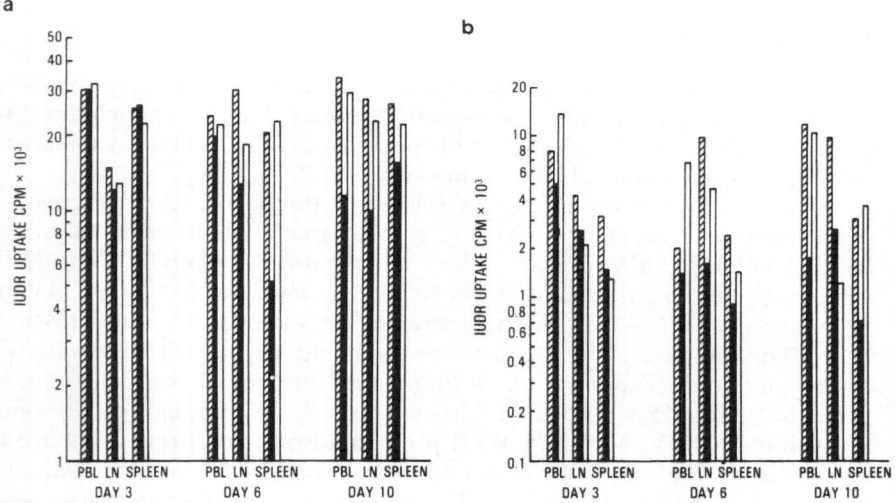

**FIGURE 1.** Peak responsiveness *in vitro* to (a) concanavalin A (Con A) and (b) anti-Ig, which are the optimal T- and B-lymphocyte mitogens, respectively, in rabbits. Rabbits with initially normal PBL mitogen responses are given Shope fibroma virus (SFV) (hatched lines) or malignant rabbit fibroma virus (MV) (closed bars) or virus-free control preparations (open bars). After sacrifice 3, 6, and 10 days later, lymphocytes from spleen were tested for mitogen responses, with [125 I]-UdR uptake used as an index of blastogenesis. Mitogen responses depicted are from the day of assay at which [125 I]-UdR incorporation was maximal. (a) Con A responses of splenic lymphocytes. (b) Anti-Ig responses of splenic lymphocytes. Comparable results were obtained with PBL and lymph node lymphocytes. (From Strayer et al.[19])

deficit in proliferation, however, is profound. Con A responses may be less than 1/100 of normal (Fig. 2a).

The mechanism by which MV alters lymphocyte proliferative responses is not yet clear. Capping and internalization of caps are not impaired.[20] Initiation of proliferation is not altered, as MV added to cultures 2 days after lectin

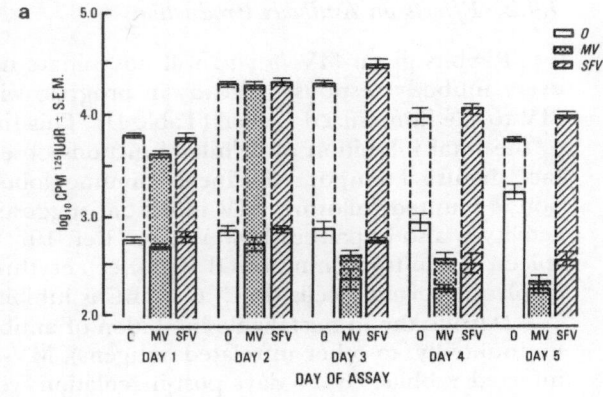

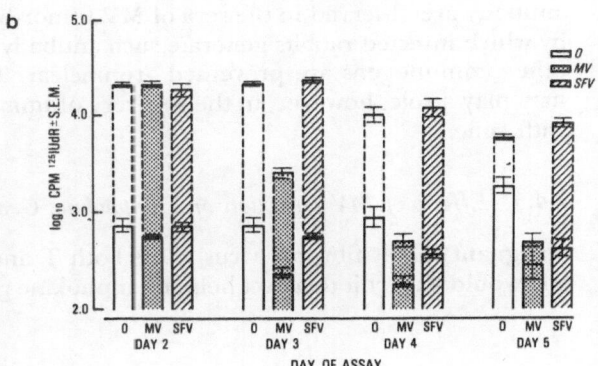

**FIGURE 2.** Effect of virus on mitogen responses of normal spleen cells. Spleen cells from normal rabbits were placed in culture with concanavalin A (Con A). Mitogens were added on day 0, with spleen cells. Virus, either malignant rabbit fibroma virus (MV) or Shope fibroma virus (SFV), was added on day 0 (a), day 1 (b), or day 2 (c), at a multiplicity of infection (MOI) of 0.1. Control cultures received an equal volume of culture medium instead of virus. Blastogenic responses of the spleen cell cultures were measured daily after viral addition by pulsing cultures with [125I]-UdR and harvesting them 24 hr thereafter. [125I]-UdR incorporation is shown as $\log_{10}$ (cpm[125I]-UdR) for unstimulated (solid bars) and mitogen-stimulated cultures ±SEM. (From Strayer et al.[18])

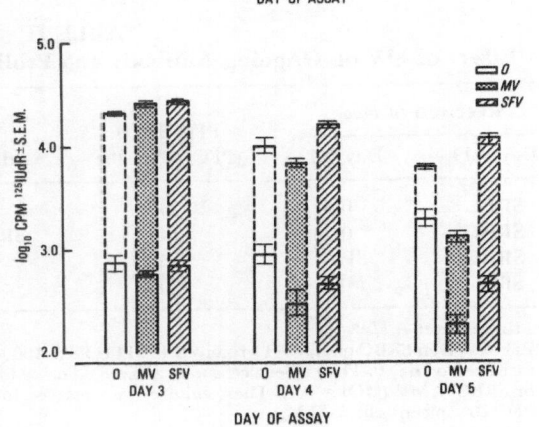

will still inhibit subsequent proliferative activity (Fig. 2b). Rather, MV seems to inhibit proliferation of infected cells in a nonspecific manner. Lymphocytes are clearly inhibited. In addition, the line of rabbit kidney epithelial cells (RK-13) used to grow MV as well as SFV and MYX stop proliferating on exposure to MV but not SFV (D. S. Strayer, unpublished results).

### 3.4.2. Effects on Antibody Production

Rabbits given MV *in vivo* will not initiate new antibody responses. However, antibody responses already in progress will not be abrogated by giving MV to the immunized animal (Table II). This finding is in contrast somewhat to the total inhibition of cellular functions observed in vaccinia infection. In fact, the usual temporal decline in immunoglobulin M (IgM) antibody production is blunted following MV infection, suggesting that somehow suppressor activity is also impaired (however, see Ref. 19). When MV is added *in vitro* to spleen cell cultures immunized with sheep erythrocytes (SRBC), the generation of plaque-forming cells (PFC) to SRBC is inhibited.[19]

Despite the unmistakable inhibition of antibody responses to SRBC (and, undoubtedly, to other unrelated antigens), MV-specific antibodies develop in infected rabbits. By 11 days post-inoculation, very high titers of neutralizing antibody are observed in the sera of MV tumor-bearing rabbits.[21] The means by which infected rabbits generate such antibody responses when responses to other immunogens are prevented are unclear. These neutralizing antibodies may play a role, however, in the recovery of immune function that is observed with time.

### 3.4.3. Effects of MV Infection on Lymphokine Generation

Malignant rabbit fibroma virus alters both T and B lymphocyte function; so one would expect it to affect helper lymphokine generation and/or action. MV

**TABLE II**
**Effect of MV on Ongoing Antibody and Proliferative Responses to SRBC[a,b]**

| Received *in vivo* | | PFC day 0 (PFC/$10^6$ cells) | Added *in vitro* | Response to SRBC[c] (PFC/$10^6$ cells) |
|---|---|---|---|---|
| Day −11 | Day −4 | | | |
| SRBC | 0 | 147 | SRBC | 10,691 ± 3500 |
| SRBC | 0 | 147 | SRBC + MV | 0 ± 0 |
| SRBC | SFV | 50 | SRBC | 4398 ± 1415 |
| SRBC | MV | 1006 | SRBC | 694 ± 284 |

[a]From Strayer *et al.*[18]
[b]Rabbits given SRBC on day −11 received $10^7$ FFU SFV, 1000 PFU MV, or nothing on day −4 and were sacrificed on day 0. Their spleen cells were assayed on day 0 for direct PFC and were cultured with SRBC or SRBC + MV (MOI = 0.1). These cultures were assayed for direct PFC on day 4.
[c]PFC/$10^6$ spleen cells ± SEM.

**TABLE III**
**Effects of MV Infection on IL-1 Generation**[a]

| Rabbit | IL-1 generation[b] | |
|---|---|---|
| | No LPS | +LPS |
| Normal | 1.80 | 3.30 |
| SFV | 2.14 | 2.79 |
| MV | 1.90 | 3.07 |

[a]Mock-infected rabbits, rabbits given SFV, or rabbits given MV were sacrificed 7 days postinfection. Peritoneal wash cells were collected and allowed to adhere to plastic plates. Plastic adherent cells were exposed to *Escherichia coli* endotoxin (LPS) or not. Supernatants were collected 2 days later and were tested for their ability to support the proliferation of D.10G4 cells in the presence of Con A.
[b]Results shown are for the proliferation of D.10G4 cells, $\log_{10}$ (cpm [$^3$H]thymidine).

is present systemically in the fixed phagocytes, so it could logically be expected to cause deficiencies in the ability of adherent cells to participate in immune responses.

Adherent cell immune function was measured by assaying interleukin-1 (IL-1) production by peritoneal macrophages from MV and control rabbits. The IL-1 responsive murine T-cell clone D.10G4 was used as a target cell line.[22] Table III shows that IL-1 production in response to endotoxin was not altered in adherent cells from rabbits bearing MV-induced tumors.[23]

The generation of and responses to interleukin-2 (IL-2) were studied as a means of dissecting the level at which a defect might occur in the T lymphocyte series. Spleen cells from MV-infected and control rabbits were stimulated with Con A and the resulting IL-2 activity measured in the culture supernatant

**TABLE IV**
**IL-2 Production in MV Tumor-Bearing Animals**[a]

| Animal | HT-2 proliferation[b] | | | |
|---|---|---|---|---|
| | No Con A$_1$ | Con A$_1$ | No Con A$_2$ | Con A$_2$ |
| Normal | 2.68 | 4.00 | 2.75 | 4.29 |
| Day 7 MV | 2.41 | 3.06 | 2.91 | 3.44 |
| Day 7 SFV | 2.70 | 3.60 | 3.12 | 4.43 |

[a]Spleen cells from normal rabbits and from rabbits sacrificed 7 days following either MV or SFV inoculation were cultured with either Con A or no mitogen. After 2 days of culture, supernates from these cultures were assayed for IL-2 activity using HT-2 cells. Target cell (HT-2) proliferation was determined by 24-hr [$^3$H]thymidine incorporation after 2 days of culture. Results shown are for two separate experiments.
[b]Proliferative activity is indicated as $\log_{10}$ (cpm [$^3$H]thymidine).

fluids using the murine IL-2 responsive cell line HT-2 (Table IV). Spleen cells from MV tumor-bearing rabbits were deficient in their ability to generate IL-2 in response to Con A. This deficiency was not absolute but resulted in much lower IL-2 generation than was observed using control rabbits.[23]

Using cytofluorographic techniques, the ability of such cells to generate cell membrane receptors for IL-2 was determined. The monoclonal antibody 7D4 recognizes both murine [24] and leporine IL-2 receptors.[21a] It was observed (Fig. 3) that Con A-stimulated spleen cells from MV tumor-bearing rabbits express receptors for IL-2 normally. This finding was somewhat surprising, as most investigators believe that T-cell activation is required to promote the expression of IL-2 receptors on target cells.[25] The immunologic defect in T lymphocytes from MV-infected rabbits is more profound than their inability to produce normal amounts of IL-2. They also respond poorly to lectins when IL-2 is added to the cultures (Table V).

Thus, the abnormality in T-cell function during infection with malignant rabbit fibroma virus is complex. It involves at least these features: (1) generalized poor ability to proliferate, (2) defective IL-2 elaboration in response to stimulation, and (3) inability to respond to added IL-2 despite normal IL-2 receptor expression.

### 3.4.4. Suppressor Lymphokines

Malignant rabbit fibroma virus infection of lymphocytes is not adequate to explain all manifestations of virus-associated immunologic dysfunction. MV also induces host cells to elaborate a substance(s) capable of inhibiting certain aspects of immune responsiveness. This activity is measurable in ultravirus (UV)-treated culture supernatants and therefore does not represent infectious virus. Suppressor supernatants have several capabilities that are different from live virus. These capabilities include a more rapid mechanism of action and the ability to turn off ongoing antibody responses.[19] The factor in question is weaker than direct virus effects and cannot alter appreciably Con A responses. It can diminish proliferative responses to less potent mitogens, however, such as SRBC. Unlike MV, which infects only rabbits, the factor(s) generated by rabbit cells in response to MV infection will suppress mouse spleen cell responses to antigen. This factor also suppresses the response of HT-2 cells to IL-2.[26]

Characterization of the suppressor factor has been hampered by its lability to freezing and thawing, and even to prolonged storage at $-70°C$. Still, some data are available regarding this substance.[26] The substance in question is not produced in the presence of indomethacin, or of either cyclic adenosine monophosphate (cAMP) or cyclic guanosine monophosphate (cGMP). Its molecular weight is $>25$ kDa. Despite the apparent involvement of prostaglandins, when serial depletions of adherent cells are performed, suppressor activity is still present in supernatants of nonadherent cells. Eliminating T lymphocytes from the MV spleen cell population eliminates the ability of the resulting pool of cells to produce suppressor activity.

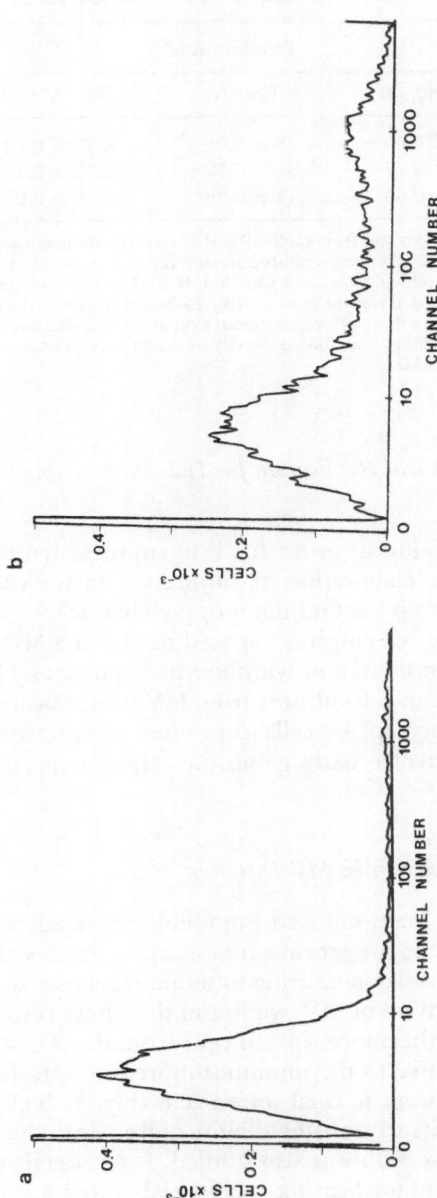

**FIGURE 3.** Labeling of spleen cells from malignant rabbit fibroma virus (MV)-infected rabbits with antibody to the receptor for interleukin-2 (IL-2). Spleen cells from rabbits given MV 7 days previously were cultured overnight with concanavalin A (Con A) and labeled the next day with the rat monoclonal antibody to the mouse IL-2 receptor, 7D4, followed by FITC-goat antirat Ig. Cells were then assayed by cytofluorography. (a) MV spleen cells cultured overnight with mitogen. (b) MV spleen cells cultured overnight without mitogen.

**TABLE V**
**Proliferative Responses of Spleen Cells from MV**
**Tumor-Bearing Rabbits with and without Added IL-2** [a]

| Cells | Proliferation [b] | | |
|---|---|---|---|
| | No mitogen | + Con A | + Con A + IL-2 |
| Normal | 4.20 ± 0.03 | 4.75 ± 0.01 | 4.73 ± 0.02 |
| Day 7 MV | 2.66 ± 0.17 | 3.01 ± 0.12 | 2.75 ± 0.15 |
| Day 11 MV | 3.14 ± 0.06 | 4.44 ± 0.08 | 4.40 ± 0.08 |

[a] Spleen cells from normal rabbits and from rabbits 7 and 11 days following intrader-
mal inoculation of 1000 PFU MV were cultured *in vitro* for 3–5 days. To these
cultures, we added either nothing, Con A, or Con A + IL-2. The ability of these
spleen cells to proliferate was determined by measuring 24-hr [³H]thymidine incorporation. Responses shown are those of a representative experiment, indicating the
proliferation of the spleen cells in question on the day of maximal responsiveness.
[b] $\log_{10}$ (cpm [³H]thymidine) ±SD.

### 3.4.5. Extent of the Need for Viral Replication for Induction of Immunologic Dysfunction

The need for MV replication in order for it to suppress lymphocyte responsiveness or to induce the elaboration of suppressor factor was studied.
Phosphonoacetic acid (PAA) is a potent inhibitor of poxvirus DNA polymerase.
When added to normal spleen cell cultures exposed to MV at a MOI of 0.1 *in
vitro*, PAA does not alter the inhibition of lymphocyte responses by MV. Similarly, PAA when added to spleen cell cultures from MV tumor-bearing rabbits
does not alter the ability of such spleen cells to produce suppressor factor. It
appears that both activities involve early genes, i.e., those expressed before
DNA replication.[27]

### 3.4.6. Recovery of Immune Function in MV Infection

Malignant rabbit fibroma virus-induced immunologic impairment is not
permanent. This is so despite the progressive tumor and gram-negative infection that afflicts infected rabbits. In measuring immune responses to mitogens
11 days after *in vivo* administration of MV, we found that these responses had
largely recovered (Fig. 4). Furthermore, spleen cells from day 11 MV tumor-
bearing rabbits were less sensitive to the immunosuppressive effects of day 7
MV-infected spleen cells than were normal spleen cells (Fig. 4). The ability of
spleen cells from day 11 rabbits to transfer their acquired resistance to MV-
induced T-cell-mediated suppression was also studied. It was determined that
spleen cells from day 11 MV tumor-bearing rabbits elaborated a substance(s)
capable of inhibiting the suppressive effects of the suppressor material produced by day 7 MV-infected spleen cells.[28] The mechanism of action of this
antisuppressive activity on the part of the former spleen cell population is not
yet clear, but it has been shown to be a T-cell product.

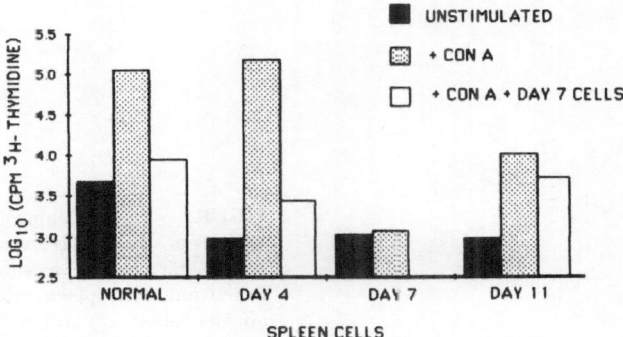

**FIGURE 4.** Adult New Zealand white rabbits were given 1000 PFU MV intradermally in the thigh on day 0. They were sacrificed 4, 7, or 11 days later, and their spleen cells cultured *in vitro* with or without concanavalin A (Con A). Their ability to respond to Con A was measured by 24-hr [³H]thymidine incorporation, and compared with the responses of normal rabbits. Parallel cultures received day 7 cells + Con A as well as day 4, day 11, or normal cells, in a ratio of 1 : 1. Background proliferation in these cultures was not different from that in cultures not receiving day 7 cells (data not shown). (From Strayer and Liebowitz.[24])

### 3.4.7. Interferon

Although vaccinia virus induces host elaboration of IFN, we have not found any evidence of IFN in the MV system. Attempts to inhibit the replication of vesicular stomatitis virus and MV in appropriate target cells, either with cell lysates or culture supernatants from MV- or SFV-infected lymphocytes have failed to detect IFN activity.

## 3.5. Malignant Rabbit Fibroma Virus Replication within the Cells of the Immune System

### 3.5.1. Characteristics of MV Replication in Lymphoid Cells Exposed to MV

The data regarding the immunosuppressive effects of adding MV to lymphocyte cultures suggest that MV infects both B and T lymphocytes. MV replication in lymphocytes was studied in two ways. First, lymphocytes from infected rabbits were studied to determine the cellular tropism of MV, its requirements for replication, and its presence in lymphocytes at selected time points during oncogenesis. In addition, MV replication in lymphocyte cultures was examined using both naive lymphocytes and those exposed to MV *in vivo*.

Malignant rabbit fibroma virus antigen was sought on lymphocyte cell membranes by double labeling and cytofluorography. Using lymphocytes from spleen cells 7 days after infection, it was determined that MV antigen was expressed primarily on T lymphocytes (Fig. 5). However, following overnight culture without added virus, a population of nonadherent non-T lymphocytes

a

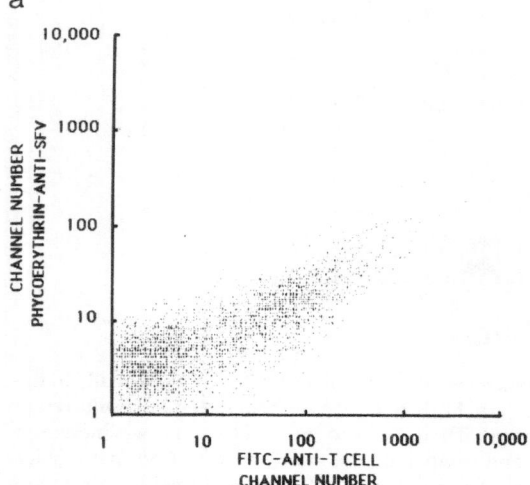

b

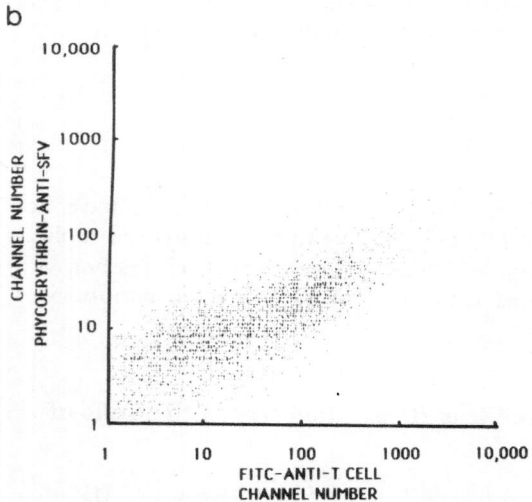

**FIGURE 5.** Cytofluorographic analysis of spleen cells from malignant rabbit fibroma virus (MV)-infected animals. Spleen cells from normal rabbits (a) and rabbits receiving 1000 PFU MV 7 days previously (b) were analyzed for virus and T-cell antigens by double-label cytofluorography. After pretreatment with normal rabbit serum, these cells were treated with the mouse monoclonal antibody 9AE10, which recognizes rabbit T cells, followed by FITC-rabbit anti-mouse Ig. This was followed by biotinylated rabbit anti-SFV, then avidin–phycoerythrin. The double label was analyzed by FACS (Becton-Dickinson). Some spillage of fluorescein fluorescence into the phycoerythrin range was noticed (a). However, analysis of the plots obtained using MV tumor-bearing rabbit spleen cells (b) shows that the population of spleen cells bearing MV antigens *in vivo* also bears T-cell antigens. Approximately 25% of spleen cells from MV-infected rabbits are positive for MV antigens. A similar number also demonstrate T-lymphocyte markers as well.

(presumably B cells) also expressed MV antigen. Insufficient fresh virus is elaborated by infected T cells after only overnight culture to infect this many B cells. Thus, the B lymphocytes are undoubtedly infected with MV *in vivo* but, for reasons as yet unknown, do not express virus antigen until given time *ex vivo* to do so.

The ability of these spleen cells to permit MV replication was also determined. Virus recovery from spleen cells of any type is very low, no matter when they are examined postinfection (maximum yield <20 PFU/10⁵ spleen cells). However, when spleen cells from MV-infected rabbits are cultured without added virus, recoverable virus titers increase approximately 10-fold daily up to 4 or 5 days of culture. In fractionating these cell populations, it was found that T lymphocytes are most supportive of such viral replication, B lymphocytes less so, and adherent cells least of all. Thymic T lymphocytes do not yield any MV using this technique.

Since Bloom et al.[12] had reported that lymphocytotropic viruses do not replicate in the absence of lymphocyte stimulation, the above results obtained without added mitogen were unusual. Additional experiments were performed to determine whether naive lymphocytes could support MV replication without mitogens. MV was added to cultures of normal spleen cells with and without Con A or LPS. Yields of MV from cultures of normal spleen cells were found to be highest when stimulation was provided. However, MV replicated well in spleen cells in the absence of added lectin (Fig. 6). This finding makes MV unique among described lymphocytotropic viruses, with the exception of human immunodeficiency virus (HIV), which also replicates in lymphocytes without added lectin.

Finally, the time course of MV infection of lymphocytes was examined. Spleen cells from rabbits 4, 7, and 11 days following inoculation were examined for expression of virus antigen and ability to yield MV, to support replication of added MV, and to affect other cells' capacity to replicate MV. Lymphocytes from rabbits 4 and 7 days postinfection both expressed MV antigen and yielded high titers of MV following culture *in vitro* without added virus or mitogen. Lymphocytes from day 11 rabbits showed no evidence of MV in this assay. The latter did not alter the ability of MV to replicate in the former.

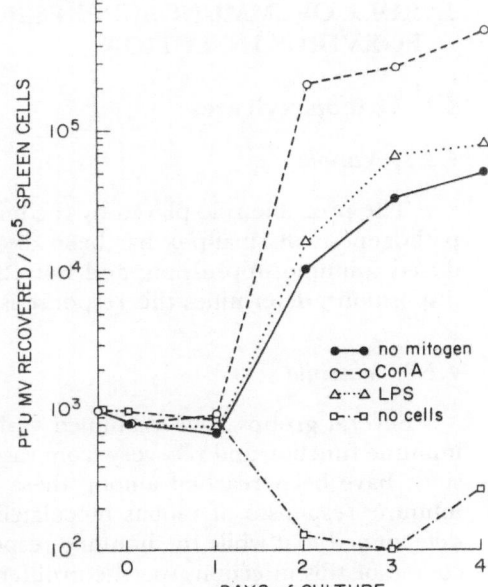

**FIGURE 6.** Growth curve of malignant rabbit fibroma virus (MV) in spleen cells. MV was added to normal spleen cells at 0.001 multiplicity of infection (MOI). No mitogen, Con A, or LPS was added to these cultures. After sequential days, *in vitro* cultures were frozen and thawed. Titers of recoverable MV are shown.

However, when MV was added to cultures of normal, day 4, day 7, and day 11 spleen cells, lymphocytes from day 11 rabbits supported MV replication much less than did other spleen cells.[21]

These findings suggest that a population of lymphocytes infected with MV is lost between 7 and 11 days postinfection *in vivo*. While MV has not been found to cause gross differences in viability in whole spleen cell cultures *in vitro*,[29] we cannot rule out the possibility that MV destroys a small population of particularly susceptible cells. Such depletion of a population of lymphocytes that preferentially replicate MV is supported by the poor replication of MV in day 11, as compared with normal, spleen cells.

The means by which these lymphocytes would be eliminated is not clear. By day 11 postinfection, rabbits have very high titers of neutralizing antibody in their sera. Macrophages from such animals have been observed by fluorescence photography ingesting MV-infected lymphocytes.[21] In addition, cytotoxic T cells (CTL) might develop. While T cells from MV-infected rabbits do not proliferate in response to virus antigen 4 and 7 days following inoculation *in vivo*,[30] they do so by day 11.[28,31]

Immunologic recovery from MV infection seems to be a complex phenomenon. In light of the high titers and rise coincident with elimination of circulating virus infected cells, antibody to virus probably participates. Direct virus toxicity to, and subsequent loss of, the pool of lymphocytes most permissive to viral replication is possible. Cell-mediated immunity to virus is also likely to be important. The role of the antisuppressive T-cell products remains to be decided. However, several different processes undoubtedly participate.

## 4. ROLE OF IMMUNOSUPPRESSION IN THE PATHOGENESIS OF POXVIRUS INFECTION

### 4.1. Orthopoxviruses

#### 4.1.1. Variola

The probable role played by the immune response to variola virus in the pathogenesis of smallpox has been discussed. The extent to which virus-induced immunosuppression, and not other host factors such as genetic predispositions, determines this response is unknown.

#### 4.1.2. Vaccinia

Several groups have examined various factors involved in the course of immune function and recovery from vaccinia virus infection. Different conclusions have been reached among these investigators. In their studies of the immune responses of rabbits inoculated with vaccinia virus, Tompkins *et al.* determined that while the immune responses to Con A were decreased in the course of this infection, specific proliferative responses to virus antigen were observable. The appearance of specific lymphoproliferative responses to vaccinia corresponded temporally to macrophage resistance to viral replication. On rechallenge, both phenomena were found to be more pronounced.[10]

Resistance to vaccinia in hamsters has been studied by the same group. They determined that vaccinia virus induces two sets of lymphocytes capable of killing virus-infected cells. One of these, found principally in the spleen, consists of lymphocytes that are capable of killing target cells infected with other viruses as well as vaccinia, and resistant to anti-T-cell antibody. These are most likely natural killer (NK)-like cells.[32] A population of CTL is also found in the peritoneal cavity and is considered important in clearing vaccinia infection.[33]

### 4.1.3. Ectromelia

Ectromelia virus produces a lethal disseminated disease in some strains of mice. Studies of the effects of this virus on the immune system are in their earliest stages (M. Buller, personal communication). However, some work has been done on factors involved in the recovery from ectromelia infection.

Both macrophages and CTL have been implicated in the genetic and induced ability of mice to resist ectromelia infection both *in vivo* and *in vitro*. Cohen *et al.*[34] noted that *C. parvum*-primed macrophages block ectromelia replication. They observed defective release of virus DNA in macrophages so treated and therefore ascribe the resistance conferred by priming to the production of IFN. Others have determined that both macrophages and T lymphocytes play an important role in recovery from ectromelia infection.[35] They have correlated resistance to ectromelia virus with CTL as well as IFN production.[36]

## 4.2. Leporipoxviruses

### 4.2.1. Shope Fibroma Virus

Shope fibroma virus is not lymphocytotropic, and is not immunosuppressive per se. However, neonates are inordinately susceptible to the oncogenic effects of SFV. They develop disseminated tumors. From these, the neonatal rabbits follow a protracted course and may or may not recover. Neutralizing antibody appears to be present in animals that die of progressive tumor and those that recover. Adult rabbits receiving SFV and those neonates that go on to reject their tumors do develop both delayed-type sensitivity to SFV and the ability to produce migration inhibitory factor. Neonates that succumb to disseminated tumor develop neither of these. Thus, it is likely that CTL are important in the recovery from SFV infection, both in neonates and in adults.[37]

### 4.2.2. Malignant Rabbit Fibroma Virus

It is mainly, then, for the oncogenic leporipoxvirus, MV, that some hypotheses can be made. Clearly, MV infection involves immunologic dysfunction. Other aspects of MV infection, such as myxosarcoma and epithelial hyperplasia, are poorly understood. The interaction of MV-induced immunologic dysfunction and these nonimmune phenomena is even more speculative. However, the profound immunosuppression engendered by MV surely facilitates virus spread and the development of opportunistic gram negative infection in host rabbits. An important and as yet unanswered question is whether, in light of the immunologic recovery, MV tumor-bearing animals treated for their gram-negative infection would eventually recover not only from their pasteurellosis but from their tumors as well.

The study of poxvirus biology to date has concentrated on orthopox viruses for the most part, primarily on the manner in which they infect and replicate in target cells. Most of the work on the immunomodulatory effects of poxviruses has been with malignant fibroma virus. The complexity of poxviruses as a group suggests that a wealth of information is to be found in studying their effects on the immune system. These studies are just in their infancy.

ACKNOWLEDGMENTS.    The studies reported here that were performed by us involved the participation of Dr. Lynnette Corbeil, Dr. Julian Leibowitz, Dr. Stewart Sell, and Dr. Eileen Skaletsky and also of Miss Patricia Sharp, Mrs. Jan Hurt, Jan Dombrowski, and Kenneth Korber. The American Cancer Society has been most generous in supporting our work, through grant IM-358. Initial studies were supported by grant CA16136 from the National Institutes of Health. I greatly appreciate personal communications with Dr. Grant McFadden and Dr. Mark Buller. Dr. Julian Leibowitz was kind enough to review and comment on this chapter.

# REFERENCES

1. Dales, S., and B. G. T. Pogo, *Biology of Poxviruses*, Springer-Verlag, Vienna (1984).
2. Moss, B., Replication of poxviruses, in: *Virology* (B. N. Fields, D. M. Knipe, R. M. Chanock, J. L. Melnick, B. Roizman, and R. E. Shope, eds.), pp. 685–703, Raven, New York (1985).
3. Blomquist, M. C., L. T. Hunt, and W. C. Barker, Vaccinia virus 19-kilodalton protein: Relationship to several mammalian proteins, including two growth factors, *Proc. Natl. Acad. Sci. USA* **81**:7363–7367 (1984).
4. Eppstein, D. A., Y. V. Marsh, A. B. Schreiber, S. R. Newman, G. H. Todaro, and J. J. Nestor, Jr., Epidermal growth factor receptor occupancy inhibits vaccinia virus infection, *Nature (Lond.)* **318**:663–665 (1985).
5. Chen, H. R., and W. C. Barker, Similarity of vaccinia 28K, V-*erb*-B and EGF receptors, *Nature (Lond.)* **316**:219–220 (1985).
6. Matthews, R. E. F., Classification and nomenclature of viruses, *Intervirology* **17**:42–46 (1982).
7. Obom, K., and B. G.-T. Pogo, Characterization of the transformation properties of Shope fibroma virus, *Virus Research* **9**:33–48 (1988).
8. Fenner, F., Poxviruses, in: *Virology* (B. N. Fields, D. M. Knipe, R. W. Chanock, J. L. Melnick, B. Roizman, and R. E. Shope, eds.), pp. 661–684, Raven, New York (1985).

9. Freed, E. R., J. D. Richard, and M. R. Escobar, Vaccinia necrosum and its relationship to impaired immunologic responsiveness, *Am. J. Med.* **52**:411–420 (1972).

10. Fulginiti, V. A., C. H. Kempe, W. E. Hathaway, D. S. Perlman, O. F. Serber, Jr., J. J. Eller, J. J. Joyner, and A. Robinson, Progressive vaccinia in immunologically deficient individuals, *Birth Defects* **4**:129–145 (1968).

11. McLaren, C., H. Cheng, D. L. Spicer, and W. A. F. Tompkins, Lymphocyte and macrophage responses after vaccinia virus infections, *Infect. Immun.* **14**:1014–1021 (1976).

12. Bloom, B. R., A. Senik, G. Stoner, G. Ju, M. Nowakowski, S. Kano, and L. Jimenez, Studies on the interactions between viruses and lymphocytes, *Cold Spring Harbor Symp. Quant. Biol.* **41**:73–83 (1976).

13. Buchmeier, N. A., S. R. Gee, F. A. Murphy, and W. E. Rawls, Abortive replication of vaccinia virus in activated rabbit macrophages, *Infect. Immun.* **26**:328–338 (1979).

14. Avila, F. R., R. M. Schultz, and W. A. F. Tompkins, Specific macrophage immunity to vaccinia virus: Macrophage–virus interaction, *Infect. Immun.* **6**:9–16 (1972).

14a. Ferrante, A., D. E. O'Keefe, and Y. H. Thong, Induction of suppressor cells in mice following vaccinia virus infection, *Med. Microbiol. Immunol.*, **168**:227– (1980).

15. Block, W., C. Upton, and G. McFadden, Tumorigenic poxviruses: Genomic organization of malignant rabbit virus, a recombinant between Shope fibroma virus and myxoma virus, *Virology* **140**:113–124 (1985).

16. Strayer, D. S., E. Skaletsky, and S. Sell, S., Strain differences in Shope fibroma virus: An immunolopathologic study, *Am. J. Pathol.* **116**:342–358 (1984).

17. Strayer, D. S., G. Cabirac, S. Sell, and J. L. Leibowitz, Malignant rabbit fibroma virus: Observations on the culture and histopathologic characteristics of a new virus-induced rabbit tumor, *J. Natl. Cancer Inst.* **71**:91–104 (1983).

18. Strayer, D. S., and S. Sell, Immunohistology of malignant rabbit fibroma virus—A comparative study with rabbit myxoma virus, *J. Natl. Cancer Inst.* **71**:105–116 (1983).

19. Strayer, D. S., S. Sell, E. Skaletsky, and J. L. Leibowitz, Immunologic dysfunction during viral oncogenesis. I. Nonspecific immunosuppression caused by malignant rabbit fibroma virus, *J. Immunol.* **131**:2595–2600 (1983).

20. Strayer, D. S., E. Skaletsky, G. F. Cabirac, P. A. Sharp, L. B. Corbeil, S. Sell, and J. L. Leibowitz, Malignant rabbit fibroma virus causes secondary immunosuppression in rabbits, *J. Immunol.* **130**:399–404 (1983).

21. Strayer, D. S., and V. L. Leibowitz, Virus–lymphocyte interactions during the course of immunosuppressive virus infection. *J. Gen. Virol.* **68**:463–471 (1987).

22. Kaye, J., S. Porcelli, J. Tite, B. Jones, and C. A. Janeway, Jr., Both a monoclonal antibody and antisera specific for determinants unique to individual cloned helper T cell lines can substitute for antigen and antigen-presenting cells in the activation of T cells, *J. Exp. Med.* **158**:836–856 (1983).

23. Strayer, D. S., M. Horowitz, and J. L. Leibowitz, Immunosuppression in viral oncogenesis. III. Effects on virus infection on interleukin-1 and interleukin-2 generation and responsiveness, *J. Immunol.* **137**:3632–3638 (1986).

24. Malek, T. R., G. Ortega, J. P. Jackway, C. Chan, and E. M. Shevach, The murine IL-2 receptor. II. Monoclonal anti-IL-2 receptor antibodies function as specific inhibitors of T cell function *in vitro*, *J. Immunol.* **133**:1976–1982 (1984).

25. Waldmann, T. A., W. J. Leonard, J. M. Depper, M. Kronke, C. B. Thompson, R. Kozak, and W. C. Greene, Structure, function, and expression of the receptor for interleukin-2 on normal and malignant lymphocytes, *Cancer Cells* **3**:221–226 (1985).

26. Strayer, D. S., K. Korber, and J. Dombrowski, Immunosuppression during viral oncogenesis. IV. Generation of soluble virus-induced immunologic suppressor molecules, *J. Immunol.* **140**:2051–2059 (1988).

27. Strayer, D. S., E. Skaletsky, and J. L. Leibowitz, Effects of inhibition of virus replication on immunosuppression induced by malignant rabbit fibroma virus, *Clin. Exp. Immunol* **66**:25–36 (1986).

28. Strayer, D. S., and J. L. Leibowitz, Reversal of virus-induced immune suppression, *J. Immunol.* **136**:2649–2653 (1986).

29. Strayer, D. S., E. Skaletsky, and J. L. Leibowitz, *In vitro* growth of two related leporipox-viruses in lymphoid cells, *Virology* **145**:330–334 (1985).

30. Skaletsky, E., P. A. Sharp, S. Sell, and D. S. Strayer, Immunologic dysfunction during viral oncogenesis. II. Inhibition of cellular immunity to viral antigens by malignant rabbit fibroma virus, *Cell. Immunol.* **86**:64–74 (1984).

31. Strayer, D. S., and J. Dombrowski, Immunosuppression during viral oncogenesis. V. Resistance to virus-induced immunosuppressive factor, *J. Immunol.* (in press).

32. Yang, H., C. A. Cain, M. C. Woan, and W. A. F. Tompkins, Evaluation of hamster natural cytotoxic cells and vaccinia-induced cytotoxic cells for Thy 1.2 homologue by using a mouse monoclonal anti-Thy-1.2 antibody, *J. Immunol.* **129**:2239–2243 (1982).

33. Yang, H., and W. A. F. Tompkins, Nonspecific cytotoxicity of vaccinia-induced peritoneal exudates in hamsters is mediated by Thy 1.2 homologue-positive cells distinct from NK cells and macrophages, *J. Immunol.* **131**:2545–2550 (1983).

34. Cohen, D. A., R. E. Morris, and H. C. Bubel, Aborative ectromelia virus infection in peritoneal macrophages activated by *Corynebacterium parvum*, *J. Leukocyte Biol.* **35**:179–192 (1984).

35. Tsuru, S., H. Kitani, M. Seno, M. Abe, Y. Zinnaka, and K. Nomoto, Mechanism of protection during the early phase of a generalized viral infection. I. Contribution of phagocytes to protection against ectromelia virus, *J. Gen. Virol.* **64**:2021–2026 (1983).

36. Sakuma, T., T. Suenaga, I. Yoshida, and M. Azuma, Mechanisms of enhanced resistance of *Mycobacterium bovis* BCG-treated mice to ectromelia virus infection, *Infect. Immun.* **42**:567–573 (1983).

37. Tompkins, W. A. F., R. M. Schultz, and G. V. S. V. Rama Rao, Depressed cell-mediated immunity in newborn rabbits bearing fibroma virus-induced tumors, *Infect. Immun.* **7**:613–619 (1973).

# 10

# Reovirus-Induced Immunosuppression

## CARLO GARZELLI and TAKASHI ONODERA

## 1. INTRODUCTION

### 1.1. Basic Properties of Reoviruses

Reoviruses are members of *Reoviridae*, a family of double-stranded (ds) RNA viruses that infect vertebrates, insects, arthropods, and plants.[1] Mammalian reoviruses are icosahedral virions, 76 nm in diameter, with a double capsid consisting of a central core of 52 nm surrounded by an outer protein shell. The viral genome consists of 10 fragments of ds RNA with molecular weight ranging from 0.5 to $2.7 \times 10^6$ kDa. There are three large segments (L1, L2, L3), three medium segments (M1, M2, M3), and four small segments (S1, S2, S3, S4). Viral replication occurs in the cytoplasm of infected cells, in which each genome fragment, corresponding to a single gene, is transcribed by a viral transcriptase into a messenger RNA (mRNA) molecule and subsequently translated into a primary polypeptide. Viral assembly takes place within cytoplasmic inclusions, and infected cells are ultimately destroyed with release of the progeny virus.

Mammalian reoviruses can be divided into three serotypes (types 1, 2, and 3) on the basis of neutralization and hemagglutination inhibition assays. The type-specific antigen is the sigma-1 protein, encoded by the S1 gene, and serves

CARLO GARZELLI • Institute of Microbiology, University of Pisa, 56100 Pisa, Italy.    TAKASHI ONODERA • Laboratory of Immunology, National Institute of Animal Health, Kodaira, Tokyo 187, Japan as the viral hemagglutinin.[2] The presence of a segmented genome has made it possible to isolate a number of recombinant viral clones containing reassorted genome segments derived from either type 1 or 3 reoviruses, [3] providing an excellent model for studying different aspects of viral pathogenesis at the genetic and molecular level.[4,5]

## 1.2. Clinical Manifestations Associated with Reovirus Infections

Serologic studies have shown that infection with each of the three serotypes of mammalian reoviruses is extremely common. Reoviruses can be isolated from feces and respiratory secretions of healthy persons and children with mild respiratory or intestinal diseases, but a precise relationship with clinical illness in humans is not established. In this sense, respiratory enteric orphan viruses, i.e., reoviruses, are indeed still orphan of disease.

Conversely, experimental infection of suckling mice with reoviruses causes a serious disease characterized by an acute phase, in which the virus replicates in cells of many organs, producing lesions notably in the liver, pancreas, central nervous system (CNS), heart, intestine, lung, spleen, salivary glands, and skeletal muscle. At least in some cases, the viral hemagglutinin, which appears to be solely responsible for the cell and tissue tropism, is a major factor in determining the pattern of disease: in the CNS, reovirus type 1 infects primarily ependymal cells, often leading to hydrocephalus, whereas type 3 infects neurons, resulting in severe encephalitis[4]; similarly, reovirus type 1 infects the anterior pituitary, while type 3 infects the intermediate and posterior pituitary.[6]

In a proportion of animals, the acute phase of illness is usually followed by a chronic, apparently virus-free, disease expressed as a runting syndrome characterized by retarded growth, oily hair, alopecia, and steatorrhea. The precise mechanisms of reovirus-induced runting are not clear, but it is thought to be immunologically mediated. In reovirus type 1 infection, autoimmunity to endocrine tissues and hormones seems to be important[7]; in fact, autoantibodies reactive with the anterior pituitary and growth hormone, with the pancreatic islets and insulin, and with the thyroid and thyroglobulin (Fig. 1) are detectable in the serum of infected mice. Monoclonal autoantibodies with a similar pattern of reactivity have also been isolated from hybridomas obtained by fusing spleen cells from reovirus type 1-infected mice with mouse myeloma cells.[8] Autoantibodies to thymic lymphocytes[9] and to splenic T lymphocytes (C. Garzelli, personal observations) have also been found in the serum of infected mice. Further evidence that autoimmunity contributes to reovirus type 1-induced runting and polyendocrinopathy is the observation that the severity of both disorders can be appreciably reduced if the animals are given immunosuppressive treatments before infection.[9]

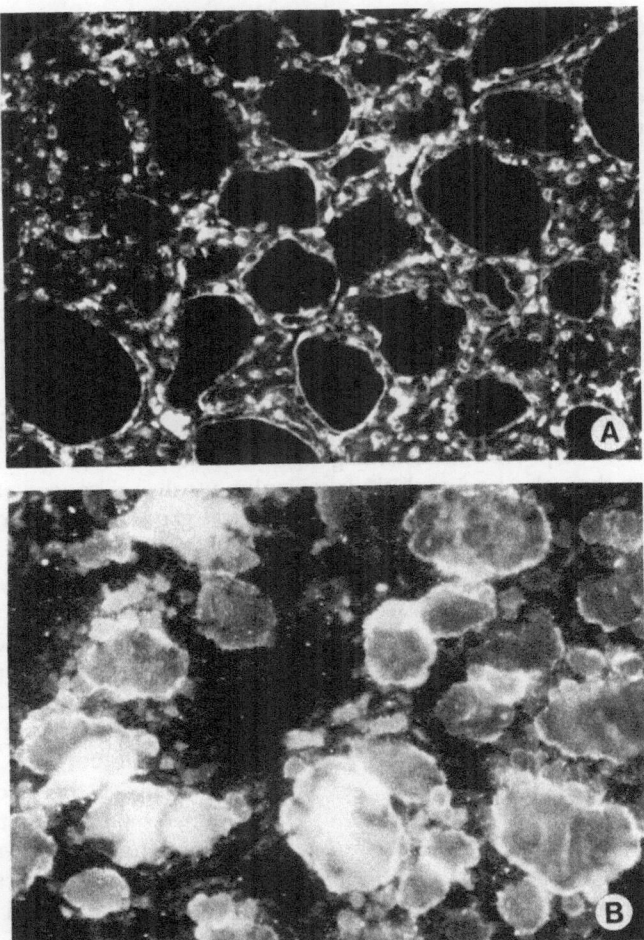

**FIGURE 1.** Autoantibodies to thyroid antigens in the serum of newborn SJL/J mice infected with reovirus type 1 (Lang strain) 3 weeks previously, as determined by indirect immunofluorescence. (A) Bouin's fixed section of thyroid tissue of normal mice incubated with serum of infected mouse. Antiacinar cells and antinuclear antibodies are seen. (B) Methanol-fixed section of thyroid tissue of normal mice incubated as above, showing the typical floccular puffy pattern of staining, characteristic of antithyroglobulin antibodies.

## 2. MODULATION OF IMMUNITY BY REOVIRUSES

### 2.1. Interaction of Reoviruses with Cells of the Immune System

Reoviruses bind to the surface of murine and human lymphocytes. In particular, reovirus type 3 binds to a subset of murine and human T and B cells, whereas reovirus type 1 binds minimally and only to human lymphocytes.[10] Murine and human T lymphocytes bearing the receptor for reovirus type 3 belong predominantly to the suppressor/cytotoxic subset, in that they express the Lyt 2,3 and T8 antigens, respectively.[11] One documented functional consequence of reovirus type 3 binding to lymphocytes is the generation of suppressor T cells, which inhibit the proliferative response of murine spleen cells to mitogens.[12] This effect, as well as the capabilities of lymphocyte binding, is a property of the viral hemagglutinin. However, reovirus types 1 and 3 seem unable to grow in either normal or mitogen-stimulated lymphocytes.

Much less is known about the interactions of reoviruses with macrophages. *In vitro*, both reovirus type 1 and 3 infect and grow in the mouse macrophage-like cell line P388D$_1$. Non-neutralizing antibodies directed to virus surface proteins other than the hemagglutinin, or subneutralizing concentrations of antibodies to the viral hemagglutinin enhance viral growth in these cells.[13]

### 2.2. Suppression of Immune Responsiveness by Reoviruses

Experimental infection of adult mice with reoviruses has practically no significant consequences regarding immune system function, but intraperitoneal infection of newborn mice with reovirus type 1 has been shown to cause strong alterations of immune responsiveness. The most striking immune defect associated with reovirus type 1 infection is undoubtedly the development of a vast array of autoantibodies, most of which are directed toward endocrine tissues and hormones.[7,8] This effect seems to be a property of the S1 gene product, i.e., the viral hemagglutinin. Recombinant viral clones containing the S1 gene segment from type 1, but not those containing the same segment from type 3, reproduce the ability of the parental type 1 virus to induce the production of autoantibodies to hormones, such as growth hormone and thyroglobulin[7] (T. Onodera and C. Garzelli, unpublished observations).

Infection of newborn mice with reovirus type 1 is also followed by a marked suppression of the antibody response to sheep erythrocytes (SRBCs), as measured by the splenic plaque-forming cells (PFC) assay.[14] The PFC produced are usually less than 1% of that detected in uninfected control animals. In addition, reovirus types 2 and 3 cause reduction of antibody production in newborn mice, but the immunodepression is usually slight and transient (T. Onodera and C. Garzelli, unpublished observations). The S1 gene segment of reovirus type 1 is also required for the immunosuppression, in that recombinant viruses containing the S1 gene segment from reovirus type 1, but not from type 3, have immunosuppressive properties comparable to that of the type 1 parental virus.[14]

Other immunologic abnormalities observed in reovirus type 1-infected mice are listed in Table I. At least in severely runted mice, the lymphoid organs show histopathologic alterations; the thymus is usually smaller than normal, and atrophy of the cortex and increased numbers of Hassal's bodies can be observed (Fig. 2a); the spleen, although larger in size, shows depletion of lymphocytes in the thymus-dependent and follicular areas (Fig. 2b). Such changes are probably due to a selective depletion of particular subsets of T lymphocytes, since in the spleen of most infected animals the proportion of cytotoxic/suppressor Lyt 2.2$^+$ cells is decreased, whereas the numbers of helper Lyt 1.2$^+$ cells are not altered (Fig. 3). The autoantibodies to T lymphocytes, present in the serum of type 1-infected mice, could be responsible for these changes. A further important immunologic abnormality seen in reovirus type 1-infected mice is a polyclonal activation of B cells, as evidenced by increased numbers of splenic PFC spontaneously producing immunoglobulins (Ig) or antibodies to SRBCs. Serum levels of IgM and IgG are also generally higher (T. Onodera and C. Garzelli, unpublished observations).

On the basis of these findings, it is difficult to delineate a possible picture of events that might account for reovirus type 1-induced immunosuppression; however, several obvious possibilities immediately come to mind. First, the virus might destroy cells of the immune system through cytolytic infection of lymphocytes; reportedly, reovirus type 1 does not bind to lymphocytes of adult mice *in vitro*,[10] but there are no data dealing with lymphoid cells of newborn mice *in vivo*. A variety of factors could influence the induction and expression of receptors to reovirus and the susceptibility of lymphoid cells to the infection, as already shown in other cases of virus–lymphocyte interactions.[15] A second possibility is also that the cells of the immune system are destroyed by virus-induced anti-lymphocyte autoantibodies, which might lead to hyporesponsiveness to antigenic stimuli. A third possibility is that the endocrine abnormalities induced by infection of endocrine organs, such as the anterior pituitary, might influence immune responsiveness in a negative manner.[16]

## TABLE I
**Immunologic Abnormalities in Newborn Mice Infected with Reovirus Type 1**

Histopathologic changes in the thymus and spleen
Suppression of antibody responses to foreign antigens
Decreased numbers of suppressor/cytotoxic lymphocytes
Polyclonal B-cell activation
Autoantibodies to thymocytes
Autoantibodies to endocrine tissues (pituitary, pancreatic islets, thyroid, gastric mucosa) and hormones (growth hormone, insulin, thyroglobulin)

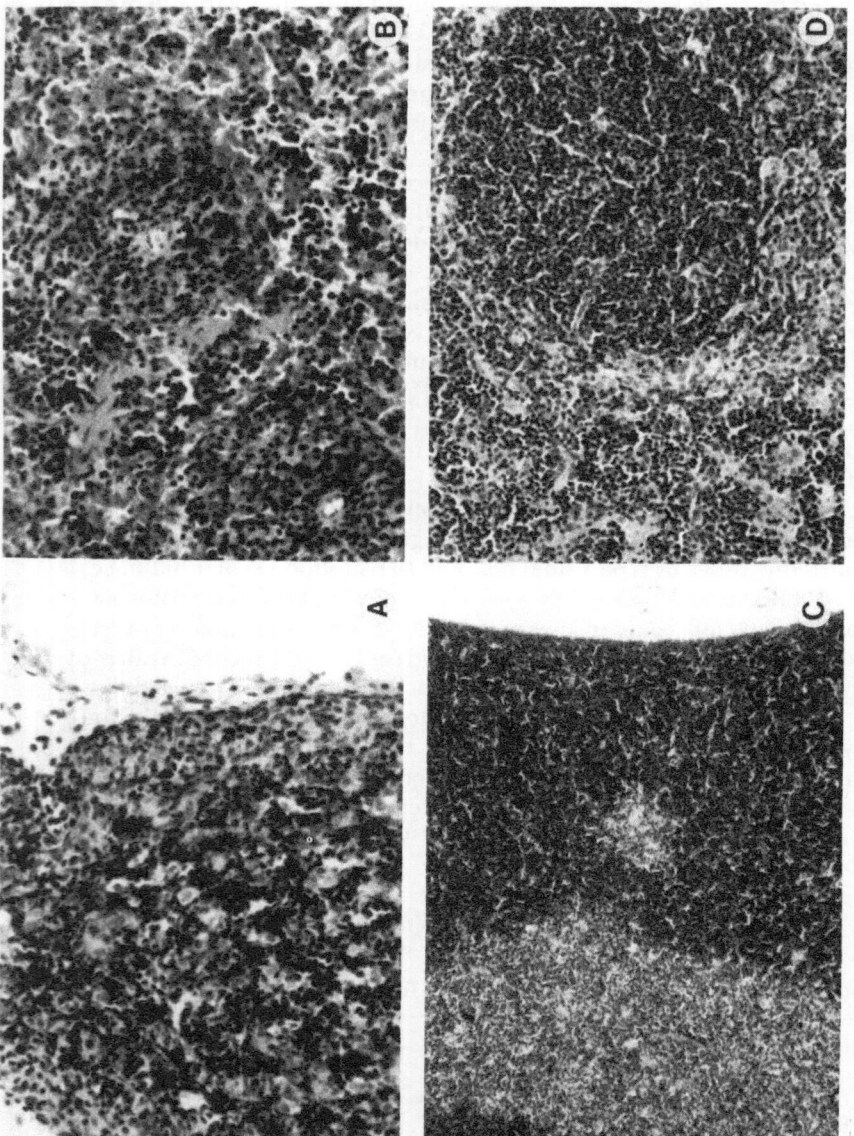

**FIGURE 2.** Histologic changes in lymphoid organs of newborn SJL/J mice infected with reovirus type 1 (Lang strain) 3 weeks previously. (A) Thymus. Atrophy of the thymus cortex and increased numbers of Hassal's bodies in the medulla. (B) Spleen. Atrophy of the spleen and lymphocyte depletion from thymus-dependent and follicular areas; fibrosis and increased numbers of phagocytes can be observed in parafollicular areas. (C) Thymus, and (D) spleen sections from uninfected 3-week-old control mice. Sections stained with hematoxylin and eosin.

**FIGURE 3.** T-lymphocyte subpopulations in the spleen of newborn SJL/J mice infected with reovirus type 1 (Lang strain) 3 weeks previously, as detected by complement-dependent cytotoxicity assayed by Trypan blue dye exclusion. (●) Lyt 1.2$^+$ helper/inducer lymphocytes. (○) Lyt 2.2$^+$ cytotoxic/suppressor lymphocytes. Data expressed as specific cytotoxicity detected in spleen cell suspensions from a single animal. Most of reovirus type-1 infected animals show decreased numbers of Lyt 2.2$^+$ suppressor/cytotoxic cells.

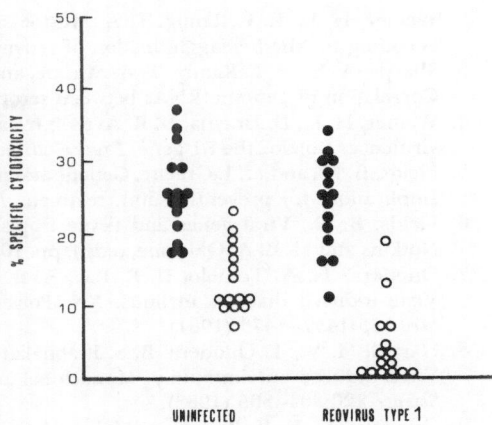

## 3. SIGNIFICANCE OF VIRUS-INDUCED IMMUNOMODULATION IN VIRAL PATHOGENESIS

The available information does not permit delineation of the contribution of viral immunodepression to reovirus-induced pathogenesis. In experimental reovirus type 1 infection of newborn mice, virus-induced immunosuppression is probably irrelevant to the infectious phase of the disease, since the virus rapidly disappears from the body, even if the animals have been treated with immunosuppressive agents before infection. Therefore, immunosuppression would appear to represent an epiphenomenon of infection rather than a means by which the virus avoids or delays immune elimination and successfully spreads through the body. By contrast, the observation that the same molecule, i.e., the viral hemagglutinin, is required for the induction of both autoimmunity and immunosuppression could suggest that, at least in the reovirus type 1 model, the two effects are closely related or even interdependent. Although the coexistence of both immunosuppression and autoimmunity is not unprecedented, the mechanisms leading to the paradoxical occurrence of an antibody response to many autoantigens and the inability of making an immune response to newly presented foreign antigens remain unexplained.

ACKNOWLEDGMENTS. We thank Dr. A. L. Notkins for introducing us to this area of research and Dr. A. Toniolo for critical revision of the manuscript. Part of this work was supported by funds from the Italian Ministry of Public Education.

## REFERENCES

1. Matthews, R. E. F., Fourth report of the International Committee on taxonomy of viruses. Classification and Nomenclature of Viruses, *Intervirology* **17**:1–199 (1982).

2. Weiner, H. L., R. F. Ramig, T. A. Mustoe, and B. N. Fields, Identification of the gene encoding for the hemagglutination of reovirus, *Virology* **86:**581–584 (1978).
3. Sharpe, A. H., R. F. Ramig, T. A. Mustoe, and B. N. Fields, A genetic map of reovirus. I. Correlation of genome RNAs between serotypes 1, 2, and 3, *Virology* **84:**63–74 (1978).
4. Weiner, H. L., D. Drayna, D. R. Averill, Jr., and B. N. Fields, Molecular basis of reovirus virulence: Role of the S1 gene, *Proc. Natl. Acad. Sci. USA* **74:**5744–5748 (1977).
5. Fields, B. N., and M. I. Greene, Genetic and molecular mechanisms of viral pathogenesis: Implications for prevention and treatment, *Nature (Lond.)* **300:**19–23 (1982).
6. Fields, B. N., Viral genes and tissue tropism, in: *Concepts in Viral Pathogenesis* (A. L. Notkins and M. B. A. Oldstone, (eds.), pp. 102–108, Springer-Verlag, New York (1984).
7. Onodera, T., A. Toniolo, U. R. Ray, A. B. Jenson, R. A. Knazek, and A. L. Notkins, Virus-induced diabetes mellitus. XX. Polyendocrinopathy and autoimmunity, *J. Exp. Med.* **153:**1457–1473 (1981).
8. Haspel, M. V., T. Onodera, B. S. Prabhakar, M. Horita, H. Suzuki, and A. L. Notkins, Virus-induced autoimmunity: Monoclonal antibodies that react with endocrine tissues, *Science* **220:**304–306 (1983).
9. Onodera, T., U. R. Ray, K. A. Melez, H. Suzuki, A. Toniolo, and A. L. Notkins, Virus-induced diabetes mellitus: Autoimmunity and polyendocrine disease prevented by immunosuppression, *Nature (Lond.)* **297:**66–68 (1982).
10. Weiner, H. L., K. A. Ault, and B. N. Fields, Interaction of reovirus with cell surface receptors. I. Murine and human lymphocytes have a receptor for the hemagglutinin of reovirus type 3, *J. Immunol.* **124:**2143–2148 (1980).
11. Epstein, R. L., R. Finberg, M. L. Powers, and H. L. Weiner, Interaction of reovirus with cell surface receptors. IV. The reovirus type 3 receptor is expressed predominantly on murine Lyt-2,3$^+$ and human T8$^+$ cells, *J. Immunol.* **133:**1614–1617 (1984).
12. Fontana, A., and H. L. Weiner, Interaction of reovirus with cell surface receptors. II. Generation of suppressor T cells by hemagglutinin of reovirus type 3, *J. Immunol.* **125:**2660–2664 (1980).
13. Burstin, S. J., M. W. Brandriss, and J. J. Schlesinger, Infection of a macrophage-like cell line, P338D$_1$ with reovirus. Effects of immune ascitic fluids and monoclonal antibodies on the neutralization and on enhancement of viral growth, *J. Immunol.* **130:**2915–2919 (1983).
14. Garzelli, C., T. Onodera, U. R. Ray, and A. L. Notkins, The S1 gene from reovirus type 1 is required for immunosuppression, *J. Infect. Dis.* **152:**640–643 (1985).
15. Morishima, T., P. R. McClintock, L. C. Billups, and A. L. Notkins, Expression and modulation of virus receptors on lymphoid and myeloid cells: Relationship to infectivity, *Virology* **116:**605–618 (1982).
16. Besedovski, H. O., A. Del Ray, and E. Sorkin, Neuroendocrine immunoregulation, in:*Immunoregulation* (N. Fabris, E. Garaci, J. Hadden, and N. A. Mitchison, eds.), pp. 315–339, Plenum, New York (1983).

# 11

# Immunosuppression by Avian Infectious Bursal Disease Virus and Mouse Hepatitis Virus

JAGDEV M. SHARMA, J. M. DUPUY, and
L. LAMONTAGNE

## 1. INTRODUCTION

This chapter briefly discusses two viruses that infect diverse species of animals but that share an important similarity in that both viruses are lymphotropic and cause profound immunosuppression in their respective hosts. Infectious bursal disease (IBD) of chickens, also referred to as Gumboro disease, is an economically important disease of commercial chickens. In unprotected chickens, the IBD virus (IBDV) rapidly destroys the lymphocyte population in the bursa of Fabricius, the principal organ that regulates humoral immunity in the chicken. Continued economic loss due to IBD in the field and recent general interest in viral immunosuppression have stimulated renewed efforts in understanding the characteristics of the immunosuppressive effects of this disease. The mouse

JAGDEV M. SHARMA • Regional Poultry Research Laboratory, U. S. Department of Agriculture, Agricultural Research Service, East Lansing, Michigan 48823. *Present address:* Department of Veterinary Pathobiology, College of Veterinary Medicine, University of Minnesota, St. Paul, Minnesota 55108. J. M. DUPUY and L. LAMONTAGNE • Immunology Research Center, Institute Armand-Frappier, University of Quebec, Quebec, Canada. *Present address for J.M.D.:* 5 Boulevard des Belges, Lyon 6906, France. *Present address for L.L.:* Department of Biological Sciences, University of Quebec at Montreal, Montreal, Quebec H3C 3P8, Canada.

hepatitis virus (MHV), also a common infection in laboratory mouse colonies, causes a debilitating disease accompanied by severe immunosuppression. The influence of MHV on immune functions of the host seems to be related to a close interaction between virus particles and host lymphoid cells.

This discussion is not intended to be a comprehensive review of IBDV and MHV. Only the important features of the infections are discussed, with emphasis on the influence these viruses have on the immune capabilities of the host.

## 2. IMMUNOSUPPRESSION BY AVIAN INFECTIOUS BURSAL DISEASE VIRUS

### 2.1. A Characterization of the Virus and the Disease

Infectious bursal disease virus is widespread in the environment and infects most commercial populations of chickens early in life. The virus nucleocapsid is a naked icosahedron with 32 capsomeres and a diameter of 55–63 nm.[1] The IBDV genome is double-stranded RNA.[2] Recent molecular cloning studies with an Australian isolate of IBDV have demonstrated that the genome has a large segment of 3400 bp and a small segment of 2900 bp.[3] The large segment codes for five proteins of molecular weights 52, 41, 32, 28, and 16, kDa, respectively, whereas the small segment codes for a single protein of 90 kDa. Several viral structural proteins identified from purified virus preparations have been examined for their immunogenic potential.[4]

In the laboratory, IBDV can be propagated in embryonated chicken eggs. Best virus yields may be obtained by inoculating 9- to 10-day-old embryos from IBDV-free flocks by the dropped chorioallantoic membrane route.[5] Some isolates of IBDV have been adapted to cell cultures of avian and mammalian origin. Chick embryo fibroblast cells are used most frequently for *in vitro* studies with cell-culture-adapted IBDV.[6]

The chicken is the most common natural host of IBDV, although natural infection may also occur in other avian species, particularly turkeys. The IBDV isolates may be classified into serotypes 1 and 2. The two serotypes cross-react by the immunofluorescent test but not by the virus neutralization test.[7,8] Most isolates of chicken origin fall into serotype 1 and most isolates of turkey origin into serotype 2, although there is no strict species restriction of the two serotypes. Under natural conditions, turkeys exposed to IBDV do not develop clinical disease or detectable immunosuppression.

Cosgrove[9] first reported the disease in chickens. Chickens acquire infection from contaminated premesis; there is no evidence for vertical transmission. The virus replicates in B lymphocytes and one of the first detectable lesions is necrosis of lymphoid elements in the bursa of Fabricius.[10] Bursal necrosis is accompanied by inflammatory changes. Other lymphoid organs such as spleen and thymus also experience transient lymphoid cell depletion. Bursal degeneration is permanent. Predilection of IBDV for B lymphocytes was also demonstrated *in vitro*. Established lymphoid cell lines of B but not of T cells were susceptible to infection with IBDV.[11]

The age of the chicken at the time of infection with IBDV seems to determine the nature of the ensuing disease. In chickens younger than 3 weeks, IBDV does not cause high mortality, although the chickens develop bursal atrophy and severe immunosuppression. In older chickens, clinical disease occurs and is characterized by sudden onset, variable but often high mortality, and rapid recovery of survivors.

The principal economic concern with IBDV in commercial chicken flocks is the effect of this disease on immune competence of young chickens. Because infection occurs soon after hatching, under natural conditions the infected chickens respond poorly to vaccines used routinely to protect against common viral infections. Infected chickens also become vulnerable to opportunistic infections. The nature of IBDV-induced immunosuppression in humoral and cellular responses is described below.

## 2.2. Influence on Circulating B and T Lymphocytes

Several attempts have been made to study the effect of IBDV on circulating B and T lymphocytes.[12-14] In general, infection with IBDV reduced the number of circulating B cells. The depression in B-cell numbers was more pronounced in chickens exposed to the virus in ovo or at the time of hatching than in those in which exposure was delayed until the birds were 3 weeks of age or older. The reduction in circulating B cells was detected within 1 week after virus inoculation and persisted through the observation period of 8 weeks.

The influence of IBDV on circulating T lymphocytes was variable. In one study,[14] T-cell numbers were reduced below control levels if infection occurred at the time of hatching but were increased if the infection was delayed until birds were 3 weeks of age. In another study,[13] this relationship of age at the time of infection with numbers of circulating T cells was the reverse of the previous findings.

## 2.3. Influence on Antibody Production

Infectious bursal disease virus severely compromises the ability of chickens to mount antibody responses against a variety of infectious and noninfectious antigens including viral, bacterial, and protozoan antigens.[13,15-21] The age at which chickens become exposed to the virus has a profound effect on the degree of B-cell immunosuppression. Infection during the first 2 weeks of age results in much more severe immunodepression than does infection at older ages.[22] Both primary and secondary antibody responses may be reduced.[13,16] B-cell immunosuppression following infection with IBDV during the early posthatching period is probably persistent, although the duration of immunosuppression has not been well established.

Infection with IBDV also affects serum immunoglobulin (Ig) levels. Serum IgM levels generally dropped following IBDV infection; whereas IgG levels varied depending on the age of the chicken at the time of infection.[13,16] The IgG levels measured at eight weeks of age were lower in virus infected chickens than in age-matched control chickens if infection occurred before or at the time of hatching but the levels were elevated if infection occurred at one week of age or older.[13] IBDV also caused a defect in the IgM that was produced. Ivanyi and Morris [16] noted that chickens infected with IBDV exclusively produced IgM as a 7S monomer. Further, the IgM of infected chickens lost the MI[a] allotypic marker normally present on chicken IgM.[23]

## 2.4. Influence on Cellular Immune Functions

Circumstantial evidence strongly indicates that IBDV may compromise cell-mediated immune functions. For example, infection with IBDV results in (1) extensive histologic lesions in the thymus and the virus replicates to high titers in the thymus [10,24]; (2) reduction in circulating T cells[14]; (3) poor efficacy of Marek disease vaccine that appears to protect mainly via cell-mediated immunity[25]; and (4) exacerbation of disease conditions in which defense by cellular immune mechanisms if important.[19,26] Despite compelling indications that IBDV may influence cell-mediated immunity, relatively meager efforts have been devoted to study this influence.

There have been conflicting reports on the ability of young chickens exposed to IBDV to reject allogeneic skin grafts; in one study, the rejection was delayed,[27] while in the others it was not.[12,18] The evidence that IBDV may influence cellular immunity comes from *in vitro* studies. Most efforts have been directed toward delineating the mitogenic response of T cells,[28–31] although other cellular functions have also been examined.[28,31,32]

Preparations of T cells obtained from IBDV-exposed chickens respond poorly to mitogens such as phytohemagglutinin (PHA) and concanavalin A (Con A). Infection at the time of hatching as well as at 3–4 weeks of age influenced mitogenic response. The depression in the mitogenic response was transient, although the time when it occurred following viral inoculation varied. When whole blood cultures were used, the T-cell responsiveness was reduced during the first 2 weeks postinfection,[28,30,31] although in one study,[28] maximum reduction occurred 6–7 weeks after viral infection. In assays conducted with peripheral blood leukocytes fractionated on Ficoll-Hypaque[30] or spleen cells,[31] the mitogenic hyporesponsiveness was consistently transient and occurred during the first 1–2 weeks of viral infection followed by complete recovery of responsiveness.

Peripheral blood leukocytes from IBDV-infected chickens were also deficient in mounting a mixed lymphocyte reaction when cocultured with allogeneic stimulator cells.[28] A reduced mixed lymphocyte response was detected in chickens exposed to IBDV at the time of hatching or at 3 weeks of age; chickens infected at the time of hatching were more severely affected than were those infected at 3 weeks. Interestingly, unlike the mitogenic response that was affected transiently, the defect in mixed lymphocyte reaction was persistent and was detectable until the birds were 10 weeks of age, the longest interval between infection and testing. The ability of cells from virus-infected chickens to serve as stimulator cells in the mixed lymphocyte reaction assay has not been examined.

The natural killer (NK) cell activity of spleen effector cells from virus-exposed chickens was compared with that of the effector cells obtained from age-matched normal chickens.[31] No consistent differences in the activity between the two groups were noted. Similarly, IBDV did not cause detectable alteration of phagocytic activity of circulating phagocytes.[32]

## 2.5. Influence on Soluble Immune Factors

Little is known about the effect of IBDV on soluble mediators of immunity. Interferon-$\beta$ ($IFN_\beta$) was detected in a variety of tissues and serum following inoculation with IBDV at 1 day or 3 weeks of age.[33] The role played by $IFN_\beta$ in regulating immune functions is unknown. Currently, avian lymphokines are being actively studied; it should be of interest to examine possible modulation of these by IBDV.

## 2.6. Mechanism of Immunosuppression

### 2.6.1. B-Cell Immunity

It is likely that one of the major reasons for depressed antibody synthesis in chickens exposed to IBDV is that the virus selectively infects and lyses B lymphocytes. The observation that infection during the early posthatching period is more immunodepressive than infection after 3 weeks of age indicates that B-cell precursors within the confines of the bursa may be more susceptible to the cytopathic effects of IBDV than are mature B cells in circulation. Indeed, when Ivanyi and Morris[16] delayed infection with IBDV from less than 6 up to 42 hr after hatching, they noted progressively decreasing proportions of birds with immune deficiency in anti-sheep erythrocyte (anti-SRBC) antibody responses. Decreased number of circulating B cells following IBDV infection of neonates[12–14] may also indicate intrabursal destruction of B-cell precursors resulting in reduced peripheralization of B cells to the circulation. Selective susceptibility of B cells with IgM but not IgG receptors[11] further suggests that the virus is more cytopathic for B cells during early stages of differentiation before the switch from IgM to IgG expression occurs.

The mechanism by which IBDV induces the production of altered IgM, i.e., monomeric IgM that fails to polymerize and loses allotypic marker MI[a], is not known.[16] The virus may destroy IgM-producing cells that may be replaced by a population of B cells with functional impairment expressed by production of altered IgM.

Other mechanisms may also be involved in the suppression of B-cell function. For example, IBDV may compromise antibody production by damaging helper T cells or other accessory cells such as macrophages that play an important role in generating B-cell responses to certain antigens. In addition to destroying B cells, IBDV may also stimulate the appearance of suppressor cells that may participate in inhibiting antibody responses.[34]

### 2.6.2. T-Cell Immunity

The mechanism of hyporesponsiveness of T cells to mitogen stimulation has been examined.[29,31] We noted that spleen cells of chickens undergoing acute infection with IBDV responded poorly to PHA but that their response was restored to near normal levels if the responder cells were pretreated with carbonyl iron.[31] Thus, the spleen cell response was being inhibited by suppressor cells that could be removed by carbonyl iron treatment. The suppressor cells shared several characteristics with macrophages, i.e., the suppressor cells were adherent to plastic, were phagocytic, and resisted treatment with antithymocyte and antibursa cell sera. Suppressor cells isolated from spleen of IBDV-infected chickens were able to inhibit the mitogenic response of spleen cells of normal virus-free chickens. We recently confirmed the presence of suppressor cells in IBDV spleens using Con A as a T-cell mitogen and have shown that addition of exogenous conditioned medium with high interleukin-2 (IL-2) activity was ineffective in restoring the mitogenic response of spleen cells of IBDV-infected chickens.[35]

The above observations suggested that reduced mitogenic response of lymphocytes in IBDV-infected chickens was not due to lack of functional T cells but to the presence of suppressor cells. Other mechanisms of T-cell immunosuppression may be involved as well. Confer and MacWilliams[29] suggested that the mitogenic hyporesponsiveness of whole blood cells from IBDV-infected chickens was associated with increased number of circulating large immature lymphocytes incapable of mitogen-induced blastogenesis.

## 3. IMMUNOSUPPRESSION BY MOUSE HEPATITIS VIRUS

### 3.1. Biology of the Virus

Mouse hepatitis viruses are classified as coronaviruses.[36] They are pleomorphic or rounded enveloped particles with a diameter of 60–220 nm, surrounded by a fringe or layer of typical club-shaped spikes. Their genome consists of single-stranded polyadenylated RNA of positive polarity. Viruses are released by internal budding into cytoplasmic vesicles derived from the endoplasmic reticulum.

The antigenicity of coronaviruses is related to three major antigens.[37,38] Surface glycoproteins of murine coronaviruses are responsible for the induction of neutralizing, complement-fixing, and hemagglutination-inhibiting antibodies.[39–41] Hybridization with MHV-specific complementary DNA (cDNA) showed a close relationship among murine strains MHV-A59, MHV-3, and JHM.[42] Oligofingerprinting demonstrated genomic variations in MHV strains that could be related to neurovirulence.[42–44] These variations did not seem to correlate with the serologic relationships of these viruses.

The coronavirus genome is a positive single-stranded infectious molecule of RNA containing about 18,000 nucleotides.[45] $T_1$ oligonucleotide mapping indicated that no extensive sequence reiteration occurred in the coronavirus genome[43,46,47] and that the 3' end of the genomic RNA was polyadenylated and formed a 3'-coterminal nested sequence.[48,49] Synthesis of each of the intracellular RNAs is initiated independently and is not processed from a large precursor protein.[50,51] Subgenomic RNAs, as well as genomic RNA, contained the 5'-cap structure.[52] These RNA molecules were also found to act as individual mRNAs and to be translated into single proteins of a size corresponding to the coding capacity of the unique 5'-terminal sequences not present in the next smallest RNA.[53-55]

The nucleocapsid protein possesses a molecular weight of 50–60kDa and is nonglycosylated.[56] A cyclic adenosine monophosphate(cAMP)-independent protein kinase is associated with the virion, but it is not yet known whether the enzyme is virally coded or is a sequestered host cell enzyme.[57] A high homology of amino acid sequences has been found between nucleocapsid proteins from neuropathogenic JHM and nonpathogenic A59, although two regions of lower homology are present.[58]

The virion possesses a lipid envelope containing matrix and peplomer proteins. All coronavirions have a glycoprotein of 20–30kDa. A small glycosylated portion of the molecule is peripheral to the lipid membrane,[59] whereas a second strong hydrophobic domain is thought to correspond to a portion of the molecule integrated into the lipid membrane.[60] A stable complex between this protein and the viral RNA could be formed *in vitro*, suggesting a third domain in the protein which is internal to the lipid membrane and responsible for the interaction with viral nucleocapsids.[60,61]

Peplomer proteins are glycoproteins of molecular weight 80–200 kDa with one or two major species derived from a single primary translation product but modified by post-translational cleavage.[62] Glycosylation of the MHV peplomer protein was inhibited by tunicamycin. The lack of reabsorption and cell fusion observed in tunicamycin-treated cells suggested that the peplomer protein plays a role in the reception of virions on cell surfaces and in the induction of cell fusion.[60] In addition, trypsin treatment of A59 virus, which cleaves the 180 kDa to two 90-kDa subunit polypeptides, greatly increases the capacity of the virus to cause cell fusion.[63]

## 3.2. Pathogenesis of Mouse Coronaviruses

The MHV-JHM strain, the first murine coronavirus to be isolated,[64] is a neurotropic virus causing acute and chronic demyelinating diseases. Infection with other mouse hepatitis viruses results in hepatitis, encephalomyelitis and/or enteritis. It is impossible, however, to classify virus strains according to target organs since several organs can be affected. The virulent strains $MHV_2$ and $MHV_3$ and the less virulent $MHV_1$, $MHV_5$ and $MHV_{A59}$ can cause hepatitis in newborns and adult mice, whereas $MHV_S$ induces enteritis in newborns.

$MHV_3$ is the most virulent strain of MHV. The severity and the type of infection however are related to age, immune resistance, and genetic factors.

Most strains of mice display a full susceptibility, leading to death within a few days. The A/J strain is unique, as it is the only full resistant strain with 100% survival of infected adult animals. Other mouse strains are semisusceptible, and animals surviving the acute disease develop chronic manifestations with progressive neurologic involvement.[65]

### 3.2.1. Acute MHV$_3$ Infection

In some strains of mice, e.g., C57BL/6, DBA/2, BALB/c, or NZB, parenteral administration of 10 $LD_{50}$ of MHV$_3$ always leads to fulminant hepatitis and death. Peritoneal macrophages, and liver Kupffer cells are the major sites of viral replication. Infectious virions are disseminated to all organs during the viremic phase.[66] By contrast, full resistance to MHV$_3$ is observed in the A/J mouse strain even after the administration of large doses of virus ($10^7$ $LD_{50}$). Histopathologic studies showed an absence of lesions. During the first 4 days of infection, virus was recovered from the liver of resistant as well as susceptible strains of mice. In A/J strain mice, viral titers were consistently $<10^3$ $LD_{50}$, whereas in susceptible DBA/2 mice, titers greater than $10^4$ were always found. In the resistant mouse strain, infectious virus was cleared from the liver, brain and serum within 7 days, whereas virus continued to replicate in susceptible animals until death.[65]

### 3.2.2. Persistent MHV$_3$ Infection

In contrast to full susceptibility or resistance, other mouse strains, such as C3H or hybrid animals resulting from a cross between susceptible and resistant parents, exhibit an intermediate sensitivity. In this type (semisusceptibility), MHV$_3$ causes a chronic disease with paralysis, virus persistence, and immunodepression. The virus could be recovered from brain, liver, spleen, lymph nodes, or peritoneal macrophages for several months.[67] This type of disease and the clinical outcome vary greatly, however, with age, immune resistance, and genetic factors. Immunologic immaturity as observed in mice during the first 2 weeks of life, or T-cell deprivation following neonatal thymectomy or treatment with antilymphocyte antiserum, induce a full susceptibility to MHV$_3$ infection in mice of the normally resistant A/J strain.[68] Similarly, genetic factors are of a primary importance in MHV infection. The first evidence of an association of host genes with resistance was reported for MHV$_2$.[69] Genetic study of MHV$_3$ infection indicated that at least two recessive genes are involved in resistance to acute and chronic diseases and showed that the genes involved in both diseases are different. The capacity to resist the development of paralysis is conferred to heterozygote as well as to homozygote mice by H-2$^f$ or H-2$^q$ alleles, indicating that resistance to paralysis is H-2 linked.[70] It was also shown that such an action was mostly mediated through the expression of mouse class I antigens of the major histocompatibility complex.[71]

## 3.3. MHV₃-Induced Immunosuppression

MHV$_3$ infection in mice induces a marked nonspecific immunosuppression during the acute as well as the chronic phase of the disease. Sequential determination of Ig levels in chronically infected mice revealed a progressive decrease of all Igs during the first 3 months of infection. At the end of that period, most animals were severely hypogammaglobulinemic, and some suffered from infections.[72] In addition, a marked decrease of the antibody response against T-cell-dependent and -independent antigens was observed in semisusceptible mice chronically infected with MHV$_3$. In these experiments, it was observed that, when tested 40 and 80 days postinfection, primary and secondary plaque-forming cell (PFC) responses and serum antibody titers against SRBC T-dependent antigen and lipopolysaccharide (LPS) T-independent antigen were markedly diminished in paralyzed as well as in nonparalyzed mice.[72]

## 3.4. Immune Functions of MHV₃-Infected Mice

It was noteworthy that chronically infected mice, in spite of their hypogammaglobulinemia, developed anti-MHV$_3$ complement-fixing antibodies. Such antibodies were detectable 2–4 weeks following viral infection, showed an important rise by day 60 and reached a maximum level by days 100–130 after which titers decreased in surviving animals. No correlation was observed, however, between the onset of paralysis and the increase of MHV$_3$ antibody, in spite of the fact that bound Igs associated with antigens were found in chronic plexus vessels.[73] In addition, most of the anti-MHV$_3$ antibody produced in chronically infected mice was of the IgM class.

Antiviral antibodies are capable of modifying a lytic acute infection into a subacute and persistent infection or even to prevent the induction of the chronic disease.[74] The effectiveness of humoral immunity seems to depend on the titer, avidity, and neutralizing capacity of the antibody. In rats infected with the strain JHM, cerebrospinal antibodies seem to occur only in animals expressing a JHM-induced disease with viral replication in the central nervous system (CNS). Nevertheless, localized production of IgG was unable to protect rats from either the acute or the chronic neurologic disease.[75]

Specific cell-mediated immune reactions against MHV$_3$ antigens were detected in chronically infected mice and protection, using lymphoid cells, could be transferred from MHV$_3$-paralyzed animals into susceptible newborns. Similarly, a specific cellular immunity was observed during the course of MHV-JHM infection in rats. Wege et al.[76] showed that spleen cells obtained from diseased animals not only proliferated in the presence of basic myelin proteins in vitro but adoptive transfer of such cells was followed by the occurrence of experimental allergic encephalomyelitis-like lesions in the CNS. Such results suggest that MHV-JHM replication in the CNS may lead to alteration of myelin, and/or to cell membrane changes by insertion of viral proteins that may trigger, in turn, an immune response against myelin.

## 3.5. Mechanism of Immunosuppression

The nonspecific immunodepression always observed in $MHV_3$-infected animals after antigenic challenge could be related to a direct effect of the virus on one or several components of the immune response. This was tested by studying the PFC response in lethally X-irradiated normal $F_1$ hybrids reconstituted with T, B, or spleen cells originating from $MHV_3$-infected or -noninfected syngeneic animals. These experiments clearly showed that the immunocompetence of T and B lymphocytes originating from $MHV_3$-infected mice was normal. In addition, macrophage functions, as tested by phagocytosis of yeast particles in vitro and by the in vivo uptake of radiolabelled SRBC, were not different in infected and control animals.[77]

### 3.5.1. Lymphocyte Depletion

A striking feature always observed during the chronic phase of mouse hepatitis, associated or not with paralysis, is the marked decrease of cell numbers in the lymphoid organs. This is observed in bone marrow, spleen, thymus, lymph nodes, peripheral blood and peritoneal exudates. Differential enumeration, however, of T and B lymphocytes in the spleen at different times postinfection did not demonstrate important variations.

The mechanism whereby $MHV_3$ infection exerts a suppressive effect is related neither to a quantitative deficiency of lymphocytes nor to a dietary insufficiency for the following reasons: (1) healthy virus carrier mice are immune deficient in spite of normal body weight and a lymphocyte count similar to that of noninfected control mice; (2) T-cell functions appeared normal when tested in $MHV_3$-paralyzed mice; (3) T cells, B cells, or macrophages, originating from $MHV_3$-infected paralyzed mice, have the capacity to reconstitute immune responses fully in lethally irradiated syngeneic animals[78]; (4) in spite of a decrease in the total number of bone marrow cells, the number of spleen colony-forming cells increased sharply after infection and reached maximum levels by day 50; and (5) LPS-stimulated bone marrow or spleen B cells obtained from $MHV_3$-infected mice, synthesized in vitro IgM and IgG in amounts comparable to that produced by B cells originating from noninfected controls. In these experiments, it was not possible to detect suppressor cells or factors.[72]

### 3.5.2. Viral Replication in Immunocompetent Cells

Inhibition of lymphocyte activation and/or proliferation by $MHV_3$ seems to be a possible factor in alteration of immune functions. Previous work indicated indeed that lymphocytes supported $MHV_3$ replication, which in turn interfered with cellular metabolism. Viral replication in macrophage-depleted T lymphocytes was demonstrated by a progressive increase of viral titers in culture supernatants and further evidenced by dot immunobinding analysis.[77] Infection of lymphocytes with virulent $MHV_3$, however, results in a strong inhibition of the lymphoproliferative response of cells stimulated with mitogens or with allogeneic cells. Viral infection of either stimulator cells or responder cells in mixed lymphocyte reactions indicated that the inhibitory effect was due to a direct contact of infectious viral particles with the proliferating cells. Since virus can be regularly recovered from brain, liver, spleen, lymph nodes, and peritoneal macrophages of MHV$_3$-infected mice during the evolution of the disease (up to 7 months postinfection), it is likely that persistent viral replication in lymphocytes and macrophages is responsible, at least in part, for immunosuppression and for some of the immunopathologic effects seen during the course of the chronic disease.

### 3.5.3. Genetic Influence

Since the direct interaction between MHV$_3$ and lymphoid cells appears to be a major factor for immunosuppression, the intrinsic capacity of cells to restrict viral replication was studied. Persistent MHV$_3$ infection can readily be induced in vitro in fibroblast and lymphocyte cultures.[79,80] Therefore, work was performed to see whether genetically controlled virus persistency could be detected at the in vitro level. A carrier state was established in vitro using short-term progeny passages in cells originating from various mouse strains exhibiting different sensitivities. Results showed a correlation between pehnotypic expression of in vivo sensitivity and the capacity, which was not H-2 linked, of macrophages or lymphocytes to restrict viral replication. This indicates that resistance to MHV$_3$ may be the result of restriction of viral replication in macrophages and lymphoid cells. Such restriction of MHV replication in macrophages from resistant strain mice also has been observed with other serotypes.[69,81] Macrophages genetically resistant to MHV$_2$ were converted in vitro to susceptible macrophages by lymphokines present in the supernatant fluid from allogeneic mixed lymphocyte cultures,[82] by spleen cells from cortisone-treated mice,[83] or by silica treatment.[84] By contrast, genetically susceptible macrophages were converted into resistant cells by Con A administration.[85]

In addition, resistance to viral infection displayed by other target cells, such as hepatocytes[86] and fibroblasts,[79] should minimize pathologic damage and therefore ensure the survival of infected mice and the development of an adequate immune response, leading to total elimination of virus. By contrast, susceptible mice infected with the virus can neither restrict viral replication nor resist virus-induced cellular injuries. Dissemination of the infection thus leads to extensive pathologic lesions and death. The intermediate behavior, displayed by infected cells originating from semisusceptible mouse strains, seems to be related to an incomplete restriction of viral replication leading to virus persistence, cell lysis, and subsequent immunodepression. Resistance mechanisms, genetically determined and expressed at macrophage and lymphocyte levels, by controlling or not controlling viral replication, therefore seem to play a major role in MHV$_3$-induced immunosuppression.

## 4. SUMMARY

Two viral diseases with strong immunosuppressive effects were discussed. Avian IBDV replicates in B lymphocytes and causes clinical disease and associated immunosuppression. Infected chickens fail to produce antibody against a variety of antigens and show reduced T-cell response to mitogenic stimulation *in vitro*. The mechanism of immunosuppression is not entirely clear, although lysis of B cells or B-cell precursors by IBDV is likely the cause of failure of the antibody response. The mitogenic hyporesponsiveness of T cells appears to be mediated by virus-activated suppressor macrophage-like cells.

Mouse hepatitis virus, a coronavirus that causes an acute or chronic disease in laboratory mice, also induces immunosuppression. Infected animals develop persistent hypogammaglobulinemia and show decreased antibody response against T-dependent and T-independent antigens. The virus replicates in lymphocytes and macrophages, and the reduced immune responsiveness likely results from direct interaction of the virus with cells of the immune system. Suppressor cells or factors do not seem to be involved.

## REFERENCES

1. Hirai, K., N. Kato, A. Fujiura, and S. Shimakura, Further morphological characterization and structural proteins of infectious bursal disease virus, *J. Virol.* **32**:323–328 (1979).
2. Muller, H., C. Scholtissek, and H. Becht, The genome of infectious bursal disease virus consists of two segments of double stranded RNA, *J. Virol.* **31**:584–589 (1979).
3. Azad, A. A., S. A. Barrett, and K. J. Fahey, The characterization and molecular cloning of the double-stranded RNA genome of an Australian strain of infectious bursal disease virus, *Virology* **143**:35–44 (1985).
4. Fahey, K. J., I. J. O'Donnell, and A. A. Azad, Characterization by Western blotting of the immunogens of infectious bursal disease virus, *J. Gen. Virol.* **66**:1479–1488 (1985).
5. Hitchner, S. B., Infectivity of infectious bursal disease virus for embryonating eggs, *Poultry Sci.* **49**:511–516 (1970).
6. Lukert, P. D., and R. B. Davis, Infectious bursal disease virus: Growth and characterization in cell cultures, *Avian Dis.* **18**:243–250 (1974).
7. McFerran, J. B., M. S. McNutty, E. R. McKillop, T. J. Conner, R. M. McCracken, D. S., Collins, and G. M. Allan, Isolation and serological studies with infectious bursal disease virus from fowl, turkeys and ducks: Demonstration of a second serotype, *Avian Pathol.* **9**:395–403 (1980).
8. Jackwood, D. J., Y. M. Saif, and J. H. Hughes, Characteristics and serologic studies of two serotypes of infectious bursal disease virus in turkeys, *Avian Dis.* **26**:871–882 (1982).
9. Cosgrove, A. S., An apparently new disease of chickens—Avian nephrosis, *Avian Dis.* **6**:385–389 (1962).
10. Cheville, N. F., Studies on the pathogenesis of Gumboro disease in the bursa of Fabricius, spleen and thymus of the chicken, *Am. J. Pathol.* **51**:527–551 (1967).
11. Hirai, K., and B. W. Calnek, In vitro replication of infectious bursal disease virus in established lymphoid cell lines and chicken B lymphocytes, *Infect. Immun.* **25**:964–970 (1979).
12. Hudson, L., M. Pattison, and N. Thantrey, Specific B lymphocyte suppression by infectious bursal agent (Gumboro virus) in chickens, *Eur. J. Immunol.* **5**:675–679 (1975).

13. Hirai, K., K. Kunihiro, and S. Shimakura, Characterization of immunosuppression in chickens by infectious bursal disease virus, *Avian Dis.* **23**:950–965 (1979).

14. Sivanandan, V., and S. K. Maheswaran, Immune profile of infectious bursal disease. I. Effect of infectious bursal disease virus on peripheral blood T and B lymphocytes of chickens, *Avian Dis.* **24**:715–725 (1980).

15. Allan, W. H., J. T. Farraghar, and G. A. Cullen, Immunosuppression by infectious bursal agent in chickens immunized against Newcastle disease, *Vet. Rec.* **90**:511–512 (1972).

16. Ivanyi, J., and R. Morris, Immunodeficiency in the chicken. IV. An immunological study of infectious bursal disease, *Clin. Exp. Immunol.* **23**:154–165 (1976).

17. Giambrone, J. J., C. S. Eidson, and S. H. Kleven, Effect of infectious bursal disease on the response of chickens to *Mycoplasma synoviae*, Newcastle disease virus and infectious bronchitis virus, *Am. J. Vet. Res.* **38**:251–253 (1977).

18. Giambrone, J. J., J. P. Donahoe, D. L. Dawe, and C. S. Eidson, Specific suppression of the bursa-dependent immune system of chicks with infectious bursal disease virus, *Am. J. Vet. Res.* **38**:581–583 (1977)

19. Anderson, W., W. M. Reid, and P. D. Lukert, Influence of infectious bursal disease on the development of immunity to *Eimeria tenella*, *Avian Dis.* **21**:637–641 (1978).

20. Hopkins, I. G., K. R. Edwards, and D. H. Thornton, Measurement of immunosuppression in chickens caused by infectious bursal disease vaccines using *Brucella abortus* strain 19, *Res. Vet. Sci.* **27**:260–261 (1979).

21. Pejkovski, C., F. G. Develaar, and B. Kouwenhoven, Immunosuppressive effect of infectious bursal disease virus on vaccination against infectious bronchitis, *Avian Pathol.* **8**:95–106 (1979).

22. Faragher, J. T., W. H. Allen, and C. J. Wyeth, Immunosuppressive effect of infectious bursal agent on vaccination against Newcastle disease, *Vet. Rec.* **95**:385–388 (1974).

23. Ivanyi, J., Polymorphism of chicken serum allotypes, *J. Immunogenet.* **2**:87–107 (1975).

24. Kaufer, I., and E. Weiss, Significance of bursa of Fabricius as target organ in infectious bursal disease of chickens, *Infect. Immun.* **27**:364–367 (1980).

25. Sharma, J. M., Effect of infectious bursal disease virus on protection against Marek's disease by turkey herpesvirus vaccine, *Avian Dis.* **28**:629–640 (1984).

26. McDonald, L. R., T. Karlson, and W. M. Reid, Interaction of infectious bursal disease and coccidiosis in layer replacement chickens, *Avian Dis.* **24**:999–1005 (1980).

27. Panigraphy, B., L. K. Misra, S. A. Naqi, and C. F. Hall, Prolongation of skin graft survival in chickens with infectious bursal disease, *Poultry Sci.* **56**:1745 (1977).

28. Sivanandan, V., and S. K. Maheswaran, Immune profile of infectious bursal disease. III. Effect of infectious bursal disease on the lymphocyte responses to phytomitogens and on mixed lymphocyte reaction of chickens, *Avian Dis.* **25**:112–121 (1981).

29. Confer, A., and P. S. MacWilliams, Correlation of hematological changes and serum and monocyte inhibition with the early suppression of phytohemagglutinin stimulation of lymphocytes in experimental infectious bursal disease, *Can. J. Comp. Med.* **46**:169–175 (1982).

30. Confer, A. W., W. T. Springer, S. M. Shane, and J. F. Donovan, Sequential mitogen stimulation of peripheral blood lymphocytes from chickens inoculated with infectious bursal disease virus, *Am. J. Vet. Res.* **452**:2109–2113 (1981).

31. Sharma, J. M., and L. F. Lee, Effect of infectious bursal disease on natural killer cell activity and mitogenic response of chicken lymphoid cells: Role of adherent cells in cellular immune suppression, *Infect. Immun.* **42**:747–754 (1983).

32. Santivatr, D. S. K. Maheswaran, J. A. Newman, and B. S. Pomeroy, Effect of infectious bursal disease virus infection on the phagocytosis of *Staphylococcus aureus* by mononuclear phagocytic cells of susceptible and resistant strains of chickens, *Avian Dis.* **25**:303–311 (1981).

33. Gelb, J., C. S. Eidson, O. J. Fletcher, and S. H. Kleven, Studies on interferon induction by

infectious bursal disease virus (IBDV). II. Interferon production in White Leghorn chickens infected with an attenuated or pathogenic isolant of IBDV, *Avian Dis.* **23**:634–645 (1979).

34. Blaese, R. M., A. V. Muchmore., I. Koski., and N. J. Dooley, Infectious agammaglobulinemia: Suppressor T cells with specificity for individual immunoglobulin classes, *Adv. Exp. Biol.* **88**:155–159 (1977).

35. Sharma, J. M., and T. Fredericksen, Mechanism of T cell immunosuppression by infectious bursal disease virus of chickens, in: *Avian Immunology.* Vol. II W. T. Weber and D. L. Ewert, eds., pp. 283–294, Liss, New York (1987).

36. Tyrrell, D. A. J., J. D. Almeida, D. M. Berry, C. H., Hamre, M. S. Hofstad, L. Malluci, and K. McIntosh, Coronaviruses, *Nature (Lond.)*, **220**:650 (1968)

37. Hajer, I., and J. Storz, Antigens of bovine coronavirus strain LY-138 and their diagnostic properties, *Am. J. Vet. Res.* **39**:441–444 (1978).

38. Yaseen, S. A., and M. Johnson-Lussenburg, Antigenic studies on coronavirus. I. Identification of the structural antigens of human coronavirus strain 229E, *Can. J. Microbiol.* **27**:334–342 (1981).

39. Garwes, D. I., M. H. Lucas, D. A. Higgins, B. V. Pike, and S. F. Cartwright, Antigenicity of structural components from porcine transmissible gastroenteritis virus, *Vet. Microbiology* **3**:179–190 (1979).

40. McNaughton, M. R., H. J. Hasony, and S. Reed, Antibody to virus components in volunteers experimentally infected with human coronaviruses 229E group viruses, *Infect. Immun.* **31**:845–849 (1981).

41. Schmidt, O. W., and G. E. Kenny, Immunogenicity and antigenicity of human coronavirus 229E and OC43, *Infect. Immun.* **32**:1000–1006 (1981).

42. Weiss, S. R., and J. L. Leibowitz, Comparison of the RNAs of murine human coronaviruses, in: *Biochemistry and Biology of Coronaviruses* (V. Ter Meulen, S. Siddell, and H. Wege, eds.), pp. 245–259, Plenum, New York (1981).

43. Lai, M. M. C., and S. A. Stohlman, Comparative analysis of RNA genome of mouse hepatitis virus, *J. Virol.* **38**:661–670 (1981).

44. Wege, H., J. R. Stephenson, M. Koga, and V. Ter Meulen, Genetic variation of neurotropic and non-neurotropic murine coronaviruses, *J. Gen. Virol.* **54**:67–74 (1981).

45. Robb, J. A., and C. W. Bond, Coronaviridae, in: *Comprehensive Virology* (H. Fraenkel-Conrat and R. R. Wagner, eds.), pp. 193–247, Plenum, New York (1979).

46. Lomniczi, B., and I. Kennedy, Genome of infectious bronchitis virus, *J. Virol.* **24**:99–107 (1977).

47. Lai, M. M. C., and S. A. Stohlman, Genomic structure of mouse hepatitis virus: Comparative analysis by oligonucleotide mapping, in: *Biochemistry and Biology of Coronaviruses* (V. Ter Meulen, S. Siddell, and H. Wege, eds.), pp. 69–82, Plenum, New York (1981).

48. Yogo, Y., N. Hirano, H. Shibuta, and M. Matumoto, Polyadenylate in the virion RNA of mouse hepatitis virus, *J. Biochem. (Tokyo)* **82**:1103–1108 (1977).

49. McNaughton, M. R., and M. H. Madge, The genome of human coronavirus strain 229E, *J. Gen. Virol.* **39**:497–504 (1978).

50. Jacobs, L., W. J. M. Spaan, M. E. Horzinek, and B. A. M. Van der Zeijst, Synthesis of subgenomic mRNA of mouse hepatitis virus is initiated independently: Evidence from UV transcription mapping, *J. Virol.* **39**:401–406 (1981).

51. Stern, D. F., and B. M. Sefton, Synthesis of coronavirus mRNAs: Kinetics of inactivation of infectious bronchitis virus RNA synthesis by UV light, *J. Virol.* **42**:755–759 (1982).

52. Lai, M. M. C., C. D. Patton, and S. A. Stohlman, Replication of mouse hepatitis virus: Negative-stranded RNA and replicative form RNA are a genome length, *J. Virol.* **44**:487–492 (1982).

53. Rottier, P. J. M., M. C. Horzinek, and B. A. M. Van der Zeijst, Viral protein synthesis in mouse hepatitis virus strain A59-infected cells: Effect of Tunicamycin, *J. Virol.* **40**:350–357 (1981).

54. Leibowitz, J. L., S. R. Weiss, E. Paavola, and C. W. Bond, Cell-free translation of murine coronavirus RNA, *J. Virol.* **43**:905–913 (1982).

55. Siddell, S. G., Coronavirus JHM: Coding assignment of subgenomic mRNAs, *J. Gen. Virol.* **64**:113–125 (1983).

56. Siddell, S. G., H. Wege, and V. Ter Meulen, The structure and replication of coronaviruses, *Curr. Topics Microbiol. Immunol.* **99**:131–163 (1982).

57. Siddell, S. G., S. G. Barthel, and V. Ter Meulen, Coronavirus JHM: A virion-associated protein kinase, *J. Gen. Virol.* **52**:235–243 (1981).

58. Armstrong, J., S. Smeekens, and P. Rottier, Sequence of the nucleocapsid gene from murine coronavirus MHV-A59, *Nucleic Acid Res.* **11**:883–891 (1983).

59. Sturman, L. S., Characterization of a coronavirus. I. Structural proteins' effects on preparative conditions on the migration of protein in polyacrylamide gels, *Virology.* **77**:637–649 (1977).

60. Sturman, L. S, The structure and behaviour of coronavirus A59 glycoprotein, in: *Biochemistry and Biology of Coronaviruses* (V. Ter Meulen, S. Siddell, and H. Wege, eds.), pp. 1–18, Plenum, New York (1981).

61. Sturman, L. S., K. V. Holmes, and J. Behnke, Isolation of coronavirus envelope glycoproteins and interaction with the viral nucleocapsids, *J. Virol.* **33**:449–462 (1980).

62. Sturman, L. S., and K. V. Holmes, Characterization of a coronavirus. II. Glycoproteins of the viral envelope: Tryptic peptide analysis, *Virology* **77**:650–660 (1977).

63. Sturman, L. S., and K. V. Holmes, Proteolytic cleavage of peplomeric glycoprotein E2 of MHV yields two 90K subunits and activates fusion, in: *Molecular Biology and Pathogenesis of Coronaviruses* (P. M. J. Rottier, B. A. M. Van der Zeijst, W. J. M. Spaan, and M. C. Horzinek, eds.), pp. 25–35, Plenum, New York (1984).

64. Cheever, F. S., J. B. Daniels, A. M. Pappenheimer, and O. T. Bailey, A murine virus (JHM) causing disseminated encephalomyelitis with extensive destruction of myelin, *J. Exp. Med.* **90**:181–194 (1949).

65. Le Prevost, C., E. Levy-Leblond, J. L. Virelizier, and J. M. Dupuy, Immunopathology of mouse hepatitis virus type 3 infection. I. Role of humoral and cell-mediated immunity in resistance mechanisms, *J. Immunol.* **114**:221–225 (1975).

66. Piazza, M., G. Pane, and F. De Ritis, The fate of MHV-3 after intravenous injection into susceptible mice, *Arch. Ges. Virusforsch.* **22**:472–475 (1967).

67. Le Prevost, C., J. L. Virelizier, and J. M. Dupuy, Immunopathology of mouse hepatitis virus type 3 infection. III. Clinical and mitogenic observations of a persistant virus infection, *J. Immunol.* **115**:640–643 (1975).

68. Dupuy, J. M., E. Levy-Leblond, and C. Le Prevost, Immunopathology of mouse hepatitis virus type 3 infection. II. Effect of immunosuppression in resistant mice, *J. Immunol.* **114**:226–230 (1975).

69. Bang, F. B., and A. Warwick, Mouse macrophages as host cells for the mouse hepatitis virus and the genetic basis of their susceptibility, *Proc. Natl. Acad. Sci. USA* **46**:1065–1075 (1960).

70. Levy-Leblond, E., D. Oth, and J. M. Dupuy, Genetic study of mouse sensitivity to MHV-3 infection: Influence of the H-2 complex, *J. Immunol.* **112**:1359–1362 (1979).

71. Oth, D., D. Pekovic, V. Cainelli-Gebara, and J. M. Dupuy, Expression of H-2K antigens in brain lesions, and influence of H-2K gene on susceptibility to paralysis, in $MHV_3$ infected mice, in: *Genetic Control of Host Resistance to Infection and Malignancy* (E. Skamene and P. Kongshavn, eds.), pp. 135–140, Liss, New York (1985).

72. Leray, D., C. Dupuy, and J. M. Dupuy, Immunopathology of mouse hepatitis virus type 3 infection. IV. MHV3-induced immunosuppressions, *Clin. Immunol. Immunopathol.* **23**:1457–1465 (1982).

73. Virelizier, J. L., A. D. Dayan, and A. C. Allison, Neuropathological effects of persistent infection of mice by mouse hepatitis virus, *Infect. Immun.* **12**:1127–1140 (1975).

74. Levy, G., R. Shaw, J. L. Leibowitz, and E. Cole, The immune response to mouse hepatitis

virus infection: Genetic variation in antibody response and disease, in: *Molecular Biology and Pathogenesis of Coronaviruses* (P. J. M. Rottier, B. A. M. Van der Zeijst, W. J. M. Spaan, and M. C. Horzinek, eds), pp. 345–364, Plenum, New York (1984).

75. Sorensen, O., S. Beushausen, S. Puchalski, S. Cheley, R. Anderson, M. Coulter-Mackie, and S. Dales, *In vivo* and *in vitro* models of demyelinating disease. VIII. Genetic, immunologic and cellular influences on JHM virus infection of rats, in: *Molecular Biology and Pathogenesis of Coronaviruses* (P. J. M. Rottier, B. A. M. Van der Zeijst, W. J. M. Spaan, and M. Horzinek, eds.), pp. 279–298, Plenum, New York (1984).

76. Wege, H., R. Watanabe, and V. Ter Meulen, Virological and immunological aspects of coronavirus induced subacute demyelinating encephalomyelitis in rats, in: *Molecular Biology and Pathogenesis of Coronaviruses* (P. J. M. Rottier, B. A. M. Van der Zeijst, W. J. M. Spaan, and M. Horzinek, eds.), pp. 259–270, Planum, New York (1984).

77. Krzystyniak, K., and J. M. Dupuy, Immunodepression of lymphocyte response in mouse hepatitis virus 3 infection, *Biomed. Pharmacol.* **37:**68–74 (1983).

78. Dupuy, J. M., C. Dupuy, and D. Decarie, Genetically determined resistance to mouse hepatitis virus 3 is expressed in hematopoietic donor cells in radiation chimeras, *J. Immunol.* **133:**1609–1613 (1984).

79. Lamontagne, L., and J. M. Dupuy, Natural resistance of mice to mouse hepatitis virus type 3 infection is expressed in embryonic fibroblast cells, *J. Gen. Virol.* **65:**1165–1171 (1984).

80. Lamontagne, L. and J. M. Dupuy, Persistent infection with mouse hepatitis virus 3 in mouse lymphoid cell lines, *Infect. Immun.* **44:**716–723 (1984).

81. Taguchi, F., N. Hirano, Y. Kiuchi, and K. Fujiwara, Difference in response to mouse hepatitis virus among susceptible mouse strains, *Jpn. J. Microbiol.* **20:**293–302 (1976).

82. Weiser, W., and F. B. Bang, Macrophages genetically resistant to mouse hepatitis virus converted *in vitro* to susceptible macrophages, *J. Exp. Med.* **143:**690–695 (1976).

83. Taylor, C. E., W. Y. Weiser, and F. B. Bang, *In vitro* macrophage manifestation of cortisone-induced decrease in resistance to mouse hepatitis virus, *J. Exp. Med.* **153:**732–737 (1981).

84. Taguchi, F., A. Yamada, and K. Fujiwara, Resistance to highly virulent mouse hepatitis virus acquired by mice after low-virulence infection: Enhanced antiviral activity of macrophages, *Infect. Immun.* **29:**42–49 (1980).

85. Weiser, W. Y., and F. B. Bang, Blocking of *in vitro* and *in vivo* susceptibility to mouse hepatitis virus, *J. Exp. Med.* **146:**1467–1472 (1977).

86. Arnheiter, H., T. Baechi, and O. Haller, Adult mouse hepatocytes in primary monolayer culture express genetic resistance to mouse hepatitis virus type 3, *J. Immunol.* **129:**1275–1281 (1982).

12

# Picornavirus-Induced Immunosuppression

CARLO GARZELLI, FULVIO BASOLO,
DONATELLA MATTEUCCI, BELLUR S. PRABHAKAR,
and ANTONIO TONIOLO

## 1. INTRODUCTION

### 1.1. Basic Properties of Picornaviruses

Picornaviridae are small RNA viruses with a naked ether-resistant icosahedral capsid, 22–30 nm in diameter. The capsid is composed of four different polypeptides ($VP_1$–$VP_4$), three of 20–40 kDa, one of 5–10 kDa. One molecule of each makes up the capsid structural subunits and the capsid comprises 60 capsomeres. In addition, a single copy of a smaller polypeptide ($VP_g$) is covalently linked to the 5' end of the genome.

Virus particles contain a single molecule of infectious linear single-stranded RNA of approximately $2.5 \times 10^6$ daltons, which serves as mRNA for

*Abbreviations:* CAV, coxsackievirus group A; CBV, coxsackievirus group B; CNS, central nervous system; EMC, encephalomyocarditis virus; FMDV, foot-and-mouth disease virus; FPV, feline picornavirus; HAV, hepatitis A virus; IFN, interferon; $IFN_\alpha$, interferon-α; $IFN_\beta$, interferon-β; mRNA, messenger RNA; PGE, prostaglandin $E_1$; PHA, phytohemagglutinin; PWM, pokeweed mitogen; SRBC, sheep red blood cells.

CARLO GARZELLI • Institute of Microbiology, University of Pisa, 56100 Pisa, Italy. FULVIO BASOLO • Institute of Pathological Anatomy, University of Pisa, 56100 Pisa, Italy. DONATELLA MATTEUCCI • Institute of Hygiene, University of Pisa, 56100 Pisa, Italy. BELLUR S. PRABHAKAR • Laboratory of Oral Medicine, National Institute of Dental Research, National Institutes of Health, Bethesda, Maryland 20205. ANTONIO TONIOLO • Institute of Microbiology, University of Sassari, 07100 Sassari, Italy.

viral protein synthesis and subsequently as template for viral RNA replication. The genome codes for about 10–12 structural and nonstructural proteins, which derive from the cleavage of a single giant polypeptide. Replication occurs exclusively in the cytoplasm in association with intracellular membranes.[1] Animal picornaviruses are divided into four genera, sharing the basic properties described above, i.e., *Enterovirus, Cardiovirus, Rhinovirus, and Aphthovirus*. These genera are distinguished on the basis of sensitivity to acid, buoyant density of the viral particle in CsCl, and clinical manifestation of infected host. The distinguishing features and the representative members of picornaviruses genera are shown in Table I.

In general, picornavirus infection rapidly inhibits host-specific macromolecular synthesis and redirects the ribosomal system for the production of viral proteins; the infectious cycle is usually short and the viral progeny is released by cell lysis. These agents are therefore highly cytopathic, although several examples of productive infection without lysis of host cells have been reported; in all these circumstances, picornaviruses seem capable of producing persistent infections.

The persistence of picornaviruses under certain conditions in infected hosts has been well documented in various animal species infected with FMDV,[2] in the chronic neurologic disease produced in mice by Theiler virus,[3] as well as in HAV infection of various cell lines.[4] Sporadic reports also indicate that FPV can produce chronic inapparent infections in cats,[5] and a few cases of polymyositis/dermatomyositis in humans have been·tentatively attributed to persistent CAV infections.[6] In addition, it has been shown that exposure of lymphoid cell lines to certain enteroviruses often results in chronic productive infection.[7–9] Although of obvious pathogenetic relevance, the molecular basis of picornavirus persistence and the apparent lack of cytopathology in infected cells are, however, unclear.

Picornaviruses species are defined on the basis of neutralization assays, but newer immunologic, biochemical, and genetic techniques show that possibly all

## TABLE I
### Distinguishing Features of Animal Picornaviruses[a]

| Genus | Primary members | Optimal growth temperature (°C) | Infectivity at pH 3 | Density in CsCl |
|---|---|---|---|---|
| Enterovirus | Poliovirus<br>Coxsackievirus A and B<br>Echovirus<br>Enterovirus 68–72 | 36–37 | Stable | 1.33–1.35 |
| Cardiovirus | Encephalomyocarditis virus | 36–37 | Stable | 1.34 |
| Rhinovirus | >100 Serotypes | 33–34 | Labile | 1.38–1.42 |
| Aphthovirus | FMDV | 36–37 | Labile | 1.43–1.45 |

[a]New members included in the picornavirus family are HAV and Theiler virus.

species undergo considerable evolutionary variation. Changes apparently result from mutation or antigenic drift,[10,11] and variants often differ from prototype strains with regard to pathogenicity.[12,13] It is therefore suspected, although not proved, that the extraordinarily wide spectrum of clinical manifestations caused, for instance, by members of the enterovirus group is due to the existence of intratypic variants endowed with different cell tropism. Thus, even if the immune system is not commonly regarded as a primary target for picornaviruses, it is definitely possible that natural infections with these agents produce, at least in some cases, untoward effects on immunocompetent cells.

## 1.2. Clinical Manifestations Associated with Picornavirus Infections

Picornaviruses may cause an extraordinarily wide spectrum of diseases in humans and animals (Table II). For instance, enteroviruses are associated with poliomyelitis, encephalitis, meningitis, carditis, encephalomyocarditis neonatorum, pleurodynia, herpangina, various exanthems, acute hemorrhagic conjunctivitis, respiratory illness, hepatitis, pancreatitis, and several other clinical entities. Rhinoviruses typically infect the upper respiratory tract, causing the common cold. EMC, the representative member of cardioviruses, typically infects the CNS and heart in rodents, while some variants are selectively tropic for pancreatic beta cells causing diabetes. Aphthoviruses cause epizootic of foot-and-mouth disease in ungulates.

This is an oversimplified view of the real situation; in the course of each particular clinical syndrome, several different cell types apart from those constituting the main target organs, can support virus replication and are consequently damaged. For example, unidentified cells in tonsils and in gut-associated lymphoid tissues are infected by poliovirus and play a key role in the pathogenesis and epidemiology of poliomyelitis.

It should also be borne in mind that genetic and phenotypic host factors control susceptibility to picornavirus infections. Some have been shown to favor the development of severe disease in humans: very young or very old age, chronic undernutrition, corticosteroid treatment, physical exertion, hypoxia, cold, irradiation, tonsillectomy, pregnancy, and adrenal-related endocrine changes.[14]

## 2. MODULATION OF IMMUNITY BY PICORNAVIRUSES

The interactions between picornaviruses and immune system are of great interest, since these viruses are important pathogens and it is known that virus–immunocyte interplay greatly influences viral pathogenesis. Unfortunately, information on the immunomodulatory activity of these agents is relatively scarce. This is probably because the host range of most of these agents is usually fairly restricted and thus it is difficult to set up experimental models to study immunologic parameters in infected hosts. The best known model of

**TABLE II**
**Main Clinical Manifestations Associated with Infections by Members**
**of Picornaviruses**

| Virus | Primary host | Target | Clinical syndrome |
|---|---|---|---|
| Enterovirus | | | |
|   Poliovirus | Human | CNS | Paralysis |
|   Coxsackievirus A | Human | CNS | Meningitis |
| | | | Paralysis |
| | | | Encephalitis |
| | | Respiratory tract | Respiratory illness |
| | | | Herpangina |
|   Coxsackievirus B | Human | CNS | Meningitis |
| | | | Paralysis |
| | | | Encephalitis |
| | | Digestive system | Pancreatitis |
| | | | Hepatitis |
| | | Heart, muscle | Myocarditis |
| | | | Pericarditis |
| | | | Pleurodynia |
| | | Respiratory tract | Respiratory illness |
|   Echovirus | Human | CNS | Meningitis |
| | | | Paralysis |
| | | | Encephalitis |
| | | Respiratory tract | Respiratory illness |
|   Hepatitis A virus | Human | Liver | Hepatitis |
| Cardiovirus | | | |
|   EMC virus | Rodents | CNS | Encephalitis |
| | | Heart | Myocarditis |
| Rhinovirus | | | |
|   (>100 serotypes) | Human | Upper respiratory tract | Common cold |
| Aphthovirus | | | |
|   FMDV (7 types) | Cloven-footed animals | Mucosa, skin and internal organs | Foot-and-mouth disease |

picornavirus-induced immunosuppression is represented by the experimental infection of mice with CBV-3 (described in detail in Section 3). Here, we summarize published data on the modulation of immune functions by some members of the picornavirus group.

## 2.1. Antibody Production

Early observations indicated that patients infected with members of CVA or CVB were susceptible to the development of unusually severe paralytic poliomyelitis [15,16] or protozoan infections.[17] Experimental studies in adult mice infected with CBV or with selected members of CAV showed that types A-15, B-1, B-3, and B-6 suppressed the antibody response to unrelated antigens (e.g., poliovirus-1, SRBC, or both). In addition, CBV-3 suppressed the development of cell-mediated contact sensitivity to oxazolone.[18] When the CBV-3-induced suppression of antibody response was analyzed in details using spleen cell cultures from infected mice, it was found that these agents could interfere with the development of the primary antibody response to thymus-dependent antigens, but not with that against thymus-independent antigens.[19] Unfortunately, no studies have dealt so far with the effects of other picornaviruses on antibody production.

## 2.2. Lymphocytes

It is known that in humans polioviruses replicate in lymphoid organs, such as tonsils and Peyer patches. Early studies on the interactions of polio-1 with cells of the immune system showed that human leukocyte cultures can support replication of this agent.[20] However, freshly isolated lymphocytes or polymorphonuclear cells appeared to resist infections, and virus was replicated only in a small proportion of cells tentatively identified as monocytes.[20] Further studies showed that poliovirus replication was enhanced after PHA stimulation of cultured leukocytes[21] and that PHA-stimulated lymphocytes became susceptible to infection during blastogenesis.[22] The capability of polioviruses to infect lymphocytes productively was later confirmed by the finding that exposure of several spontaneously transformed human lymphoid cell lines to attenuated or virulent strains of polio-1 and -2 led to a sharp increase in virus titers followed by the establishment of a persistent infection.[8] More recently, it was shown that large amounts of polio-1 to -3 could be grown in Epstein–Barr virus-transformed human lymphoblastoid B-cell lines isolated from patients with infectious mononucleosis.[7]

From a functional point of view, polioviruses infections impairs PHA-induced proliferation of human lymphocytes[23,24] but, although viral replication can take place in these cells, no evidence has been produced for a direct damage of lymphocytes in these cultures. It is possible that the alteration of lymphocyte blastogenesis induced by these agents is indeed mediated by virus effects on macrophages. Related viruses that suppress leukocyte mitogenesis include HAV in humans and mengovirus in mice.[25,26] The six CBVs are easily replicated in human lymphoid cell lines of B and T lineage, although unable to replicate in freshly isolated human or mouse lymphocytes and to reduce the mitogenic response of mouse spleen cells[27] (unpublished observations). CBV-1 through -5, in addition, consistently establish persistent productive infections in these lines, probably replicating in a minority of cells at any given time, while CBV-6 usually kills infected cultures.[9] Upon infection *in vivo*, CBVs do not appear to alter lymphocyte functions directly, although surface receptors for these and other picornaviruses are present in various types of murine lymphoid cells.[28] CBV-3, which is strongly immunosuppressive in experimental animals, does not influence either the relative proportion of Ig+ and Thy 1.2+ cells or helper T-cell function in infected mice; it fails to alter primary antibody responses when added *in vitro* to murine spleen cells.[19,29] Thus, it is unlikely that picornavirus-induced immunologic defects derive from a direct attack of virus on lymphocytes.

## 2.3. Phagocytic Cells

A few experimental data indicate that certain macrophages are permissive cells for replication of some picornaviruses: for instance, polioviruses can replicate in human monocytes[20,23,24] and in PHA-stimulated murine macrophages of genetically susceptible mice.[30] It was also shown that the administration of antibody against $IFN_\alpha/IFN_\beta$ to normal mice renders most peritoneal macrophages susceptible to EMC *in vitro*, while macrophages from untreated mice are intrinsically resistant[31]; this observation clearly indicates that the production of trace amounts of endogenous IFN exerts a great influence on the susceptibility of lymphoreticular cells to picornaviruses. Poliovirus infection of human mononuclear cell cultures can alter some macrophage functions, such as the enhancing effect of lymphocyte blastogenesis after stimulation with either PHA, PWM, tuberculin, or allogeneic cells.[23,24]

In spite of these findings, no morphologic proof of picornavirus replication within macrophages has been produced so far, nor is it clear whether these agents are cytopathic for these cells. Although a functional defect of antigen-presenting cells has been demonstrated in the spleen of adult mice infected with CBV-3, electron microscopy and infectious center assay failed to clarify whether spleen macrophages are infected by CBV-3 *in vivo* or whether these cells are merely carrying engulfed virus particles as previously suggested.[29] Early studies had in fact shown that spleen or peritoneal macrophages may carry absorbed CBV-3 for a considerable time after infection and that the fate of macrophage-associated virus components was dependent both on the strain and the age of infected mice.[32,33]

Other studies indicate that macrophages often represent early targets in experimental picornavirus infections: HAV was detected by immunofluorescence in hepatic Kupffer cells of marmosets before hepatitis became clinically manifested,[34] and an isolate of FPV was seen to replicate in alveolar macrophages of kittens before virus spreading to epithelial cells of alveoli and bronchioles occurred.[35] Some enteroviruses (echo-9, in particular) have been shown to alter surface properties of human polymorphonuclear cells. In particular, it was shown that virus–cell membrane interaction was followed by the dissolution of the functional receptor units governing the chemotactic response of the cell. In addition, exposure to these viruses *in vitro* appeared to increase the binding of polymorphs to endothelial cells.[36–38] Although the genesis of these alterations is obscure, they may contribute to the reduction of natural resistance seen in infants with disseminated enteroviral infections. It has also been shown that the incubation of human polymorphs with rhinovirus-16 *in vitro* makes these cells less responsive to the physiologic mediators controlling the release of lysosomal components (e.g., catecholamine, histamine, $PGE_1$). Viral interference of this normal homeostatic granulocyte response may accentuate inflammation and, perhaps, increase virus-induced tissue injury. This effect appeared to occur within 1 hr of exposure to both live or inactivated virus, and it was shown that virus-induced abnormalities were produced at some intracellular step distal to the stimulation of the agonist receptor.[39] It might very well be that similar effects are produced in lymphoid tissues by other picornaviruses and result in alterations of immunoregulation without obvious damage to lymphoid cells.

### 2.4. Interferon System

Although to a variable extent, the replication of all picornaviruses can be inhibited by IFN. Their ability to induce IFN in infected cells varies greatly not only with the species, but often depending on each particular strain. In fact, while many enteroviruses are very good IFN inducers, some highly cytocydal variants (e.g., the D variant of EMC, some isolates of poliovirus, mengovirus) cause little IFN response, and HAV is noted for its inability to cause any detectable IFN response.[40] To our knowledge, no studies have been performed to establish whether these agents can impair host defenses by specifically blocking the IFN system. What has been shown, instead, is that the administration of exogenous IFN or IFN inducers may exacerbate tissue injury, especially when given a few days after infection.[41,42] In addition, it has been shown that when peripheral blood lymphocytes from normal donors are exposed to certain HAV-carrier cells, they first produce $IFN_\alpha$ and subsequently lyse infected targets.[4] Thus, it appears that, at least in certain picornavirus infections, the IFN system can enhance virus-induced damage.

To check whether the systemic lymphoid atrophy produced in mice by CBV-3 infection[29] was partly mediated by the elevated IFN levels seen in both spleen and thymus, we treated infected animals with potent antibody to $IFN_\alpha/IFN_\beta$. This treatment not only failed to alleviate lymphoid involution but actually increased the mortality rate, showing that this pathologic effect of obscure origin is not directly caused by the production of endogenous IFN (unpublished observations).

There are indications that low levels of endogenous IFN help maintain some cells such as macrophages in a picornavirus-resistant state.[31] By contrast, it has been shown that many different cell lines, of both lymphoid and nonlymphoid origin, persistently infected with HAV or CBV do not produce detectable levels of IFN.[4,9] As a rule, however, all these lines are rapidly cured by the administration of exogenous IFN.[9,43] Thus, it appears that small defects of the IFN response will increase the host susceptibility to chronic picornavirus infections.

## 3. EXPERIMENTAL INFECTION OF MICE WITH COXSACKIEVIRUS B-3

Within a few days of CBV-3 infection, adult mice of susceptible strains lose much of their immunocompetence against newly presented antigens. In general, secondary responses are unaffected, while primary responses are reduced to one half to one fifth of normal levels. CBV-3-induced immunodeficiency is accompanied by the development of progressive atrophic changes in all lymphoid organs, but infected mice become hyporesponsive well before lymphoid atrophy is detected hystologically. It appears, therefore, that immunosuppression is a primary consequence of infection. Table III summarizes the effects of CBV-3 infection on various immunologic parameters. Several conclusions can be drawn from these results:

1. B lymphocytes are not a major target of immunosuppression, since spleen cell cultures from infected mice respond well to T-independent

### TABLE III
**Immunosuppression by CBV-3 Infection in BALB/c Mice[a,b]**

| Parameter | Response | Restoration |
|---|---|---|
| Induction of cell-mediated immunity *in vivo* | Reduced | Unknown |
| Ab response to T-dependent Ag *in vivo* | Reduced | Unknown |
| Ab response to T-dependent Ag *in vitro* | Reduced | Addition of normal macrophages Substitution of normal T cells for infected T cells |
| Ab response *in vitro* of normal spleen cells + infected cells | Reduced | Lysis of infected T cells with Thy 1.2 Ab + C' |
| Ab response *in vitro* of normal non-adherent spleen cells + infected adherent cells | Reduced | Unknown |
| Ab response to T-independent Ag *in vitro* | Normal | — |
| Ab response *in vitro* of infected B cells + macrophages and T cells from normal mice | Normal | — |
| Ab response *in vitro* of normal spleen cells + infectious virus | Normal | — |
| Mitogenic response of infected spleen cells to: | | |
| Con A, PWM | Normal | — |
| *E. coli* LPS | Increased | Unknown |

[a]From Bendinelli *et al.*[(18,19)] and Toniolo *et al.*[(27)]
[b]*Abbreviations:* Ab, antibody; Ag, Antigen(s); C', complement; —, not applicable; Con A, concanavalin A; PWM, pokeweed mitogen; LPS, lipopolysaccharide.

antigens and, if supplemented with non-B cells from normal mice, are normally also responsive to T-dependent antigens.

2. Helper T-cell functions are also probably unaffected, since supplementation with antigen-presenting cells from normal spleen restores the response of infected cultures;

3. Significant numbers of nonspecific suppressor T cells are activated during infection, although to a variable extent;

4. Antigen-presenting cells are certainly involved, since spleen macrophages obtained from infected mice cannot restore the response of macrophage-depleted uninfected spleen cultures and, conversely, normal spleen macrophages can make infected spleen cells capable of responding. This notion is reinforced by electron microscopy showing that spleen macrophages of infected mice are crammed with phagosomes and replicating internal membranes (a possible sign of picornavirus infection);

5. Immune deficiency does not appear to derive from direct virus damage to lymphocytes, since these cells are resistant to infection *in vitro*, remain normally viable *in vivo*, and respond normally to different mitogens in cultures. Moreover, the direct addition of virus to cultured lymphoid cells does not prevent the development of primary antibody responses *in vitro*.

These observations suggest that immune deficiency primarily derives from damage to antigen-presenting cells and from the simultaneous activation of T-suppressor lymphocytes; loss of viable lymphocytes from lymphoid organs certainly contributes to the effects seen *in vivo*. As shown in Table IV, the loss of lymphoid elements occurring early in the course of infection is apparently irreversible and can be prevented only by the immediate treatment of mice with antiviral antibody. Since CBV-3 infection does not produce either necrotic changes in lymphoid organs or selective pathologic effects in T- or B-dependent areas, and since lymphocyte viability is preserved during infection, it is speculated that the lymphoid atrophy is caused by the selective depletion of unidentified trophic cells in target organs, to a diminished production of precursor cells and/or to alterations of lymphocyte trafficking.

Thus far, we have not had any evidence of direct virus effect on lymphoid organs, since both the infectious center assay (done with mechanically prepared cell suspensions) and immunofluorescence staining of frozen tissues (with polyclonal antibodies) failed to detect significant numbers of CBV-3-infected cells in lymphoid organs. Recently, however, direct immunofluorescence with monoclonal antibody to CBV-3 has begun to delineate an entirely different picture. In fact, early production of viral components was seen in scattered cells of thymus, spleen, and lymph nodes. Figures 1–3 show a random distribution of infected cells in different lymphoid organs obtained 2 days postinfection. Virus-positive cells do not resemble lymphocytes and are usually polygonal in shape, and viral antigen distribution is confined to the cytoplasm. Although these cells have not been identified, they are morphologically similar

## TABLE IV
### Characteristics of the Systemic Lymphoid Atrophy Induced by CBV-3 Infection in Adult Mice[a]

Induced by most, but not all, CBV-3 isolates in susceptible mice

Genetically restricted to certain mouse strains

Independent of sex

Characterized by a progressive loss of lymphocytes without noticeable necrosis in all areas of thymus and spleen and by the rapid development of fibrosis

Viability of lymphocytes from thymus, spleen, and lymph nodes unaffected (no evidence for *in vivo* or *in vitro* replication of virus in lymphocytes)

Relative proportions of B- and T-spleen lymphocytes unchanged

Rare unidentified cells distributed at random in all areas of lymphoid organs are virus-positive from the very early phase of infection

Prevented by treatment with antibody to CBV-3 on the day of infection (partial protection possible if antibody is given at later time)

Not prevented by adrenalectomy, transfusion of normal spleen cells during infection, antibody to $IFN_\alpha/IFN_\beta$

Exacerbated by immunopotentiating agents

---

[a]Bendinelli *et al.*,[19] Toniolo *et al.*,[27] and Matteucci *et al.*[29]

to certain types of antigen-presenting cells and to the epithelial cells found in thymus. It is of interest that these lymphoid elements represent an immediate target for CBV-3. In fact, viral antigens can be detected in thymus just 12 hr postinfection, and the retraction of cytoplasmic projection of dendritic epithelial cells represents one of the earliest consequences of infection in this organ. Small numbers of cells (usually less than 1%) are infected in all lymphoid organs at any given time from day 1 to day 6 postinfection, but it is not clear whether CBV-3 kills infected cells and new targets are continuously available or whether persistent infection occurs in the same cells without obvious cytopathologic effect. In any case, infection is accompanied by a progressive loss of lymphocyte, suggesting that the increased peripheral request caused by infection in different organs is not balanced by sufficient production and/or entry of new cells into lymphoid organs. Previous studies showed that significant numbers of virus-infected cells are present in the bone marrow of infected mice,[29] and it is known that lymphocytes enter lymphoid organs through high endothelial venules to which they adhere by means of specific receptors.[44] Therefore, one might speculate that systemic infection directly or indirectly (e.g., release of IFN, proteolytic enzymes), can alter the expression of these interactive molecules, subverting the structural organization of lymphoid organs. Provocative data in this direction have been already presented with regard to echo-9 infection of human endothelial cells.[38] Finally, since immunopotentiating agents exacerbate CBV-3-induced lymphoid atrophy, it has been proposed that autoreactivity may play some pathogenetic role in picornavirus-induced immunodeficiency,[29] as in the case of other pathologic effects attributable to these agents.[3,45]

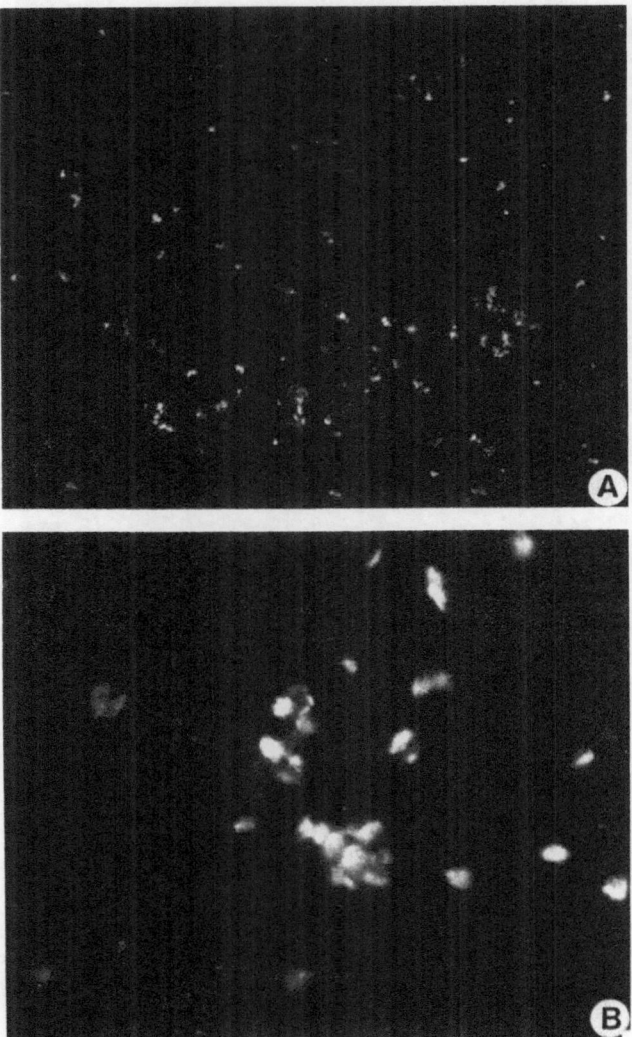

**FIGURE 1.** Frozen sections of thymus of an adult BALB/c mouse 2 days after intraperitoneal infection with CBV-3, as detected by direct immunofluorescence with anti-CBV-3 monoclonal antibody. (A) Small clusters of infected cells are occasionally evident. (×160) (B) Cytoplasmic fluorescence in a small cluster of infected cells; large numbers of thymocytes are clearly virus negative. Note the random distribution of infected cells in both the cortical and medullary areas. (×500)

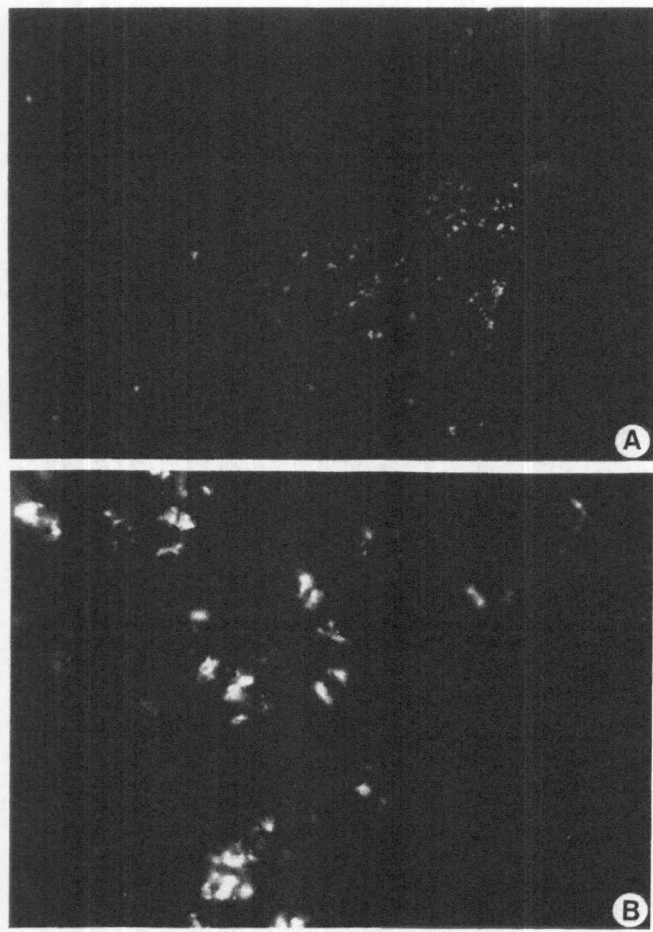

**FIGURE 2.** Frozen sections of spleen of an infected mouse stained by direct immunofluorescence (see legend to Fig. 1). (A) (×160); (B) (×500). Infected cells are more frequently distributed along connective vascular structures and, as in the thymus, lymphocytes are largely negative.

## 4. MECHANISMS OF IMMUNOSUPPRESSION

It is premature to attempt to delineate a coherent picture of the events leading from picornavirus infection to immunologic defects on the basis of the many disparate observations reported. A few conclusions, however, seem possible:

1. Picornaviruses as a group do not seem particularly trophic for lymphocytes, although the possibility that these agents cause productive and

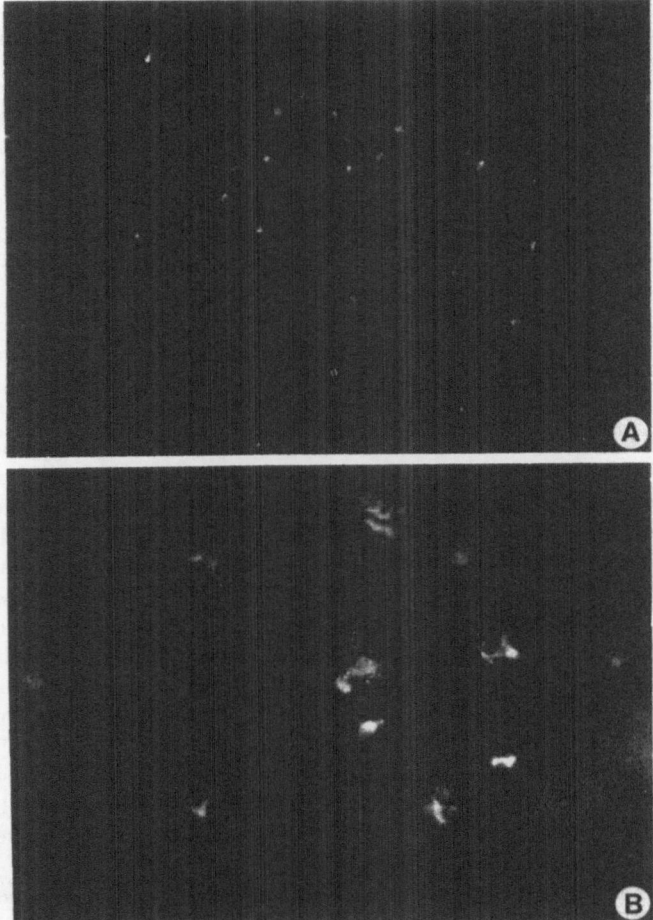

**FIGURE 3.** Frozen sections of inguinal lymph node of an infected mouse stained by direct immunofluorescence (see legend to Fig. 1). (A) (×160); (B) (×500). Lower numbers of virus-positive cells are present in lymph nodes as compared with thymus and spleen; the strong cytoplasmic granular fluorescence shown in (B) suggests ongoing viral replication rather than the mere presence of engulfed viral components.

persistent infections of these cells has been demonstrated both *in vivo* and *in vitro*.

2. Phagocytic cells, and perhaps other types of antigen-presenting cells, are frequently targets for these viruses.

3. At least in some cases, infection of trophic cells and of vascular endothelial cells[46] probably contributes in altering the natural microenvironment of lymphoid organs.

4. The variable and ill-defined metabolic and hormonal changes occurring during systemic picornavirus infections undoubtedly contribute to immunosuppression.

Our present knowledge is therefore insufficient for devising specific treatments to counteract the immunodepressive effects of these viruses, with the exception of the early therapeutic use of antiviral antibody. It should be recalled here that the administration of non-specific immunostimulating drugs, IFN or IFN-inducers, can actually exacerbate virus-induced pathologic changes and immunodeficiency in experimental animals.[29,41,42,45]

## 5. SIGNIFICANCE OF VIRUS-INDUCED IMMUNOMODULATION IN VIRAL PATHOGENESIS

Clinical observations indicate that serious enterovirus infections are not particularly common among children with T-cell immune-deficiency syndrome, while children with Bruton X-linked agammaglobulinemia or severe combined immune deficiency frequently develop severe disease. The occurrence of persistent picornavirus infections has been reported in these patients,[46-51] as well as in mice given treatments suppressing antibody production. The conclusion that humoral defenses play a key protective role against picornaviruses has also been reached in mice infected with various enteroviruses, FMDV, or EMC.[33,52,53] It has also been shown that the availability of endogenous or exogenous IFN during the initial phase of infection clearly reduces clinical manifestations and virus titers in target organs.[13,42,54] In addition, the protective effect of macrophages and NK cells has been suggested in some models, although not clearly defined.[55,56] Theoretically, these cells should destroy virus-producing cells if enough viral antigen is expressed on the cell surface, but usually this is not the case of picornaviruses. The study of different models of chronic picornaviral infection has actually disclosed that cytotoxic T-lymphocytes can frequently become autoreactive and destroy normal uninfected cells.[3,59-61]

Thus, it appears that in order to invade the host, picornaviruses would need means to avoid the activation of the IFN system and to inhibit or delay the induction of specific antibody response. As a rule, aggressive strains are poor IFN inducers,[13] and several picornaviruses have been shown to interfere with the physiology of antigen-presenting cells, depressing and delaying the development of primary responses.[19,23,24] These properties may well have evolved in order to favor extensive virus synthesis in the early phase of infection. As a consequence, because of their rapid replication rate, these agents might spread into the host to a significant degree in the absence of first line defenses, such as natural cytotoxicity and IgM antibody. However, the capability of inducing these defects may also be beneficial: damage to antigen-presenting cells together with lowered IFN responses can probably inhibit the activation of cytotoxic T cells, preventing the development of detrimental autoreactivity.

ACKNOWLEDGMENTS.   This work was supported in part by the Italian National Research Council, special project Control of Infectious Diseases. We thank Dr. M. Bendinelli for introducing us to this area of research and for his continuous encouragement and support.

## REFERENCES

1. Matthews, R. E. F., Fourth report of the International Committee on taxonomy of viruses. Classification and Nomenclature of viruses, *Intervirology* **17**:1–199 (1982).
2. Gustafson, D. P., Foot and mouth disease, in: *Disease Transmitted from Animals to Man* (W. H. Hubbert, W. Mc Cullock, and P. R. Schnurrenberger, eds.), pp. 859–865, Charles C Thomas, Springfield, Illinois (1975).
3. Friedman, A., and Y. Lorch, Theiler's virus infection: A model for multiple sclerosis, *Prog. Med. Virol.* **31**:43–83 (1985).
4. Kurane, I., L. N. Binn, W. H. Bancroft, and F. A. Ennis, Human lymphocyte responses to hepatitis A virus-infected cells: Interferon production and lysis of infected cells, *J. Immunol.* **135**:2140–2144 (1985).
5. Povey, R. C., R. C. Wardley, and H. Jessen, Feline picornavirus infection: the *in vivo* carrier state, *Vet. Rec.* **92**:224–229 (1973).
6. Gyorkey, F., G. A. Cabral, P. K. Gyorkey, G. Uribe-Botero, G. R. Dreesman, and J. L. Melnick, Coxsackievirus aggregates in muscle cells of polymyositis patient, *Intervirology* **10**:69–77 (1978).
7. Bjare, U., Propagation of large quantities of poliovirus in human lymphoblastoid cells grown in serum-free medium, *J. Virol. Methods* **9**:259–268 (1984).
8. Carp, R. I., Persistent infection of human lymphoid cells with poliovirus and development of temperature-sensitive mutants, *Intervirology* **15**:49–56 (1981).
9. Matteucci, D., M. Paglianti, A. M. Giangregorio, M. R. Capobianchi, F. Dianzani, and M. Bendinelli, Group B coxsackieviruses readily establish persistent infections in human lymphoid cell lines, *J. Virol.* **56**:651–654 (1985).
10. Prabhakar, B. S., M. A. Menegus, and A. L. Notkins, Detection of conserved and nonconserved epitopes on Coxsackievirus B4: Frequency of antigenic change, *Virology* **146**:302–306 (1985).
11. Rozhon, E. J., A. K. Wilson, and B. Jubelt, Characterization of genetic changes occurring in attenuated poliovirus 2 during persistent infection in mouse central nervous systems, *J. Virol.* **50**:137–144 (1984).
12. Nottay, B. K., O. M. Kew, M. H. Hatch, J. T. Heyward, and J. F. Obijeski, Molecular variation of type 1 vaccine-related and wild polioviruses during replication in humans, *Virology* **108**:405–423 (1981).
13. Yoon, J. W., P. R. Mc Clintock, T. Onodera, and A. L. Notkins, Virus-induced diabetes mellitus. XVIII. Inhibition by a non diabetogenic variant of encephalomyocarditis virus, *J. Exp. Med.* **152**:878–892 (1980).
14. Moore, M., and D. M. Morens, Enteroviruses, including polioviruses, in: *Textbook of Human Virology* (R. B. Belshe ed.), pp. 407–483, PSG Publishing Company, Littleton, Massachusetts (1984).
15. Melnick, J. L., and A. S. Kaplan, Dual antibody response to coxsackie and poliomyelitis viruses in patients with paralytic poliomyelitis, *Proc. Soc. Exp. Biol. Med.* **74**:812–815 (1950).
16. Melnick, J. L., A. S. Kaplan, E. Zabin, G. Contreras, and N. W. Larkum, An epidemic of paralytic poliomyelitis characterized by dual infections with poliomyelitis and coxsackie viruses, *J. Exp. Med.* **94**:471–492 (1951).

17. Sebastiani, A., G. Fontana, and A. Balestrieri, Su due casi di polmonite pneumocistica del lattante associata ad infezione da virus Coxsackie B1, *Arch. Ital. Sci. Med. Trop. Parass.* **47**:191–198 (1966).

18. Bendinelli, M., A. Ruschi, M. Campa, and A. Toniolo, Depression of humoral and cell-mediated immune responses by coxsackieviruses in mice, *Experientia*, **31**:1227–1229 (1975).

19. Bendinelli, M., D. Matteucci, A. Toniolo, A. M. Patane, and A. M. Pistillo, Impairment of immunocompetent mouse spleen cell functions by infection with coxsackievirus B3, *J. Infect. Dis.* **146**:797–805 (1982).

20. Gresser, I., and C. Cheny, Multiplication of poliovirus type I in preparation of human leukocytes and its inhibition by interferon, *J. Immunol.* **92**:889–895 (1964).

21. Willems, F. T. C., J. L. Melnick, and W. E. Rawls, Viral inhibition of the phytohemagglutinin response of human lymphocytes and application to viral hepatitis, *Proc. Soc. Exp. Biol. Med.* **130**:652–661 (1969).

22. Willems, F. T. C., J. L. Melnick, and W. E. Rawls, Replication of poliovirus in phytohemagglutinin-stimulated human lymphocytes, *J. Virol.* **3**:451–457 (1969).

23. Soontiens, F. J. C. J., and J. Van Der Veen, Evidence for a macrophage-mediated effect of poliovirus on the lymphocyte response to phytohemagglutinin, *J. Immunol.* **111**:1411–1419 (1973).

24. Van Loon, A. M., J. Th. M. Van Der Logt, and J. Van Der Veen, Poliovirus-induced suppression of lymphocyte-stimulation: A macrophage-mediated effect, *Immunology* **37**:135–143 (1979).

25. Mella, B., and D. J. Lang, Leucocyte mitosis: Suppression in vitro associated with infectious hepatitis, *Science* **155**:80–81 (1967).

26. Olson, G. B., Effect of various viruses on the responsiveness of mouse lymphocytes to phytohemagglutinin stimulation, *Infect. Immun.* **7**:438–444 (1973).

27. Toniolo, A., D. Matteucci, F. Basolo, and M. Bendinelli, The immune system in experimental coxsackievirus-B3 infection, in: *Viruses Immunity and Immunodeficiency* (A. Szetivanyi and H. Friedman, eds.), pp. 101–105, Plenum, New York (1986).

28. Morishima, T., P. R. McClintock, L. C. Billups, and A. L. Notkins, Expression and modulation of virus receptors on lymphoid and myeloid cells: relationship to infectivity, *Virology* **116**:605–618 (1982).

29. Matteucci, D., A. Toniolo, P. G. Conaldi, F. Basolo, Z. Gori, and M. Bendinelli, Systemic lymphoid atrophy in coxsackievirus B3-infected mice: Effects of virus and immunopotentiating agents, *J. Infect. Dis.* **51**:1100–1108 (1985).

30. Kantoch, M., and H. Dobrowolska, Studies on the inheritance of the susceptibility to poliovirus of phytohemagglutinin-transformed macrophages, *Acta Virol. (Praha)* **13**:153–155 (1969).

31. Belardelli, F., F. Vignaux, E. Proietti, and I. Gresser, Injection of mice with antibody to interferon renders peritoneal macrophages permissive for vesicular stomatitis virus and encephalomyocarditis virus, *Proc. Natl. Acad. Sci. USA* **81**:602–606 (1984).

32. Gauntt, C. J., M. D. Trousdale, D. R. L. LaBadie, R. E. Paque, and T. Nealon, Properties of coxsackievirus B3 variants which are amyocarditic or myocarditic for mice, *J. Med. Virol.* **3**:207–220 (1979).

33. Rager-Zisman, B. R., and A. C. Allison, The role of antibody and host cells in the resistance of mice against infection by Coxsackie B-3 virus, *J. Gen. Virol.* **19**:329–338 (1973).

34. Mathiesen, L. R., S. M. Feinstone, R. H. Purcell, and J. O. Wagner, Detection of hepatitis A antigen by immunofluorescence, *Infect. Immun.* **18**:524–530 (1977).

35. Kahn, D. E., and J. H. Gillespie, Feline viruses: Pathogenesis of picornavirus infection in the cat, *Am. J. Vet. Res.* **32**:521–531 (1971).

36. Bultmann, B. D., O. Haferkamp, H. J. Eggers, and H. Gruler, Echo 9 virus-induced order-disorder transition of chemotactic response of human polymorphonuclear leucocytes: Phenomenology and molecular biology, *Blood Cells* **10**:79–106 (1984).

37. Bultmann, B. D., P. Allmendinger, R. U. Raus, I. Melzner, O. Haferkamp, H. J. Eggers, and H. Gruler, F-Met-Leu-Phe and echo 9 virus interaction with human granulocytes. Changes of cell membrane structure, *Am. J. Pathol.* **116**:46–55 (1984).

38. Kirkpatrick, C. J., B. D. Bultmann, and H. Gruler, Interaction between enteroviruses and human endothelial cells in vitro. Alterations in the physical properties of endothelial cell plasma membrane and adhesion of human granulocytes, *Am. J. Pathol.* **118**:15–25 (1985).

39. Busse, W. W., C. L. Anderson, E. C. Dick, and D. Warshauer, Reduced granulocyte response to isoproterenol, histamine, and prostaglandin $E_1$ after *in vitro* incubation with rhinovirus 16, *Am. Rev. Respir. Dis.* **122**:641–646 (1980).

40. Vallbracht, A., P. Gabriel, J. Zahn, and B. Flehmig, Hepatitis A virus infection and the interferon system, *J. Infect. Dis.* **152**:211–213 (1985).

41. Gould, C. L., K. G. McMannama, N. J. Bigley, and D. J. Giron, Exacerbation of the pathogenesis of the diabetogenic variant of encephalomyocarditis virus in mice by interferon, *J. Interferon Res.* **5**:33–37 (1985).

42. Lutton, C. W., and C. J. Gauntt, Ameliorating effect of IFN-beta and anti-IFN-beta on coxsackievirus B3-induced myocarditis in mice, *J. Interferon Res.* **5**:137–146 (1985).

43. Vallbracht, A., L. Hofman, K. G. Wurster, and B. Flehmig, Persistent infection of human fibroblasts by hepatitis A virus, *J. Gen. Virol* **65**:609–615 (1984).

44. Siegelman, M., M. W. Bond, M. W. Gallatin, T. St. John, H. T. Smith, V. A. Fried, and I. L. Weissman, Cell surface molecule associated with lymphocyte homing is a ubiquitinated branched-chain glycoprotein, *Science* **231**:823–829 (1986).

45. Gudvangen, R. J., P. S. Duffey, R. E. Paque, and C. J. Gauntt, Levamisole exacerbates coxsackievirus B3-induced murine myocarditis, *Infect. Immun.* **43**:1157–1165 (1983).

46. Friedman, H. M., E. J. Macarack, R. R. MacGregor, J. Wolfe, and N. A. Kefalides, Virus infection of endothelial cells, *J. Infect. Dis.* **143**:266–273 (1981).

47. Bodensteiner, J. B., H. H. Morris, J. T. Howell J. T., and S. S. Schochet, Chronic ECHO type 5 virus meningoencephalitis in X-linked hypogammaglobulinemia: treatment with immune plasma, *Neurology* (NY), **29**:815–819 (1979).

48. Wilfert, C. M., R. H. Buckley, T. Mohanakumar, J. F. Griffith, S. L. Katz, J. K. Whisnant, P. A. Eggleston, M. Moore, E. Treadwell, M. N. Oxman, and F. S. Rosen, Persistent and fatal central nervous system echovirus infections with agammaglobulinemia, *N. Engl. J. Med.* **296**:1485–1489 (1977).

49. Mease, P. J., H. D. Ochs, and R. J. Wedgwood, Successful treatment of echovirus meningoencephalitis and myositis-fasciitis with intravenous immune globulin therapy in a patient with X-linked agammaglobulinemia. *N. Engl. J. Med.* **304**:1278–1281 (1981).

50. Webster, A. D. B., J. H. Tripp, A. R. Hayward, A. D. Dayan, R. Doshi, E. E. Macintyre, and A. J. Tyrrell, Echovirus encephalitis and myositis in primary immunoglobulin deficiency, *Arch. Dis. Child.* **53**:33–37 (1978).

51. Bardelas, J. A., J. A. Winkelstein, D. S. Y. Seto, T. Tsai, and A. G. Rogal, Fatal ECHO 24 infection in a patient with hypogammaglobulinemia. Relationship to dermatomyositis-like syndrome, *J. Pediatr.* **90**:396–399 (1977).

52. Rozhon, E. J., A. K. Wilson, and B. Jubelt, Characterization of genetic changes occurring in attenuated poliovirus 2 during persistent infection in mouse central nervous system, *J. Virol.* **50**:137–144 (1984).

53. Borca, M. V., F. M. Fernandez, A. M. Sadir, and A. A. Schudel, Reconstitution of immunosuppression mice with mononuclear cells from donors sensitized to foot-and-mouth disease virus (FMDV), *Vet. Microbiol.* **10**:1–11 (1984).

54. Greenberger, S. B., M. W. Harmon, R. B. Couch, P. E. Johnson, S. Z. Wilson, C. C. Dacso, K. Bloom, and J. Quarles, Prophylactic effect of low doses of human leukocytes interferon against infection with rhinovirus, *J. Infect. Dis.* **145**:542–546 (1982).

55. Chaturvedi, U. C., H. O. Tandon, and A. Mathur, Control of *in vivo* and *in vitro* spread of coxsackievirus B4 infection by sensitized spleen cells and antibody, *J. Infect Dis.* **138**:181–190 (1978).

56. Woodruff, J. F., Lack of correlation between neutralizing antibody production and suppression of coxsackievirus B-3 replication in target organs: Evidence for involvement of mononuclear inflammatory cells in host defense, *J. Immunol.* **123:**31–36 (1979).
57. Huber, S. A., L. P. Job, and J. F. Woodruff, Sex-related differences in the pattern of coxsackievirus B-3 induced immune spleen cell cytotoxicity against virus-infected myofibers, *Infect. Immun.* **32:**68–73 (1981).
58. Bruserud, O., M. Stenersen, and E. Thorsby, T lymphocyte responses to Coxsackie B4 and mumps virus. II. Immunoregulation by HLA-DR3 and DR4 associated restriction elements, *Tissue Antigens* **26:**179–192 (1985).
59. Woodruff, J. F., Viral myocarditis. A review, *Am. J. Pathol.* **101:**425–483 (1980).
60. Huber, S. A., L. P. Job, and J. F. Woodruff, In vitro cultures of coxsackievirus group B, type 3 immune spleen cells on infected endothelial cells and biological activity of the cultured cells *in vivo*, *Infect. Immun.* **43:**567–573 (1984).
61. Huber, S. A., L. P. Job, K. R. Auld, and J. F. Woodruff, Sex-related differences in the rapid production of cytotoxic spleen cells active against uninfected myofibers during coxsackievirus B-3 infection, *J. Immunol.* **126:**1336–1340 (1981).

# 13

# Arenaviruses

## KATHRYN E. WRIGHT and WILLIAM E. RAWLS

## 1. INTRODUCTION

Over the years, three phenomena have focused attention on members of the arenaviruses family. Soon after the discovery of lymphocytic choriomeningitis virus (LCMV), it became apparent that the virus was capable of establishing a persistent infection in its natural host, *Mus musculus*.[1] Persistence of virus in animals infected *in utero* or in the newborn period is a feature shared by all *Arenaviridae* and probably represents a mechanism for virus survival in nature.[2] The mechanisms responsible for this viral persistence have been extensively investigated. Second, acute disease due to LCMV in adult mice can be abrogated by treatments that immunosuppressed the host. Subsequent studies have demonstrated that much of the pathology associated with Arenaviridae-induced diseases has an immune basis. These include the acute central nervous system (CNS) disease attributed to T lymphocytes as well as renal disease in chronically infected animals attributable to immune complex formation.[3,4] Finally, arenaviruses produce severe hemorrhagic diseases in humans, an incidental host. These include Argentine hemorrhagic fever caused by Junin virus, Bolivian hemorrhagic fever caused by Machupo virus, and Lassa fever caused by Lassa virus. The virulence of these viruses in humans has hampered investigation of the pathophysiology of the diseases they produce, although considerable information is available from clinical studies and from experimental infections using avirulent mutants in susceptible laboratory animals. Despite the fact that virus modulation of host immunity could contribute to all three phenomena, there is only limited evidence that this is the case.

KATHRYN E. WRIGHT • Scripps Clinic and Research Foundation, La Jolla, California 92037.    WILLIAM E. RAWLS • Department of Pathology, McMaster University, Hamilton, Ontario, Canada L8N 3Z5.    *Present address for K. E. W.:* Department of Microbiology and Immunology, University of Ottawa, Ottawa, Ontario, Canada K1H 8M5.

## 2. BASIC PROPERTIES OF THE VIRUSES

### 2.1. Morphology

Arenaviridae are characterized by a segmented single-stranded RNA genome, pleomorphic virions containing electron-dense granules, and an envelope with regular club-shaped peplomers.[5] Thirteen species have been identified on the basis of the natural host and serology.[6] These are listed in Table I. Viruses from the Americas form one group (the Tacaribe complex), while those from African rodents and LCMV comprise a second group (the LCMV–Lassa complex) on the basis of serologic relatedness. All arenaviruses are indistinguishable by electron microscopic examination.[7] On thin sections, the virus particles are usually 110–130 nm in diameter and contain electron-dense granules that resemble cellular ribosomes.[8] The nucleoprotein–RNA complex is not regularly ordered within the virion, and the envelope is acquired by budding through the plasma membrane.

### 2.2. Biochemical Properties

The genetic information coding for the proteins of the Arenaviridae is contained in two single-stranded RNA molecules. These have been designated L and S for the large and small segments, which have sedimentation values of 31S and 23S, respectively.[9] In addition, 28S, 18S, and 4–6S RNA can be read-

## TABLE I
### Arenavirus Species

| Virus | Natural host | Geographic distribution |
| --- | --- | --- |
| LCMV-Lassa complex | | |
| LCMV | *Mus musculus* | Europe and Americas |
| Lassa[a] | *Mastomys sp.* | West Africa |
| Mopeia | *Mastomys natalensis* | Mozambique |
| Mobala | *Praomys jacksoni* | Central African Republic |
| Tacaribe complex | | |
| Tacaribe | *Artibeus jamaicensis* | Trinidad |
| | *Artibeus literatus* | |
| Junin[a] | *Calomys musculinus* | Argentina |
| Machupo[a] | *Calomys calosus* | Bolivia |
| Amapari | *Oryzemys gaeldi* | Brazil |
| | *Neacomys quianae* | |
| Parana | *Oryzomys buccinatus* | Paraguay |
| Tamiami | *Sigmodon hispidus* | Florida (United States) |
| Pichinde | *Oryzomys albiqularis* | Colombia |
| Latino | *Calomys calosus* | Bolivia |
| Flexal | *Oryzomys sp.* | Brazil |

[a]Viruses for which humans may be an incidental host.

ily isolated from purified virions; these molecules appear to arise from host cell ribosomes packaged within the virion during the budding process. No function has been assigned to these entrapped ribosomes, since studies using cells with a temperature-sensitive ribosomal subunit protein indicate that the functional integrity of the ribosome is not needed for viral replication.[10]

The genome of an arenavirus has been shown to code for three primary gene products. The nucleoprotein, with a molecular weight of 61–72 kDa, is encoded in a negative sense on the 3' end of the S RNA segment. The 5' end of the S RNA segment encodes a molecule that is the precursor for glycoproteins inserted in the virion envelope. Interestingly, the glycoprotein gene appears to be encoded in a positive sense; the fact that two genes are encoded in the same segment with different polarity has been termed ambisense.[11] The precursor glycoprotein of 72–79 kDa gives rise to two virion glycoproteins in most,[12,13] but not all, arenaviruses.[14] The L RNA segment is thought to code for a large protein of ~200 kDa, which may be the RNA polymerase reported in virions of two members of the family.[13] Thus, although additional minor polypeptides have been found associated with purified virions, the basic structure of the Arenaviridae appears to be an RNA–nucleoprotein–polymerase core surrounded by an envelope containing two virus-specified glycoproteins.

## 2.3. Pathogenesis

### 2.3.1. Human Diseases

Human diseases from the arenaviruses are thought to be acquired through the respiratory or gastrointestinal (GI) tract or through abrasions in the skin from contaminated soil, water, or food.[15] In the case of LCMV, the contamination of the environment can usually be traced to persistently infected house mice, pet mice or hamsters, or laboratory animals. LCMV disease in humans is usually mild, and so few fatalities have been recorded that little information is available on the nature of the pathology produced by the virus in humans.

Argentine hemorrhagic fever and Bolivian hemorrhagic fever are diseases with similar features. These diseases are characterized by symptoms of a high unremitting fever, malaise, headache, muscular pains, nausea, and leukopenia appearing 7–14 days after exposure. In severe cases, the symptoms become more pronounced with hemorrhagic manifestations: petechiae and bleeding from the gums, nose, stomach, intestine, and uterus. There may also be signs of CNS involvement. Acute disease can last for up to 3 weeks. Death is usually attributable to uremic coma, hypotension, and shock secondary to plasma loss. Lassa fever is encountered in Western Africa. Early symptoms are similar to those for the South American hemorrhagic fevers, but with marked pharyngitis. Petechiae and subcutaneous hemorrhages may occur later; death, occurring in a high percentage of hospitalized cases during the second week, is associated with cardiovascular collapse.

The mechanisms whereby these viruses cause disease in humans are not well understood. Virus can be recovered from the blood, throat, and some-times urine during the acute illness. In hemorrhagic fever victims, the spleen and lymph nodes are most often positive for virus. A consistent finding is swell-ing of the endothelium of capillaries and arterioles in all organs, in the absence of inflammatory reactions.[15] A role for complement has been hypothesized early in infection,[16] and it has been suggested that high levels of interferons (IFN) produced as a result of lymphocyte destruction may damage phagocytic cells and activate complement, leading to the observed damage. In Lassa fever, a direct cytopathic effect may be responsible for major lesions in the liver, lungs, spleen, lymph nodes, and intestinal mucosa. Antibodies detectable by complement fixation, immunofluorescence, or virus neutralization develop in infected humans. In patients infected with Lassa virus, antibodies detectable by immunofluorescence usually appear about 10 days after the onset of symp-toms, while antibodies detected by the other two assays do not usually appear until about 18 days. Antibodies also appear late in the South American hemor-rhagic fevers, and peak titers are not observed until 30–60 days postinfection. Since antibodies arise so late after infection, it has been postulated that they are unlikely to play a role in eliminating the primary infection; favorable results after transfer of immune sera have been reported only for Argentine hemor-rhagic fever.[15]

### 2.3.2. Animal Diseases

Lymphocytic choriomeningitis virus is the best studied of the arenaviruses. LMCV produces a wide spectrum of disease in mice, depending on the strain of virus, dose and route of inoculation, as well as the strain and the age of the host. When virus is given by any route to neonatal mice, all organs become infected, although the primary targets are cells of the lymphoreticular system. These animals survive and remain persistently infected.

Intracranial inoculation of LCMV into adult mice results in a disease that kills virtually 100% of the animals in 8–10 days. Widespread infiltration of lymphocytes into most organs, including the CNS, is the major pathologic find-ing. The absence of similar infiltrates in immunologically immature mice that survived infections indicates that this pathology is not a direct effect of the virus. It has been demonstrated that immunosuppression prevents death and lymphocyte infiltration in the CNS of LCMV-infected adult mice without alter-ing viral replication.[17] Furthermore, adoptive transfer of LCMV immune T lymphocytes, but not immune serum, into infected immunosuppressed mice has been shown to induce the disease.[18–20] Thus, this form of arenavirus pathology is immunologically mediated.

The outcome of LCMV infection when the virus is administered extra-neurally varies with the strain of mouse and the virus used. Generally, intra-peritoneal or intravenous inoculation of LCMV does not produce disease.[3] The virus replicates in many organs, including spleen, thymus and lymph nodes, liver, kidney, and brain, and can be demonstrated in macrophages, B cells, and T cells.[21–23] Titers in the spleen and blood peak at day 5 after infection and decline rapidly thereafter. These mice are protected from intra-cranial challenge with LCMV, and the protective immunity that develops by day 4 is mediated by T lymphocytes that are also involved in viral clear-ance.[24,25]

Members of the Tacaribe complex of arenaviruses have been studied less extensively than LCMV, but the few studies that have been carried out reveal similarities to the pathogenesis of LCMV. In their natural hosts, these viruses clear slowly or establish persistence after neonatal exposure, whereas older ani-mals tend only to undergo transient infection.[2,26] This pattern may be differ-ent in rodents other than natural hosts.[27] For example, Pichinde virus, which was originally isolated from a South American rodent, is fatal for neonatal mice and hamsters but causes inapparent infection in adult mice and most strains of hamster.[28,29] The virus is fatal for adults of one inbred strain of Syrian ham-ster,[29,30] and an adapted strain of Pichinde virus has been developed that is fatal for guinea pigs.[31] There is no evidence that disease is immunologically mediated in these instances, and the major pathologic findings are extensive lesions in the kidneys and liver.[31,32] Wild-type Junin virus causes a fatal dis-ease in guinea pigs similar to human hemorrhagic fever, and this disease does not appear to be immunologically mediated.[33] However, Tamiami, Tacaribe, and Junin viruses cause a fatal neurologic disease in neonatal or suckling mice that resembles the CNS disease of LCMV.[27,33–36] Nude mice were found to be resistant to lethal infection with Junin virus, whereas immunocompetent controls were not,[37] and thymectomized mice resisted infection with Ma-chupo, Tacaribe, and Junin viruses.[38] An attenuated strain of Junin virus, but not wild-type virus, causes a disease in guinea pigs that appears to be immuno-logically induced.[39] The arenaviruses appear to replicate preferentially in cells of the reticuloendothelial system and lymphoid tissues,[2,40,41] with the exception of Tacaribe virus.[35] Both Tamiami and Junin viruses have been found in megakaryocytes in the bone marrow,[40,41] while Junin virus has been demonstrated in macrophages in the peritoneal cavity,[42] and an attenuated strain of Junin virus has been found chiefly in plastic adherent cells identified as dendritic cells.[39]

## 3. MODULATION OF IMMUNITY

### 3.1. Antibody Production

#### 3.1.1. Responses to Unrelated Antigens

Relatively few studies of the effect of arenavirus infections in antibody responses to unrelated antigens have been reported, and most of these deal with LCMV infection in mice. In adult mice, acute infection by footpad inocu-lation was found to suppress the humoral responses to sheep red blood cells (SRBC) and human serum albumin.[43] The suppressed response lasted for up to 2 months postinfection. A suppressed response to SRBC was also observed in neonatal mice acutely infected with LCMV, but the suppression lasted for a shorter period than was observed in adult mice. In another study of adult mice infected intraperitoneally with LCMV, antibody responses to thymus (T)-dependent antigens were found to be suppressed; this suppression lasted more than 2 months postinfection.[44] By contrast, adult mice persistently infected with LCMV were found to have normal humoral responses to SRBC and keyhole limpet hemocyanin[43,45,46] and to other viruses.[47,48]

We were able to locate only one other report of immunosuppression by an arenavirus in the literature. Guinea pigs infected with Junin virus were tested for humoral responses to SRBC and were found to be suppressed.[49] Reduced antibody production appeared to be due to reduced numbers of antibody-producing B cells.

### 3.1.2. Responses to Viral Antigens

Antibody responses to the proteins of the arenaviruses vary according to virus strains, host, and age at infection. The role of modulation by the infecting virus in the different patterns of responsiveness is unclear. Most experimental studies have dealt with LCMV in mice, since the virus produces persistence following neonatal infection, and this is a convenient model with which to explore the role of immunity in persistence. Adult mice inoculated with virus develop antibodies detectable by neutralization of virus or by complement fixation; these antibodies appear 2–3 weeks postinfection. Interestingly, neutralizing antibodies are readily detected postinfection with a viscerotropic strain (WE strain) of virus but not after infection with a neurotropic strain (Armstrong strain) of virus.[50] Whether these phenomena represent a possible modulation of the antibody response to certain epitopes on the glycoproteins of virus by neurotropic variants is unknown.

Early investigations failed to demonstrate antibodies to LCMV in sera of persistently infected animals.[17] Antibodies in these studies were sought using neutralization assays. On the basis of these observations, it was postulated that the mice were tolerant to viral antigens, possibly by deletion of B-cell clones capable of recognizing viral epitopes. Subsequently, it was demonstrated that persistently infected mice synthesize antiviral antibodies that participate in immune-complex formation.[51] In addition, low levels of neutralizing antibodies have been reported,[52] providing evidence that humoral tolerance to LCMV antigens does not exist in these mice. The virus-specific antibodies to LCMV appear to differ with respect to isotype between acutely and persistently infected mice.[53] More recent studies showed relatively few cells making antibodies to LCMV in the spleens of persistently infected mice as compared with acutely infected mice. In addition, the number of antibody-producing cells observed was dependent on the age and strain of the mice.[54] The spleens of a mouse strain that developed immune-complex disease contained appreciable numbers of antibody-producing cells, while the spleens of persistently infected house mice did not contain detectable antibody-producing cells. Thus, there appears to be humoral hyporesponsiveness to viral antigens in the persistently infected mice, and the degree of hyporesponsiveness may depend on the genetic background of the animal.

## 3.2. T Lymphocytes

### 3.2.1. Responses to Unrelated Antigens

Reports of suppressed T-cell responses following arenavirus infections again are restricted primarily to studies with LCMV. Adult mice infected in the footpad with LCMV were found to be more susceptible to ectromelia virus than were control mice, and increased susceptibility correlated with a reduced footpad response to ectromelia.[43] Lehmann-Grube and colleagues[55] showed a slight inhibition of skin graft rejection in LCMV-infected adult mice, and lymphocytes from adult mice infected intravenously displayed less cytotoxic activity against allogeneic targets in vitro than did lymphocytes from uninfected mice.[44,56] In addition, reduced mitogenic responses of T cells obtained from acutely infected adult mice[57] and from humans infected with the M-P variant of LCMV have been observed.[58] As with humoral responses, cellular responses to nonviral antigens appear to be normal in persistently infected mice. Such mice showed no differences in their ability to reject skin grafts over normal mice.[59] Of the Tacaribe complex, Junin virus has been found to depress the delayed hypersensitivity skin reaction to purified protein derivative (PPD) markedly in acutely infected guinea pigs.[60]

### 3.2.2. Responses to Viral Antigens

T-lymphocyte responses to viral antigens have been readily demonstrated in adult laboratory animals infected with arenaviruses. Adult mice acutely infected with LCMV develop delayed-type hypersensitivity (DTH), which has been demonstrated by the footpad swelling response as well as cytotoxic T lymphocytes (CTL) capable of lysing virus-infected target cells in vitro.[61,62] The same has been found for mice infected with Pichinde virus, and a secondary CTL response can be demonstrated using this virus–host combination.[63] Exceptions have been recorded, as exemplified by the development of a CTL response in the absence of a DTH response in adult mice infected with Junin virus.[36,64] The basis of the lack of a DTH response to Junin virus in mice is unknown. A lack of footpad swelling following footpad inoculation of Pichinde virus in the inbred MHA strain of hamsters, but not other hamster strains, was noted. This lack of responsiveness was inherited as a recessive trait and could be attributed to the induction of splenic suppressor cells that inhibited the DTH response.[65] However, T-cell responses to arenaviruses in adult rodents are usually vigorous and often contribute to the immunopathology of the diseases produced by these viruses.

Substantial data exist suggesting that viral persistence associated with neonatal infection is attributable to the absence of effector T cells to viral antigens. Virus-specific T-cell responses have been difficult to demonstrate in mice persistently infected with LCMV[66,67]; however, the transfer of immune T cells to persistently infected mice results in a rapid clearance of virus.[68] It has been postulated that the lack of T-cell responses associated with neonatal LCMV infection is due to tolerance resulting from selective inactivation of virus-specific T-cell precursors by LCMV.[69,70] Alternate possibilities include inhibition of effector T-cell responses by suppressor cells in persistently infected mice[71] or suppression by large amounts of virus.[72] This latter observation is supported by recent findings that the suppressive action of spleen cells from persistently infected mice on virus-specific CTL responses is due to the presence of variant LCMV that evolved during the course of infection.[73] The common theme of all these hypotheses is that the virus in the newborn mouse selectively modulates the T-cell responses to prevent the development of effector T cells that normally clear virus in acutely infected adult mice.

### 3.2.3. T-Cell Numbers

Arenavirus infections are generally associated with reduced numbers of lymphocytes in various organs during the acute phase of the infection.[60] Again, the effects of LCMV in mice have been most carefully studied. In adult mice infected with LCMV, a leukopenia with marked reduction in lymphocytes has been observed.[74] A reduction in bone marrow lymphocytes was also noted in this study, and lymphocytes in both compartments return to normal numbers by 30 days postinfection. Other studies have shown preferential loss of lymphocytes in T-dependent areas of lymphoid organs and the thymus within 1 week of infection.[23,75,76] Cortical thymocytes appeared particularly susceptible, and the bone marrow of infected mice showed a reduced ability to repopulate the thymi of irradiated recipients, suggesting that T-cell progenitors rather than mature T cells were lacking.[76] Acutely infected mice were found to be less able to recover from irradiation, and this was attributed to reduced numbers of hematopoietic stem cells, as assessed in vivo (CFU-S) or in vitro (CFC).[77,78] Similar changes could be observed in the absence of irradiation in acutely infected but not persistently infected mice.[74] It is not clear whether stem cells detected in these assays served as lymphoid precursors, but apparently some T lymphocyte precursors were diminished by infection.

## 3.3. Macrophages

Macrophages serve as one of the major targets for arenaviruses,[21,22,32,39,40,42] but there are only scattered data on alterations in macrophage function postinfection. Macrophages infected in vitro with LCMV showed no defect in either phagocytic function[43,46,47] or in hydrolytic enzymes.[79] However, alterations in β-galactosidase activity[80] and the uptake of neutral red[81] by macrophages of LCMV-infected mice have been noted, suggesting that the virus may alter lysosomal functions. In addition, it has been observed that mice infected with LCMV clear colloidal carbon more slowly at 6 days postinfection than do control mice.[82] The reduced mitogenic response of T and B lymphocytes from acutely infected mice can be reversed by the addition of normal macrophages to the cultures.[57] However, this may not be indicative of specific macrophage defects in infected mice, since the macrophages could serve to improve the viability of the lymphocytes in culture.[83]

Infection of murine macrophages with Pichinde virus interfered with the responsiveness of cells to macrophage growth factor (MGF) such that infected macrophages failed to proliferate when cultured with MGF.[84] Whether this defect interferes with immune responses to viral infection was not established, although activated macrophages did not appear to have direct antiviral effects. However, activated macrophages might contribute to resistance through other mechanisms, such as antigen presentation, or by the release of monokines necessary for the generation of virus-specific responses.

An attenuated strain of Junin virus was found to differ from a virulent strain in its failure to infect macrophages *in vitro*, suggesting that virulence may be due, at least in part, to interference by virus with macrophage functions.[42] The only function examined in this system, however, has been the phagocytic ability of macrophages infected with the virus *in vitro*, which was normal.[85]

### 3.4. Natural Killer Cells and Soluble Factors

Infection of mice with LCMV and Pichinde virus and hamsters with the latter virus resulted in increased nonspecific cytotoxic (NK) activity in the spleen.[30,63,86,87] After LCMV infection, enhanced NK activity is thought to result from IFN elicited by the infection.[86] NK cells do not appear to play a significant role in resistance to the arenaviruses,[87,89] although susceptibility to fatal Pichinde virus in an inbred strain of hamsters correlated with increased NK activity.[30] Nor are IFN production or NK activity negatively affected by infection with these viruses. Examination of interleukins after infection with arenaviruses has not been carried out, but two groups have examined production of colony-stimulating factor (CSF) in mice acutely infected with LCMV. One group found increased CSF activity during the time of suppressed CFU and CFC[74]; the other found reduced amounts of CSF.[90]

## 4. MECHANISMS OF VIRAL SUPPRESSION

Two quite different types of reduced immune responsiveness are associated with arenavirus infections. The immunosuppression observed in adult animals acutely infected with virus appears to affect responses to antigens unrelated to the virus but generally not to virus-specific antigens. However, in persistently infected animals, there appears to be reduced immune responsiveness only to viral antigens. Different mechanisms have been postulated for these two phenomena.

It has been postulated that the suppressed responses of both T and B lymphocytes in acutely infected animals is due to a defect in macrophage function brought about by infection of such cells with virus.[57] However, functions of macrophages thought to be important in the reduction of specific immune responses, such as monokine production and antigen presentation, have not been examined in infected mice. Infection of macrophages with Pichinde virus does alter DNA replication, but again it is not clear that this alters macrophage functions important in the generation of immune responses.[84]

Another defect proposed to account for suppressed humoral and cell-mediated responses to unrelated antigens is reduced numbers of T cells.[60,75,76,90] Thomsen and co-workers[76] established that only antibody responses to T-dependent antigens were reduced in infected mice. They hold that the reduced numbers of functional T cells is the result of viral interference with the differentiation of stem cells of the T lineage. While this might be the case, evidence that arenaviruses are directly cytolytic to lymphocytes at different maturational stages is lacking, and in adult mice there is no evidence that LCMV infects significant numbers of bone marrow cells or T lymphocytes.[22,23,91]

LCMV induces IFN which peaks at days 2–4 postinfection,[74,92] a period during which hematopoiesis is first suppressed. This correlation has led to the suggestion that IFN may have a suppressive effect on stem cells in the marrow and spleen,[74] since injection of mice with IFN or IFN inducers results in similar changes. Others have speculated that NK cells activated by IFN induced by infection mediate lysis of bone marrow cells.[86] NK cells are able to lyse bone marrow cells and are thought to have a regulatory role in normal hematopoiesis.[93,94] IFN may also have a direct effect on immunocompetent cells. When present before immunization, it has been shown to suppress humoral and cell-mediated responses to antigens and proliferative responses of lymphocytes to mitogens.[95] IFN could act directly on macrophages or B or T lymphocytes to prevent the proliferation necessary for the generation of a specific response. Arguing against virus-induced IFN as the sole mechanism of immunosuppression is the temporal pattern of the IFN response, which is transient as compared with the suppressed hematopoiesis of bone marrow cells and reduced humoral responses, which last for some time after infection.[42,72] Thus, the precise mechanisms responsible for immunosuppression following acute infection of adult animals by arenaviruses remain to be elucidated.

The mechanisms by which arenaviruses persist following neonatal infection are also not yet understood. It is clear that persistently infected animals need not be completely tolerant to viral antigens, since some of the viremic animals make antibodies to viral antigens.[3] However, a number of investigators have failed to demonstrate cell-mediated responses to viral antigens. Three mechanisms have been postulated to explain this lack of response to viral antigens by the effector T-cell compartment. These include the induction of virus-specific suppressor cells, a direct immunosuppressive effect of the virus or inactivation of precursor CTL capable of recognizing viral antigens during neonatal infection. Evidence from cell-transfer experiments suggesting the existence of suppressor cells for cell-mediated responses to LCMV in persistently infected mice has been reported[71]; however, the experimental design was such that this suppression may have been due to the presence of virus in the transferred cells.[72] This possibility has been further supported by the recent demonstration of variant viruses arising in persistently infected mice; the variant virus was capable of suppressing the generation of CTL responses to LCMV upon transfer to uninfected adult mice.[73] Thus, infection in the neonatal period may give rise to virus variants capable of specifically suppressing CTL responses to viral antigens. Why the variant virus itself does not induce a CTL response is unknown. The virus may not express epitopes recognized by CTL precursors, or the virus may interfere with appropriate antigen presentation on T cells, as has been demonstrated for other antigens.[96] Alternatively, the virus may replicate to high levels in organs of the immune system, so that local high concentrations of virus or antigens alone may inhibit responses.[72,73]

Both B and T lymphocytes from mice persistently infected with LCMV have been shown to contain infectious virus,[22,69,70] and further studies demonstrated that immature but not mature T cells could be infected with the virus both *in vivo* and *in vitro*.[70,91] It was suggested that T cells are permissive for infection by LCMV only at a particular stage during differentiation and that they become infected via their antigen-specific receptors. Once infected, they would lose their functional capacity, explaining the apparent tolerance.[91] Ahmed and colleagues[73] postulated that the variant virus isolated from persistently infected mice might prevent CTL induction by infecting these cells and again interfering with their differentiation into effector cells. Other factors have been noted that might account for the absence of virus-specific responses in persistently infected animals, but none has been examined in detail. The defects in both cellular and humoral responses may reside in the macrophage, so that viral antigens are not appropriately displayed for immune recognition, or there may be a selective failure of macrophages or T-helper cells to synthesize growth factors after stimulation with LCMV antigens, as has been documented for *Mycobacterium*.[97] A last suggestion is that CTL effectors are themselves lysed after infection, so that they are deleted before they carry out their effector functions.[72]

## 5. SIGNIFICANCE OF VIRUS-INDUCED MODULATION IN VIRAL PATHOGENESIS

Immunosuppression to nonviral antigens has been repeatedly demonstrated in adult animals acutely infected with arenaviruses; however, evidence is lacking that this immunosuppression plays a significant role in the pathogenesis of the virus-induced disease. On the contrary, there is ample evidence indicating that the immune responses to viral antigens contribute to the pathology of the virus-induced disease, as exemplified by the immunologically mediated CNS disease caused by LCMV in mice.[3] Conversely, the failure of virus-specific T-cell responses, particularly CTL, to develop in neonatal animals infected with arenaviruses appears to be crucial in the establishment of viral persistence. This phenomenon appears to represent a highly specific suppression of the immune system. There is ample evidence that CTL can clear virus when transferred to persistently infected mice.[25,62,98] The variant viruses described by Ahmed and co-workers[73] that induce persistent infections in immunocompetent mice induce normal humoral responses but fail to induce CTL. These data strengthen the argument that the suppressed or reduced CTL responses contribute to persistence of the virus. Further investigation of virus–host interactions in this system should provide a greater understanding of the importance of virus-induced suppression of immune responses in the pathogenesis of viruses capable of establishing persistent infections.

## REFERENCES

1. Traub, E., Persistence of lymphocytic choriomeningitis virus in immune animals and its relation to immunity, *J. Exp. Med.* **63**:847–861 (1936).
2. Johnson, K. N., P. A. Webb, and G. Justines, Biology of Tacaribe–complex viruses, in: *Lymphocytic Choriomeningitis Virus and Other Arenaviruses* (F. Lehmann-Grube, ed.), pp. 241–258, Springer-Verlag, New York (1973).
3. Buchmeier, M. J., R. M. Welsh, F. J. Dutko, and M. B. A. Oldstone, The virology and immunobiology of lymphocytic choriomeningitis virus infection, *Adv. Immunol.* **30**:275–331 (1980).
4. Lehmann-Grube, F., Portrait of viruses: Arenaviruses, *Intervirol.* **22**:121–145 (1984).
5. Howard, C. R., and P. R. Young, Variation among new and old world arenaviruses, *Trans. R. Inc. Trop. Med. Hyg.* **78**:299–306 (1984).
6. Wulff, H., J. V. Lange, and P. A. Webb, Interrelationships among arenaviruses measured by indirect immunofluorescence, *Intervirol.* **9**:344–350 (1978).
7. Murphy, F. A., P. A. Webb, K. M. Johnson, and S. G. Whitfield, Morphological comparison of Machupo with lymphocytic choriomeningitis virus: Basis for a new taxonomic group, *J. Virol.*, **4**:535–541 (1969).
8. Murphy, F. A., and S. G. Whitfield, Morphology and morphogenesis of arenaviruses, *Bull. WHO* **52**:409–419 (1975).
9. Rawls, W. E., and W. C. Leung, Arenaviruses, in: *Comprehensive Virology*, Vol. 14 (H. Fraenkel-Conrat and R. R. Wagner, eds.), pp. 157–192, Plenum, New York (1979).
10. Leung, W. C., and W. E. Rawls, Virion-associated ribosomes are not required for the replication of Pichinde virus, *Virology* **81**:174–176 (1977).
11. Auperin, D., V. Romanowski, M. Galinski, and D. H. L. Bishop, Sequencing studies of Pichinde arenavirus S RNA indicate a novel coding strategy in ambisense viral S RNA, *J. Virol.* **52**:897–908 (1984).
12. Buchmeier, M. J., and M. B. A. Oldstone, Protein structure of lymphocytic choriomeningitis virus: Evidence for a cell-associated precursor of the virion glycopeptides, *Virology* **99**:111–120 (1979).
13. Harnish, D. G., W. C. Leung, and W. E. Rawls, Characterization of polypeptides immunoprecipitable from Pichinde virus-infected BHK-21 cells, *J. Virol.* **38**:840–848 (1981).
14. Gard, G. P., A. C. Vezza, D. H. L. Bishop, and R. W. Compans, Structural proteins of Tacaribe and Tamiami virions, *Virology* **83**:84–95 (1977).
15. Casals, J., Arenaviruses, in: *Viral Infections of Humans*, 2nd ed. (A. Evans, ed.), pp. 127–150, Plenum, New York (1984).
16. de Bracco, M. M. E., M. T. Rimoldi, P. M. Cossio, A. Rabinovich, J. I. Maistegui, G.

Carballa, and R. M. Arana, Argentine hemorrhagic fever. Alterations of the complement system and anti-Junin-virus humoral response, *N. Engl. J. Med.* **299**:216–220 (1978).

17. Hotchin, J., The biology of lymphocytic choriomeningitis infection. Virus-induced immune disease, *Cold Spring Harbor Symp. Quant. Biol.* **27**:479–499 (1962).

18. Gilden, D. H., G. A. Cole, and N. Nathanson, Immunopathogenesis of acute central nervous system disease produced by lymphocytic choriomeningitis virus. II. Adoptive immunization of virus carriers, *J. Exp. Med.* **135**:874–889 (1972).

19. Doherty, P. C., and R. M. Zinkernagel, Capacity of sensitized thymus-derived lymphocytes to induce fatal lymphocytic choriomeningitis is restricted by the H-2 gene complex, *J. Immunol.* **114**:30–33 (1975).

20. Allan, J. E., and P. C. Doherty, Immune T cells can protect or induce fatal neurological disease in murine lymphocytic choriomeningitis, *Cell. Immunol.* **90**:401–407 (1985).

21. Doyle, M. V., and M. B. A. Oldstone, Interactions between viruses and lymphocytes. I. *In vivo* replication of lymphocytic choriomeningitis virus in mononuclear cells during both chronic and acute viral infections, *J. Immunol.* **121**:1262–1269 (1978).

22. Popescu, M., J. Löhler, and F. Lehmann-Grube, Infectious lymphocytes in lymphocytic choriomeningitis virus carrier mice, *J. Gen. Virol.* **42**:481–492 (1979).

23. Lohler, J., and F. Lehmann-Grube, Immunopathologic alterations of lymphatic tissues of mice infected with lymphocytic choriomeningitis virus. I. Histopathologic findings, *Lab. Invest.* **44**:193–204 (1981).

24. Lehmann-Grube, F., U. Assmann, C. Loliger, D. Moskophidis, and J. Löhler, Mechanism of recovery from acute virus infection. I. Role of T lymphocytes in the clearance of lymphocytic choriomeningitis virus from spleens of mice, *J. Immunol.* **134**:608–615 (1985).

25. Byrne, J. A., and M. B. A. Oldstone, Biology of cloned cytotoxic T lymphocytes-specific for lymphocytic choriomeningitis virus. VI. Migration and activity *in vivo* in acute and persistent infection, *J. Immunol.* **136**:698–704 (1986).

26. Webb, P. A., K. M. Johnson, C. J. Peters, and G. Justines, Behavior of Machupo and Latino viruses in calomys callosus from two geographic areas of Bolivia, in: *Lymphocytic Choriomeningitis Virus and Other Arenaviruses* (F. Lehmann-Grube, ed.), pp. 313–322, Springer-Verlag, Berlin (1973).

27. Gilden, D. H., H. M. Friedman, C. O. Kyj, R. A. Roosa, and N. Nathanson, Tamiami virus-induced immunopathological disease of the central nervous system, in: *Lymphocytic Choriomeningitis Virus and Other Arenaviruses* (F. Lehmann-Grube, ed.), pp. 287–297, Springer-Verlag, New York (1973).

28. Trapido, H., and C. Sanmartin, Pichinde virus, a new virus of the Tacaribe group from Colombia, *Am. J. Trop. Med. Hyg.* **20**:631–641 (1971).

29. Buchmeier, M. J., and W. E. Rawls, Variation between strains of hamsters in the lethality of Pichinde virus infections, *Infect. Immun.* **16**:413–421 (1977).

30. Gee, S. R., M. A. Chan, D. A. Clark, and W. E. Rawls, Role of natural killer cells in Pichinde virus infection of Syrian hamsters, *Infect. Immun.* **31**:919–928 (1981).

31. Jahrling, P. B., R. A. Hesse, J. B. Rhoderick, M. A. Elwell, and J. B. Moe, Pathogenesis of a Pichinde virus strain adapted to produce lethal infection in guinea pigs, *Infect. Immun.* **32**:872–880 (1981).

32. Murphy, F. A., M. J. Buchmeier, and W. E. Rawls, The reticuloendothelium as the target in a virus infection: Pichinde virus pathogenesis in two strains of hamsters, *Lab. Invest.* **37**:502–515 (1977).

33. Weissenbacher, M. C., L. B. De Guerrero, and M. C. Boxaca, Experimental biology and pathogenesis of Junin virus infection in animals and man, *Bull. WHO* **52**:507–515 (1975).

34. Walker, D. H., H. Wulff, J. V. Lange, and F. A. Murphy, Comparative pathology of Lassa virus infection in monkeys, guinea-pigs and *Mastomys natalensis*, *Bull. WHO* **52**:523–534 (1975).

35. Borden, E. C., and N. Nathanson, Tacaribe virus infection of the mouse: An immunopathologic disease model, *Lab. Invest.* **30**:465–473 (1974).

36. Barrios, H. A., S. N. Rondinone, J. L. Blejer, O. A. Giovaniello, and N. R. Nota, Development of specific immune response in mice infected with Junin virus, *Acta Virol. (Praha)* **26**:156–164 (1982).

37. Weissenbacher, M. C., M. A. Calello, C. J. Quintans, H. Panisse, N. M. Woyskowski, and V. H. Zanndi, Junin virus infection in genetically athymic mice, *Intervirology* **19**:1–5 (1983).

38. Besuschio, S. C., M. C. Weissenbacher, and G. A. Schmunis, Different histopathological response to arenavirus infection in thymectomized mice, *Arch. Ges. Virusforsch.* **40**:21–28 (1973).

39. Laguens, R. M., M. M. Avila, S. R. Samoilovich, M. C. Weissenbacher and R. P. Laguens, Pathogenicity of an attenuated strain (XJC1₃) of Junin virus. Morphological and virological studies in experimentally infected guinea pigs, *Intervirology* **20**:195–201 (1983).

40. Murphy, F. A., W. Winn, D. H. Walker, M. R. Flemister, and S. G. Whitfield, Early lymphoreticular viral tropism and antigen persistence. Tamiami virus infection in the cotton rat, *Lab. Invest.* **34**:125–140 (1976).

41. Carballal, G., M. Rodriguez, M. J. Frigerio, and C. Vasquez, Junin virus infection of guinea-pigs: Electron microscope studies of peripheral blood and bone marrow, *J. Infect. Dis.* **135**:367–373 (1977).

42. Laguens, M., J. G. Chambo, and R. P. Laguens, *In vivo* replication of pathogenic and attenuated strains of Junin virus in different cell populations of lymphatic tissue, *Infect. Immun.* **41**:1279–1283 (1983).

43. Mims, C. A., and S. Wainwright, The immunodepressive action of lymphocytic choriomeningitis virus in mice, *J. Immunol.* **101**:717–724 (1968).

44. Bro-Jørgensen, K., F. Güttler, P. N. Jørgensen, and M. Volkert, T lymphocyte function as the principal target of lymphocytic choriomeningitis virus-induced immunosuppression, *Infect. Immun.* **11**:622–629 (1975).

45. Bro-Jørgensen, K., and M. Volkert, Defects in the immune system of mice infected with lymphocytic choriomeningitis virus, *Infect. Immun.* **9**:605–614 (1974).

46. Oldstone, M. B. A., A. Tishon, J. M. Chiller, W. O. Weigle, and F. J. Dixon, Effect of chronic viral infection on the immune system. 1. Comparison of the immune responsiveness of mice chronically infected with LCM virus with that of non-infected mice, *J. Immunol.* **110**:1268–1278 (1973).

47. Haas, V. H., Some relationships between lymphocytic choriomeningitis (LCM) virus and mice, *J. Infect. Dis.* **94**:187–198 (1954).

48. Traub, E., Observations on immunological tolerance and "immunity" in mice infected congenitally with the virus of lymphocytic choriomeningitis, *Arch. Ges. Virusforsch.* **10**:303–314 (1961).

49. Parodi, A. S., N. R. Nota, L. B. deGuerrero, M. J. Frigerio, M. Weissenbacher, and E. Rey, Inhibition of immune response in experimental hemorrhagic fever (Junin virus), *Acta Virol. (Praha)* **11**:120–125 (1967).

50. Kimming, W., and F. Lehmann-Grube, The immune response of the mouse to lymphocytic choriomeningitis virus. I. Circulating antibodies, *J. Gen. Virol.* **45**:703–710 (1979).

51. Oldstone, M. B. A., and F. J. Dixon, Lymphocytic choriomeningitis; production of antibody by "tolerant" infected mice, *Science* **158**:1193–1195 (1967).

52. Hotchin, J., L. Benson and E. Sikora, The detection of neutralizing antibody to lymphocytic choriomeningitis virus in mice, *J. Immunol.* **102**:1128–1135 (1969).

53. Thomsen, A. R., M. Volkert and O. Marker, Different isotype profiles of virus-specific antibodies in acute and persistent lymphocytic choriomeningitis virus infection in mice, *Immunology* **55**:213–223 (1985).

54. Moskophidis, D., and F. Lehmann-Grube, The immune response of the mouse to lym-

phocytic choriomeningitis virus. IV. Enumeration of antibody-producing cells in spleens during acute and persistent infection, *J. Immunol.* **133**:3366–3370 (1984).

55. Lehmann-Grube, F., I. Niemeyer, and J. Lohler, Lymphocytic choriomeningitis of the mouse. IV. Depression of the allograft reaction, *Med. Microbiol. Immunol.* **158**:16–25 (1972).

56. Güttler, F., K. Bro-Jørgensen, P. N. Jørgensen, Transient impaired cell-mediated tumor immunity after acute infection with lymphocytic choriomeningitis virus, *Scand. J. Immunol.* **4**:327–336 (1975).

57. Jacobs, R. P., and G. A. Cole, Lymphocytic choriomeningitis virus-induced immunosuppression: A virus-induced macrophage defect, *J. Immunol.* **117**:1004–1009 (1976).

58. Wilson, J. D., H. E. Webb, N. M. Molomut, and M. Padnos, Depression of PHA response in patients during therapeutic infection with MP virus, *Intervirology* **2**:41–47 (1973/1974).

59. Holterman, O. A., and J. A. Majde, An apparent histoincompatibility between mice chronically infected with LCMV and their uninfected syngeneic counterparts, *Transplantation* **11**:20–29 (1971).

60. Carballal, G., J. R. Oubiña, S. N. Rondinone. B. Elsner, and M. J. Frigerio, Cell-mediated immunity and lymphocyte populations in experimental Argentine hemorrhagic fever (Junin virus), *Infect. Immun.* **34**:323–327 (1981).

61. Tosolini, F. A., and C. A. Mims, Effect of murine and viral strain on the pathogenesis of lymphocytic choriomeningitis infection and a study of footpad responses, *J. Infect. Dis.* **123**:134–144 (1971).

62. Zinkernagel, R. M., and P. C. Doherty, MHC-restricted cytotoxic T cells: Studies on the biological role of polymorphic major transplantation antigens determining T-cell restriction-specificity, function, and responsiveness, *Adv. Immunol.* **27**:51–177 (1979).

63. Walker, C. M., W. E. Rawls, and K. L. Rosenthal, Generation of memory cell-mediated immune responses after secondary infection of mice with Pichinde virus, *J. Immunol.* **132**:469–474 (1984).

64. Barrios, H. A., S. N. Giovanniello, S. N. Rondinone, O. E. Competella, and N. R. Nota, Passive transfer protection against Junin virus in cyclophosphamide-suppressed mice, *Acta Virol. (Praha)* **28**:343 (1984).

65. Chan, M., D. Clark, and W. E. Rawls, Pichinde virus-specific cell-associated suppression of primary footpad swelling in an inbred strain of Syrian hamsters, *J. Immunol.* **130**:925–931 (1983).

66. Cole, G. A., R. A. Prendergast, and C. S. Henney, *In vitro* correlates of LCM virus-induced immune response, in: *Lymphocytic Choriomeningitis Virus and Other Arenaviruses* (F. Lehmann-Grube, ed.), pp. 61–71, Springer-Verlag, Berlin (1973).

67. Marker, O., and M. Volkert, *In vitro* measurement of the time course of cellular immunity to LCM virus in mice, in: *Lymphocytic Choriomeningitis Virus and Other Arenaviruses* (F. Lehmann-Grube, ed.), pp. 207–216, Springer-Verlag, Berlin (1973).

68. Volkert, M., K. Bro-Jørgensen, O. Marker, B. Rubin, and L. Trier, The activity of T and B lymphocytes in immunity and tolerance to the lymphocytic choriomeningitis virus in mice, *Immunology* **29**:455–464 (1975).

69. Cihak, J., and F. Lehmann-Grube, Immunological tolerance to lymphocytic choriomeningitis virus in neonatally infected virus carrier mice: evidence supporting a clonal inactivation mechanism, *Immunology* **34**:265–275 (1978).

70. Popescu, M., and D. H. Ostrow, Multiplication of lymphocytic choriomeningitis virus in thymocytes during its persistence in mice, *J. Gen Virol.* **61**:293–298 (1982).

71. Zinkernagel, R. M., and P. C. Doherty, Indications of active suppression in mouse carriers of lymphocytic choriomeningitis virus, in: *Immunological Tolerance* (D. H. Katz and B. Benacerraf, eds.), pp. 403–411, Academic, New York (1974).

72. Dunlop, M. B. C., and R. V. Blanden, Mechanisms of suppression of cytotoxic T-cell

responses in murine lymphocytic choriomeningitis virus infection, *J. Exp. Med.* **145:**1131–1143 (1977).

73. Ahmed, R., A. Salmi, L. D. Butler, J. M. Chiller, and M. B. A. Oldstone, Selection of genetic variants of lymphocytic choriomeningitis virus in spleens of persistently infected mice. Role in suppression of cytotoxic T lymphocyte responses and viral persistence, *J. Exp. Med.* **60:**521–540 (1984).

74. Bro-Jørgensen, K., and S. Knudtzon, Changes in hemopoiesis during the course of acute LCM virus infection in mice, *Blood* **49:**47–57 (1977).

75. Hanaoka, M., S. Suzuki, and J. Hotchin, Thymus-dependent lymphocytes: Destruction by lymphocytic choriomeningitis virus, *Science* **163:**1216–1219 (1969).

76. Thomas, A. R., K. Bro-Jørgensen, and B. L. Jensen, Lymphocytic choriomeningitis virus-induced immunosuppression: Evidence for viral interference with T-cell maturation, *Infect. Immun.* **37:**981–986 (1982).

77. Bro-Jørgensen, K., and M. Volkert, Haemopoietic defects in mice infected with lymphocytic choriomeningitis virus. 1. The enhanced x-ray sensitivity of virus infected mice, *Acta Pathol. Microbiol. Scand. B* **80:**845–852 (1972).

78. Bro-Jørgensen, K., and M. Volkert, Haemopoietic defects in mice infected with lymphocytic choriomeningitis virus. 2. The viral effect upon the function of colony-forming stem cells, *Acta Pathol. Microbiol. Scand. B* **80:**853–862 (1972).

79. Schwartz, R., J. Löhler, and F. Lehmann-Grube, Infection of cultivated mouse peritoneal macrophages with lymphocytic choriomeningitis virus, *J. Gen. Virol.* **39:**565–570 (1978).

80. Yarborough, D. J., O. T. Meyer, A. M. Dannenberg, Jr., and B. Pearson, Histochemistry of macrophage hydrolases. III. Studies on β-galactosidase, β-glucuronidase and aminopeptidases with inodyl and naphthyl substrates, *J. Reticuloendothel. Soc.* **4:**390–408 (1967).

81. Allison, A., Lysosomes in virus-infected cells, in: *Perspectives in Virology*, Vol. V (V. M. Pollard, ed.), pp. 29–62, Academic, New York (1967).

82. Gledhill, A. W., D. L. J. Bilbey, and J. S. F. Niven, Effect of certain murine pathogens on phagocytic activity, *Br. J. Pathol.* **46:**433–442 (1965).

83. Bro-Jørgensen, K., The interplay between lymphocytic choriomeningitis virus, immune function and hemopoiesis in mice, *Adv. Viral Res.* **22:**327–369 (1978).

84. Friedlander, A. M., P. B. Jahrling, P. Merrill, and S. Tobery, Inhibition of mouse peritoneal macrophage DNA synthesis by infection with the arenavirus Pichinde, *Infect. Immun.* **43:**283–288 (1984).

85. Gonzalez, P. H., J. S. Lampuri, C. E. Coto, and R. P. Laguens, *In vitro* infection of murine macrophages with Junin virus, *Infect. Immun.* **35:**356–358 (1982).

86. Welsh, R. M., Jr., Cytotoxic cells induced during lymphocytic choriomeningitis virus infection of mice. I. Characterization of natural killer cell induction, *J. Exp. Med.* **148:**163–181 (1978).

87. Gee, S. R., D. A. Clark, and W. E. Rawls, Differences between Syrian hamsters strains in natural killer cell activity induced by infection with Pichinde virus, *J. Immunol.* **123:**2618–2626 (1979).

88. Welsh, R. M., and R. Kiessling, Natural killer cell response to lymphocytic choriomeningitis virus in beige mice, *Scand. J. Immunol.* **11:**363–367 (1980).

89. Bukowski, J. F., B. A. Woda, S. Habu, K. Okumura, and R. M. Welsh, Natural killer cell depletion enhances virus synthesis and virus-induced hepatitis *in vivo*, *J. Immunol.* **131:**1531–1538 (1983).

90. Silberman, S. L., R. P. Jacobs, and G. A. Cole, Mechanisms of hemopoietic and immunological dysfunction induced by lymphocytic choriomeningitis virus, *Infect. Immun.* **19:**533–539 (1978).

91. Lehmann-Grube, F., F. Tijerina, W. Zeller, U. C. Chaturvedi, and J. Löhler, Age-dependent susceptibility of murine T lymphocytes to lymphocytic choriomeningitis virus, *J. Gen. Virol.* **64:**1157–1166 (1983).

92. Merigan, T. C., M. B. A. Oldstone, and R. M. Welsh, Interferon production during lymphocytic choriomeningitis virus infection of nude and normal mice, *Nature (Lond.)* **268**:67–68 (1977).

93. Kiessling, R., P. S. Hochman, O. Haller, G. M. Shearer, H. Wigzell, and G. Cudkowicz, Evidence for a similar or common mechanism for natural killer cell activity and resistance to haematopoietic grafts, *Eur. J. Immunol.* **7**:655–663 (1977).

94. Riccardi, C., A. Santoni, T. Barlozzari, and R. B. Herberman, *In vivo* reactivity of mouse natural killer (NK) cells against normal bone marrow cells, *Cell. Immunol.* **60**:136–143 (1981).

95. deMaeyer, E., Interferon and delayed-type hypersensitivity to viral antigen, *J. Infect. Dis.* **133**:A63–A65 (1976).

96. Fink, P. J., I. L. Weissman, and M. J. Bevan, Haplotype-specific suppression of cytotoxic T cell induction by antigen inappropriately presented on T cells, *J. Exp. Med.* **157**:141–154 (1983).

97. Hoffenbach, A., P. H. Lagrange, and M. A. Bach, Influence of dose and route of *Mycobacterium lepraemurium* inoculation on the production of interleukin 1 and interleukin 2 in C57 Bl/6 mice, *Infect. Immun.* **44**:665–671 (1984).

98. Anderson, J., J. A. Byrne, R. Schreiber, S. Patterson, and M. B. A. Oldstone, Biology of cloned cytotoxic T lymphocytes specific for lymphocytic choriomeningitis virus: Clearance of virus and *in vitro* properties, *J. Virol.* **53**:552–560 (1985).

# 14

# Togavirus-Induced Immunosuppression

## UMESH C. CHATURVEDI

## 1. INTRODUCTION

The viruses included in the Togaviridae family are commonly known as arboviruses that replicate in the tissues of arthropods and are transmitted to the vertebrate host by the bite of blood-sucking arthropods. The Togaviridae include alpha- and flaviviruses. The flaviviruses were recently separated from Togaviridae into a family, the Flaviviridae[1]; however, they have been included for discussion in the present chapter. Arbovirus infections are common in the tropical and subtropical areas of the world. They produce febrile illness with or without a rash, encephalitis, or hemorrhagic manifestations. Some are serious public health problems and have produced extensive epidemics. Eight of the 25 known alphaviruses and 26 of 60 flaviviruses can cause human disease.[2]

In the past, few virologists were interested in the study of the immunopathogenesis and other aspects of togavirus infection due to the slow pace of work, limited appropriate technology, and other inherent problems in their study. Thus, in this large group of viruses, immunopathologic studies have been concentrated on only a few, including Venezuelan equine encephalitis virus (VEEV), Sindbis virus (SV), and Semliki Forest virus (SFV), among the alphaviruses, and yellow fever virus (YFV), Japanese encephalitis virus (JEV), and dengue virus (DV) in the flavivirus group; other togaviruses include lactate dehydrogenase-elevating virus (LDV) and rubella virus (RV).

Suppression of immune responses during viral infection has been known from the time of the observations of von Pirquet in 1908 that tuberculin skin

UMESH C. CHATURVEDI • Department of Microbiology, King George's Medical College, Lucknow, India 226 003.

reactivity is lost in children infected with measles. Since then, a number of viral infections have been shown to be associated with depressed immune responses, and the possible mechanisms have been discussed.[3] An excellent description of the effects of togavirus infection on the immunologic parameters of humans can be found in a review by Halstead.[4] Animal models of viral infection, especially mice, provide the best understood phenomena in virus immunity and pathogenesis. This is also true for togaviruses. This chapter summarizes the effects of togavirus infection on the immune system, with special reference to animal models, followed by a description of the mechanisms of immunosuppression, using studies on DV infection of mice.

## 2. BASIC PROPERTIES OF THE VIRUSES

Togaviruses are enveloped spherical particles, 40–70 nm in diameter, with surface projections protruding from the envelope. The genome is a single molecule of positive single-stranded RNA. The capsid consists of 3–4 polypeptides, of which one or more are glycosylated.[5]

The genomic RNA of flaviviruses is a polycistronic single-stranded molecule having a 5' cap but lacking a poly (A) tail at the 3' end.[6–8] No subgenomic messenger RNA (mRNA) has been detected in cells infected by any flavivirus, and it has been proposed that internal initiation of protein synthesis occurs on the genomic mRNA.[9–11] The naked 42S RNA genome of alphaviruses is infectious and, in certain members the genome, is capped at its 5' end and polyadenylated at its 3' end.[5] The complete nucleotide sequence of several flaviviruses was recently elucidated.[12]

## 3. ANTIGENIC RELATIONSHIPS

Togaviruses have been grouped on the basis of serologic cross-reactivity of their antigens by hemagglutination inhibition (HI), complement fixation (CF), and neutralization (NT) tests. Alphaviruses share a group-reactive nucleoprotein antigen that does not react in HI or NT tests. Two other proteins are envelope glycoproteins and show type-specific reactivity. The flaviviruses have divergent antigenic and biologic characteristics but share the basic property of having three virion polypeptides, one of which is a glycoprotein.[13] They have genus-specific, species-specific, and subgroup-specific antigens. The major antigens are an envelope (E) glycoprotein and the nucleocapsid (C) protein. The E glycoprotein reacts in HI, CF, and NT tests.[14,15] Antigenic cross-reactivity among togaviruses has also been shown in tests for cell-mediated immunity (CMI) by cross-protection using adoptively transferred immune spleen cells[16] and by the activity of cytotoxic T lymphocytes (CTL).[17] Antigenic cross-reactivity among togaviruses is complex and creates problems in clinical diagnosis as well as in seroepidemiology.

## 4. HOST-DEFENSE MECHANISMS

Host defenses are active against both extracellular and intracellular virus. Mechanisms that combat extracellular viruses include antibody and phagocytic macrophages that may combine with and inactivate the virus. Mechanisms responsible for the elimination of intracellular virus include (1) lysis of virus-infected cells bearing viral antigenic membrane surface antigens by CTL, acting directly or through their mediators; antibody plus macrophages; (2) antibody plus complement; (3) antibody plus T cells; and (4) antibody plus K cells. Cytolysis of virus-infected cells can also occur nonspecifically by activated macrophages and natural killer (NK) cells. The replication of intracellular virus can be inhibited nonspecifically by interferon (IFN). The effect of togaviruses on some of these mechanisms is described later.

### 4.1. Antibodies

Antibodies are important in the recovery from most acute viral infections. They may terminate the primary infection, limit viremic spread, and prevent disease, reinfection, and mucosal entrance.[18-20] The mechanisms by which circulating antibodies contribute to recovery against viral infection may include prevention of adsorption to cells,[21] lysis of the virus,[22] altered intracellular handling of virus–antibody complexes,[23] and lysis of virus-infected cells.[24,25] In a number of togavirus infections, adoptively transferred antibodies provide protection by limiting viremia, preventing virus spread to susceptible target organs.[26-38] JEV infection in humans is associated with persistence of virus-specific immunoglobulin M (IgM) antibody from 6 months to 2 years.[39-41] Persistence of IgM in such cases has been correlated with the increased virulence of the viral infection.[41] By contrast, Mathur et al.[34] showed that JEV-specific IgM antibodies last for 2 weeks in mice and are responsible for protection against the virus. Long-lasting virus-specific IgM and IgG memory is found in mice latently infected with JEV, which facilitates elimination of the reactivated virus.[42]

Properties of antibodies such as NT or HI have generally been associated with protective immunity, but the role of non-neutralizing antibodies is unclear. With the availability of monoclonal antibodies of known specificity, it has become possible to investigate the protective phenomenon more precisely. Studies on VEEV, SV, and SFV[32,33,35] have shown that protection was provided both by the neutralizing as well as by some non-neutralizing antibodies, while only neutralizing antibodies provided protection against challenge with tick-borne encephalitis (TBEV) and St. Louis encephalitis (SLEV) viruses.[43-45] Recently, Gould et al. reported that monoclonal antibodies against both the 54-kDa structural envelope protein and 48-kDa nonstructural protein of yellow fever virus can protect mice against infection, but the ability to protect depends on the neurovirulence of the virus.[38] There was no correlation between the ability of monoclonal antibody to neutralize the virus in vitro and protect mice in vivo. Immunization of mice with dengue 2 virus-specified nonstructural protein NSI provides significant protection against intracerebral challenge with the homologous virus but not against heterologous dengue 1 virus.[43] It may be possible to have a broad-spectrum cross-protection among flaviviruses by stimulating an antibody response to appropriate epitopes of the envelope protein.[46]

*Enhancing Antibodies*

Antiviral antibodies that increase the infectivity of many togaviruses have been termed enhancing antibodies. Halstead[47] presented an extensive review on the enhancement of viral infections by such antibodies. This phenomenon has been studied mostly using the mouse macrophage-like cell line P388D1. Enhancement by specific non-neutralizing antiviral IgG requires the presence of Fc receptors on the cell but not complement.[48] Similar enhancement via $C_3b$ receptors of macrophages can be produced by virus-specific IgM antibody and complement.[49] Although the prototype model is DV, other togaviruses that show antibody-dependent enhancement are West Nile virus (WNV), YFV, JEV, Murray-Valley encephalitis virus, and Kunjin and Getah viruses.[47] Using monoclonal antibodies, it has been shown that most of the antibodies that were cross-reactive within the flavivirus group mediate immune enhancement, while type-specific antibodies, whether NT or HI, did not.[47,50,51] For the first time, *in vivo* enhancement of virus virulence has been reported using monoclonal antibodies against YFV that may provide an opportunity to correlate this unusual phenomenon *in vivo* with that *in vitro*.[52] In DV[53] and TBEV[54] infections, the enhancement is epitope-specific, as not all monoclonal antibodies that bind to the envelope protein of the virus are able to enhance viral replication. Clearly, the production of future vaccines using recombinant DNA technology will require the exclusion of epitopes that elicit enhancing antibody and those that are not involved in the protective immune response.

## 4.2. Cell-Mediated Immune Response

Cell-mediated immunity promotes recovery from virus infection by eliminating or restricting virus-infected cells. CMI involves different subpopulations of sensitized T cells and macrophages. Virus-induced inflammatory process and virus clearance are functions of class I-MHC-restricted T-cell populations.[55] It has been difficult to demonstrate sensitization of T lymphocytes in togavirus infection, especially CTL activity. However, there are reports of sensitization of different subpopulations of T cells by various togaviruses (Table I). The role of CTL in alphavirus infections is questionable, although these viruses bud from the cell plasma membrane, and their antigens would be available to sensitize T cells. The flaviviruses are assembled in intracytoplasmic cisternae of the cell; thus, antigens are not present on the cell surface. The major contribution of T cells to flavivirus clearance is via T-helper—B-cell interactions, leading to antibody production.[55]

**TABLE I**
**Examples of Sensitization of T Lymphocytes in Mice by Togaviruses, as Detected in Different Assays**

| Functions/virus | References |
|---|---|
| Delayed-type hypersensitivity | |
| St. Louis encephalitis | 242 |
| Semliki Forest | 243 |
| Japanese encephalitis | 229, 253, 254 |
| Dengue | 207, 208 |
| Lymphocyte activation | |
| Venezuelan equine encephalitis | 68 |
| Japanese encephalitis | 240 |
| Leukocyte migration inhibition | |
| Japanese encephalitis | 34, 95, 244, 245, 254 |
| Dengue | 246 |
| Cytotoxic T lymphocytes | |
| Tick-borne encephalitis | 247 |
| Kunjin, Bebaru, West Nile | 16, 58, 255 |
| Getah, Ross River, Semliki Forest, Sindbis, Barmah Forest | 59 |
| Helper T cells | |
| Dengue | 60, 61 |
| Suppressor T cells | |
| Dengue | 222, 227 |
| Japanese encephalitis | 228, 229, 231 |
| Inducer of macrophage procoagulant activity | |
| West Nile | 62 |
| Protection by adoptive transfer | |
| Sindbis | 248 |
| Venezuelan equine encephalitis | 27 |
| Banzi | 249 |
| Japanese encephalitis | 34, 244, 250 |

Various T-cell functions may not be type specific due to extensive cross-reactivity between togavirus antigens. Kelkar and Banerjee[56] showed cross-reactivity among flaviviruses in the leukocyte migration inhibition test. CTL, while specific for a group of viruses, do not generally show specificity within groups. Among alphaviruses, cross-reactivity is absolute.[16,57–59] CTL generated against any one member of the group will cross-react with all other members, even if they are serologically distant viruses.[58]

Induction of helper T (Th) cells for a humoral immune response has been reported in DV infection of mice.[60,61] T cells of the helper/delayed-type hypersensitivity (DTH) class, stimulated by WNV, have been assayed *in vitro* by the induction of macrophage procoagulant activity.[62] Lysis of DV-infected cells by natural cell-mediated cytotoxicity and antibody-dependent cell-mediated cytotoxicity has been reported.[63]

The protective role of CMI in togavirus infection has been studied in mice depleted of T cells by treatment with antithymocyte sera,[26,31,64] in athymic nude mice,[65-67] or by adoptive transfer of immune lymphocytes into recipient animals challenged with the virus. For example, Mathur et al.[34] reported that adoptive transfer of spleen cells from mice immunized with JEV provided protection to syngeneic mice against challenge with a lethal dose of the virus. Protection was dependent on T cells but not on B cells or macrophages. Furthermore, protection was provided only when the cells were harvested within 1–2 weeks of immunization but not later.[34] These findings are reminiscent of the observations in mice infected with attenuated VEEV.[27,68] Conversely, immune spleen cells did not protect mice against DV infection.[31]

## 5. INTERACTIONS WITH CELLS OF THE IMMUNE SYSTEM

### 5.1. Lymphoid Tissues

Both proliferative and degenerative changes may be produced in the lymphoid tissues during togavirus infection. Attenuated VEEV vaccine produces proliferation of germinal centers of spleen and lymph nodes of guinea pigs, mainly due to proliferation of reticuloendothelial cells and immature lymphoblasts.[69] Infection of germfree mice with VEEV vaccine results in an increase in spleen weight and formation of germinal centers.[70] In LDV-infected mice, increased weight of the spleen and lymph nodes, with hyperplasia of germinal centers, and reduced weight of the thymus are observed.[71-76] The increased spleen weight is due to the trapping of circulating lymphocytes,[77] and the fall in thymus weight is due to steroids, as it could be prevented by adrenalectomy.[78] A proliferative splenomegaly with increased T and B cells is observed in JEV-infected mice. Two of the cell types in which JEV replicated are the macrophages and T cells. The perifollicular area shows a large number of macrophages, with many of these cells containing immunofluorescent (IF) JEV antigen during the initial period of infection. IF antigen was also found in lymphoblast-like cells. By day 5, polymorphonuclear leukocytes become discernible, and by day 7 germinal centers appear that reach maximum size by day 15. The virus-antigen-positive cells disappear with the appearance of the germinal centers. The findings showed that splenomegaly in JEV infection was due to both proliferation and trapping of cells[79] (A. Mathur and U. C. Chaturvedi et al. unpublished data). By contrast, spleen weight is reduced markedly in DV infection of mice.[67,80] This is associated with a sharp decline in the proportion of T lymphocytes in the spleen[80] and of macrophages in both the spleen and peritoneal cavity.[81] Histologic study of the spleen of DV-infected mice shows hypocellularity, necrosis of cells, disorganization, and atrophy of follicular architecture.[82] This is similar to autopsy studies in cases of dengue hemorrhagic fever (DHF) that demonstrate necrosis and hemorrhage in the thymus-dependent areas of the lymphoid tissues.[83] A reduction in the concentration of lymphocytes in the thymus-dependent areas of lymphoid tissues of LDV-infected mice was observed by Proffitt et al.[75] and Snodgrass et al.[76] but not by Michaelides and Simms.[84]

## 5.2. Blood Leukocytes

Changes commonly occur in blood leukocytes in viral infections, the most frequent alteration being leukopenia with lymphopenia; in rare cases, leukocytosis may also be seen;[85] 65–80% of blood lymphocytes are T lymphocytes and, in cases of lymphopenia, these are affected most.

Lymphopenia is a common occurrence of VEEV, LDV, and TBEV infections.[69,85–87] Conversely, the alterations that occur in human peripheral blood leukocytes during DV infection may be leukopenia due to neutropenia[88] or leukopenia associated with lymphocytosis.[89–92] Wells et al.[93] showed that in cases of DHF, a significant lymphocytosis occurs that is associated with a loss of T lymphocytes and an increase in non-T, non-B, non-Fc receptor-bearing null cells. In a similar study, Ikeuchi et al.[94] reported decreased helper (OKT4) and suppressor (OKT8) cells in such cases. By contrast, JEV infection causes leukocytosis, with very marked neutrophilia accompanied by lymphopenia, due to diminished T lymphocytes, while B-cell numbers remain unaffected.[95]

## 5.3. Macrophages

Cells of the monocyte–macrophage system are a major component in the host defense against viral infection. In this regard, certain facts are known: (1) depletion or inhibition of macrophages decreases host resistance; (2) the resistance of macrophage-deficient or neonatal animals can be enhanced by the adoptive transfer of macrophages, especially activated ones; (3) age-related resistance can be correlated with maturity of macrophages; and (4) genetic susceptibility/resistance of an animal can be correlated with the nonpermissiveness/permissiveness of macrophages for viral replication.[96,97] Macrophages can affect viral infection by extrinsic or intrinsic mechanisms. In some viral infections, they can extrinsically affect viruses or viral replication in other surrounding cells.[97,98] Several mechanisms of antiviral activity are described for macrophages, including the secretion of arginase;[99] however, the precise mode of action is not well understood. Macrophages are also known to participate in immunopathogenetic mechanisms triggered by viruses.[100] Macrophages are permissive to some togaviruses (e.g., DV, YFV, WNV, and tick-borne flaviviruses). Cytopathic viruses, if they can replicate in macrophages, usually produce lethal infections in adult animals. Conversely, viruses that are not cytopathic but are capable of replication within macrophages result in persistent infections.[101]

The ability of YFV and other flaviviruses to produce lethal infections in genetically susceptible mice is paralleled by the capacity of cultured peritoneal macrophages from mice to support viral multiplication.[102,103] However, intraperitoneal inoculation of such viruses into suckling mice produces lethal infection, even in strains that are resistant as adults. SFV replication in mouse macrophages *in vivo* and *in vitro* is potentiated by treatment with Myocrisin.[104]

The entry of viruses into macrophages is facilitated by the phagocytic capability of these cells and may occur through virus receptors, or through Fc and complement receptors, singly or in combination.[49,105] Macrophages are the principal target for replication of DV in humans, monkeys, and mice.[67,82,106–108] DV infects macrophages through either a trypsin-sensitive virus receptor or the Fc receptor.[48] In DV-infected mice, the total number of cells in the spleen and peritoneal cavity is reduced, and the proportion of macrophages is decreased up to 75% on day 10 postinfection (p.i.).[81,109] Electron microscopy showed degenerative and necrotic changes of splenic macrophages and lymphocytes. Macrophages showed hypertrophic rough endoplasmic reticulum, osmophilic inclusions, vesicular Golgi complex, dense material on the inner nuclear membrane, and occasionally virions between the strands of endoplasmic reticulum.[82] DV antigen can be demonstrated in the splenic and peritoneal macrophages by the fluorescent antibody technique.[81,110] Macrophage functions, viz. phagocytosis of opsonized erythrocytes (EA), latex particles and neutral red dye, attachment of EA, migration from capillary tubes on glass surface, and antigen presentation,[81,111,112] are depressed in DV-infected mice. At late periods p.i., phagocytosis remained depressed, but attachment functions recovered.[108] JEV had similar effects on mouse macrophages (S. Rawat, A. Mathur, U.C. Chaturvedi, unpublished data), but human macrophages remain unaffected.[95]

## 5.4. Soluble Factors

A number of viruses induce host cells to produce a variety of soluble factors that have different immunopathologic effects. Induction of these factors by togaviruses is described briefly below.

### 5.4.1. Interferon

Interferons (IFNs) are of major interest in this context due to their immunomodulatory properties. For example, they may inhibit antibody production either through induction of suppressor T cells (Ts) or through activation of natural killer (NK) cells.[113] CMI is also affected greatly. IFNs activate macrophages, increasing their cytotoxicity, mobility, adherence, phagocytosis, and intracellular enzymatic activity. IFN also increases the expression of Fc receptors and MHC antigens on macrophages and other cells.[114]

Viruses differ in their capacity to induce IFN production. Tongaonkar and Ghosh[115] studied 37 different arboviruses in this respect and have grouped them into low, moderate, or high inducers. Most of the togaviruses are moderate inducers (100–1000 units/ml). The ability to induce IFN had no relationship with infectivity, CF antigen, incubation period in mice, or source of the virus.

The susceptibility of different viruses to IFNs also varies. Administration of 500 units of IFN had no effect on the outcome of DV infection.[116] By contrast, pretreatment of mice with a fungal IFN inducer, 6-MFA (isolated from the fungus *Aspergillus ochraceus* ATCC 28706),[117] protected the mice against challenge with SFV, JEV, WNV, and Kyasanur Forest disease virus.[117-119] Rodda and White[120] demonstrated induction of IFN by SFV and Kunjin virus in mice, which activated macrophages. 6-MFA had no effect on the outcome of DV infection, but it abrogated the virus-induced depression of attachment and phagocytosis of EA by peritoneal macrophages.[121] Adoptive transfer of peritoneal macrophages stimulated with IFN can protect mice from infection with SFV.[122] Thus, the role of IFN in recovery from togavirus infection is variable.

### 5.4.2. Suppressor Factors

Splenic T lymphocytes produce suppressor factors in DV- and JEV-infected mice. Two suppressor factors are produced in DV infection, one by the $Ts_1$ subpopulation of suppressor cells (SF), and the other by $Ts_2$ suppressor cells ($SF_2$). These highly potent, heat-labile, low-molecular-weight substances suppress DV-specific IgM antibody plaque-forming cells (PFC). $SF_2$ is prostaglandin-like and differs from SF in several properties.[123,124] Both are absorbed by anti-I-J and anti-I-A immunosorbent columns and are produced by $Thy1.2^+$, $Ly1^-2^+$, $I-J^+$, $I-A^+$, and cyclophosphamide-resistant T cells.[125-130] DV induces production of SF *in vivo* and *in vitro*,[131,132] but the production of $SF_2$ is induced only by SF.[124] The participation of these suppressor factors in the suppressor pathway is described later. The Ts cells generated in JEV infection of mice also produce two suppressor factors: one suppresses JEV-specific IgM PFC, while the other suppresses virus-specific DTH in mice.[133,134]

### 5.4.3. Cytotoxic Factors

Dengue virus induces a subpopulation of T lymphocytes ($Thy1.2^+$, $Ly1^+2^-$) in the mouse spleen to produce a cytotoxic factor (CF) that kills lymphoid cells of different animal species in a 1-hr assay at 4° or 37°C, without complement, but has no effect on different tissue culture cells or bacteria.[130,135-137] CF is a highly potent, trypsin-sensitive, heat- and pH-labile substance that separates out into two peaks on ion-exchange chromatography, both of which contain equally cytotoxic material.[138,139] *In vitro* treatment with CF kills nearly two thirds of macrophages in the culture but those that survive produce another cytotoxin ($CF_2$) *in vitro* and *in vivo*.[140-141] CF kills I-A negative macrophages and induces I-A positive macrophages to produce $CF_2$.[142] $CF_2$ differs from CF in a number of properties but kills lymphoid cells equally well, amplifying the cytotoxic effect.[108,140,143-146] The precise chemical nature of CF is not yet known, but it is not a virion protein, an Ig, or a subcellular organelle.[139,147] The production of CF is inhibited by treatment of mice with DV-induced suppressor factors.[148] CF is actively synthesized, and its production is inhibited irreversibly by treatment of mice with cycloheximide and mitomycin C.[149] These properties suggest that it might be a product of active metabolism of DV-primed T cells. The DV-induced cytotoxic pathway is described in Section 6.4.1.

### 5.4.4. Serum Protective Factor

Price et al.[149] described a protective mechanism induced during infection by group B arboviruses, i.e., production of a serum protective factor (SPF), independent of serum NT antibodies or IFN. The same group reported that in recovery from primary and secondary arbovirus infection in mice, SPF had a more significant role than did NT antibodies.[150] However, confirmation of these findings is not reported, and the role of SPF remains doubtful.

### 5.4.5. Helper Factor

Dengue virus given i.v. in low doses induces generation of Th cells in the mouse spleen that have a Thy $1.2^+$, $Ly1^+2^-$, $I-J^-$ surface phenotype and are antigen specific.[61] The DV-induced Th cells have been shown to produce a soluble helper factor that increases DV-specific IgM PFC both in vivo and in vitro. The helper factor is a potent heat-labile protein of low molecular weight having double chain; one chain binds DV antigen and the other has I-A determinant (U. C. Chaturvedi et al., unpublished data). The production and activity of Th cells in DV infection are regulated by the virus-induced Ts cells[60] and $CF/CF_2$[256,257]

### 5.4.6. Other Factors

Viruses are known to induce production of interleukin 1 (IL-1) on the basis of virus-induced fever with assays for circulating endogenous pyrogen in rabbits exposed to influenza virus[151,152] or by release of the pyrogen following exposure of rabbit blood cells.[153] Dinarello[154] presented an exhaustive review on IL-1 and its activities. Respiratory syncytial virus (RSV) has been shown to induce human macrophages to produce an IL-1 inhibitor.[155] However, the role of IL-1 or IL-1 inhibitor in togavirus infection is unclear. Effects of viruses on the production of other soluble factors are reported in this volume as well. However, no reports are available on togavirus effects on IL-2 or other factors.

## 6. SUPPRESSION OF IMMUNE RESPONSES

A number of viral infections are associated with suppression involving humoral, cellular, or both types of immune responses. This could be a mechanism of survival of viruses in the host. Immune responses to an antigen are the result of the interaction of T cells, B cells, and macrophages. Any disturbance in the function of these cells or in the transmission of signals between them can adversely affect the immune response. These effects can be produced directly as a result of replication of virus in the cell or indirectly through production of toxic substances by the infected host cells. Virus can also injure a cell functionally without histologic evidence of cell damage.[156] Various mechanisms by which viruses affect function of lymphocytes,[157] suppress immune responses,[3] or produce tolerance and suppression[158] have been described.

## 6.1. Immunosuppression by Togaviruses

Depressed immune responses have been reported during a few togavirus infections. Immunologic tolerance can be produced by VEEV,[70] lactate dehydrogenase-elevating virus,[159] and Border disease virus.[160] However, VEEV may also act as an adjuvant and enhance antibody production.[70,161] A summary of some of the togaviruses involved in suppression of CMI is presented in Table II.

## 6.2. Immunosuppression by Rubella Virus

A transient or prolonged immunosuppression is a well-established feature of rubella virus infection or vaccination.[162,163] A suppressed humoral response in such cases is indicated by hypogammaglobulinemia, low levels of total IgG and IgA, with or without elevation of IgM, low specific HI antibody titers, and depressed B-cell function in vitro.[134–168] CMI has been investigated extensively in cases of congenital rubella, experimental infection in volunteers, following vaccination, and in vitro on lymphocytes and monocytes. Rubella virus depresses skin-test reactivity,[169–173] lymphocyte responsiveness to mitogens in vitro,[163,173,174] and production of migration inhibitory factor.[173,175]

The mechanism of immunosuppression in rubella virus infection is unclear; however, it has been suggested that reduced stimulation of lymphocytes to phytohemagglutinin (PHA) may be due to direct damage of lymphocytes by the rubella virus[163] or indirectly by infecting macrophages, thereby diminishing the enhancing effect of macrophages on the PHA response of lymphocytes.[176] Rubella virus replicates in T lymphocytes and monocytes/macrophages but persists, without replication in B lymphocytes.[177,178] The depressed response could also be due to virus-induced functional damage to lymphocytes, as rubella is not cytocidal.[179] A more indirect role through the production of an inhibitory factor has been suggested by Mahler and Soren,[180] who observed greater suppression in the presence of autologous serum from patients with rubella. This could be through antigen–antibody (IgM) complexes present in the serum.[181] Immune responses are regulated by Ts cells; therefore, Arneborn et al.[182] investigated whether the suppression of CMI could be correlated with changes in different subpopulations of T lymphocytes. These workers showed that T lym-

**TABLE II**
**Examples of Togavirus-Induced Suppression of T-Cell Functions**

| Functions/virus | Antigen | References |
|---|---|---|
| *Delayed-type hypersensitivity* | | |
| Yellow fever | Tuberculin | 252 |
| Japanese encephalitis | Japanese encephalitis | 34 |
| Dengue | Dengue | U. C. Chaturvedi (unpublished data) |
| Dengue | Sheep erythrocytes | 80, 209, 210 |
| Rubella | Heterologous | 169–173 |
| *Leukocyte migration inhibition* | | |
| Dengue | Dengue | 31 |
| Rubella | — | 173, 175 |
| *Lymphocyte activation* | | |
| Japanese encephalitis | Japanese encephalitis | 240 |
| Dengue | Dengue | 212, U. C. Chaturvedi (unpublished data) |
| Rubella | PHA | 163, 173, 174 |
| *Graft-versus-host reaction* | | |
| Dengue | Allogeneic spleen cells | 80 |
| Lactate dehydrogenase | — | 70 |
| Japanese encephalitis | Allogeneic spleen cells | 240 |
| *Helper* | | |
| Dengue | Sheep erythrocyte | 80, 209 |
| | Lipopolysaccharide | 252 |
| Dengue | Dengue | U. C. Chaturvedi (unpublished data) |
| *Reduced T cells in spleen* | | |
| Dengue | — | 80 |

phocytes with Fc receptors for IgG (Ts cells) are increased after rubella vaccination, while no change was observed in the proportion of T lymphocytes with IgM-Fc receptors (Th cells). Similarly, Hyypia et al.[168] showed an increased proportion of Ts cells and decreased Th cells during rubella infections. But there is no direct evidence in their studies to show that Ts cells cause immunosuppression.

## 6.3. Immunosuppression by LDV

Lactate dehydrogenase-elevating virus causes a nonfatal lifelong persistent infection in mice without producing any significant pathologic lesions but is associated with perturbation of the immune system. The infection is characterized by splenomegaly due to trapping of lymphocytes,[71–74] leukopenia with reduced T lymphocytes,[87] and the presence of circulating immune complexes that do not bind complement.[183,184] Serum Ig are greatly increased, mainly in the IgG2a class.[185,186] Most of the increase in Ig is due to polyclonal stimulation of B cells by the virus, with very little production of antiviral antibodies.[186] CMI is also depressed, resulting in decreased skin contact sensitivity,[84,187,188] prolonged survival of skin allografts,[70,189] inhibition of graft-versus-host reaction,[70] depressed autoimmune response in NZB mice,[190] and increased growth of tumors.[74,191–193] Other effects are prevention of the induction of tolerance to aggregated gammaglobulin[194] and decreased resistance to murine malaria[195] and *Listeria monocytogenes*.[196,197]

Lactate dehydrogenase-elevating virus replicates exclusively in macrophages,[198,199] but only 5–20% of these cells are susceptible to the virus.[200] Susceptibility to LDV is related to the presence of Ia antigens on the macrophage cell surface, which act as receptors for the virus.[201,202] Macrophages are the principal antigen-presenting cells,[203] and several cellular immune responses depend on the T-lymphocyte–macrophage cooperation. In LDV infection, the antigen-presenting function of macrophages is significantly depressed.[204,205] The presence of circulating virus–antibody complexes, especially the transient appearance of complexes containing IgG1 in the early stages of infection, has also been implicated in the suppression of CMI.[192] Activation of Ts cells occurs through immune complexes as well,[206] so they are likely to be generated in LDV infection. Thus, trapping of lymphocytes in the spleen, destruction of macrophages or impairment of their functions and generation of Ts cells may contribute to perturbation of immune response in LDV infection.

### 6.4. Immunosuppression by Dengue Virus

Virus-specific CMI in DV-infected mice is depressed, as shown by the absence of a response against DV antigen by (1) the footpad swelling reaction (even after cyclophosphamide pretreatment); (2) the leukocyte migration inhibition test, and (3) the lymphocyte proliferation test; further evidence includes: (4) failure of adoptively transferred immune spleen cells to protect mice against challenge with a small dose of DV; (5) failure of antithymocyte serum treatment to potentiate DV infection; (6) failure of reconstitution of immunosuppressed mice by sensitized spleen cells to protect against DV; and (7) reduced direct graft-versus-host reactivity in Parker strain infant mice against the spleen cells of DV-infected Swiss mice[30,80] (U. C. Chaturvedi, unpublished data). By contrast, Pang *et al.*[207,208] showed a normal DTH response in cyclophosphamide-treated DV-infected mice. Adoptive transfer of serum obtained from mice 1–2 weeks after the third i.p. dose of DV failed to protect recipient mice against challenge with DV, whereas the serum obtained after 3–5 weeks conferred protection.[30,31] Suppressed humoral and CMI responses to heterologous antigens have also been reported using sheep red blood cells (SRBC)[80,209,210] and polyvinylpyrrolidone.[211]

Depressed T-cell function in human cases of dengue has been shown using [³H]thymidine uptake by peripheral blood leukocytes from immune subjects,[212] but there are no reports describing DTH or other T-cell functions in such cases. Immunosuppression in DV infection has been shown to be due to the development of an antigen-specific suppressor pathway and a nonspecific cytotoxic pathway in the spleen of mice.

### 6.4.1. The Cytotoxic Pathway in Experimental Dengue

Splenic T lymphocytes (TCF) from DV-infected mice produce a CF that kills lymphoid cells (without complement) of a variety of animals including mice but does not kill epithelial or fibroblastic tissue culture cells or bacteria. The killed cells are mainly macrophages, Th, and T effector (Te) lymphocytes.[130,135–138,147,256,257] Following exposure to CF, about one third of the macrophages survive its lethal effect and produce another cytotoxic substance $CF_2$ both in vitro and in vivo[140,141] amplifying the cytotoxic effects.

*Cytotoxic Factor Reproduces Effects of Dengue Virus* Intravenous inoculation of CF depresses DTH, leukocyte migration inhibition and the antibody PFC responses of mice to SRBC.[213] The macrophages are reduced in number, and their ability to attach, migrate, and phagocytize are depressed in such mice.[81,111,214] Similar effects are observed in DV-infected mice as well and are considered to be mediated by $CF–CF_2$.[111,216] The subcellular changes seen in macrophages and lymphocytes of DV-infected mice[78] can be reproduced by in vivo or in vitro treatment with CF.[215] The adverse effects of DV infection could be abrogated by pretreatment of mice with cycloheximide, which inhibits CF production, or by surgical removal of the spleen, the site of CF production.[216]

Human blood leukocytes treated in vitro with CF or $CF_2$ show diminished phagocytic activity and E-rosette formation by T cells.[145,217] Such adverse effects of $CF–CF_2$ are inhibited by pretreatment of blood leukocytes with plasma membrane-stabilizing substances.[144] Similar effects are produced in monkeys inoculated with CF.[218] Some of the changes produced in human blood leukocytes by in vitro treatment with $CF–CF_2$ and those by inoculating CF in monkeys are similar to those described in cases of DHF.[93] Thus, the cytotoxic factors produced in DV infection of mice are very potent substances with a variety of pathologic effects on the body, which result in pronounced immunosuppression. There is a strong possibility that similar substances are produced in DHF and other viral infections, including acquired immune-deficiency syndrome (AIDS).[219] CF is similar to lymphotoxin in a few aspects including the phenotype of its producer lymphocyte but differs from it in others.[138]

### 6.4.2. The Suppressor Pathway in Dengue

In DV infection of mice, the virus replicates in the spleen and other organs[31,220,221] and provides a strong stimulus for the induction of an antigen-specific suppressor pathway involving the generation of three types of suppressor cells. The virus induces the activation of $Ts_1$ in the mouse spleen,[222,223] producing a soluble SF[123] that is transmitted via live macrophages to recruit a second subpopulation $Ts_2$.[125,130,223] Depletion of macrophages from the cultures or in mice by silica treatment blocks the transmission of signal from $Ts_1$ to precursors of $Ts_2$.[224,225] The macrophages can only transmit the signal by direct cell to cell contact.[226] $Ts_2$ cells produce a soluble prostaglandin (PG)-like $SF_2$[124,125] that recruits the third subpopulation of Ts $(Ts_3)$[126] to mediate antigen-specific suppression of IgM antibody-forming cells. The precursors of both $Ts_2$ and $Ts_3$ are normal mouse spleen cells. The three generations of the suppressor cells and their products have been characterized,[124–130,226,227,258] and the findings are summarized in Fig. 1. Similar Ts that secrete suppressor factors can also be generated in spleen cell cultures *in vitro*.[131,132] The generation of DV-specific Th cells is inhibited by pretreatment of mice with SF or $SF_2$, but once the Th cells are produced the $SF/SF_2$ have little effect.[60,61] Suppressor cells against SRBC[210] and polyvinylpyrrolidone[211] have also been reported in DV-infected mice.

We have further investigated the suppressor phenomenon using JEV, another flavivirus. It was observed that the virus-specific protective humoral and CMI responses in JEV-primed mice disappear by the third week p.i. due to the generation of two Ts cell pathways, one for the humoral response and the other for DTH. Investigations up to the present show that both the suppressor pathways are composed of separate circuits of $Ts_1$-suppressor factor-$Ts_2$ cells, and the signal between the two suppressor cells is transmitted by live macrophages.[34,133,134,228–230] Mathur *et al.*[231] showed the presence of memory

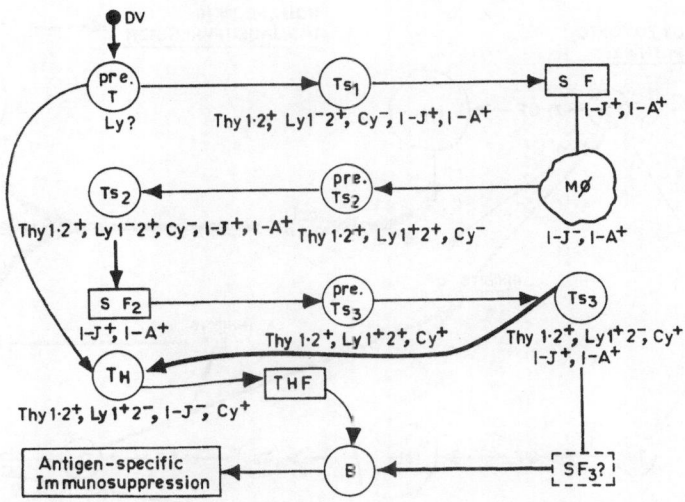

**FIGURE 1.** Characteristics of dengue virus-induced suppressor pathway. (Modified from Chaturvedi.[227])

Ts cells in mice latently infected with JEV that can be stimulated to produce secondary suppressor T cells by the reactivated virus or by exogenous virus challenge. The JEV–mouse model is unique in having two separate suppressor pathways, but the cell circuits are essentially similar to that described for DV infection.

The presence of suppressor cells has been described in a number of viral and nonviral infections, but few attempts have been made to investigate their mechanism of action. By using synthetic antigens, however, suppressor pathways have been established that have up to three generations of Ts.[232] So far, the only infectious agents for which the sequential events in the suppressor pathway have been delineated are DV and to some extent JEV. Since the suppressor pathways induced by DV, JEV, and the synthetic antigens are similar, it seems likely that they exist in other viral infections.

### 6.4.3. Mechanism of Immunosuppression in Dengue

The pathogenesis of dengue is not fully understood, but immunopathologic processes appear to be significant. Initially, immunopathogenesis was thought to be mediated by antigen–antibody immune complexes.[233] Recently, the role of enhancing antibody has been suggested, via augmentation of virus uptake by the macrophage/monocyte through Fc receptors, thereby enhancing the replication of DV.[47]

Chaturvedi and co-workers proposed a mechanism of immunosuppression in DV infection, summarized in Fig. 2. DV replicates mainly in mac-

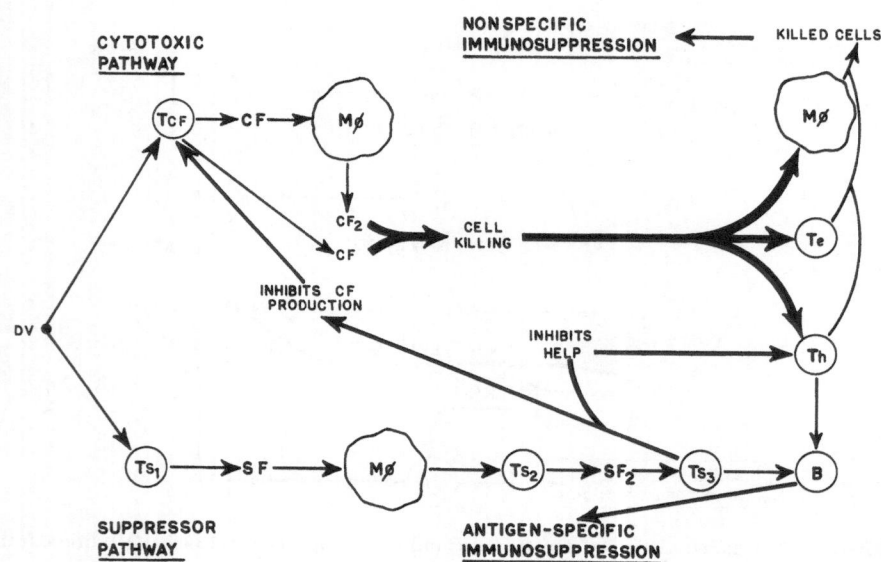

**FIGURE 2.** Mechanisms of immunosuppression in dengue viral infection.

rophages; therefore, the body makes an effort to restrict viral replication by eliminating these virus-permissive cells. This is achieved through the cytotoxic pathway by the production of CF and $CF_2$ that kill the macrophages and T cells (helper and effector); those cells that escape killing are functionally damaged. The capacity of the functionally damaged macrophages to bind recovers, but phagocytic activity does not; thus, the virus remains attached to the surface of the macrophages, where it is exposed to the action of NT antibodies. The damage to macrophages and T cells leads to depressed CMI response and T-cell-dependent B-cell functions, resulting in a nonspecific immunosuppression. The cytotoxic activity was not found in the spleen of mice inoculated with JEV or SRBC (U. C. Chaturvedi, unpublished data). Pang et al. were able to find the development of DTH against DV and no cytotoxic activity in the spleen homogenate.[207] Production of CF is greater in inbred than in outbred mice (U. C. Chaturvedi, unpublished data). Apparently, the strains of mice that do not produce sufficient CF may manifest a DTH response.

Suppressor cells are the regulators of immune response and are thus of primary importance in preventing immunopathologic consequences of overreaction by the immune system. The role of the suppressor phenomenon in controlling immunopathologic events in dengue can be crucial. The suppressor pathway suppresses antigen-specific antibody production (see Fig. 2); thus, it may prevent (1) increased replication of the virus mediated by enhancing antibody, and (2) the immunopathology mediated by immune complexes. Suppression of CMI may protect against T-cell-mediated damage. Liew and Russell[234] presented direct evidence for a beneficial effect of specific Ts for DTH during potentially lethal influenza A virus infection of mice. The induction of cytotoxic and suppressor pathways could be beneficial to the host in the DV model. Both pathways are mutually regulated. However, uncontrolled production of $CF–CF_2$ and PG-like $SF_2$ might also produce pathologic effects.

## 7. ADVERSE EFFECTS OF IMMUNOSUPPRESSION

Immunosuppression induced by a virus can be viewed as an effort to evade the host defenses and to ensure survival of viral progeny. The effect can be beneficial to the host when it prevents the immunologic destruction of cells, thereby preventing or limiting clinical illness but most often can lead to secondary bacterial, viral, or parasitic infections. Experimental immunosuppression often converts a silent abortive viral infection into a lethal infection, accompanied by increased viral titers and increased damage to cells in the target organs.[20] In the clinical setting, such examples of opportunistic viral infections are numerous.[235]

Nash[158] considered tolerance and suppression as mechanisms by which some viruses evade specific immune responses and persist in their host. An ineffective immune response results in failure to eliminate the virus. JEV is transmitted transplacentally and infects the fetus in humans[236,237] and mice,[238] favoring tolerance. JEV also produces a persistent and latent infection in mice[239,240] that can be reactivated by inducing immunosuppression.[239–241] Waning specific immunity due to Ts from the third week p.i. may result in persistence of JEV if it survives the initial 2-week period. Other togaviruses seen to produce persistent infection are Border disease[160] and LDV.[159]

## 8. CONCLUSION

The data from the mouse models described here cannot be extrapolated to humans, but they strongly indicate the possibility that in humans similar mechanisms can contribute to the outcome of infection. There is need to investigate human cases of togavirus infections along similar lines and other animal models to see whether there are basic mechanisms common to all.

Considerable progress has been made in delineation of the cytotoxic and suppressor pathways in DV infection, but much still needs to be done. Future studies will determine the chemical nature of suppressor and cytotoxic factors, their mechanism of signal transfer, feedback regulation of each step, their cell receptors, and the mechanism(s) of effector functions. Cloning of CF genes is also a definite possibility. There are glaring deficiencies in our understanding of the pathology and pathogenesis of togavirus infections. For example, we do not know why a particular organ is affected, nor why limited areas of the world are stricken more severely by certain viruses. Answers may come from experimental studies at the molecular level, by the study of genome structure, by the identification of virulence determinants, and so forth. Credence is being given to the view that various immune responses are triggered by different epitopes of a native antigen. Therefore, it may soon be possible to have a custom-tailored vaccine from which the epitopes for the cytotoxic pathway, suppressor pathway, enhancing antibody, and others are selectively eliminated, but the epitopes for protective humoral and CMI responses are preserved. Information about these aspects will help provide a better understanding of the pathogenesis and immunoregulation, rapid diagnosis techniques, and effective control of togavirus diseases.

ACKNOWLEDGMENTS.    I am grateful to Dr. Asha Mathur, Dr. Rachna Nagar, and Miss Nigar Rizvi for their help in writing this chapter. The studies described here were carried out with the financial assistance of the Indian Council of Medical Research, New Delhi, and the U.P. State government, Lucknow.

## REFERENCES

1. Westaway, E. G., M. A. Brinton, S. Y. Gaidamovich, M. C. Horzinek, A. Igarashi, L. Kaariainen, D. K. Lvov, J. S. Porterfield, P. K. Russell, and D. W. Trent, Flaviviridae, *Intervirology* **24**:183–192 (1985).
2. Simpson, D. I. H., Togaviridae, in: *Topley and Wilson's Principle of Bacteriology, Virology and Immunology* (F. Brown and G. Wilson, eds.), Vol. 4, pp. 233–254, Edward Arnold, London (1984).

3. Woodruff, J. F., and J. J. Woodruff, The effect of viral infections on the function of the immune system, in: *Viral Immunology and Immunopathology* (A. L. Notkins, ed.), pp. 393–418, Academic, New York (1975).

4. Halstead, S. B., Immunological parameters of togavirus disease syndromes, in: *The Togaviruses* (R. W. Schlesinger, ed.), pp. 107–173, Academic, New York (1980).

5. Brown, F. *Topley and Wilson's Principles of Bacteriology, Virology and Immunology* (F. Brown and G. Wilson, eds.), Vol. 4, pp. 5–13, Edward Arnold, London (1984).

6. Boulton, R. W., and E. G. Westaway, Comparison of togaviruses: Sindbis virus (group A) and Kunjin virus (group B), *Virology* 49:283–289 (1972).

7. Wengler, G., G. Wengler, and H. J. Gross, Studies on virus-specific nucleic acid synthesized in vertebrate and mosquito cells infected with flaviviruses, *Virology* 89:423–437 (1978).

8. Cleaves, G. R., and D. T. Dubin, Methylation status of intracellular dengue type 240S RNA, *Virology* 96:159–165 (1979).

9. Westaway, E. G., Strategy of flavivirus genome: Evidence for multiple internal initiation of translation of proteins specified by Kunjin virus in mammalian cells, *Virology* 80:320–335 (1977).

10. Westaway, E. G., Replication of flaviviruses, in: *The Togaviruses* (R. W. Schlesinger, ed.), pp. 531–581, Academic, New York (1980).

11. Westaway, E. G., G. Speight, and L. Endc, Gene order of translation of the flavivirus Kunjin: Further evidence of internal initiation *in vivo*, *Virus Res.* 1:333–350 (1984).

12. Pang, T., Important implications of recent insights into the molecular structure of flavivirus genomes, *Virus Inform. Newsl.* 2:70–71 (1985).

13. Russell, P. K., W. E. Brandt, and J. M. Dalrymple, Chemical and antigenic structure of flaviviruses, in: *The Togaviruses* (R. W. Schlesinger, ed.), pp. 503–529, Academic, New York (1980).

14. Qureshi, A. A., and D. W. Trent, Group B arbovirus structural and non-structural antigens. III. Serological specificity of solubilized intracellular viral proteins, *Infect. Immun.* 8:993–999 (1973).

15. Trent, D. W., Antigenic characterization of flavivirus structural proteins separated by isoelectric focusing, *J. Virol.* 22:608–618 (1977).

16. Porterfield, J. S., Antigenic characteristic and classification of togaviridae, in: *The Togaviruses* (R. W. Schlesinger, ed.), pp. 13–46, Academic, New York (1980).

17. Mullbacher, A., I. D. Marshall, and P. Ferris, Classification of Barmah Forest virus as an alphavirus using cytotoxic T cell assays, *J. Gen. Virol.* 67:295–299 (1986).

18. Murphy, B. R., and L. A. Glasgow, Factors modifying host resistance to viral infection. III. Effect of whole body X-irradiation on experimental encephalomyocarditis virus infection in mice, *J. Exp. Med.* 127:1035–1052 (1968).

19. Robinson, T. W. E., R. J. R. Curcton, and R. B. Heath, The effect of cyclophosphamide on Sendai virus infection of mice, *J. Med. Microbiol.* 2:137–145 (1969).

20. Nathanson, N., and G. A. Cole, Immunosuppression and experimental virus infection of the nervous system, *Adv. Virus Res.* 16:397–448 (1970).

21. Svehag, S. E., Formation and dissociation of virus antibody complexes with special reference to the neutralization process, *Prog. Med. Virol.* 10:1–63 (1968).

22. Berry, D. M., and J. D. Almeida, The morphological and biological effects of various antisera on avian infectious bronchitis, *J. Gen. Virol.* 3:97–102 (1968).

23. Silverstein, S., Macrophages and viral immunity, *Semin. Hematology* 7:185–214 (1970).

24. Brier, A. M., C. R. Wanlenberg, and A. L. Notkins, Immune injury of cells infected with herpes simplex virus (HSV), *Fed. Proc.* 30:353 (1971).

25. Smith, J. W., E. Adam, J. L. Melnick, and W. E. Rawls, Use of the release test to demonstrate patterns of antibody response in humans to herpes virus type 1 and 2, *J. Immunol.* 109:554–564 (1972).

26. Zisman, B., E. F. Wheelock, and A. C. Allison, Role of macrophages and antibody in resistance of mice against yellow fever virus, *J. Immunol.* 107:236–243 (1971).

27. Rabinowitz, S. G., and W. H. Adler, Host defences during primary Venezuelan equine encephalitis virus infection in mice. I. Passive transfer of protection with immune serum and immune cells, *J. Immunol.* **110**:1345–1353 (1973).
28. Camenga, D. L., N. Nathanson, and G. A. Cole, Cyclophosphamide-potentiated West Nile virus encephalitis: Relative influence of cellular and humoral factors, *J. Infect. Dis.* **130**:634–641 (1974).
29. Griffin, D. E., and R. T. Johnson, Role of the immune response in recovery from Sindbis encephalitis in mice, *J. Immunol.* **118**:1070–1075 (1977).
30. Chaturvedi, U. C., P. Tandon, and A. Mathur, Effect of immunosuppression on dengue virus infection in mice, *J. Gen. Virol.* **36**:449–458 (1977).
31. Chaturvedi, U. C., P. Tandon, A. Mathur, and A. Kumar, Host defence mechanism against dengue virus infection of mice, *J. Gen. Virol.* **39**:293–302 (1978).
32. Schmaljohn, A. L., E. D. Johnson, J. M. Dalrymple, and G. A. Cole, Non-neutralizing monoclonal antibodies can prevent lethal alphavirus encephalitis, *Nature (Lond.)* **297**:70–72 (1982).
33. Mathews, J. H., and J. T. Roehrig, Determination of the protective epitopes on the glycoproteins of Venezuelan equine encephalomyelitis virus by passive transfer of monoclonal antibodies, *J. Immunol.* **129**:2763–2767 (1982).
34. Mathur, A., K. L. Arora, and U. C. Chaturvedi, Host defence mechanism against Japanese encephalitis virus infection in mice, *J. Gen. Virol.* **64**:805–811 (1983).
35. Boere, W. A. M., B. J. Benaissa-Trouw, N. T. Harmsen, T. Erich, C. A. Kraaijeveld and H. Snippe, Mechanisms of monoclonal antibody-mediated protection against virulent Semliki Forest Virus, *J. Virol.* **54**:546–551 (1985).
36. Hunt, A. R. and J. T. Roehrig, Biochemical and biological characteristics of epitopes on the E1 glycoprotein of Western equine encephalitis virus, *Virology* **142**:334–336 (1985).
37. Buckley, A., and E. A. Gould, Neutralization of yellow fever virus studied using monoclonal and polyclonal antibodies, *J. Gen. Virol.* **66**:2523–2531 (1985).
38. Gould, E. A., A. Buckley, A. D. T. Barrett, and N. Cammack, Neutralizing (54K) and non-neutralizing (54K and 48K) monoclonal antibodies against structural and non-structural yellow fever virus proteins confer immunity in mice, *J. Gen. Virol.* **67**:591–595 (1986).
39. Ishi, K., Y. Matsunaga & R. Kono, Immunoglobulins produced in response to Japanese encephalitis virus infections of man, *J. Immunol.* **101**:770–775 (1968).
40. Yasui, H., and S. Sugiyama, Evaluation of immunoglobulin against Japanese encephalitis. Report. I. Immunoglobulin in Japanese encephalitis patients at the time of and after hospitalization, *J. Jpn. Assoc. Infect. Dis.* **43**:5–12 (1969).
41. Edelman, R., R. J. Schneider, A. Vejjajiva, R. Pornpibul, and P. Voodhikul, Persistence of virus specific IgM and clinical recovery after Japanese encephalitis, *Am. J. Trop. Med. Hyg.* **25**:733–738 (1976).
42. Kulshreshtha, R., A. Mathur, and U. C. Chaturvedi, Immunological memory in latent Japanese encephalitis virus infection, *Br. J. Exp. Pathol.* **69** (in press).
43. Schlesinger, J. J., M. W. Brandriss, and E. E. Walsh, Protection of mice against dengue 2 virus encephalitis by immunization with the dengue 2 virus non-structural glycoprotein NSI, *J. Gen. Virol.* **68**:853–857 (1987).
44. Heinz, F. X., R. Berger, W. Tuma, and C. Kunz, A topological and functional model of epitopes on the structural glycoprotein of tick-borne encephalitis virus defined by monoclonal antibodies, *Virology* **126**:525–537 (1983).
45. Mathews, J. H., and J. T. Roehrig, Elucidation of the topography and determination of the protective epitopes on the E glycoproteins of St. Louis encephalitis virus by passive transfer with monocolonal antibodies, *J. Immunol.* **132**:1533–1537 (1984).
46. Brandriss, M. W., J. J. Schlesinger, E. E. Walsh, and M. Briselli, Lethal 17D yellow fever encephalitis in mice. I. Passive protection by monoclonal antibodies to the envelope proteins of 17D yellow fever and dengue 2 viruses, *J. Gen. Virol.* **67**:229–234 (1986).

47. Halstead, S. B., Immune enhancement of viral infection, *Prog. Allergy* **31**:301–364 (1982).
48. Daughaday, C. C., W. E. Brandt, J. M. McCown, and P. K. Russell, Evidence for two mechanisms of dengue virus infection of adherent human monocytes: Trypsin sensitive virus receptors and trypsin resistant immune complex receptors, *Infect. Immun.* **32**:469–473 (1981).
49. Cardosa, M. J., J. S. Porterfield, and S. Gordon, Complement receptor mediates enhanced flavivirus replication in macrophages, *J. Exp. Med.* **158**:258–263 (1983).
50. Henchal, E. A., J. M. McCown, M. K. Gentry, J. M. Dalrymple, and W. E. Brandt, Serological characterization of monoclonal antibodies produced against dengue virus antigens, *Fed. Proc.* **40**:1065 (1981).
51. Brandt, W. E., J. M. McCown, M. K. Gentry, and P. K. Russell, Immune enhancement of dengue-2 virus replication in the U-937 human monocyte line by cross-reactive monoclonal antibodies, *Fed. Proc.* **40**:1065 (1981).
52. Barrett, A. D. T., and E. A. Gould, Antibody mediated early death *in vivo* after infection with Yellow Fever virus, *J. Gen. Virol.* **67**:2539–2542 (1986).
53. Halstead, S. B., C. N. Venkateshan, M. K. Gentry, and L. K. Larsen, Heterogeneity of infection enhancement of dengue 2 strains by monoclonal antibodies, *J. Immunol.* **132**:1529–1532 (1984).
54. Phillpotts, R. J., J. R. Stephenson, and J. S. Porterfield, Antibody-dependent enhancement of tick-borne encephalitis virus infectivity, *J. Gen. Virol.* **66**:1831–1837 (1985).
55. Doherty, P. C., T cells and virus infections, *Br. Med. Bull.* **41**:7–14 (1985).
56. Kelkar, S. D., and K. Banerjee, Cross reactions among flaviviruses in macrophage migration inhibition assay, *Acta Virol. (Praha)* **22**:337–340 (1978).
57. Müllbacher, A., and R. V. Blanden, Murine cytotoxic T-cell response to alphavirus is associated mainly with H-2D^k, *Immunogenetics* **7**:551–561 (1978).
58. Müllbacher, A., I. D. Marshall, and R. V. Blanden, Cross-reactive cytotoxic T cells to alphavirus infection, *Scand. J. Immunol.* **10**:291–296 (1979).
59. Wolcott, J. A., C. J. Wust, and A. Brown, Immunization with one alphavirus cross-primes cellular and humoral immune responses to a second alphavirus, *J. Immunol.* **129**:1267–1271 (1982).
60. Chaturvedi, U. C., M. I. Shukla, M. Pahwa, and A. Mathur, Inhibition of B and helper T lymphocytes by dengue virus-induced suppressor factor, *Indian J. Med. Res.* **82**:471–474 (1985).
61. Chaturvedi, U. C., M. Pahwa, and A. Mathur, Dengue virus-induced helper T cells, *Indian J. Med. Res.* **86**:1–8 (1987).
62. Allan, J. E., and P. C. Doherty, Stimulation of helper/delayed-type hypersensitivity T cells by flavivirus infection: Determination by macrophage procoagulant assay, *J. Gen. Virol.* **67**:39–46 (1986).
63. Kurane, I., D. Hebblewaite, W. E. Brandt, and F. A. Ennis, Lysis of dengue virus-infected cells by natural cell-mediated cytotoxicity, *J. Virol.* **52**:223–230 (1984).
64. Woodman, D. R., A. T. McManus, and G. A. Eddy, Extension of the mean time to death of mice with a lethal infection of Venezuelan equine encephalomyelitis virus by antithymocyte serum treatment, *Infect. Immun.* **12**:1006–1011 (1975).
65. LeBlanc, P. A., W. F. Scherer, and D. H. Sussdorf, Infections of congenitally athymic (nude) and normal mice with avirulent and virulent strains of Venezuelan encephalitis virus, *Infect. Immun.* **21**:779–785 (1978).
66. Bradish, C. J., R. Fitzgeorge, D. Titmuss, and A. Baskerville, The responses of nude-athymic mice to nominally avirulent togavirus infections, *J. Gen. Virol.* **42**:555–556 (1979).
67. Hotta, H., I. Murakami, K. Miyasaki, Y. Takeda, H. Shirane, and S. Hotta, Inoculation of dengue virus into nude mice, *J. Gen. Virol.* **52**:71–76 (1981).
68. Adler, W. H., and S. G. Rabinowitz, Host defences during primary Venezuelan equine

encephalomyelitis virus infection in mice. II. *In vitro* methods for the measurement and qualitation of immune response, *J. Immunol.* **110:**1354–1362 (1973).

69. Airhart, J. W., G. S. Trevino, and C. P. Craig, Alterations in immune responses by attenuated Venezuelan equine encephalitis vaccine. II. Pathology and soluble antigen localization in guinea pigs, *J. Immunol.* **102:**1228–1234 (1969).

70. Howard, R. J., C. P. Craig, G. S. Trevino, S. F. Dougherty, and S. E. Mergenhagen, Enhanced humoral immunity in mice infected with attenuated Venezuelan equine encephalitis virus, *J. Immunol.* **103:**699–707 (1969).

71. Stauber, E., D. Berger, and R. Piper, The pathogenesis of lactate dehydrogenase elevating virus in mice, *J. Comp. Pathol.* **85:**171–183 (1975).

72. Rowson, K. E. K., and B. W. J. Mahy, Lactic dehydrogenase virus, *Virol. Monog.* **13:**1–121 (1975).

73. Riley, V., and D. Spackman, Modifying effects of a benign virus on the malignant process and the role of physiological stress on tumour incidence, *Fogarty Int. Ctr. Proc.* **28:**319–336 (1974).

74. Riley, V., and D. Spackman, Melanoma enhancement by viral induced stress, in: *Pigment Cell,* Vol. 2 (V. Riley, ed.), pp. 163–173, S. Karger, Basel (1976).

75. Proffitt, M. R., C. C. Congdon, and R. L. Tyndall, The combined action of Rauscher leukemia virus and lactic dehydrogenase virus on mouse lymphatic tissue, *Int. J. Cancer* **9:**193–211 (1972).

76. Snodgrass, M. J., D. S. Lowrey, and M. G. Hanna, Changes induced by lactic dehydrogenase virus in thymus and thymus-dependent areas of lymphatic tissue, *J. Immunol.* **108:**877–892 (1972).

77. Isakov, N., M. Feldman, and S. Segal, The mechanism of modulation of humoral immune responses after infection of mice with lactic dehydrogenase virus, *J. Immunol.* **128:**969–975 (1982).

78. Santisteban, G. A., V. Riley, and M. A. Fitzmaurice, Thymolytic and adrenal cortical responses to the LDH-elevating virus, *Proc. Soc. Exp. Biol. Med.* **139:**202–206 (1972).

79. Mathur, A., M. Bharadwaj, R. Kulshreshtha, S. Rawat, A. Jain, and U. C. Chaturvedi, Immunopathological study of spleen during Japanese encephalitis virus infection in mice, *Br. J. Exp. Pathol.* **69:**423–432 (1988).

80. Tandon, P., U. C. Chaturvedi, and A. Mathur, Differential depletion of T lymphocytes in the spleen of dengue virus-infected mice, *Immunology* **37:**1–6 (1979).

81. Chaturvedi, U. C., R. Nagar, and A. Mathur, Effect of dengue virus infection on Fc-receptor functions of mouse macrophages, *J. Gen. Virol.* **64:**2399–2407 (1983).

82. Nath, P., P. Tandon, L. Gulati, and U. C. Chaturvedi, Histological and ultrastructural study of spleen during dengue virus infection of mice, *Indian J. Med. Res.* **77:**83–90 (1983).

83. Aung-Khin, M., M. M. Khin, Z. Thant, and U. M. Tin, Changes in the tissues of the immune system in dengue haemorrhagic fever. *J. Trop. Med. Hyg.* **78:**256–261 (1975).

84. Michaelides, M. C., and E. S. Simms, Immune response in mice infected with lactic dehydrogenase virus. II. Contact sensitization to DNFB and characterization of lymphoid cells during acute LDV infection, *Cell. Immunol.* **29:**285–294 (1977).

85. de Gruchy, G. C., *Clinical Haematology in Medical Practice, 3rd ed.,* pp. 359–403 ELBS and Blackwell Scientific Publications, Oxford (1976).

86. Riley, V., Lactate dehydrogenase in the normal and malignant state in mice and the influence of a benign enzyme-elevating virus, in: *Methods in Cancer Research* Vol. 4 (H. Busch, ed.), pp. 493–618, Academic London (1968).

87. Crispens, C. G., Detection and characterization of a lymphocyte proliferating factor (LPF) in plasma of SJL/J mice infected with LDH virus, *Arch. Virol.* **72:**67–74 (1982).

88. Simmons, J. S., J. H. St. John, and F. H. K. Reynolds, Experimental studies of dengue, *Philip. J. Sci.* **44:**1–8 (1931).

89. Nimmannitya, S., S. B. Halstead, S. N. Cohen, and M. R. Margiotta, Dengue and

Chikungunya virus infection in man in Thailand, 1962–64. I. Observations on hospitalized patients with haemorrhagic fever, *Am. J. Trop. Med. Hyg.* **18**:954–971 (1969).

90. Halstead, S. B., S. Nimmannitya, and S. Cohen, Observations related to the pathogenesis of dengue haemorrhagic fever. IV. Relation of disease severity to antibody response and virus recovered, *Yale J. Biol. Med.* **42**:311–328 (1970).

91. Chaturvedi, U. C., A. Mathur, A. K. Kapoor, N. K. Mehrotra, and R. M. L. Mehrotra, Virological study of an epidemic of febrile illness with haemorrhagic manifestations at Kanpur, India during 1968, *Bull. WHO* **42**:289–293 (1970).

92. Chaturvedi, U. C., A. K. Kapoor, A. Mathur, D. Chandra, A. M. Khan, and R. M. L. Mehrotra, A clinical and epidemiological study of an epidemic of febrile illness with haemorrhagic manifestations which occurred at Kanpur, India in 1968, *Bull. WHO* **43**:281–287 (1970).

93. Wells, R. A., R. McN. Scott, K. Pavanand V. Sathitsathein, U. Cheamudon, and R. P. MacDermott, Kinetics of peripheral blood leucocyte alterations in Thai children with dengue haemorrhagic fever, *Infect. Immun.* **28**:428–433 (1980).

94. Ikeuchi, H., S. Cornain, Sumarmo, Y. Funahara, A. Shirahata, N. Fujita, Y. Okumo, A. Igarashi, T. Oda, S. W. Agus, R. Dharma, T. Matsuo, and S. Hotta, Analysis of lymphocytes of dengue/DHF patients observed at Jakarta, Indonesia in 1982, in: *Proceedings of the International Conference on Dengue/Dengue Haemorrhagic Fever, Kuala Lumpur* (T. Pang and R. Pathmanathan, eds.), pp. 355–363, University of Malaya, Kuala Lumpur (1983).

95. Chaturvedi, U. C., A. Mathur, P. Tandon, S. M. Natu, S. Rajvanshi, and H. O. Tandon, Variable effect on peripheral blood leucocytes during JE virus infection of man, *Clin. Exp. Immunol.* **38**:494–498 (1979).

96. Mogensen, S. C., Role of macrophages in natural resistance to virus infection, *Microbiol. Rev.* **43**:1–26 (1979).

97. Morahan, P. S., and S. S. Morse, Macrophage virus interactions, in: *Virus–Lymphocyte Interactions: Implications for Disease*, (M. R. Proffitt, ed.) Vol. 7, pp. 17–35, Elsevier/North-Holland, Amsterdam (1979).

98. Morahan, P. S., J. R. Connor, and K. R. Leary, Viruses and the versatile macrophages, *Br. Med. Bull.* **41**:15–21 (1985).

99. Wildy, P., Inhibition of herpes virus multiplication by activated macrophages: A role for arginase, *Infect. Immun.* **37**:42–45 (1982).

100. Denman, A. M., and M. Pinder, Measurement of immunological function in man: Interaction between virus and human leucocytes, *Proc. R. Soc. Med.* **67**:1219–1222 (1974).

101. Bloom, B. R., and B. Rager-Zisman, Cell-mediated immunity in viral infections, in: *Viral Immunology and Immunopathology* (A. L. Notkins, ed.), pp. 113–133, Academic, London (1975).

102. Bang, F. B., and A. Warwick, Mouse macrophages as host cells for the mouse hepatitis virus and the genetic basis for their susceptibility, *Proc. Natl. Acad. Sci. USA* **46**:1065–1075 (1960).

103. Goodman, G. T., and H. Koprowski, Study of the mechanism of innate resistance to virus infection, *J. Cell. Comp. Physiol.* **59**:333–373 (1962).

104. Oaten, S. W., S. Jagelman, and H. E. Webb, Further studies of macrophages in relationship to avirulent Semliki Forest virus infections, *Br. J. Exp. Pathol.* **61**:150–155 (1980).

105. Halstead, S. B., and E. J. O'Rourke, Dengue viruses and mononuclear phagocytes. I. Infection enhancement by non-neutralizing antibody, *J. Exp. Med.* **146**:201–217 (1977).

106. Halstead, S. B., The pathogenesis of dengue. Molecular epidemiology in infectious diseases, *Am. J. Epidemiol.* **114**:632–648 (1981).

107. Halstead, S. B., E. J. O'Rourke, and A. C. Allison, Dengue viruses and mononuclear phagocytes. II. Identity of blood and tissue leucocytes supporting *in vitro* infection, *J. Exp. Med.* **146**:218–229 (1977).

108. Chaturvedi, U. C., L. Gulati, and A. Mathur, Further studies on the properties of dengue virus-induced macrophage cytotoxin, *Indian J. Exp. Biol.* **21**:275–279 (1983).

109. Chaturvedi, U. C., R. Nagar, L. Gulati, and A. Mathur, Macrophage functions in dengue virus infection. *Proceedings of the International Conference on Dengue/Dengue Haemorrhagic Fever, Kuala Lumpur* (T. Pang and R. Pathmanathan, eds.), pp. 343–354, University of Malaya, Kuala Lumpur (1983).

110. Hotta, H., I. Murakami, K. Miyasaki, Y. Takeda, H. Shirane, and S. Hotta, Experimental dengue virus infection using athymic nude mice, in: *Proceedings of the First ICMR Seminar on Dengue Haemorrhagic Fever (1980) Kobe, Japan* (S. Hotta, ed.), pp. 151–158, International Centre for Medical Research, Kobe University School of Medicine, Kobe (1981).

111. Gulati, L., U. C. Chaturvedi, and A. Mathur, Depressed macrophage functions in dengue virus-infected mice: Role of the cytotoxic factor, *Br. J. Exp. Pathol.* **63**:194–202 (1982).

112. Rizvi, N., U. C. Chaturvedi, R. Nagar, and A. Mathur, Macrophage functions during dengue virus infection: Antigenic stimulation of B cells, *Immunology* **62**:493–499 (1987).

113. Baron, S., F. Dianzani, and G. J. Stanton, General considerations of the interferon system, in: *The Interferon System: A Review to 1982, Part I.,* Vol. 41 (S. Baron, F. Dianzani, and G. J. Stanton, eds.), pp. 1–12, University of Texas Medical Branch at Galveston, Galveston, Texas (1982).

114. Rager-Zisman, B., and B. R. Bloom, Interferons and natural killer cells, *Br. Med. Bull.* **41**:22–27 (1985).

115. Tongaonkar, S. S., and S. N. Ghosh, Interferon induction by arboviruses, *Indian J. Med. Res.* **69**:865–873 (1979).

116. Cole, G. A., and C. L. Wisseman, Jr., Pathogenesis of type 1 dengue virus infection in suckling, weanling and adult mice. I. The relation of virus replication to interferon and antibody formation, *Am. J. Epidemiol.* **89**:669–680 (1969).

117. Maheshwari, R. K., and B. M. Gupta, A new antiviral agent designated 6-MFA from *Aspergillus flavus.* III. Amplification of anti Semliki Forest virus activity of 6-MFA by cycloheximide treatment in mice, *J. Antibiot.* **26**:335–340 (1973).

118. Maheshwari, R. K., B. M. Gupta, S. N. Ghosh, and N. P. Gupta, Antiviral agent (6-MFA) from *Aspergillus ochraceus*-sensitivity of arbovirus in experimentally infected mice, *Indian J. Med. Res.* **67**:183–189 (1978).

119. Ghosh, S. N., M. K. Goverdhan, D. Cecilia, S. Chelliah, N. Kedarnath, and B. M. Gupta, Protective effect of a fungal interferon inducer (6-MFA) against Japanese encephalitis infection in mice, *Indian J. Med. Res.* **79**:705–708 (1984).

120. Rodda, S. J., and D. O. White, Cytotoxic macrophages: A rapid non-specific response to viral infection, *J. Immunol.* **117**:2067–2072 (1976).

121. Nagar, R., U. C. Chaturvedi, and A. Mathur, Effect of a fungal interferon inducer (6-MFA) on macrophage functions during dengue virus infection, *Indian J. Med. Microbiol.* **4**:191–197 (1986).

122. Glasgow, L. A., Transfer of interferon producing macrophages: New approach to viral chemotherapy, *Science* **170**:854–856 (1970).

123. Chaturvedi, U. C., and M. I. Shukla, Characterization of the suppressor factor produced in the spleen of dengue virus-infected mice, *Ann. Immunol. Pasteur Inst.* **132C**:245–251 (1981).

124. Shukla, M. I., and U. C. Chaturvedi, Dengue virus-induced suppressor factor stimulates production of prostaglandin to mediate suppression, *J. Gen. Virol.* **56**:241–249 (1981).

125. Shukla, M. I., and U. C. Chaturvedi, Cycloheximide and mitomycin C treatment inhibits production of dengue virus-induced suppressor factor, *Indian J. Exp. Biol.* **19**:826–828 (1981).

126. Shukla, M. I., and U. C. Chaturvedi, Study of the target cell of the dengue virus-induced suppressor signal, *Br. J. Exp. Pathol.* **65**:267–273 (1984).

127. Shukla, M. I., and U. C. Chaturvedi, Differential cyclophosphamide sensitivity of T

lymphocytes of the dengue virus-induced suppressor pathway, *Br. J. Exp. Pathol.* **65**:397–403 (1984).

128. Shukla, M. I., and U. C. Chaturvedi, Duration of adoptively transferred dengue virus-induced suppressor activity, *Br. J. Exp. Pathol.* **66**:1–7 (1985).

129. Shukla, M. I., and U. C. Chaturvedi, Presence of I-region gene products on the cells and their products of the dengue virus-induced suppressor pathway, *Br. J. Exp. Pathol.* **67**:563–569 (1986).

130. Shukla, M. I., H. Dalakoti, and U. C. Chaturvedi, Ly phenotype of T lymphocytes producing dengue virus-induced immunosuppressive factors, *Indian J. Exp. Biol.* **20**:525–528 (1982).

131. Chaturvedi, U. C., N. Singh, M. I. Shukla, and A. Mathur, *In vitro* induction of suppressor T cells by dengue virus, *Ann. Virol. Pasteur Inst.* **136E**:341–351 (1985).

132. Chaturvedi, U. C., S. Ahmed, A. Mathur and A. Kumar, Characterization of suppressor factor generated *in vitro* by the dengue virus, *Indian J. Med. Microbiol.* **4**:1–6 (1986).

133. Rawat, S., A. Mathur, and U. C. Chaturvedi, Characterization of Japanese encephalitis virus-induced suppressor T cells and their products for delayed type hypersensitivity, *Br. J. Exp. Pathol.* **67**:571–579 (1986).

134. Rawat, S., A. Mathur, and U. C. Chaturvedi, Characterization of Japanese encephalitis virus-specific suppressor T cells and their product in suppression of the humoral immune response in mice, *Ann. Immunol. Pasteur Inst.* **137D**:391–401 (1986).

135. Chaturvedi, U. C., A. Bhargava, and A. Mathur, Production of cytotoxic factor in the spleen of dengue virus-infected mice, *Immunology* **40**:665–671 (1980).

136. Chaturvedi, U. C., K. R. Mathur, L. Gulati, and A. Mathur, Target lymphoid cells for the cytotoxic factor produced in the spleen of dengue virus-infected mice, *Immunol. Lett.* **3**:13–16 (1981).

137. Singh, M., L. Gulati, A. Jain, A. Mathur, and U. C. Chaturvedi, Absence of antimicrobial activity of the dengue virus-induced macrophage cytotoxin, *Indian J. Med. Microbiol.* **5**:53–56 (1987).

138. Chaturvedi, U. C., H. Dalakoti, and A. Mathur, Characterization of the cytotoxic factor produced in the spleen of dengue virus-infected mice, *Immunology* **41**:387–393 (1980).

139. Dalakoti, H., U. C. Chaturvedi, and A. Mathur, Inhibition of production of dengue virus-induced cytotoxic factor by treatment with cycloheximide and mitomycin C. *Indian J. Exp. Biol.* **20**:1–3 (1982).

140. Gulati, L., U. C. Chaturvedi, and A. Mathur, Dengue virus-induced cytotoxic factor induces macrophages to produce a cytotoxin, *Immunology* **49**:121–130 (1983).

141. Gulati, L., U. C. Chaturvedi, and A. Mathur, Production of dengue virus-induced macrophage cytotoxin *in vivo*, *Br. J. Exp. Pathol.* **67**:269–277 (1986).

142. Chaturvedi, U. C., R. Nagar, L. Gulati, and A. Mathur, Variable effects of dengue virus-induced cytotoxic factors on different subpopulations of macrophages, *Immunology* **61**:297–301 (1987).

143. Gulati, L., U. C. Chaturvedi, and A. Mathur, Characterization of the cytotoxin produced by macrophages in response to dengue virus-induced cytotoxic factor, *Br. J. Exp. Pathol.* **64**:192–197 (1983).

144. Gulati, L., U. C. Chaturvedi, and A. Mathur, Plasma membrane acting drugs inhibit the effect of dengue virus-induced cytotoxic factor on target cells, *Ann. Immunol. Pasteur Inst.* **134C**:227–235 (1983).

145. Gulati, L., U. C. Chaturvedi, and A. Mathur, Effect of dengue virus-induced macrophage cytotoxin on functions of human blood leucocytes, *Indian J. Med. Res.* **79**:709–715 (1984).

146. Gulati, L., U. C. Chaturvedi, and A. Mathur, Study of target cell receptor sites for dengue virus-induced cytotoxins, *Indian J. Med. Microbiol.* **2**:235–242 (1984).

147. Dalakoti, H., U. C. Chaturvedi, and A. Mathur, Studies on dengue virus-induced cytotoxic factor, *Indian J. Exp. Biol.* **21**:375–378 (1983).

148. Nagar, R., U. C. Chaturvedi, M. I. Shukla, A. Mathur, and A. Kumar, Suppressor factor abrogates dengue virus-induced depression of Fc-receptor functions of macrophages, *Indian J. Exp. Biol.* **23:**121–125 (1985).

149. Price, W. H., I. S. Thind, W. O'Leary, and A. H.El Dadah, A protective mechanism induced by live group B arbovirus independent of serum neutralizing antibodies or interferon, *Am. J. Epidemiol.* **86:**11–27 (1967).

150. Thind, I. S., and W. H. Price, The effect of cyclophosphamide treatment on experimental arbovirus infections, *Am. J. Epidemiol.* **90:**62–68 (1969).

151. Bennett, I. L. Jr., R. R. Wagner, and V. S. LeQuire, Pyrogenecity of influenza virus in rabbits, *Proc. Soc. Exp. Biol. Med.* **71:**132–133 (1949).

152. Atkins, E., and W. C. Huang, Studies on the pathogenesis of fever with influenza viruses. I. The appearance of an endogenous pyrogen in the blood following intravenous injection of virus, *J. Exp. Med.* **107:**383–401 (1958).

153. Atkins, E., M. Cronin, and P. Isacson, Endogenous pyrogen release from rabbit blood cells incubated *in vitro* with parainfluenza virus, *Science* **146:**1469–1470 (1964).

154. Dinarello, C. A., Interleukin-1, *Rev. Infect. Dis.* **6:**51–95 (1984).

155. Roberts, N. J., Jr., A. H. Prill, and N. Mann Thomas, Interleukin-1 and interleukin-1 inhibitor production by human macrophages exposed to influenza virus or respiratory syncytial virus. Respiratory syncytial virus is a potent inducer of inhibitor activity, *J. Exp. Med.* **163:**511–519 (1986).

156. Oldstone, M. B. A., M. Rodriguez, W. H. Daughaday, and P. W. Lampert, Viral perturbation of endocrine function: Disordered cell function leads to disturbed homeostasis and diseases, *Nature (London)* **307:**278–281 (1984).

157. Woodruff, J. F., and J. J. Woodruff, T lymphocyte interaction with viruses and virus-infected tissues, *Prog. Med. Virol.* **19:**120–160 (1975).

158. Nash, A. A., Tolerance and suppression in virus disease, *Br. Med. Bull.* **41:**41–45 (1985).

159. Riley, V., Persistence and other characteristics of the lactate dehydrogenase-elevating virus (LDH-virus), *Prog. Med. Virol.* **18:**198–213 (1974).

160. Barlow, R. M., Some interactions of virus and maternal/foetal immune mechanism in Border disease of sheep, *Prog. Brain Res.* **59:**255–268 (1983).

161. Craig, C. P., S. L. Reynolds, J. W. Airhart, and E. V. Staab, Alterations in immune responses by attenuated Venezuelan equine encephalitis vaccine. I. Adjuvant effect of VEE virus infection in guinea pigs, *J. Immunol.* **102:**1220–1227 (1969).

162. Morag, A., B. Morag, J. M. Bernstein, K. Beutner, and P. L. Ogra, *In vitro* correlates of cell-mediated immunity in human tonsils after natural or induced rubella virus infection, *J. Infect. Dis.* **131:**409–416 (1975).

163. Vesikari, T., and E. Buimovici-Klein, Lymphocyte responses to rubella virus antigen and phytohemagglutinin after administration of the RA 27/3 strain of live attenuated rubella vaccine, *Infect. Immun.* **11:**748–753 (1975).

164. Plotkin, S. A., R. M. Klaus, and J. A. Whitely, Hypogammaglobulinemia in an infant with congenital rubella syndrome; failure of 1-adamantanamine to stop virus excretion, *J. Pediatr.* **69:**1085–1091 (1966).

165. Hancock, M. P., C. C. Huntley, and J. L. Sever, Congenital rubella syndrome with immunoglobulin disorders, *J. Pediatr.* **72:**636–645 (1968).

166. Michaels, R. H., Immunologic aspects of congenital rubella, *Pediatrics* **43:**339–350 (1969).

167. Hardy, J. B., J. L. Sever, and M. R. Gilkeson, Declining antibody titres in children with congenital rubella, *J. Pediatr.* **75:**213–220 (1969).

168. Hyypia, T., J. Eskola, M. Laine, and O. Meurman, B cell function *in vitro* during rubella infection, *Infect. Immun.* **43:**589–592 (1984).

169. Debre, R., and K. Papp, Sur Ia Cuti-reaction tuberculinique au cours de la rougeole et la rubeole, *C.R. Soc. Biol* **34:**29–33 (1926).

170. Lamb, G. A., Effect of HPV-80 rubella vaccine on the tuberculin reaction, *Am. J. Dis. Child.* **118:**261 (1969).

171. Midulla, M., L. Businco, and L. Moschini, Some effects of rubella vaccination on immunologic responsiveness, *Acta Paediatr. Scand.* **61:**609–611 (1972).

172. Kauffman, C. A., J. P. Phair, C. C. Linnemann, Jr., and G. M. Schiff, Cell-mediated immunity in humans during viral infection. I. Effect of rubella on dermal hypersensitivity, Phytohemagglutinin response and T lymphocyte numbers, *Infect. Immun.* **10:**212–215 (1974).

173. Ganguly, R., C. L. Cusumano, and R. H. Waldman, Suppression of cell-mediated immunity after infection with attenuated rubella virus, *Infect. Immun.* **13:**464–469 (1976).

174. Simons, M. J., and M. G. Fitzgerald, Rubella virus and human lymphocytes in culture, *Lancet* **2:**937–940 (1968).

175. Buimovici-Klein, E., P. B. Land, P. R. Ziring, and L. Z. Cooper, Impaired cell-mediated immune response in patients with congenital rubella: Correlation with gestational age at time of infection, *Pediatrics* **64:**620–626 (1979).

176. Soontiens, F. J. C. J., and J. Van-der Veen. Evidence for a macrophage-mediated effect of poliovirus on the lymphocyte response to phytohemagglutinin, *J. Immunol.* **111:**1411–1419 (1973).

177. Chantler, J. K., and A. J. Tingle, Replication and expression of rubella virus in human lymphocyte populations, *J. Gen. Virol.* **50:**317–328 (1980).

178. Van der Logt, J. T. M., A. M. Van Loon, and J. Van-der Veen, Replication of rubella virus in human mononuclear blood cells, *Infect. Immun.* **27:**309–314 (1980).

179. Buimovici-Klein, E., and L. Z. Cooper, Immunosuppression and isolation of rubella virus from human lymphocytes after vaccination with two rubella vaccines, *Infect. Immun.* **25:**352–356 (1979).

180. Mahler, R., and L. Soren, *In vitro* effects of rubella virus strain RA 27/3 on human lymphocytes, *Acta Pathol. Microbiol. Scand. Sect. C* **85:**49–56 (1977).

181. Lee, J. C., and M. M. Siegel, A differential effect of IgM and IgG antibodies on the blastogenic response of lymphocytes to rubella virus, *Cell. Immunol.* **13:**22–31 (1974).

182. Arneborn, P., G. Biberfeld, and J. Wasserman, Immunosuppression and alterations of T-lymphocyte subpopulations after rubella vaccination, *Infect. Immun.* **29:**36–41 (1980).

183. Oldstone, M. B. A., and F. J. Dixon, Lactic dehydrogenase virus induced immune complex type of glomerulonephritis, *J. Immunol.* **106:**1260–1263 (1971).

184. McDonald, T. L., Isolation of C1q-binding virus–antibody immune complexes from lactic dehydrogenase virus (LDV) infected mice, *Immunology* **45:**365–370 (1982).

185. Michaelides, M. C., and E. S. Simms, Immune response in mice infected with lactic dehydrogenase virus. I. Antibody response to DNP-BGG and hyperglobulinaemia in BALB/C mice, *Immunology* **32:**981–988 (1977).

186. Coutelier, J. P., and J. Van Snick, Isotypically restricted activation of B lymphocytes by lactic dehydrogenase virus, *Eur. J. Immunol.* **15:**250–255 (1985).

187. Bendinelli, M., and G. L. Asherson, Depression of contact sensitivity by Friend and Riley viruses, *Boll. Ist. Sieroterap. Mil.* **50:**502–507 (1971).

188. Vincent, M. D., J. G. Potter, and S. Cooper, Immunodepression, ascites tumour and lactate dehydrogenase virus, *S. Afr. Med. J.* **52:**924–926 (1977).

189. Isakov, N., S. Segal, and M. Feldman, The immuno-regulatory characteristics of the lactic dehydrogenase virus (LDV), a common contaminant of tumours, *Cell. Mol. Biol.* **27:**83–96 (1981).

190. Oldstone, M. B. A., and F. J. Dixon, Inhibition of antibodies to nuclear antigen and to DNA in New Zealand mice infected with lactate dehydrogenase virus, *Science* **175:**784–786 (1972).

191. Michaelides, M. C., and S. Schlesinger, Effect of acute or chronic infection with lactic

dehydrogenase virus (LDV) on the susceptibility of mice to Plasmacytoma MOPC-315, *J. Immunol.* **112**:1560–1564 (1974).

192. McDonald, T. L., Blocking of cell-mediated immunity to Moloney murine sarcoma virus-transformed cells by lactate dehydrogenase virus–antibody complex, *J. Natl. Cancer Inst.* **70**:493–497 (1983).

193. Johnson, R. J., and H. S. Shin, Lack of correlation of growth attenuation of murine lymphoma caused by *in vitro* passage with loss of lactate dehydrogenase virus, *J. Natl. Cancer Inst.* **71**:1337–1341 (1983).

194. Mergenhagen, S. E., A. L. Notkins, and S. F. Dougherty, Adjuventicity of lactic dehydrogenase virus: Influence of virus infection on the establishment of immunologic tolerance to a protein antigen in adult mice, *J. Immunol.* **99**:576–581 (1967).

195. Henderson, D. C., C. E. Tosta, and N. Wedderburn, Exacerbation of murine malaria by concurrent infection with lactic dehydrogenase-elevating virus, *Clin. Exp. Immunol.* **33**:357–359 (1978).

196. Bonventre, P. F., H. C. Bubel, J. G. Michael, and A. D. Nickol, Impaired resistance to bacterial infection after tumour implant is traced to lactic dehydrogenase virus, *Infect. Immun.* **30**:316–319 (1980).

197. Isakov, N., and S. Segal, A tumour associated lactic dehydrogenase virus suppresses the host resistance to infection with *Listeria monocytogenes*, *Immunobiology* **164**:402–416 (1983).

198. Herman, G., H. G. Du Buy, and M. L. Johnson, Studies on the *in vivo* and *in vitro* multiplication of the LDH virus of mice. *J. Exp. Med.* **123**:985–998 (1966).

199. Porter, D. D., H. G. Porter, and B. B. Deerhale, Immunofluorescence assay for antigen and antibody in lactic dehydrogenase virus infection in mice, *J. Immunol.* **102**:431–436 (1968).

200. Stueckemann, J. A., D. M. Ritzi, M. Holth, M. S. Smith, W. J. Swart, W. A. Cafruny, and P. G. W. Plagemann, Replication of lactate dehydrogenase-elevating virus in macrohpages. I. Evidence for cytocidal replication, *J. Gen. Virol.* **59**:245–262 (1982).

201. Inada, T., and C. A. Mims, Mouse Ia antigens are receptors for lactate dehydrogenase virus, *Nature (Lond.)* **308**:59–61 (1984).

202. Inada, T., and C. A. Mims, Ia antigens and Fc-receptor of mouse peritoneal macrophages as determinants of susceptibility to lactic dehydrogenase virus, *J. Gen. Virol.* **66**:1469–1477 (1985).

203. Unanue, E. R., D. I. Beller, C. Y. Lu, and P. M. Allen, Antigen presentation: Comments on its regulation and mechanism, *J. Immunol.* **132**:1–5 (1984).

204. Isakov, N., M. Feldman, and S. Segal, Lactic dehydrogenase virus (LDV) impairs the antigen-presenting capacity of macrophages yet fails to affect their phagocytic activity, *Immunobiology* **162**:15–27 (1982).

205. Isakov, N., M. Feldman, and S. Segal, Acute infection of mice with lactic dehydrogenase virus (LDV) impairs the antigen-presenting capacity of their macrophages, *Cell. Immunol.* **66**:317–332 (1982).

206. Taylor, R. B., and A. Basten, Suppressor cells in humoral immunity and tolerance, *Br. Med. Bull.* **32**:152–157 (1976).

207. Pang, T., P. Y. Wong, and R. Pathmanathan, Induction and characterization of delayed-type hypersensitivity to dengue virus in mice, *J. Infect. Dis.* **146**:235–242 (1982).

208. Pang, T., S. Devi, W. P. Yeen, I. F. C. McKenzie, and Y. K. Leong, Lyt phenotype and H-2 compatibility requirements of effector cells in the delayed-type hypersensitivity response to dengue virus infection, *Infect. Immun.* **43**:429–431 (1984).

209. Nagarkatti, M., P. S. Nagarkatti, and K. M. Rao, Effect of experimental dengue virus infection on humoral and cell-mediated immune response to thymus-dependent antigen, *Int. Arch. Allergy Appl. Immunol.* **62**:361–369 (1980).

210. Wong, P. Y., S. Devi, I. F. C. McKenzie, K. L. Yap, and T. Pang, Induction and Ly

phenotype of suppressor T cells in mice during primary infection with dengue virus, *Immunology* **51**:51–56 (1984).

211. Nagarkatti, P. S., and M. Nagarkatti, Effect of experimental dengue virus infection on immune response of the host. I. Nature of changes in T suppressor cell activity regulating the B and T cell response to heterologous antigens, *J. Gen. Virol.* **64**:1441–1447 (1983).

212. Halstead, S. B., N. J. Marchette, J. S. Sung Chow, and S. Lolekha, Dengue virus replication enhancement in peripheral blood leucocytes from immune human beings, *Proc. Soc. Exp. Biol. Med.* **151**:136–139 (1976).

213. Chaturvedi, U. C., M. I. Shukla, K. R. Mathur, and A. Mathur, Dengue virus-induced cytotoxic factor suppresses immune response of mice to sheep erythrocytes, *Immunology* **43**:311–316 (1981).

214. Nagar, R., U. C. Chaturvedi, and A. Mathur, Effect of dengue virus-induced cytotoxic factor on Fc-receptor functions of mouse macrophages, *Br. J. Exp. Pathol.* **65**:11–17 (1984).

215. Chaturvedi, U. C., P. Nath, L. Gulati, A. Jain, and A. Mathur, Subcellular changes in spleen cells of mice treated with the dengue virus-induced cytotoxic factor, *Indian J. Med. Res.* **86**:284–289 (1987).

216. Nagar, R., U. C. Chaturvedi, A. Kumar, and A. Mathur, Abrogation of dengue virus-induced alteration of Fc-receptor functions of macrophages by splenectomy or drug treatment, *Indian J. Med. Res.* **81**:537–546 (1985).

217. Chaturvedi, U. C., L. Gulati, and A. Mathur, Inhibition of E-rosette formation and phagocytosis by human blood leucocytes after treatment with the dengue virus-induced cytotoxic factor, *Immunology* **45**:679–685 (1982).

218. Nagar, R., U. C. Chaturvedi, A. Mathur, and A. Kumar, Effect of dengue virus-induced cytotoxic factor on blood leucocytes of monkeys, *Indian J. Med. Res.* **84**:339–347 (1986).

219. Chaturvedi, U. C., Virus-induced cytotoxic factor in AIDS and dengue, *Immunol. Today* **7**:159 (1986).

220. Chaturvedi, U. C., A. Mathur, and R. M. L. Mehrotra, Experimentally produced cardiac injury following dengue virus infection, *Indian J. Pathol. Bacteriol.* **17**:218–220 (1974).

221. Agarwal, D. K., P. Tandon, U. C. Chaturvedi, and A. Kumar, Biochemical study of certain enzymes and metabolites of carbohydrate metabolism in the skeletal muscles of dengue virus infected mice, *J. Gen. Virol.* **40**:399–408 (1978).

222. Tandon, P., U. C. Chaturvedi, and A. Mathur, Dengue virus-induced thymus-derived suppressor cells in the spleen of mice, *Immunology* **38**:653–658 (1979).

223. Chaturvedi, U. C., M. I. Shukla, and A. Mathur, Thymus dependent lymphocytes of the dengue virus-infected mice spleen mediate suppression through prostaglandin, *Immunology* **42**:1–6 (1981).

224. Chaturvedi, U. C., M. I. Shukla, and A. Mathur, Role of macrophages in transmission of dengue virus-induced suppressor signal to a subpopulation of T lymphocytes, *Ann. Immunol. Pasteur Inst.* **133C**:83–96 (1982).

225. Shukla, M. I., and U. C. Chaturvedi, In *vivo* role of macrophages in transmission of dengue virus-induced suppressor signal to T lymphocytes, *Br. J. Exp. Pathol.* **63**:522–529 (1982).

226. Shukla, M. I., and U. C. Chaturvedi, Transmission of dengue virus-induced suppressor signal from macrophage to lymphocyte occurs by cell contact, *Br. J. Exp. Pathol.* **64**:87–92 (1983).

227. Chaturvedi, U. C., Dengue virus-induced suppressor pathway, *Curr. Sci.* **53**:971–976 (1984).

228. Mathur, A., S. Rawat, and U. C. Chaturvedi, Induction of suppressor cells in Japanese encephalitis virus-infected mice, *Br. J. Exp. Pathol.* **64**:336–343 (1983).

229. Mathur, A., S. Rawat, and U. C. Chaturvedi, Suppressor T cells for delayed-type hypersensitivity to Japanese encephalitis virus, *Immunology* **52:**395–402 (1984).
230. Mathur, A., S. Rawat, U. C. Chaturvedi, and V. S. Misra, Macrophage transmission of suppressor signal for suppression of delayed type hypersensitivity and humoral response in JEV-infected mice, *Br. J. Exp. Pathol.* **67:**171–179 (1986).
231. Mathur, A., R. Kulshreshtha, S. Rawat, and U. C. Chaturvedi, Memory suppressor T cells in latent Japanese encephalitis virus infection, *Immunology* **60:**71–74 (1987).
232. Germain, R. N., and B. Benacerraf, A single major pathway of T lymphocyte interactions in antigen-specific immune suppression, *Scand. J. Immunol.* **13:**1–10 (1981).
233. WHO Memoranda, Pathogenetic mechanisms in dengue haemorrhagic fever. Report of an International collaborative study, *Bull. WHO* **48:**117–133 (1973).
234. Liew, F. Y., and S. M. Russell, Inhibition of pathogenic effect of effector T cells by specific suppressor T cells during influenza virus infection in mice, *Nature (Lond.)* **304:**541–543 (1983).
235. Burns, W. H., and R. Saral, Opportunistic viral infections, *Br. Med. Bull.* **41:**46–49 (1985).
236. Chaturvedi, U. C., A. Mathur, A. Chandra, S. K. Das, H. O. Tandon, and U. K. Singh, Transplacental infection with Japanese encephalitis virus, *J. Infect. Dis.* **141:**712–715 (1980).
237. Mathur, A., H. O. Tandon, K. R. Mathur, N. B. S. Sarkari, U. K. Singh, and U. C. Chaturvedi, Japanese encephalitis virus infection during pregnancy, *Indian J. Med. Res.* **81:**9–12 (1985).
238. Mathur, A., K. L. Arora, and U. C. Chaturvedi, Congenital infection of mice with Japanese encephalitis virus, *Infect. Immun.* **34:**26–29 (1981).
239. Mathur, A., K. L. Arora, S. Rawat, and U. C. Chaturvedi, Persistence, latency and reactivation of Japanese encephalitis virus infection in mice, *J. Gen. Virol.* **67:**381–385 (1986).
240. Mathur, A., K. L. Arora, S. Rawat, and U. C. Chaturvedi, Japanese encephalitis virus latency following congenital infection in mice, *J. Gen. Virol.* **67:**945–947 (1986).
241. Mathur, A., K. L. Arora, and U. C. Chaturvedi, Transplacental Japanese encephalitis virus (JEV) infection in mice during consecutive pregnancies, *J. Gen. Virol.* **59:**213–217 (1982).
242. Hudson, B. W., K. Wolff, and J. C. DeMartini, Delayed type hypersensitivity responses in mice infected with St. Louis encephalitis virus: Kinetics of the response and effects of immunoregulatory agents, *Infect. Immun.* **24:**71–76 (1979).
243. Kraaijeveld, C. A., M. Harmsen, and B. B. Khader, Delayed type hypersensitivity against Semliki Forest virus in mice, *Infect. Immun.* **23:**219–223 (1979).
244. Kelkar, S. D., and K. Banerjee, Passive transfer of protection through immune cells in Japanese encephalitis virus infection in mice. A preliminary report, *Indian J. Med. Res.* **64:**1720–1721 (1976).
245. Kelkar, S. D., F. M. Rodrigues, and K. Banerjee, Development of *in vitro* correlate of cell mediated immunity following Japanese encephalitis inactivated vaccine among laboratory personnel, *Indian J. Med. Res.* **83:**104–107 (1986).
246. Nagarkatii, P. S., M. B. D'Souza, and K. M. Rao, Use of sensitized spleen cells in capillary tube migration inhibition test to demonstrate cellular sensitization to dengue virus in mouse, *J. Immunol. Methods* **23:**341–348 (1978).
247. Gajdosova, E., C. Oravec, and V. Mayer, Cell-mediated immunity in flavivirus infection. I. Induction of cytotoxic T lymphocytes in mice by an attenuated virus from tick-borne encephalitis complex and its group reactive character, *Acta Virol. (Praha)* **25:**10–18 (1981).
248. McFarland, H. F., D. E. Griffin, and R. T. Johnson, Specificity of the inflammatory response in viral encephalitis. I. Adoptive immunization of immunosuppressed mice infected with Sindbis virus, *J. Exp. Med.* **136:**216–226 (1972).

249. Jacoby, R. O., P. N. Bhatt, and A. Schwartz, Protection of mice from lethal flaviviral encephalitis by adoptive transfer of spleen cells from donors infected with live virus, *J. Infect. Dis.* **141**:617–624 (1980).

250. Zhang, Y. H., G. G. Zhou, and W. F. Yue, Cell mediated immune response to Japanese encephalitis virus. II. Protection of mice against virulent virus challenge by adoptive transfer of immune spleen cells from donor injected with an attenuated live or inactivated vaccine, *Chinese J. Microbiol. Immunol.* **2**:213–219 (1982).

251. Berkovich, S., and S. Starr, Effects of live type 1 poliovirus vaccine and other viruses on the tuberculin skin test, *N. Engl. J. Med.* **274**:67–72 (1966).

252. Nagarkatti, M., and P. S. Nagarkatti, Suppression of intrinsic B cell function in dengue-infected mice, *Experientia* **35**:1518–1519 (1979).

253. Mathur, A., K. L. Arora, and U. C. Chaturvedi, Immune response to Japanese encephalitis virus in mother mice and their congenitally infected offspring, *J. Gen. Virol.* **64**:2027–2031 (1983).

254. Mathur, A., R. Kulshreshtha, and U. C. Chaturvedi, Induction of secondary immune response by reactivated Japanese encephalitis virus in latently infected mice, *Immunology* **60**:481–484 (1987).

255. Kesson, A. M., R. V. Blanden, and A. Müllbacher, The primary *in vitro* murine cytotoxic T cell response to the flavivirus, West Nile, *J. Gen. Virol.* **68**:2001–2006 (1987).

256. Khanna, M., U. C. Chaturvedi and A. Mathur. Abrogation of helper T cells by dengue virus-induced cytotoxic factor. *Curr. Sci.* **57**:411–414 (1988).

257. Khanna, M., U. C. Chaturvedi, and A. Mathur. Cytotoxic effects of dengue virus-induced cytotoxin on helper T cells. *Indian J. Med. Microbiol.* **6**:46–50 (1988).

258. Chaturvedi, U. C., N. Srivastava, and A. Mathur, Presence of receptors for antigen and antibody on dengue virus-induced suppressor T cells and their products. *Curr. Sci.* **57**:691–696 (1988).

# 15

# Rhabdoviruses

## Effect of Vesicular Stomatitis Virus Infection on the Development and Regulation of Cell-Mediated and Humoral Immune Responses

MAN-SUN SY and ROBERT FINBERG

## 1. BASIC PROPERTIES OF VESICULAR STOMATITIS VIRUS

Vesicular stomatitis virus (VSV) belongs to the family of rhabdoviruses. This group of viruses includes VSV, rabies virus, Marburg virus, and several other viruses that replicate in anthropoids as well as in other mammals.[1,2] The natural hosts of these viruses are listed in Table I. One of the major morphologic features of this group of viruses is their bullet-shaped virion. These bullet-shaped rods are cylinders approximately 170 nm in length and 70 nm in diameter. Each particle consists of a nucleoprotein (RNA) core helically wound around a central axial hollow. This is enclosed in an envelope consisting of a lipid bilayer and associated structural proteins. In the VSV group, there are distinct strains, including Indiana, New Jersey, Brazil, Argentina, Cocal, Piry, and Chandipura. All the strains except Piry and Chandipura share some common antigens; those two strains appear to be totally unrelated antigenically to others.[3] According to the nomenclature of Wagner et al.,[4] proteins of the Indiana serotype are designated L, G, N, NS, and M and have molecular weights of 190, 69, 50, 45, and 29 kDa, respectively.[5]

MAN-SUN SY • Department of Pathology, Harvard Medical School, Boston, Massachusetts 02115.    ROBERT FINBERG • Laboratory of Infectious Diseases, Dana Farber Cancer Institute, Boston, Massachusetts 02115.

## TABLE I
### Natural Hosts of Rhabdoviruses

| Virus | Host |
|---|---|
| VSV, Hart Park virus, Flanders virus, Kern Canyon virus | Infect a variety of mammals, fish, anthropods, and plants |
| Rabies virus (street rabies virus) | Infects a variety of mammals (i.e., fox, skunk, and dog); the only member of the group known to infect and produce disease in humans naturally |
| Marburg virus | Simian virus, only accidentally infects humans |

Pathogenetic studies of VSV infections in young hamsters and mice have shown that viral replication, particularly in thoracic and abdominal organs, precedes and contributes to viremia, which then results in central nervous system (CNS) invasion and encephalitis. In older infected animals, the degree of necrosis of CNS tissues coincides with low titers in extraneural organs and high titers in the brain and spinal cord.[6]

There is very little information regarding the effect of rhabdoviruses except VSV on host-immune responses *in vivo*. Wiktor *et al.*[7] reported that mice lethally infected with rabies virus, or street rabies virus, failed to generate virus-specific cytotoxic T-cell responses. Moreover, infection with street rabies virus suppresses the development of a primary T-cell-mediated immune response against other unrelated viruses. This indicates that some of the rhabdoviruses may be able to induce immunosuppression in experimental animals.

## 2. MODULATION OF HOST-IMMUNE RESPONSES BY VESICULAR STOMATITIS VIRUS

### 2.1. Introduction

Infection of humans or animals by a variety of viruses has been shown to result in either suppression or enhancement of immune responses.[8,9] The outcome of an immunologic response depends critically on the participation and interaction of various subpopulations of lymphocytes and monocytes.[10,11] Viral infections may affect the normal immunologic responsiveness of the host by interfering with any of the lymphoid cell types involved in an immune response. Some of the possible targets of virus-induced immune dysfunction and their possible mechanisms are listed in Table II.

Our current understanding of the immune responses in the murine system has demonstrated a fascinating complexity within lymphoid subpopulations.[12,13] It is known that the level of an immunologic response to a foreign antigen is determined by both the positive and negative influence of regulatory lymphocytes. Helper T cells are required for the optimal elicitation of both humoral and cell-mediated immune responses, while suppressor T cells (Ts) dampen these responses. Moreover, antigen-specific suppression of immune responses has been shown to involve the interaction of several T-cell subsets and specialized antigen-presenting cells.[14,15] Therefore, it is obvious that immunologic dysfunction, due to viral infections, may reflect changes in the physiologic or immunologic properties of B lymphocytes, T lymphocytes, and/or macrophages, and the consequence of lymphocyte infection for immune reactivity would depend on many factors, including type(s) and number of cells infected and the nature of the viruses.

This chapter discusses the effect of VSV infection on host immunity. The emphasis is not on the host-immune response against the virus per se, but rather on the effects of viral infection on the development and regulation of immune responses to some well-characterized antigens. It is possible that immune dysfunction may not be attributable to the direct effect of viral infection, but rather to the elimination of some regulatory T cells. In the case of helper T cells, this may result in immune suppression. For example, infection with human immune deficiency virus (HTLV-III/LAV) in some patients results in complete elimination of helper T cells and may lead to the development of acquired immune deficiency syndrome (AIDS).[16,17] In a situation in which Ts are selectively eliminated, this may result in the development of autoimmunity. It has been postulated that some of the autoimmune diseases observed in experimental models or in humans may have a viral etiology.[8,9]

## 2.2. Interferon and Anti-VSV Activity

The interferons, i.e., IFN$_\alpha$, IFN$_\beta$, and IFN$_\gamma$, were first defined by their ability to prevent infection of mammalian cells by viruses. While by definition all IFNs have antiviral properties, only one, IFN$_\gamma$, produced by mitogenic or antigenic stimulation of T lymphocytes, is a lymphokine.[18,19]

**TABLE II**
**Possible Targets of Virus-Induced**
**Immune Dysfunction**

Abnormalities in macrophage function
Abnormalities in T-cell function
Abnormalities in B-cell function

Possible mechanisms
  Direct
    Receptor blockade
    Cell destruction—lysis
  Indirect
    Inhibition of macromolecular synthesis
    Virus-induced factors—interferon
    Virus-induced suppressor cells
    Alternation of lymphocyte traffic

This lymphokine has multiple effects on the immune response to the virus because it has primary effects on the immune system as well. An important effect of $IFN_\gamma$ lies in its ability to increase expression of class II (Ia) major histocompatibility complex (MHC) antigens on the surface of antigen-presenting cells.[20] Since class II antigen-bearing cells (accessory cells) are necessary for the stimulation of T cells, the stimulation of these antigens is important in activating an immune response.[11] In certain situations (e.g., in mice exposed to UV irradiation) a functional deficiency of class II-bearing cells leads to a lack of immune responses.[21] Similarly, patients with AIDS are characterized as having a paucity of class II-bearing cells. This defect has been reversed *in vitro* by the addition of $IFN_\gamma$.[22]

Furthermore, we previously demonstrated that virus-specific helper T cells are capable of stimulating nonimmune cells (macrophages or fibroblasts) to produce $IFN_\alpha/IFN_\beta$.[23] Since the presence of class II MHC antigen-presenting cells is required for stimulation of helper T cells, $IFN_\gamma$ in particular is vitally important for stimulating the release of antiviral substances. In addition, specific cytolytic T cells (CTL) can be demonstrated to release $IFN_\gamma$ upon contact with the target cells to which they are sensitized.[23] However, whether this is the actual mechanism of action of T cells in combatting viral infections is uncertain. Alternatively, the virus-specific T cells may serve to lyse virus-infected cells, or recruit other non-immune cells to act as effector cells against the virus.

### 2.3. Effect of VSV Infection on Macrophage Activation

Interferon-$\gamma$ also may have the most important function in the activation of macrophages during VSV infection. In addition to being a modulator of MHC class II antigen expression, it is able to activate macrophages; i.e., $IFN_\gamma$ displays macrophage activation factor (MAF) activity. Several studies suggest that the injection of viruses in animals can lead to macrophage activation, which in turn leads to enhanced viral clearance.[24,25]

Activated macrophages may be important in the elimination of viruses. In some instances, the macrophage may be the crucial factor in determining whether an infection is lethal for the host. Hirsch *et al.*[24] demonstrated that macrophages obtained from adult mice were able to protect neonatal mice from lethal herpes simplex virus (HSV) infection. The population of macrophages that was protective was obtained by intraperitoneal inoculation of mice with proteose–peptone. These elicited macrophages were shown to be activated in that they ingested more virus, produced more IFN, and destroyed virus more effectively than did resting macrophages. Significantly, the suckling mouse did not respond to proteose-peptone stimulation with such a population of cells. These neonatal mice were extremely susceptible to HSV infection. Thus, it is suggested that these activated macrophages are responsible for the fact that viral infection is mild in adults and lethal in neonates. Mogensen[25] demonstrated that macrophages appear to be important in resistance to viruses in a number of experimental systems.

Belardelli *et al.*[26] demonstrated that whereas VSV can only multiply in a small percentage of peritoneal macrophages, the injection of mice with anti-IFN antibodies prior to harvesting peritoneal macrophages allowed the virus to replicate to a high titer. This finding suggests that IFNs may have a dual role in this situation, protecting macrophages against destruction by virus as well as activating them to ingest and eliminate other viral particles.

In summary, there is evidence that virus-stimulated interfering substances (IFNs) may be important in protecting macrophages from lytic infection. Macrophages are themselves essential for stimulating T cells. However, only those with class II antigens on their surface are capable of such stimulation. $IFN_\gamma$ released by virus-specific T cells is capable of increasing expression of class II MHC antigens on the surface of antigen-presenting cells. Virus-specific helper T cells are, in turn, capable of stimulating the release of not only $IFN_\gamma$ but of $IFN_\alpha / IFN_\beta$ as well.

## 2.4. Effect of VSV Infection on T- and B-Cell-Mediated Immune Response *in Vivo*

Bloom and colleagues[27,28] were the first to analyze the interaction between VSV and lymphocytes. Their results clearly indicate that VSV infects only activated T cells. Infectious VSV is produced by activated T cells, but not by B cells. Further studies showed that both Lyt-1+ and Lyt-2+3+ T cells are capable of producing VSV virus upon activation.

### 2.4.1. Delayed-Type Hypersensitivity

The induction and elicitation of delayed-type hypersensitivity (DTH) in the murine system depends critically on both Lyt-1+ T lymphocytes and Ia-bearing antigen-presenting cells.[11] Effects of infection with VSV on the DTH reaction to heterogeneous erythrocytes and the hapten, dinitrofluorobenzene (DNFB) in mice has been investigated by Katsura and colleagues.[30–33] These investigators found that infection at the time of immunization results in an enhancement of the DTH reaction specific for the immunizing antigen. These results clearly suggest that the target of VSV infection is not the DTH effector T cell. Therefore, a subpopulation of regulatory T cells responsible for down-regulation of DTH may be the target of VSV infection. In contrast to these studies, VSV given at the time of antigenic challenge not only failed to enhance the DTH response but inhibited the elicitation of the DTH reaction. The fact that elicitation of the DTH reaction was inhibited in VSV-infected mice may indicate that some of the DTH effector cells are sensitive to VSV or that the Ia-bearing antigen-presenting cells may be damaged by the virus. The latter possibility is supported by findings showing that under some conditions, VSV may be able to replicate in macrophages.[34]

More recently, we investigated whether VSV infection can also alter the DTH reaction to another simple hapten azobenzene arsonate (ABA).[35] Using our experimental protocol, we were unable to detect any effect of VSV infection either in the intensity or duration of the ABA-specific DTH reaction (Fig. 1). These observations were confirmed independently by employing two different *in vitro* assays. First, VSV-infected mice did not develop a higher level of ABA-specific cytotoxic T-cell activity than the untreated control mice. The development of ABA-specific CTL requires the participation of both Lyt-1+ helper T cells and Lyt-2+ CTL. Second, VSV infection failed to show a significant enhancement in T-cell-dependent ABA-specific proliferation. The cells proliferating *in vitro* are mainly Lyt-1+ helper T cells. The precise reason why VSV infection failed to enhance the T-cell-mediated immune response to ABA is unclear. It is possible that the route of administration of antigen may be responsible for the discrepancies (e.g., the fact that the hapten was conjugated with syngeneic cells).

### 2.4.2. Antibody Responses

Infection of mice by VSV has been shown to enhance the primary antibody response to sheep red blood cells (SRBC) *in vivo*.[32] A greater response can be observed in both IgM and IgG levels, regardless of the immunizing dose of

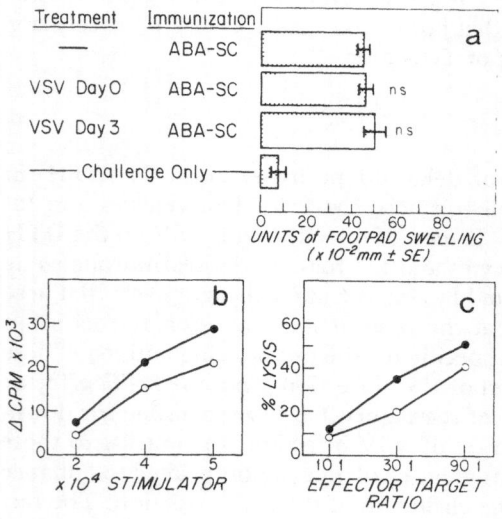

**FIGURE 1.** Effect of VSV infection in the induction of T-cell-mediated immune response to ABA-conjugated spleen cells (ABA-SC). (a) Delayed-type hypersensitivity (DTH). Normal A/J mice were immunized subcutaneously with $3 \times 10^7$ ABA-SC. Within 1 hr after immunization, some of the mice were also infected with $1 \times 10^7$ PFU of VSV (i.p.). Another group of mice were injected with $1 \times 10^7$ PFU of VSV (i.p.) 3 days after immunization. All mice, including the controls were challenged in the footpad with 30 μl of the ABA diazonium salt 6 days after immunization; the increase in footpad swelling was measured 24 hr later. Each group consisted of at least four mice. (b) *In vitro* proliferation. Normal A/J mice were immunized subcutaneously with $3 \times 10^7$ ABA-SC (●). Within 1 hr after immunization, some of the immunized mice were also given $1 \times 10^7$ PFU of VSV (i.p.) (○). Five days later, their draining lymph nodes were removed and cultured *in vitro* with ABA-SC and pulsed [³H]thymidine for the last 24 hr of the culture. (c) *In vivo* priming for the generation of cytotoxic T cells. Normal A/J mice were immunized subcutaneously with $3 \times 10^7$ ABA-SC (●). Within 1 hr after immunization, some of the immunized mice were also given $1 \times 10^7$ PFU of VSV (i.p.) (○). Five days later, their spleens were removed and cultured *in vitro* for the induction of ABA-specific cytotoxic T cells. ABA-specific cytotoxic T cells were quantitated using a 4-hr Cr-release assay. (From Nowakowski *et al.*[34])

SRBC. However, the enhancement effect depends critically on the time of viral infection. Marked augmentation can be observed only when VSV and SRBC were given on the same day. Injection of VSV 2 days before or 3 days after immunization failed to produce an enhancement. Moreover, a reduction of antibody production is observed in animals given VSV 4 days after antigen administration. This provides strong evidence that the effect of VSV infection on the host-immune response depends critically on the dynamics and homeostasis of various lymphoid subpopulations. Related to these studies is the observation of Goodman-Snikoff and co-workers. These investigators reported that the glycoprotein (G protein) of VSV is a T-cell-independent B-cell mitogen and polyclonal activator.[36]

### 2.4.3. Effect of VSV Infection on Regulatory T Cells

The elegant studies of Katsura and colleagues were the first to suggest that the targets of VSV infection may be Ts.[30-33] These investigators found that the generation of T cells capable of replicating VSV is observed in antigen-primed cultures of lymphocytes taken from animals immunized with the challenge antigen. These VSV-replicating cells do not represent all the activated T cells, but may represent a subpopulation of activated T cells. It was further shown that both the antibody response and DTH reaction were strongly augmented by the infection with VSV, suggesting that the generation of Ts was prohibited.[30-33] Because most of the earlier studies on the effect of viral infection on immune responses used complex antigens (i.e., SRBC), we believe that the simple haptenic system (ABA) and the information it provides on well-characterized subpopulations of T cells permits a more fruitful analysis of the effect of VSV viral infections on lymphocyte subpopulations and eventually of their effect on the development and regulation of various aspects of immune responses.

In examining the T-cell-mediated murine immune responses to ABA conjugated cells,[37,38] we have identified three effector T-cell subpopulations. These include $T_{DH}$ (T cells that mediated DTH), T prol (T cells that proliferate in vitro, which are probably helper T cells), and CTL. In addition, we have described three distinct subpopulations of suppressor T cells (Ts-1, Ts-2, and Ts-3) that interact sequentially to mediate suppressor activities (Table III). Our inability to observe an enhancement of T-cell-mediated immune responses to ABA in animals infected with VSV does not rule out the possibility that VSV may interfere with Ts interactions. Thus, we investigated whether VSV infection will render animals insensitive to tolerance induction or to suppression mediated by various subpopulations of antigen-specific suppressor T cells, i.e., Ts-1, Ts-2, or Ts-3.[35] Our results clearly demonstrated that the induction of immunologic tolerance to ABA was not affected by the in vivo infection of VSV (Fig. 2). Similarly, the induction of ABA-specific $Ts_1$ (Table IV) and Ts-2 (Fig. 3), was also unaffected by VSV infection. Interestingly, VSV infection renders mice insensitive to suppression mediated by Ts factors ($TsF_1$ and $TsF_2$) (Fig. 4). These findings provide direct evidence that VSV infection can indeed modu-

**TABLE III**
**Suppressor T Cells Involved in Regulating T-Cell-Mediated Immune Responses to Azobenzene Arsonate**[a]

| T suppressor | Inducer/route | Cell phenotype of suppressor | Receptor specificity |
|---|---|---|---|
| Ts$_1$ | ABA-conjugated syngeneic spleen cell (intravenous)<br>Anti-idiotypic antibodies (intravenous) | Lyt-1$^+$ | Anti-antigen (binds ABA) |
| Ts$_2$ | TsF$_1$ suppressor (T-cell factor obtained from Ts$_1$)<br><br>Idiotype-conjugated spleen cells (intravenous) | Lyt-1$^+$2$^+$3$^+$ | Anti-idiotype (binds anti-ABA antibodies of the appropriate strain of mice) |
| Ts$_3$ | ABA-conjugated spleen cell (subcutaneous) | Lyt-2$^+$3$^+$ | Anti-antigen |

[a] ABA, azobenzene arsonate.

late an immune response by interfering with some cellular component in the suppressor T-cell pathway required for the manifestation of suppressor activities. Since neither Ts-1 nor Ts-2 cells are affected by VSV, the most likely candidate for the target of VSV infection in the ABA system is therefore the Ts-3 cell. This is supported by our observation that it is possible to bypass the VSV effect with normal Lyt-2$^+$ T cells adoptively transferred to the host after the virus has been cleared (Table V). Exactly how the VSV eliminates the action of Ts has not been clearly established. Although IFN has been shown to interfere with the action of these cells,[39] we do not believe that this mediator is involved in such phenomena because other viruses (i.e., reovirus type 3) that are excellent IFN inducers had no effect on the action of ABA-specific Ts. In addition, only live VSV, but not UV-irradiated VSV, is capable of reversing the Ts effect. It is also unlikely that VSV binding to its target cell via a specific receptor is important for suppression, since VSV binds equally well to Lyt-1$^+$ and Lyt-2$^+$ T cells.[29]

Why does VSV affect only some T-cell subpopulations while it seems to bind to all? It is possible that different T-cell subpopulations may have different activation requirements. For example, activation of some T-cell subpopulations may require DNA synthesis, whereas others do not. Some T-cell subpopulations may require a longer period of time before they can be activated, whereas other T cells can be activated relatively quickly. These kinds of differences may provide the explanation for the differential in the sensitivity of different T-cell subpopulations to VSV infection.

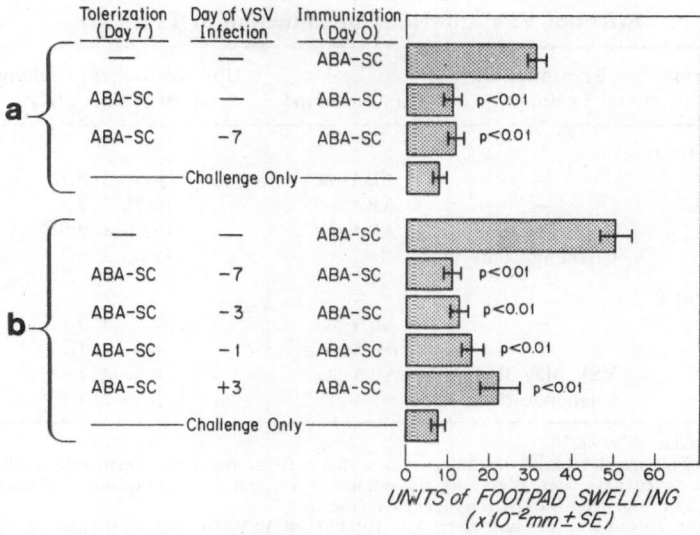

**FIGURE 2.** Effect of VSV infection on the induction of immunologic tolerance to ABA-SC. Normal A/J mice were injected i.v. with $5 \times 10^7$ ABA-SC on day 7 to induce ABA-specific unresponsiveness. At various times after i.v. injection of ABA-SC, some of the mice were also given $1 \times 10^7$ PFU of VSV (i.p.) All the mice and their appropriate controls were immunized subcutaneously with $3 \times 10^7$ ABA-SC on day 0 Footpad challenges were done on day 6 as described and the increase in footpad swelling measured at 24 hours post challenge. (a,b) Two different experiments. (From Nowakowski *et al.*[34])

## 2.5. Effect of VSV Infection *in Vitro*

### 2.5.1. Inhibition of Lymphocyte Mitogenic Responses

It is known that viruses can exert inhibitory effects on the mitogenic responses of lymphocytes *in vitro*. One effect of the virus could be its direct lytic effect on lymphocytes. Another mechanism is the *in vitro* activation of suppressor cells that possess the capacity to inhibit mitogen-driven lymphocyte proliferation.[40–42] It is also possible that the effect of the virus may be totally nonspecific. Wainberg and Israel[43] reported that *in vitro* addition of VSV virus can completely abolish the mitogenic response of murine spleen cells to the T-cell mitogens phytohemaglutinin (PHA) or concanavalin A (Con A). Both infectious and UV-inactivated virus can inhibit *in vitro* mitogenesis. Moreover, these investigators have shown that the observed inhibition is truly nonspecific, since it can also be induced by membrane vesicles that approximate virus size and by crude pelleted material derived from supernatant fluids of normal or virus-infected cells. In these studies, the targets of these inhibitory

**TABLE IV**
**Effect of VSV Infection on Induction of $Ts_1$[a]**

| Inducer of Ts[b] | Treatment of Ts donors[c] | Immunization[d] | Units of footpad swelling ($\times 10^{-2}$ mm $\pm$SE)[e] |
|---|---|---|---|
| **Experiment 1** | | | |
| — | — | ABA-SC | 38.5 ± 2.1 |
| ABA-SC | — | ABA-SC | 16.75 ± 2.8 |
| ABA-SC | VSV (day 0) | ABA-SC | 12.50 ± 2.0 |
| — | Challenge only | — | 12.00 ± 0.7 |
| **Experiment 2** | | | |
| — | — | ABA-SC | 22.5 ± 3.2 |
| ABA-SC | — | ABA-SC | 8.5 ± 1.7 |
| ABA-SC | VSV (day 3) | ABA-SC | 8.5 ± 1.1 |
| — | Challenge only | — | 5.25 ± 1.7 |

[a] From Nowakowski et al.[34]
[b] To induce Ts, normal A/J mice were injected i.v. with $5 \times 10^7$ azobenzene arsonate spleen cells (ABA-SC). Seven days later, they were the donors of Ts, and $5 \times 10^7$ spleen cells were transferred i.v. into groups of naive syngeneic recipients.
[c] Some of the Ts donors were also given $1 \times 10^7$ PFU of VSV (Indiana) at the day of Ts induction (day 0) or 3 days later (day 3).
[d] Within 2 hr of receiving Ts, all recipients were immunized subcutaneously with $3 \times 10^7$ ABA-SC.
[e] Five days after immunization, all experimental groups including the controls were challenged in the footpad. Increase in footpad swelling was measured 24 hr postchallenge.

materials were not addressed, and the suppressive effect of the virus may be due to a mechanical alteration of the cells, perhaps the cell membrane. The importance of this effect *in vivo* is not clear.

### 2.5.2. Inhibition of Antigen-Specific Proliferative Response

The *in vitro* addition of VSV has been shown to inhibit mixed lymphocyte reactions *in vitro*.[44] Pre-infection of responding cells with the virus also causes a complete abrogation of mixed lymphocyte reactions. In these experiments, resting lymphocytes are infected before being stimulated in an MLR, and the virus plaque-forming cells and thymidine incorporation are measured. Under these conditions, in which virus plaque-forming cells are generated, there is a complete elimination of proliferation.[21] It has been suggested that lymphocyte activation for replication of VSV is an early event, precedes DNA synthesis, and probably occurs maximally during $G_0$ of the cell cycle.[21] In investigating the effect of VSV infection on ABA-specific proliferative responses, we have also found that *in vitro* addition of VSV results in complete abrogation of ABA-specific proliferative responses. However, this differs from studies reported by Wainberg and Israel.[43] In the ABA-system, involving *in vitro*-primed T cells and *in vitro* restimulation, UV-inactivated VSV is totally ineffective. Moreover, reovirus type 3 used as another control failed to inhibit *in vitro*

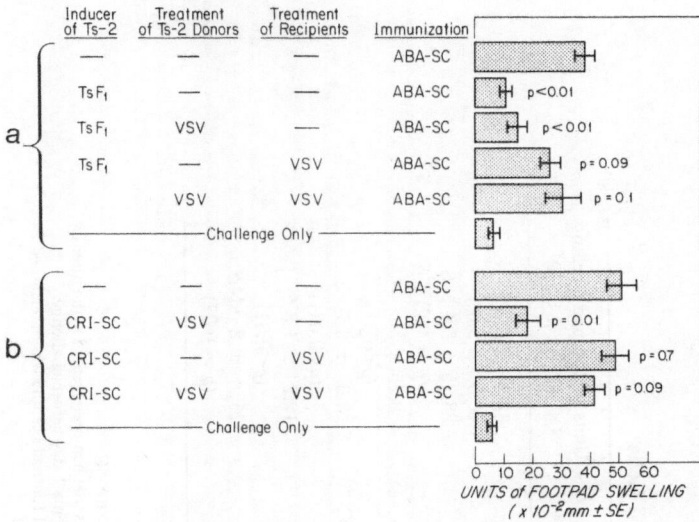

**FIGURE 3.** Effect of VSV infection on the induction and expression of $Ts_2$. Normal A/J mice were injected with $2 \times 10^7$ cell equivalents of $TsF_1$ (i.v.) for 5 successive days or $5 \times 10^7$ cross-reactive idiotype (CRI-SC), to serve as donors of $Ts_2$. Some of these mice were also given $1 \times 10^7$ PFU of VSV. In the case of $Ts_2$ induced with $TsF_1$, they were given VSV only on the first day of $TsF_1$ injection. Seven days later, these mice were served as donors of $Ts_2$. Five $\times 10^7$ spleen cells from these mice were transferred into different groups of naive recipients. All recipients and respective controls were immunized subcutaneously with $3 \times 10^7$ ABA-SC. In addition, some of the recipients were also given $1 \times 10^7$ PFU of VSV. Footpad challenge was done 6 days after immunization and the increase in footpad swelling measured 24 hr later. (a,b) Two different experiments. (From Nowakowski *et al.*[31])

**FIGURE 4.** Failure of $TsF_1$ and $TsF_2$ to suppress the development of ABA-specific delayed-type hyper-sensitivity (DTH) in VSV-infected mice. Normal A/J mice were immunized with $3 \times 10^7$ ABA-SC (s.c.) to induce ABA-specific DTH reaction. Within 2 hr after immunization some of the animals were also given $1 \times 10^7$ PFU of VSV; $10^8$ cell equivalents of $TsF_1$ were administered i.v. into different groups of immunized animals over a 5-day period, beginning on the day of immunization. For $TsF_2$, $7 \times 10^7$ cell equivalents of

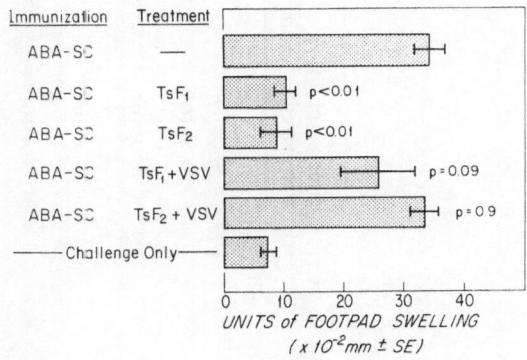

$TsF_1$ were given only on the day of challenge. All mice were challenged with the diasonium salt on day 6; 24 hr later, the degree of footpad swelling was determined.

## TABLE V
### Targets of VSV Infection: Lyt 2+3+ T Cells[a]

| Number of Ts$_1$ transferred[b] | Ts$_1$ recipients[c] | Immunization[d] | Reconstitution[e] | Units of footpad swelling ($\times 10^{-2}$ mm ±SE)[f] |
|---|---|---|---|---|
| — | — | ABA-SC | — | 38.0 ± 5.7 |
| 5 × 10$^7$ | — | ABA-SC | — | 10.0 ± 3.0 |
| 5 × 10$^7$ | VSV | ABA-SC | — | 28.25 ± 3.4 N.S. ($p = 0.2$) |
| 5 × 10$^7$ | VSV | ABA-SC | 5 × 10$^7$ NSC | 9.25 ± 3 ($p<0.01$) |
| 5 × 10$^7$ | VSV | ABA-SC | 5 × 10$^7$ NSC + C' | 8.0 ± 0.7 ($p<0.01$) |
| 5 × 10$^7$ | VSV | ABA-SC | 5 × 10$^7$ NSC + αLy 1.2 + C' | 12.2 ± 1.8 ($p<0.01$) |
| 5 × 10$^7$ | VSV | ABA-SC | 5 × 10$^7$ NSC + αLy 2.2 + C' | 25 ± 2 N.S. ($p = 0.07$) |

[a]From Nowakowski et al.[(34)]

[b]Ts$_1$ were induced by i.v. injection of 5 × 10$^7$ ABA-SC 7 days earlier.

[c]One × 10$^7$ PFU of VSV (Indiana) was given (i.p.) to the recipients of Ts$_1$.

[d]On the day of transfer and within 2 hr, after immunization with 3 × 10$^7$, ABA-SC.

[e]To reconstitute infected animals, five × 10$^7$ normal spleen cells or spleen cells that had been treated with either Ly 1.2 + C' or anti-Ly 2.2 + C' were injected i.v. into VSV-infected Ts recipients 1 day before challenge.

[f]Five days after immunization, all animals were challenged in the footpad, and footpad swelling was determined 24 hr later.

proliferation. *In vitro* proliferation of antigen-primed T cells requires the participation of both T cells and antigen-presenting cells (which can be either B cells or macrophages). Therefore, attempts were made to address whether the *in vitro* effect of VSV is due to a direct effect of virus on T cells or to the antigen-presenting cells. We attempted to pulse T cells or macrophages with VSV and then mix these cells together to elicit an *in vitro* response. Pulsing of either T cells or macrophages with VSV results in complete elimination of *in vitro* proliferative responses. We are unable to rule out the possibility that the observed effect is simply due to a carryover of the virus from T cells to the macrophage or from macrophage to T cells, a problem common to studies of the *in vitro* effect of live viruses.

### 2.5.3. Effect of VSV Infection on the Generation of Cytotoxic T Cells

We have also investigated the effect of VSV infection on the generation of hapten-specific and virus-specific CTL. The *in vitro* addition of VSV results in dramatic reduction in ABA-specific CTL. Moreover, the addition of VSV *in vitro* inhibits the generation of reovirus specific CTL (Table VI). Interestingly, the *in vitro* addition of VSV into cultures obtained from VSV-primed animals, results in the generation of VSV-specific CTL. Therefore, VSV appears to be

**TABLE VI**
**Inhibition of Virus-Specific Cytotoxic T-Lymphocyte**
**Responses by VSV**

| *In vivo* immunization[a] | *In vitro* addition[b] | % Specific lysis of reovirus-infected L cells[c] | | |
|---|---|---|---|---|
| | | Effector/target ratio | | |
| | | 100 : 1 | 33 : 1 | 11 : 1 |
| Experiment 1 | | | | |
| Reo 3 | Reo 3 | 70 | 67 | 45 |
| Reo 3 | Reo 3 + VSV (day 0) | 0 | 1 | 6 |
| Reo 3 | Reo 3 + VSV (day 1) | 2 | 3 | 3 |
| Reo 3 | Reo 3 + VSV (day 2) | 3 | 3 | 4 |
| Experiment 2 | | | | |
| Reo 3 | Reo 3 | 40 | 37 | 35 |
| Reo 3 | Reo 3 + VSV (day 0) | 10 | 8 | 5 |
| Reo 3 | Reo 3 + VSV (day 1) | 3 | 7 | 4 |
| Reo 3 | Reo 3 + VSV (day 4) | 34 | 36 | 23 |

[a]Normal A/J mice were injected i.p. with $1 \times 10^7$ PFU of reovirus type 3 (Reo 3); 7 days later, the spleens from these animals were removed and single-cell suspension prepared and cultured *in vitro*.
[b]One $\times 10^5$ PFU of reovirus or VSV (Indiana) was added at the beginning of a 5-day culture.
[c]Specific killing: percentage lysis of reovirus infected L cells minus percentage lysis of uninfected L cells.

unable to inhibit CTL activity against the virus, but will inhibit the generation of CTL against another determinant. It is known that the generation of virus-specific CTL requires *in vivo* priming. Thus, if an animal has been primed with VSV *in vivo*, its activated lymphocytes or monocytes may be able to eliminate some of the virus. Host responses may reduce the number of viral particles *in vitro*, rendering them unable to cause the inhibitory effect. Kinetic studies show that the addition of VSV is most effective in inhibiting anti-ABA CTL on day 0 or 1–2 days after culture initiation; 4 days after the initiation of culture, VSV fails to cause any significant effect (Table VI).

### 2.5.4. Antibody Responses

It has been reported that after the inoculation of VSV into the culture, the *in vitro* secondary anti-SRBC antibody response of spleen cells taken more than 3 days after immunization, was markedly enhanced.[30] The timing of immunization appears to be critical since no augmentation was observed in the culture of day 3 or earlier spleen cells. The optimal enhancement appears to be in the culture of day 7 spleen cells. In investigating the effect of VSV infection on a T-dependent hapten–carrier system, we found *in vitro* addition of VSV results in significant inhibition of both IgG- and IgM-specific for the hapten ABA. The effect appears to be specific for VSV, since addition of reovirus type 3 causes only minimal inhibition (Table VII). Similar results were obtained when we investigated the effect of VSV infection using the T-cell-independent antigen TNP–*Brucella abortus* (Table VIII). Replication of virus appears to be essential for the inhibitory effect, since UV irradiation of virus completely abrogated its effect.

In summary, we have provided evidence indicating that infection of animals with a rhabdovirus, VSV, can result in the functional elimination of suppressor T-cell subpopulations. This renders the animal unable to downregulate

### TABLE VII
### Effect of *in Vitro* Addition of VSV on ABA-Specific IgG and IgM Production

| Antigen[a] | In vitro addition[b] | PFU/1 × 10⁶ cells | |
|---|---|---|---|
| | | IgM | IgG |
| ABA-KLH | — | 990 ± 25 | 6500 ± 210 |
| ABA-KLH | VSV 1 × 10⁴ PFU | 220 ± 86 | 2000 ± 177 |
| ABA-KLH | VSV 1 × 10⁵ PFU | 90 ± 8 | 1035 ± 82 |
| ABA-KLH | Reo 3 1 × 10⁴ PFU | 1000 ± 77 | 6190 ± 54 |
| ABA-KLH | Reo 3 1 × 10⁵ PFU | 900 ± 22 | 5000 ± 710 |

[a]A/J mice were primed i.p. with 100 µg of ABA-KLH in CFA; 7 days later, spleen cells were removed and cultured in presence of ABA-KLH for 5 days.
[b]VSV (Indiana) and Reo virus type 3 were added at the beginning of culture.

**TABLE VIII**
**Inhibition of the Generation of TNP-Specific PFC Response by Live VSV**

| Antigen[a] | VSV[b] | VSV addition | PFC/$1 \times 10^6$ spleen cells |
|---|---|---|---|
| TNP–*Brucella* | — | — | $1796 \pm 271$ |
| TNP–*Brucella* | $1 \times 10^4$ PFU | Day 0 | $87 \pm 14$ |
| TNP–*Brucella* | $1 \times 10^4$ PFU | Day 2 | $34 \pm 8$ |
| TNP–*Brucella* | $1 \times 10^4$ PFU UV-irradiated | Day 0 | $1376 \pm 209$ |
| TNP–*Brucella* | $1 \times 10^4$ PFU UV-irradiated | Day 2 | $1677 \pm 87$ |

[a]Spleen cells from normal A/J mice were cultured *in vitro* in the presence of TNP–*Brucella abortus*
[b]VSV (Indiana) was used.

its immune response. The precise mechanism whereby VSV can alter the immunologic function of some, but not all. T-cell subpopulations is not clear. In contrast to *in vivo* infection, *in vitro* addition of VSV results in inhibition of a variety of immunologic functions. This includes mitogenic response, CTL response, and antibody response. The precise mechanism(s) of these *in vitro* inhibitory effects also remain to be determined.

# REFERENCES

1. Prevec, L., Physiological properties of vesicular stomatitis virus and some related Rhabdoviruses, in: *Viruses, Evolution and Cancer, Basic Consideration* (E. Kurstak and K. Maramorosch, eds.), pp. 677–695, Academic, New York (1974).
2. Howatson, A. F., Vesicular stomatitis and related viruses, *Adv. Virus Res.* **16:**195–256 (1970).
3. Cartwright, B., and F. Brown, Serological relationship between different strains of vesicular stomatitis virus, *J. Gen. Virol.* **16:**391–398 (1972).
4. Wagner, R. R., L. Prevec, F. Brown, F. Summer, F. Sokol, and R. MaCleod, Classification of Rhabdovirus protein: A proposal, *J. Virol.* **10:**1228–1230 (1972).
5. Kang, C. Y., and L. Prevec, Proteins of vesicular stomatitis virus. III. Intracellular synthesis and extracellular appearance of virus specific proteins, *Virology* **46:**678–690 (1971).
6. Murphy, F. A., Evolution of Rhabdovirus tropisms, in: *Viruses, Evolution and Cancer, Basic Consideration* (E. Kurstak and K. Maramorosch, eds.), pp. 696–702, Academic, New York (1974).
7. Wiktor, T. J., P. C. Doherty, and H. Koprowski, Suppression of cell-mediated immunity by street rabies virus, *J. Exp. Med.* **145:**1617–1622 (1977).
8. Proffitt, M. R. (ed.), *Virus–Lymphocyte Interactions Implication for Disease*, Elsevier/North-Holland, New York (1979).
9. Notkins, A. L. (ed.), *Viral Immunology*, Academic, New York (1975).
10. Cantor, H., and I. Weissman, Development and function of subpopulation of thymocytes and T lymphocytes, *Prog. Allergy* **20:**1–50 (1970).
11. Klein, J., Immunology: *The Science of Self–Non-Self Discrimination*, Wiley, New York (1982).

12. Benacerraf, B., and R. N. Germain, Specific suppressor responses to antigen under I region control, *Fed. Proc.* **38**:2053–2057 (1979).

13. Cantor, H., and R. K. Gershon, Immunological circuits: Cellular composition, *Fed. Proc.* **38**:2058–2063 (1979).

14. Germain, R. N., and B. Benacerraf, Hypothesis: A single major pathway of T lymphocyte interactions in antigen-specific immune suppression, *Scand. J. Immunol.* **13**:1–10 (1981).

15. Tada, T., and K. Okumura, The role of antigen-specific T cell factors in the immune response, *Adv. Immunol.* **28**:1–80 (1980).

16. Broder, S. M., and R. C. Gallo, A pathogenic retrovirus (HTLV-III) linked to AIDS, *N. Engl. J. Med.* **311**:1291–1297 (1984).

17. Klatzman, D., F. Barre-Sinoussi, M. T. Nugeyre, C. Dauguet, F. Vilmer, C. Griscelli, F. Brun-Vezinet, C. Rouzioux, J. C. Gluckman, J. C. Chermann, and L. Montagnier, Selective tropism of lymphadenopathy associated virus (LAV) for the helper/inducer T lymphocytes, *Science* **225**:59–63 (1984).

18. Nathan, C. F., W. H. Murray, M. E. Wiebe, and R. Y. Rubin, Identification of interferon-γ as the lymphokine that activates human macrophage oxidative metabolism and antimicrobial activity, *J. Exp. Med.* **158**:670–689 (1983).

19. Ennis, F. A., and A. Meager, Immune interferon produced to high levels by antigenic stimulation of human lymphocytes with influenze virus, *J. Exp. Med.* **154**:1279–1289 (1981).

20. Vignaux, F., and I. Gresser, Differential effects of interferon on the expression of H-2K, H-2D, and Ia antigen on mouse lymphocytes, *J. Immunol.* **118**:721–723 (1977).

21. Letvin, N. L., R. S. Kauffman, and R. Finberg, T lymphocytes immunity to reovirus: Cellular requirements for generation and role in clearance of primary infections, *J. Immunol.* **127**:2334–2339 (1981).

22. Heagy, W., T. Strom, V. Kelley, R. Finberg, Decreased expression of human Class II antigen on monocytes in AIDS patients: Increased expression with interferon-γ, *J. Clin. Invest.* **74**:2089–2096 (1984).

23. Ertl, H. C. J., E. Brown, and R. Finberg, Sendai virus specific T cell clones. II. Induction of interferon production of Sendai virus-specific T helper cell clones, *Eur. J. Immunol.* **12**:1051–1053 (1982).

24. Hirsch, M. S., B. Zisman, and A. C. Allison, Macrophages and age-dependent resistance to herpes simplex virus in mice, *J. Immunol.* **104**:1160–1165 (1970).

25. Mogensen, S. C., Role of macrophages in natural resistance to virus infections, *Microbiol. Rev.* **43**:1–26 (1979).

26. Belardelli, F., F., Vignaux, E. Prioetti, and I. Gresser, Injection of mice with antibody to interferon renders peritoneal macrophages permissive for vesicular stomatitis virus and encephalomyocarditis, *Proc. Natl. Acad. Sci. USA*, **81**:602–606 (1984).

27. Bloom, B. R., L. Jimenez, and P. I. Marcus, A plaque assay for enumerating antigen-sensitive cells in delayed-type hypersensitivity, *J. Exp. Med.* **132**:16–30 (1970).

28. Kano, S., B. R. Bloom, and M. I. Howe, Enumeration of activated thymus-derived lymphocytes by the virus plaque assay, *Proc. Natl. Acad. Sci. USA* **70**:2299–2303 (1970).

29. Bloom, B. R., A. Senik, G. Stoner, G. Ju, M. Nowakowski, S. Kano, and L. Jimenez, Studies on the interactions between viruses and lymphocytes, *Cold Spring Harbor Symp. Quant. Biol.* **43**:73–83 (1976).

30. Minato, N., and Y. Katsura, Virus replicating T cells in the immune response of mice. I. Virus plaque assay of the lymphocytes reactive to sheep erythrocytes, *J. Exp. Med.* **145**:390–404 (1977).

31. Minato, N., and Y. Katsura, Virus replicating T cells in the immune response of mice. II. Characterization of T cells capable of replicating vesicular stomatitis virus, *J. Exp. Med.* **148**:837–849 (1978).

32. Katsura, Y., M. Takaoki, Y. Kono, and N. Minato, Augmentation of delayed-type hyper-

sensitivity by vesicular stomatitis virus infection in mice, *J. Immunol.* **125**:1459–1462 (1980).

33. Takahashi, C., and Y. Katsura, Hapten-specific virus replicating T cells: Analysis of the functional role in contact sensitivity, *J. Immunol.* **124**:2721–2727 (1980).

34. Nowakowski, M., J. D. Feldman, S. Kano, and B. R. Bloom, The production of vesicular stomatitis virus by antigen or mitogen-stimulated lymphocytes and continuous lymphoblastoid lines, *J. Exp. Med.* **137**:1042–1059 (1973).

35. Sy, M.-S., M. Tsurufuji, R. Finberg, and B. Benacerraf, Effect of vesicular stomatitis virus (VSV) infection on the develop ment and regulation of T cell-mediated immune responses, *J. Immunol.* **131**:30–36 (1983).

36. Goodman-Snitkoff, G., R. J. Mannino, and J. J. McSharry, The glycoprotein isolated from vesicular stomatitis virus is mitogenic for mouse B lymphocytes, *J. Exp. Med.* **153**:1489–1502 (1981).

37. Greene, M. I., M. J. Nelles, M.-S. Sy, and A. Nisonoff, Regulation of immunity to azobenzenearsonate hapten, *Adv. Immunol.* **32**:253–300 (1982).

38. Benacerraf, B., M. I. Greene, M.-S. Sy, and M. E. Dorf, Suppressor T cell circuits, *Ann. NY Acad. Sci.* **392**:300–308 (1982).

39. Knop, J., R. Stremmer, C. Neumann, E. deMaeyer, and E. Macher, Interferon inhibits the suppressor T cell response of delayed-type hypersensitivity, *Nature (Lond.)* **296**:757–758 (1982).

40. Soontiens, F. J. C. J., and J. Van der Veen, Evidence for a macrophage-mediated effect of poliovirus on the lymphocyte response to phytohemagglutinin, *J. Immunol.* **111**:1411–1419 (1973).

41. Toy, S. T., and E. F. Wheelock, *In vitro* depression of cellular immunity by Friend leukemic spleen cells, *Cell. Immunol.* **17**:57–73 (1975).

42. Kirchner, H., R. B. Herberman, M. Glacer, and D. H. Lavrin, Suppression of *in vitro* lymphocyte stimulation in mice bearing primary Moloney sarcoma virus-induced tumors, *Cell. Immunol.* **13**:32–51 (1974).

43. Wainberg, M. A., and E. Israel, Viral inhibition of lymphocyte mitogenesis. I. Evidence for the nonspecificity of the effect, *J. Immunol.* **124**:64–70 (1980).

44. Kano, S., B. R. Bloom, D. C. Shreffler, D. Schendel, and F. H. Bach, Studies of the mixed lymphocyte reaction by the virus plaque assay, *Cell. Immunol.* **20**:229–240 (1975).

# 16

# Virus-Induced Immunosuppression
## Influenza Virus

NORBERT J. ROBERTS, JR., and FRANK DOMURAT

## 1. INTRODUCTION

Influenza virus infection is common and a significant cause of morbidity and mortality, the latter usually due to secondary (bacterial) complications.[1-3] Several recent reviews are available that discuss in depth one or more particular aspects of the anti-influenza virus immune response,[4,5] including aspects of the virus that might play a role in the development of leukocyte dysfunction.[3,6] The review by Sweet and Smith[3] is especially pertinent to any considerations of influenza virus pathogenicity. Concepts of viral immunosuppression, with attention to viruses other than influenza, are reviewed in other chapters in this volume, as well as elsewhere.[7-11]

It is becoming apparent from both *in vitro* and *in vivo* challenge studies that immune responses after influenza virus infection are a complex group of events in which non-virus-specific parameters are suppressed, while virus-specific responses develop.[12-14] Most of the data available are in regard to influenza A infection. A detailed discussion of an individual function (e.g., cytotoxic cell activity) is beyond the scope of the current review, the focus of which is the interaction of influenza virus with the immunologic system to produce, directly or indirectly, effects that may be broadly related to immunosuppression. Thus, those effects of influenza virus infection on leukocytes that result in depressed cell functions, production of immunomodulatory signals, or that contribute or lead to virus-induced disease are emphasized, particularly noting more recent

NORBERT J. ROBERTS, JR., and FRANK DOMURAT • Infectious Diseases Unit, Department of Medicine, University of Rochester School of Medicine, Rochester, New York 14642.

observations. Analysis of the potential beneficial versus detrimental roles of the various virus-induced immunologic alterations is included, in light of the recognized common course of influenza infection.

## 2. BASIC PROPERTIES OF THE VIRUS

Type A, B, and C influenza viruses are structurally similar yet antigenically distinguished orthomyxoviruses that are significant human pathogens and are continuing targets of immunologic investigation.[2,15,16] Influenza C remains largely an enigma, especially with regard to effects on immunologic systems; most of the data concern influenza A and, to a much lesser extent, influenza B. The viral genome (RNA) contains eight single-stranded segments coding for 10 or more gene products.[3,15,17] There does not appear to be one single gene responsible for pathogenicity; an optimal combination of all RNA segments is required to produce a highly pathogenic virus strain.[3,18] The viral envelope hemagglutinin (HA) glycoprotein is required for attachment and initiation of infection of cells.[2,5,15,18] While the HA, after cleavage, is clearly responsible for adsorption to cell surfaces, the function of the neuraminidase (NA) is less clearly established.[15] However, these viral proteins appear to confer subtype and strain specificity in the development of the host immunologic response. Furthermore, the epidemic and sporadic nonepidemic potential of the virus is established by the antigenic lability of the HA and NA glycoproteins, evading neutralization by the antibodies possessed at the time by host populations.[5] Influenza virus types B and C are generally believed to undergo less antigenic variation than the A strains.[16] Internal antigens, nucleoprotein (NP), matrix (M), and the three polymerase (P) proteins can elicit immune responses but are shared within but not between influenza types, with only minor variations between subtypes.[5] At least three virus-coded nonstructural proteins are found in infected cells but have unknown functions.[15]

## 3. MODULATION OF IMMUNITY BY THE VIRUS

### 3.1. Antibody Production

Influenza virus induces homotypic immunity. Successful protection against infection, as well as reduction in disease severity, correlate in the host with systemic anti-HA antibody titers,[19–21] which correlate with *in vitro* virus neutralization. Although antibody titers against NA are similarly important, the significance of antibody responses against influenza nucleoprotein and matrix protein is unclear. The presence of local anti-HA antibody at the mucosal surfaces of the upper and lower respiratory tract is most important in host protection.[22] Virus-specific immunoglobulins (IgA and IgG) function at this level to prevent viral attachment and to limit the cell-to-cell spread of infectious progeny. Furthermore, although absence of antibodies in human and animal models does not preclude recovery, the incidence of fatal pneumonia and respiratory reinfection is increased within this context.[19]

The systemic primary antibody response is initially IgM, followed by IgG.[23,24] Although rising titers of IgA are less predictable in the serum, IgA predominates in nasal secretions. The primary nasal response also includes significant titers of IgM. Administration of live attenuated vaccine to healthy adult volunteers leads to a secondary systemic antibody response, predominantly IgG.[25] Similarly, a brisk IgA response in nasal secretions follows within 1 week of intranasal challenge with inactivated virus, demonstrating a memory response.[26] While IgA is detected in nasal secretions, bronchoalveolar lavage reveals the presence of IgG.[27] In mice challenged with influenza virus via small-particle aerosols, IgA, IgG, and IgM, all specific for the infecting virus, are detected in bronchoalveolar washings.[28] Although nasal wash neutralizing antibody titers drop markedly within 1 or 2 years following infection, the homotypic immunity reflected by systemic anti-HA antibody may persist for decades. However, the duration of heterotypic immunity (protection against influenza A subtype variants) is considerably less.[26] In addition, most persons immunized with an inactivated virus vaccine demonstrate a more rapid decline in antibody production as measured by *in vitro* stimulation tests.[29]

*In vitro* studies of influenza virus challenge demonstrate both serotype-specific and cross-reactive antibody production. Viral stimulation of human peripheral blood leukocytes shows that most antibodies produced are directed against the stimulating strain.[30] Cross-reactive antibodies against different subtypes or different strains within a subtype also appear with specificity for shared determinants on the HA, nucleoprotein, or matrix proteins. Furthermore, heteroclitic antibodies appear against heterologous strains of virus and appear to be determined by the prior exposure of the donor.[30] At least in murine systems, HA expressed on isolated virions is recognized only by a minority of antibodies.[31] Most antibodies require the co-recognition of histocompatible antigenic determinants.

Although influenza virus (especially $H_2$ and $H_6$ subtypes) can be mitogenic[32,33] in the presence of class II immune response antigens, there remains an absolute T-cell dependence for the production of anti-influenza antibody.[34] Furthermore, a soluble antigen-specific helper factor from influenza-specific T cells is also genetically restricted.[34] Sensitized populations of helper T lymphocytes recognize both subtype-specific and cross-reactive determinants of the intact virion. Help occurs cross-reactively for all subtypes of a virus strain.[34] There is a greater T-cell dependence exhibited by serospecific antibody responses than that exhibited by heterologous antibody responses.[21] However, T-helper clones may recognize and react with different antigenic determinants, or even entirely different proteins, in facilitating a subsequent B-cell response. For example, matrix protein-specific T-helper clones assist with anti-HA production if intact virus is used as the source of stimulating determinants.[35] Despite the HLA-restricted T-cell dependence for antibody production, B-cell memory may develop and be triggered independent of Th cells.[36] Immunologic memory is retained in the form of specific B memory cells that are readily detected in most primed adults.[37]

## 3.2. T Lymphocytes

In addition to B cells and the T-helper cells (Th) with which they interact, other subpopulations of lymphocytes figure prominently in the immune response to influenza virus. These include cytotoxic T lymphocytes (Tc), their precursors or memory cells, as well as their helpers (TcH) and suppressors (TcS). In addition, delayed-type hypersensitivity (DTH) lymphocytes (Td) recognize influenza antigens, modulated in turn by their respective helpers and suppressors. The latter may be further characterized as suppressors of either Td induction (afferent) or Td expression (efferent).[5,34,38–42]

Influenza virus-induced Tc are restricted in target cell killing of infected cells by histocompatibility genes, predominantly class I.[43] Activity against influenza A and influenza B is mutually exclusive.[44] However, unlike the anti-HA antibody response, Tc are noted for their reactivity across influenza virus subtypes.[45] Nonetheless, the presence of these cells in the host cannot prevent infection.[34] They do protect against a lethal outcome and appear to contain the viral infection.[46,47] Thus, T-cell-deficient mice exhibit more frequent dissemination of influenza virus, including to the central nervous system (CNS).[48] Other known cytotoxic cell systems demonstrate Tc development independent of accessory cells and soluble factors other than interleukin-2 (IL-2).[49] However, the generation of influenza virus-specific Tc requires both accessory cells and antigen in a class II-histocompatible presentation. Tc appear within the first week of infection and usually persist for 2–4 weeks.[50] With antecedent priming to different strains of influenza virus, Tc may appear in the lungs within 2 days after challenge. Lung virus titers are inversely proportional to the number of Tc.[46,47] Infectious virus demonstrates greater efficiency in the induction of such cells than that afforded by inactivated virus.[34] Although their numbers increase in the lung and diminish in the circulation during episodes of pneumonia, influenza virus-specific Tc and their precursors appear in blood, cervical lymph nodes, and spleen.[48]

Cytotoxic lymphocytes as mature cells figure prominently in the recovery from infection of the host. However, Tc are relatively short-lived compared with Tc memory cells, which are not themselves cytotoxic. Memory cells may be increased 100-fold by infection and may persist for extended intervals.[51] The presence of cytotoxic cells and cytotoxic memory cells is associated with a rapid decline in influenza virus titers after challenge despite the absence of humoral immunity.[5] After primary challenge, T memory cells remain for up to 6 weeks in the lung, and for years in the reticuloendothelial system.[34] The estimated half-life of cytotoxic T memory cells is approximately 3 years.[52]

The mechanism for efferent immune recognition of influenza virus by Tc is under intense scrutiny, since these cells facilitate host recovery in a manner commonly cross-reactive across several influenza virus subtypes. By contrast, exposure to current vaccines exploits the induced homotypic immunity against a specific hemagglutinin antigen. Induction of influenza virus-specific T-cell activity in humans appears to correlate primarily with HLA-DR-plus-antigen on accessory cells, although DP and DQ also serve to restrict antigen presentation.[53] Although most Tc are cross-reactive for virus subtypes, a minority recognize only homologous subtype virus. Subtype-specific Tc may recognize unique epitopes, while cross-reactive Tc recognize common epitopes of the HA molecule. The latter can be demonstrated by blocking with monoclonal antibodies.[45,54,55] In addition, influenza virus nucleoprotein has received much attention recently as a major cross-reactive epitope for cytotoxic cells, despite the fact that it is not a viral surface antigen.[56,57] However, marked individual variation in cross-reactive cytotoxic activity within human and murine systems suggests that other influenza A virus cross-reactive antigens remain to be determined.[58] Killed virus, inactivated vaccines, and some HA subunit vaccines induce little cross-reactive Tc activity or cytotoxic T-cell memory. This has been ascribed to the lack of integration of viral antigen into the cell membrane of the infected cells that must co-express histocompatibility-restricted antigens.[51]

It has been established more recently, however, that infection is not required and that purified nucleoprotein alone may prime *in vivo* or recall *in vitro* cross-reactive cytotoxic lymphocytes.[59,60] Although traditionally regarded as recognizing only class I-restricted targets, recent investigations establish that Tc may lyse class II-restricted targets.[56] It appears that class I Tc recognize conformationally dependent antigen epitopes. Alternatively, class II Tc may further recognize sequential determinants. Hence, inactivated influenza or purified HA will sensitize cells as targets for class II cytotoxic lymphocytes. The latter efficiently eliminate infectious virus from the respiratory tract, despite the absence of class II determinants on respiratory epithelium. The role of soluble mediators in this regard needs further investigation.[34]

There also exist class II helper-amplifier cells for cytotoxic T lymphocytes (TcH).[51] They appear in the lungs within 3 days after exposure and may be equally cross-reactive in antigen recognition. Thus, the T-helper cells achieve peak activity quickly and may persist for 3 weeks, exhibiting specificity for the homologous virus, within the virus subtype, or between strains of the same type of virus.[38,61] Determinants are different from those of B cells, however.[62]

Delayed-type hypersensitivity lymphocytes (Td) appear 5–7 days after influenza challenge and are cross-reactive in their recognition of virus subtypes.[34,63] Similarly, Td helpers (TdH) peak 7–9 days after challenge in a secondary, recall response. The presence of Td cells is directly proportional to the degree of pulmonary consolidation and the dose of virus exposure. However, their presence is inversely proportional to the degree of Tc activity.[63,64] DTH activity is associated *in vitro* with lymphocyte proliferation after exposure to purified viral antigen and appropriate MHC restriction elements.[63] Td cells may be class I or class II histocompatibility antigen restricted in their recognition of infectious virus. However, Td cells exhibit class II restriction alone in the recognition of inactivated virus.[34] Class II Td cells have been associated with increased mortality in murine models.[34] Furthermore, natural infection leads to increased numbers of DTH suppressor cells (TdS) within 2–7 weeks.[41,63] Td suppressors correlate with decreased mortality and a fall in Td activity. This occurs less predictably following immunization with inactivated virus.[63] It appears that helper or suppressor epitopes may associate preferentially with immune response gene products which, respectively, facilitate help or suppression of the Td response.

In regard to viral immunosuppression, several groups have reported depression of mitogen- or antigen-stimulated human lymphocyte proliferative responses and DTH skin-testing responses in patients with influenza virus infection.[12,65-67] Initially, influenza virus was believed to affect lymphocyte proliferative responses to mitogens and specific antigens directly. However, it has been shown more recently that influenza virus-infected human lymphocyte proliferative responsiveness is preserved. The depressed lymphocyte proliferative responses are due to virus-induced alterations in peripheral blood monocyte–macrophage accessory cell function, which is required for such responses.[14,70,71] Nonetheless, proliferative responses to the challenging virus are concomitantly stimulated.[72]

### 3.3. Macrophages

Several aspects of monocyte–macrophage function are altered by influenza virus infection *in vivo* and *in vitro*, including accessory cell function, chemotaxis, phagocytosis, and bactericidal activity. *In vivo*, volunteers exposed to infectious virus have shown decreased leukocyte proliferation, as seen in natural infections, whereas volunteers exposed to inactivated (vaccine) virus showed no alteration of leukocyte proliferative activity.[12] Similar results, and more extensive studies, using *in vitro* infection have suggested that such alterations in immune responsiveness are due to influenza virus interactions with monocytes–macrophages. *In vitro*, exposure to infectious virus but not to inactivated influenza virus results in depressed mitogen- and nonviral antigen-stimulated human peripheral blood lymphocyte proliferation due to altered monocyte–macrophage accessory cell function, with lymphocyte function preserved.[14,71] However, the accessory cell function of autologous human alveolar macrophages does not appear to be altered by exposure to the virus.[73] Furthermore, the depressed peripheral blood mononuclear leukocyte proliferative responses to mitogens and nonviral antigens after *in vitro* infection is associated concomitantly with active leukocyte proliferation in response to the challenging virus itself, suggesting selectivity in the alteration of macrophage accessory cell function.[72]

In animal models, influenza virus infection has been shown to depress *in vitro* macrophage chemotaxis and *in vivo* macrophage accumulation at sites of insult,[74] as well as *in vivo* phagocytosis[75] and killing[75,76] of microorganisms, including bacteria. Others, however, have reported that mouse alveolar macrophages, exposed to the virus, did not show depressed phagocytosis or bactericidal activity despite evidence (by surface-expressed hemagglutinating activity) of viral infection.[77] Different results could be due to differences in timing of measurements of the various functions after exposure to the virus.[6]

Chemotaxis of human monocytes is depressed after *in vivo*[78] and *in vitro*[79,80] influenza virus infection. Influenza virus infection of human monocytes and macrophages *in vitro* has also been shown to result in depressed phagocytosis[14,79] and microbicidal activity,[79] although phagocytosis of staphylococci was enhanced after exposure to virus at a low (0.1) multiplicity of infection.[14]

Concomitant with these alterations in monocyte–macrophage function, exposure to influenza virus results in active production, by the macrophages, of potentially immunomodulatory factors, such as interleukin-1 (IL-1) and interferon (IFN).[13,81−83] Furthermore, monocytes–macrophages are not permissive for influenza virus, rapidly taking up and effectively inactivating the infectious virus upon encounter.

## 3.4. Natural Killer Cells and ADCC

Splenic leukocytes, possibly both adherent monocytes–macrophages and nonadherent natural killer (NK) cells, from mice that were antibody negative for influenza virus have been shown to be spontaneously cytotoxic for influenza virus-infected fibroblast target cells.[84] Shortly after intratracheal infection of mice with influenza virus, NK activity has been stimulated in the lungs,[85,86] but not in the spleen.[85] NK cell activity has been shown to increase in humans shortly after influenza virus infection,[87] and in vitro addition of influenza virus to human lymphocytes did not abrogate NK cell activity or antibody-dependent cell-mediated cytotoxicity (ADCC) in standard assays.[66]

Influenza virus-induced anti-HA antibody has been reported to be produced in vitro after recent in vivo natural infection and has been shown to support cytotoxic responses of leukocytes to virus-infected target cells (ADCC) during that postchallenge period.[88] Other studies have shown that the ADCC could be mediated by lymphocytes, monocytes, and neutrophils (human cord blood-derived as well as adult).[89]

## 3.5. Miscellaneous Cells

### 3.5.1. Polymorphonuclear Leukocytes

Increased polymorphonuclear leukocytes (PMNL) circulating in the peripheral blood have been demonstrated early after influenza virus infection,[90] and there is evidence that PMNL can participate in antiviral defense.[91] Various aspects of PMNL function have been reported to be affected, often adversely, with in vivo infection or in vitro exposure to influenza virus. For example, human PMNL random locomotion,[92] chemokinesis,[92] chemotaxis,[93,94] phagocytosis,[93] lysosome–phagosome fusion,[95] and bactericidal activity[95] have all been reported to be depressed after infection. In vivo exposure to influenza virus can result in depressed rat neutrophil exudation in response to an inflammatory stimulus.[96]

Chemiluminescence by human PMNL, measured because of its association with the ability to generate an oxidative respiratory burst and with the bactericidal capacity of the cells, has been shown to be depressed (in response to standard stimuli) after exposure to the virus, although the virus itself induces chemiluminescence.[97,98] The virus-induced activation of chemiluminescence has been associated with serum factors, notably being correlated with serum HA-inhibiting antibody titer,[99] possibly implicating immune complex formation in the response. However, others have reported that human PMNL respond to influenza virus with increased oxygen consumption, generation of chemiluminescence, and production of superoxide in the absence of serum.[100]

### 3.5.2. Fibroblasts and Other Cells

Influenza virus infection may affect immune defenses by infecting ciliated and mucus-secreting cells as well as cells lining the alveolar spaces,[3,6,101] leading to decreased nonspecific (often mechanical) defenses and an increased burden on leukocyte functions. Influenza virus-infected fibroblasts have been shown to be more susceptible to macrophage-mediated cytostasis than are normal fibroblasts[102] and would be expected to be susceptible to cytotoxic anti-influenza leukocyte effector cells' responses.

## 3.6. Soluble Factors

### 3.6.1. Interferons

Influenza virus has been demonstrated to be an effective inducer of interferon-$\alpha$ ($IFN_\alpha$) production both in vivo[103] and in vitro.[13,104] Interferon is commonly found in the serum and nasal washes of patients infected with influenza virus.[103,105–107] IFN has also been detected in the lung lavage fluid of mice challenged intranasally with influenza virus and in the supernatant fluids of cultured leukocytes obtained by lavage.[108] Furthermore, IFN induced by influenza virus can inhibit influenza viral replication.[109] Influenza virus has recently been shown to be virtually as sensitive as the standard reference assay virus (vesicular stomatitis virus, chosen in part for sensitivity) to the antiviral effects of influenza virus-induced human macrophage-derived IFN.[81]

### 3.6.2. Interleukins

Influenza virus has long been recognized to be an effective inducer of IL-1 production, the latter detected in early animal model studies by endogenous pyrogen activity.[110–112] In vitro influenza virus infection of human macrophages induces production of IL-1 as well as of IL-1-inhibitor(s), with the net activity being an enhancement of mouse thymocyte proliferation in the standard IL-1 assay.[83] It remains to be determined whether the IL-1 and IL-1 inhibitor have similar activities when assayed for effects on human lymphocyte proliferative response to mitogens and nonviral antigens, as well as influenza virus antigens specifically. IL-1 inhibitors may be major candidates for mediators of virus-induced immunosuppression, but data to substantiate such a role are currently unavailable.

### 3.6.3. Other Factors

Influenza virus infection of human peripheral blood-derived macrophages and, to a lesser extent, alveolar macrophages *in vitro*, has been shown to induce cellular production of a factor(s) that stimulate human fibroblast proliferation.[113] In those studies, exposure to the virus itself (with a multiplicity of infection equivalent to that for the macrophages) did not alter fibroblast proliferation, suggesting that effects were due to the macrophage-derived factors rather than to viral or subviral components. Of potential relevance is the observation that primary cultures of human lung cells, exposed *in vitro*, could be persistently infected with influenza virus, and showed release of virus following subcultivation, i.e., during a period of greater proliferative activity.[114]

## 4. MECHANISMS OF IMMUNOSUPPRESSION

### 4.1. Viral Replication within the Immune System

Influenza virus infection of PMNL is abortive, with added virus effectively inactivated.[115] Similarly, both monocytes–macrophages and lymphocytes can be infected by the virus, but the infection is again abortive.[3,116–122] No infectious virus is released by the cells, and virus added to the cells is rapidly inactivated. Nonetheless, influenza virus-specific proteins, including HA and NA, are produced and expressed on the surface of the infected leukocytes.[118,123] Such surface-expressed viral proteins can serve as target structures for developing antiviral responses.[124]

The cell-expressed viral proteins do appear to be a result of actual infection, being newly synthesized by the monocytes–macrophages and lymphocytes and not merely input viral proteins processed and presented by monocytes to lymphocytes.[125] Earlier studies also suggested that peripheral blood leukocyte cultures that had not been transformed (induced to proliferate) with a mitogen did not support the synthesis of viral proteins.[126–128] However, more recent studies have established that both resting and proliferating human lymphocytes can be infected by influenza virus, although in each case the infection is dependent on the presence of monocytes–macrophages.[118,125,129]

### 4.2. Potentially Immunosuppressive Virus-Induced Host Responses

Various local and systemic features of the immunologic response to influenza virus infection could contribute to virus-induced immunosuppression. However, virus-induced changes in one compartment, such as the peripheral blood, may be quite unrepresentative of changes in another, such as the lung. For example, influenza virus infection is associated with an acute influx of monocytes, lymphocytes, and neutrophils into the lavageable lung compartment.[130,131] It should be noted that such recruitment of monocytes–macrophages occurs despite a demonstrated decrease in monocyte chemotaxis in response to other stimuli after *in vivo* or *in vitro* influenza virus infection.

Cellular immune function is suggested to be an integral part of host defense against influenza virus infection by the finding of lymphocyte infiltration during influenza pneumonia.[130,131] A major component of the lymphocytic response has been shown to be the generation, or recruitment, and action of antiviral Tc. However, DTH has also been detected in assays using cells from lung, hilar lymph nodes, and spleen of hamsters following intratracheal influenza virus inoculation[86] and using cells recovered from the lungs of mice after intranasal challenge with infectious influenza virus.[64] The cells were shown to produce DTH responses on transfer to naive animals. Notably for consideration of virus-induced immunosuppression, suppressor T lymphocytes were also generated during influenza virus infection of the mice.[64] The cells inhibited the development of virus-specific class II MHC antigen-restricted DTH responses but did not interfere with protective class I MHC-restricted responses (such as that of Tc) or neutralizing antibody production.[41,132,133] Thus, the cells that can suppress the delayed hypersensitivity response may play a role in the control of, as well as contribution to, the immunopathologic process.

In contrast to the responses to challenge with infectious virus, after exposure of mice to inactivated influenza virus, suppressor cells were induced that inhibited class I MHC-restricted T-cell responses, such as cytotoxic activity, which may contribute to the immunopathology.[40,132] Other studies have suggested that recognition of external (HA and NA) as well as internal (M and NP) viral proteins can contribute to both generation and function (by inhibition of proliferation to virus or viral proteins) of murine influenza virus-specific suppressor cells.[134]

Also in regard to potential viral immunosuppression, it should be noted that, after previous challenge, exposure to the same subtype of virus can cause a drop in total cytotoxic lymphocyte response. By contrast, exposure to viruses of different subtypes will cause an increase in (presumably cross-reactive) cytotoxic activity. This observation may be related either to the presence of suppressor cells and/or neutralizing antibody.[40,45]

Although the acute inflammatory response to influenza virus challenge has been recognized for some time, only recently have experimental data emerged to suggest that a more chronic inflammatory response may ensue after infection. Studies have shown, using labeled inactivated virus preparations, that viral antigens can persist in the lungs within antigen-presenting cells, and can stimulate lymphocytes.[135] The persisting immunogenic stimulus could lead to more effective establishment of local immunity or, conversely, to prolonged immunopathology. Thus, after infection of mice, pulmonary leukocytosis may persist for as long as a year, with histologic evidence of patchy interstitial pneumonia, deposition of collagen in the affected areas, and marked hyperplasia of bronchial-associated lymphoid tissue.[136] Although infectious virus could not be recovered after 9 days, viral antigens persisted at high concentrations in the lungs.

Immunoglobulin-secreting cells and antiviral antibodies, specifically, have also been detected, by immunofluorescence assays, in the lungs of mice long after aerosol challenge with influenza virus.[137,138] Cells producing IgA and IgM were detected early (3 days) after challenge, soon followed by IgG-producing cells. The cells appeared in two principal locations, i.e., along major airways and in consolidated lesions within the lung parenchyma. Cells producing IgG and IgA persisted after the decline in IgM-producing cells, at approximately 30 days. Virus-binding antibodies, specifically, were shown still to be present for all classes on day 33 after challenge.

Such data, combined with the demonstration of influenza virus-induced production of macrophage-derived fibroblast-stimulating factors, [113] raise the possibility that influenza virus infections may play a role in the development of interstitial lung disease. However, interstitial fibrosis is an uncommon recognized sequela of influenza virus infection, and other factors may play an as-yet unrecognized role in enhancing or limiting the clinical development of such a process.[113] Hence, the chronic alveolitis in mice, due to influenza virus infection, did not result in progressive generalized interstitial fibrosis.[136]

Various additional aspects of the immune response, which can be common with such acute as well as chronic inflammation, could contribute to the pathologic manifestations of influenza virus infection. Immune complexes of detectable size are induced by influenza virus infection (of mice) during the interface between antigen excess and antibody excess conditions, and the influenza virus-induced immune complexes have been reported to suppress murine alveolar macrophage phagocytosis.[139] Such suppressive effects may contribute to the phagocytic dysfunction that occurs 7–10 days after influenza pneumonia in the animals.

Neutrophils, although not extensively recruited by an influenza virus challenge (compared with bacterial challenge), could be the source of various proteinases, produced in the process of responding to and eliminating the offending material. Normal lung tissues and both recruited and resident immunocompetent cells may suffer unwarranted damage due to the same proteases.

Several soluble immunomodulatory factors, induced by influenza virus, have reasonably been considered as possible contributors to viral immunosuppression. For example, IFN has well-documented antiproliferative activity in many systems, including demonstrated ability to depress mitogen-stimulated lymphocyte proliferation.[140,141] However, several lines of evidence suggest that IFN production by peripheral blood-derived monocytes–macrophages[13] per se cannot explain the monocyte–macrophage-mediated depression of mitogen-stimulated human lymphocyte proliferation.[14] For example, IFN production was induced by exposure of human macrophages to either infectious or inactivated influenza virus, but only macrophages exposed to infectious virus exhibited depression of accessory cell support for lymphocyte proliferation, and the depression was not clearly related to the titer of IFN produced.[73] Furthermore, neutralization of IFN, by addition of appropriate antisera to cultures, did not return macrophage-supported lymphocyte proliferative responses to normal after exposure to the virus. Similarly, prostaglandin production, induced by exposure to the virus, has warranted consideration as an immunosuppressive factor. However, evidence also suggests that prostaglandin production per se cannot account for altered monocyte–macrophage support of mitogen-stimulated lymphocyte proliferation after exposure to influenza virus. Responses of influenza virus-exposed human leukocytes were not returned to normal by addition of indomethacin, catalase, or antiprostaglandin antibody to the cultures.[73] Immunomodulatory factors remain strong candidates as mediators of virus-induced immunosuppression, notably the influenza virus-induced human macrophage-derived IL-1 inhibitor, which has recently been described.[83] Such a factor(s) could suppress mitogen-stimulated mouse thymocyte proliferation, although concomitant IL-1 production resulted in net enhancement of the thymocyte response. The specificity or nonspecificity of the inhibitor, especially in regard to human lymphocyte proliferation, warrants further investigation. It is possible that several of the factors induced by influenza virus produce their effects by actions on certain cells (e.g., even the producing macrophages themselves), which in turn mediate immunosuppression.

# 5. SIGNIFICANCE OF VIRUS-INDUCED IMMUNOMODULATION IN VIRAL PATHOGENESIS

## 5.1. Immunologic Participation in Observed Pathology

A vast array of local, recruited, and systemic immunological responses can play a concerted and integrated role in anti-influenza virus defense, including intrinsic antiviral activity of different leukocyte populations (uptake and inactivation of the virus), as well as various extrinsic antiviral responses, such as neutralizing antibody production, ADCC, antibody-independent specific and nonspecific (or natural) cell-mediated cytotoxicity, and production of various antiviral and immunomodulatory factors, such as IFN and IL.

The same immunologic responses that ensure recovery from the viral challenge may contribute to the characteristics or extent of clinical expression of the viral disease. This has been demonstrated by several groups of investigators, using various protocols to render animals immunodeficient, resulting in prolonged viral replication and decreased lung pathology.[142,143] In the absence of immunologic responses, death of the animals commonly results from virus dissemination and involvement of other sites, such as the CNS.[144] For example, mice treated with nontoxic doses of antilymphocyte globulin showed decreased lung consolidation and even improved survival after challenge with influenza virus under controlled experimental conditions.[143] By contrast, others have shown murine mortality rates to increase after cyclophosphamide-induced immunosuppression, with no significant delay in mortality.[144] Other studies have used analyses of active cells present in the challenged animals' lungs, or passive transfer of effector cells, or comparisons of influenza virus-induced pneumonia in normal and lymphocyte-deficient animals.[145–147] Overall, there is evidence that specific T-lymphocyte responses contribute not only to anti-influenza virus defense, but also to the severity of lung pathology.[64]

Indications of systemic immunosuppression may coexist with evidence of an active local and recruited immune response to the virus challenge itself. The most prominent resident lung leukocyte population in the unchallenged state is the alveolar macrophage.[73] However, after influenza virus challenge of mice, it has been shown clearly that numerous additional leukocytes are recruited to the lungs, with the predominant population being monocytes–macrophages, but with large numbers of lymphocytes, as well as a lesser percentage of PMNL.[130,131] Patients challenged with influenza virus have long been known to demonstrate a peripheral blood lymphocytopenia soon after infection.[90] The animal models have shown that this peripheral or systemic decrease in lymphocytes can correlate with an increase in the number of those cells at the primary site of virus challenge. Thus, the mean number of lymphocytes in the peripheral blood was decreased after challenge of mice with influenza virus, whereas the mean number of lymphocytes in the lungs was increased, with peak lymphocyte numbers reached in the lung when pneumonia was most marked and peripheral blood lymphocyte counts had decreased by 50%.[148] The Tc in the lung were shown to be specific for the HA of the infecting virus. Tc were also detected in cervical lymph nodes, spleen, and peripheral blood of the infected mice, and the cytotoxic activity decreased with resolution of the pneumonia. Such data appear to indicate that specific anti-influenza virus Tc in the major infected organ, the lung, are part of the immunologic and pathologic response to the virus. Similar observations have been made by others.[4,46] Virus-specific cytotoxic cells have been detected in the lung but not in the hilar lymph nodes or spleen of hamsters following intratracheal influenza virus inoculation.[86] Further studies indicated that transfer of induced Tc could protect mice challenged with lethal influenza viruses, reducing lung virus titers and enhancing recovery from the infection and resolution of pneumonia.[47,146,149]

Other investigators have examined local accumulation of anti-influenza antibody-producing B lymphocytes in pertinent tissues. For example, after influenza virus challenge of ferrets, the number of antibody-producing cells (measured by a standard plaque-forming cell assay) increased in the lung wash, as well as lymph nodes draining the upper and lower respiratory tract.[150] Of note, T-lymphocyte proliferative responses to the challenge virus were also measured and found to develop downstream from the major site of infection, but not in the lung wash cell population itself.

In addition to contributing to viral pathogenesis, influenza virus-induced effects on immunological functions may also contribute to secondary manifestations of influenza virus infection. Such observations do not directly demonstrate a role for influenza virus-induced immunomodulation in viral pathogenesis. However, secondary bacterial pulmonary infections are often cited as evidence of potentially deleterious immunosuppression produced by influenza virus. Influenza virus infection has long been recognized to be associated with secondary bacterial infections of both humans and laboratory animals; such sequelae include bacterial pulmonary infections relatively uncommon in non-influenzal settings.[151] Furthermore, such virus-induced immunomodulation can also be manifested by local effects that are not fully reflected by determinations of systemic immune functions after influenza virus challenge. For example, pulmonary, but not splenic or liver, bactericidal activity against staphylococci (administered intravascularly) was depressed by influenza virus infection of mice, and more staphylococci persisted in the lungs of influenza virus-infected animals.[76] Defense against challenge with airborne staphylococci was also suppressed by the viral infection. It should be noted that others have shown that pulmonary infection of mice with influenza virus can depress or enhance resistance to a second (bacterial) infection depending on the timing of bacterial challenge after virus infection.[152]

## 5.2. Genetic Contributions

Both viral and host genetic factors can affect the degree to which virus-induced immunomodulation might contribute to the pathogenesis of observed viral diseases, and this has been documented in particular in the case of influenza virus. Different influenza virus subtypes or strains can differ in the magnitude of effects on some immunologic functions and in the degree of sequelae potentially arising due to those effects. Variability in such observations may be related to differences in immunologic functions measured, or species exposed. For example, H1N1 and H3N2 strains of virus did not differ in ability to alter human macrophage accessory cell function or induce interferon production after in vitro infection.[71] Seropositivity to the influenza virus subtype or specific strain did not appear to modify the infectious virus-induced alterations of human macrophage function. However, H1N1 and H3N2 strains have been shown to have differential effects (H3N2 greater than H1N1) on PMNL chemiluminescence, clearance of pneumococci from nasal washings, and development of pneumococcal otitis media after in vivo influenza virus infection of chinchillas.[153] In rats, it has been shown that different influenza virus strains differ in ability to agglutinate thoracic duct lymphocytes.[154]

The influence of histocompatibility (HLA, class I, and class II) determinants on host interaction with influenza viruses has been well established with measurements of several aspects of the immune response. For example, genetic constitution, measured by assays using patients with varying HLA determinants, can affect the presence or magnitude of anti-influenza cytotoxic[155] or antibody responses.[156,157] Primed human lymphocyte responses to influenza virus may be restricted to class II HLA determinants different, in an individual, from those involved in responses to alternate viruses.[158] Genetic susceptibility or resistance of a host to influenza virus infection has been shown in animal models to correlate with macrophage susceptibility or resistance to virus replication, and to correlate in turn with host survival.[159]

## 5.3. Immunofocusing and the Selectivity of Immunomodulation

The altered leukocyte functions may actually represent a focusing of activity by the cells, which results in the depression of some immunologic responses combined with the activation or augmentation of other responses deemed more beneficial to the host in the setting of acute influenza virus challenge. Thus, several observations suggest that many features of influenza virus infection commonly viewed as indications of virus-induced immunosuppression warrant reconsideration, as features indicative of a virus-induced but host-directed immunofocusing rather than viral suppression of defense.[13]

First, influenza virus infection commonly resolves without significant residual disease, and with emergence of homotypic immunity. Such a course suggests either that the observed depressions of immune responses are either not deleterious to any significant degree or that they are in fact actually beneficial to the host. The depressed immune responses are so consistently observed in the setting of the infection that it can reasonably be suggested that the changes have emerged through natural selection, as opposed to inconstant features, or features never seen and supposedly selected against (as deleterious) as the species evolves. It is, however, possible that the consistent alterations of immune response in fact represent the maximum adverse effects which can be determined by the virus, as it passes from individual to individual, without resulting in extinction of the host. The relative constancy of immunological alterations across various species which have been studied in depth also argues for such a concept of potential benefit to the host of the virus-induced immunomodulations.

Second, even a single type of leukocyte, for which virus-induced depression of an immunological function has been demonstrated, can be shown to have preservation and even activation or augmentation of an alternate function. The latter function may reasonably be presumed to be of greater benefit to the host in the setting of the acute influenza virus challenge. For example, *in vitro* influenza virus infection of human peripheral blood-derived macrophages results in their decreased accessory cell support for mitogen-stimulated lymphocyte proliferation,[14] yet concomitantly induces active macrophage production of IFN.[13] Even macrophage accessory cell function is selectively altered. Thus, concomitant with decreased support of lymphocyte proliferative responses to mitogens and nonviral antigens, the macrophages fully support proliferative responses to the virus itself.[72] Lymphocyte blastogenic (proliferative) responses to influenza antigens have been detected in cultures established early after infectious virus challenge of volunteers.[12] Furthermore, the addition of (infectious but not inactivated) influenza virus to human leukocytes *in vitro* resulted in decreased mitogen-stimulated Ig synthesis.[66] By contrast, there is extensive evidence for production of antiviral antibodies in the setting of influenza virus infection, and such specific anti-influenza antibody production is macrophage dependent.[160] There would seem to be a relative focusing even of each of several functional activities of the macrophages after exposure to influenza virus.

Third, even a response that has been shown to be depressed systemically after influenza virus infection, using standard assays, may not be inappropriately depressed in regard to the immediate and/or local viral challenge. For example, in the mouse model, influenza virus infection results in a significant decrease in monocyte–macrophage chemotaxis using standard assays,[74] yet is associated with an effective local (pulmonary) recruitment, manifested by a marked increase in lavageable monocytes–macrophages after challenge.[130,131]

Fourth, even certain of the immunomodulatory effects of influenza virus infection commonly considered immunosuppressive could have consequences potentially beneficial to the host. The consequences would only have to have net adaptive value to be conserved through natural selection (i.e., the species would benefit, although an individual host encountering a second challenge may be at a disadvantage). For example, depressed macrophage accessory cell function for various (nonviral) lymphocyte proliferative responses might be inferred to have net adaptive value, since proliferating lymphocytes, while not productively infected, may support the synthesis of influenza viral proteins better than do resting lymphocytes.[126] Human lung cells, persistently infected with influenza virus, showed release of virus following subcultivation, i.e., during a period of greater proliferative activity.[114] Thus, an active host suppression of nonviral specific proliferation (i.e., other than that of anti-influenza virus-specific lymphocytes) could play a role in decreasing host tissue permissiveness for viral replication, whether productive or abortive in the various tissues.

The interactions of influenza virus and host immunologic systems would appear to reflect a relationship that allows persistence of both the pathogen and the host. It is as yet unclear whether any aspects of immunosuppression induced by influenza virus truly potentiate the persistence of the virus itself.[3] While influenza virus remains a cause of substantial morbidity and mortality, further examination of immunoregulation in the setting of influenza virus challenge is likely to reveal much information regarding the constitution of a generally effective immune defense.

ACKNOWLEDGMENTS. This work was supported by grant AI 15547 from the National Institute of Allergy and Infectious Diseases and by a Wilmot Cancer Research Fellowship awarded to Frank Domurat.

# REFERENCES

1. Davenport, F. M., Influenza viruses, in: *Viral Infections of Humans: Epidemiology and Control* (A. S. Evans, ed.), pp. 372–396, Plenum, New York (1982).
2. Kilbourne, E. D., Influenza as a problem in immunology, *J. Immunol.* **120:**1447–1452 (1978).
3. Sweet, C., and H. Smith, Pathogenicity of influenza virus, *Microbiol. Rev.* **44:**303–330 (1980).
4. Ennis, A., Some newly recognized aspects of resistance against and recovery from influenza, *Arch. Virol.* **73:**207–217 (1982).
5. Mitchell, D. M., A. J. McMichael, and J. R. Lamb, The immunology of influenza, *Br. Med. Bull.* **41:**80–85 (1985).
6. Couch, R. B., The effects of influenza on host defenses, *J. Infect. Dis.* **144:**284–291 (1981).

7. Friedman, H., S. Specter, and M. Bendinelli, Influence of viruses on cells of the immune response system, in: *Host Defenses to Intracellular Pathogens* (T. K. Eisenstein, P. Actor, and H. Friedman, eds.), pp. 463–474, Plenum, New York (1983).

8. Kirn, A., and F. Keller, How viruses may overcome non-specific defences in the host, *Philos. Trans. R. Soc. Lond. B* **303**:115–122 (1983).

9. Mims, C. A., Immunopathology in virus disease, *Philos. Trans. R. Soc. Lond. B* **303**:189–198 (1983).

10. Nash, A. A., Tolerance and suppression in virus diseases, *Br. Med. Bull.* **41**:41–45 (1985).

11. Ohmann, H. B., and L. A. Babiuk, Viral infections in domestic animals as models for studies of viral immunology and pathogenesis, *J. Gen. Virol.* **66**:1–25 (1986).

12. Dolin, R., B. R. Murphy, and E. A. Caplan, Lymphocyte blastogenic responses to influenza virus antigens after influenza infection and vaccination in humans, *Infect. Immun.* **19**:867–874 (1978).

13. Roberts, N. J., Jr., R. G. Douglas, Jr., R. L. Simons, and M. E. Diamond, Virus-induced interferon production by human macrophages, *J. Immunol.* **123**:365–369 (1979).

14. Roberts, N. J., Jr., and R. T. Steigbigel, Effect of *in vitro* virus infection on response of human monocytes and lymphocytes to mitogen stimulation, *J. Immunol.* **121**:1052–1058 (1978).

15. Murphy, B. R., and R. G. Webster, Influenza viruses, in: *Virology* (B. N. Fields, D. M. Knipe, R. M. Chanock, J. L. Melnick, B. Roizman, and R. E. Shope, eds.), pp. 1179–1239, Raven, New York (1985).

16. Palese, P., and J. F. Young, Variation of influenza A, B, and C viruses, *Science* **215**:1468–1474 (1982).

17. Webster, R. G., W. G. Laver, G. M. Air, and G. C. Schild, Molecular mechanisms of variation in influenza viruses, *Nature (Lond.)* **296**:115–121 (1982).

18. Rott, R., Molecular basis of infectivity and pathogenicity of myxovirus: Brief review, *Arch. Virol.* **59**:285–298 (1979).

19. Couch, R. B., and J. A. Kasel, Immunity to influenza in man, *Annu. Rev. Microbiol.* **37**:529–549 (1983).

20. Morris, J. A., J. A. Kasel, M. Saglam, V. Knight, and F. A. Loda, Immunity to influenza as related to antibody levels, *N. Engl. J. Med.* **274**:527–535 (1966).

21. Virelizier, J.-L., Host defenses against influenza virus: The role of anti-hemagglutinin antibody, *J. Immunol.* **115**:434–439 (1975).

22. Ramphal, R., W. Fischlschweiger, J. S. Shands, and P. A. Small, Jr., Murine influenzal tracheitis: A model for the study of influenza and tracheal epithelial repair, *Am. Rev. Respir. Dis.* **120**:1313–1324 (1979).

23. Murphy, B. R., D. L. Nelson, P. F. Wright, E. L. Tierney, M. A. Phelan, and R. M. Chanock, Secretory and systemic immunological response in children infected with live attenuated influenza A virus vaccines, *Infect. Immun.* **36**:1102–1108 (1982).

24. Virelizier, J.-L., Humoral immune response to viruses in humans, in: *Human Immunity to Viruses* (F. A. Ennis, ed.), pp. 71–79, Academic, New York (1983).

25. Burlington, D. B., M. L. Clements, G. Meiklejohn, M. Phelan, and B. R. Murphy, Hemagglutinin-specific antibody responses in immunoglobulin G, A and M isotypes as measured by enzyme-linked immunosorbent assay after primary or secondary infection of humans with influenza A virus, *Infect. Immun.* **41**:540–545 (1983).

26. Wright, P. F., B. R. Murphy, M. Kervina, E. M. Lawrence, M. A. Phelan, and D. T. Karzon, Secretory immunological response after intranasal inactivated influenza A virus vaccinations: Evidence for immunoglobulin A memory, *Infect. Immun.* **40**:1092–1095 (1983).

27. Waldman, R. H., P. F. Jurgensen, G. N. Olsen, R. Ganguly, and J. E. Johnson III, Immune response of the human respiratory tract. I. Immunoglobulin levels and influenza virus vaccine antibody response, *J. Immunol.* **111**:38–41 (1973).

28. Scott, G. H., and R. J. Sydiskis, Responses of mice immunized with influenza virus by aerosol and parenteral routes, *Infect. Immun.* **13**:696–703 (1976).
29. Mitchell, D. M., P. Fitzharris, R. A. Knight, and G. C. Schild, Kinetics of specific *in vitro* antibody production following influenza immunization, *Clin. Exp. Immunol.* **48**:491–498 (1982).
30. Yarchoan, R., and D. L. Nelson, Specificity of *in vitro* anti-influenza virus antibody production by human lymphocytes: Analysis of original antigenic sin by limiting dilution cultures, *J. Immunol.* **132**:928–935 (1984).
31. Wylie, D. E., L. A. Sherman, and N. R. Klinman, Participation of the major histocompatibility complex in antibody recognition of viral antigens expressed on infected cells, *J. Exp. Med.* **155**:403–414 (1982).
32. Butchko, G. M., R. B. Armstrong, W. J. Martin, and F. A. Ennis, Influenza A viruses of the H2N2 subtype are lymphocyte mitogens, *Nature (Lond.)* **271**:66–67 (1978).
33. Scalzo, A. A., and E. M. Anders, Influenza viruses as lymphocyte mitogens. I. B cell mitogenesis by influenza A viruses of the H2 and H6 subtypes is controlled by the I-E/C subregion of the major histocompatibility complex, *J. Immunol.* **134**:757–760 (1985).
34. Ada, G. L., and P. D. Jones, The immune response to influenza virus infection, in: *Options for the Control of Influenza* (A. P. Kendal and P. A. Patriarca, eds.), pp. 107–124, Liss, New York (1986).
35. Fischer, A., S. Nash, P. C. L. Beverly, and M. Feldmann, An influenza virus matrix protein-specific human T cell line with helper activity for *in vitro* anti-hemagglutinin antibody production, *Eur. J. Immunol.* **12**:844–849 (1982).
36. Virelizier, J.-L., P. Postlethwaite, G. C. Schild, and A. C. Allison, Antibody responses to antigenic determinants of influenza virus hemagglutinin. I. Thymus dependence of antibody formation and thymus independence of immunological memory, *J. Exp. Med.* **140**:1559–1570 (1974).
37. Callard, R. E., G. W. McCaughan, J. Babbage, and R. L. Souhami, Specific *in vitro* antibody responses by human blood lymphocytes: Apparent nonresponsiveness of PBL is due to a lack of recirculating memory B cells, *J. Immunol.* **129**:153–156 (1982).
38. Ashman, R. B., and A. Mullbacher, A T helper cell for anti-viral cytotoxic T-cell responses, *J. Exp. Med.* **150**:1277–1282 (1979).
39. Ennis, F. A., W. J. Martin, M. W. Verbonitz, and G. M. Butchko, Specificity studies on cytotoxic thymus-derived lymphocytes reactive with influenza virus-infected cells: Evidence for dual recognition of H-2 and viral hemagglutinin antigens, *Proc. Natl. Acad. Sci. USA* **74**:3006–3010 (1977).
40. Leung, K. N., R. B. Ashman, H. C. J. Ertl, and G. L. Ada, Selective suppression of the cytotoxic T cell response to influenza virus in mice, *Eur. J. Immunol.* **10**:803–810 (1980).
41. Liew, F. Y., and S. M. Russell, Delayed-type hypersensitivity to influenza virus: Induction of antigen-specific suppressor T cells for delayed-type hypersensitivity to haemagglutinin during influenza virus infection in mice, *J. Exp. Med.* **151**:799–814 (1980).
42. Reiss, C. S., and S. J. Burakoff, Specificity of the helper T cell for the cytolytic T lymphocyte response to influenza viruses, *J. Exp. Med.* **154**:541–546 (1981).
43. Biddison, W. E., and S. Shaw, Differences in HLA antigen recognition by human influenza virus-immune cytotoxic T cells, *J. Immunol.* **122**:1705–1709 (1979).
44. Effros, R. B., P. C. Doherty, W. Gerhard, and J. Bennink, Generation of both cross-reactive and virus-specific T-cell populations after immunization with serologically distinct influenza A viruses, *J. Exp. Med.* **145**:557–568 (1977).
45. Greenspan, N. S., D. H. Schwartz, and P. C. Doherty, Role of lymphoid cells in immune surveillance against viral infection, in: *Advances in Host Defense Mechanisms*, Vol. 2: *Lymphoid Cells* (J. I. Gallin and A. S. Fauci, eds.), pp. 101–141, Raven, New York (1983).
46. Yap, K. L., and G. L. Ada, Cytotoxic T cells in the lungs of mice infected with influenza A virus, *Scand. J. Immunol.* **7**:73–80 (1978).

47. Yap, K. L., G. L. Ada, and I. F. C. McKenzie, Transfer of specific cytotoxic T lymphocytes protects mice inoculated with influenza virus, *Nature (Lond.)* **273**:238–239 (1978).
48. Ogra, P. L., J. C. Cumella, and R. C. Welliver, Immune response to viruses, in: *Immunology of the Lung and Upper Respiratory Tract* (J. Bienenstock, ed.), pp. 242–263, McGraw-Hill, New York (1984).
49. Erard, F., P. Corthesy, and M. Nabholz, Induction of cytolytic T lymphocytes by lectin or alloantigen: Lack of requirement for accessory cells or differentiation factors other than interleukin-2, in: *Cellular and Molecular Biology of Lymphokines* (C. Sorg, A. Schimpl, and M. Landy, eds.), pp. 105–110, Academic, Orlando, Florida (1985).
50. Cambridge, G., J. S. MacKenzie, and D. Keast, Cell-mediated immune response to influenza virus infections in mice, *Infect. Immun.* **13**:36–43 (1976).
51. Askonas, B. A., and P. M. Taylor, Murine cytotoxic T-cells in influenza, in: *Human Immunity to Viruses* (F. A. Ennis, ed.), pp. 137–149, Academic, New York (1983).
52. McMichael, A. J., F. M. Gotch, D. W. Dongworth, A. Clark, and C. W. Potter, Declining T-cell immunity to influenza, 1977–82, *Lancet* **2**:762–764 (1983).
53. Lamb, J. R., and D. D. Eckels, Human T cell recognition of influenza viral antigens and class II MHC gene products, in: *Options for the Control of Influenza* (A. P. Kendal and P. A. Patriarca, eds.), pp. 423–434, Liss, New York (1986).
54. Askonas, B. A., and R. G. Webster, Monoclonal antibodies to hemagglutinin and to H-2 inhibit the cross-reactive cytotoxic T cell populations induced by influenza, *Eur. J. Immunol.* **10**:151–156 (1980).
55. Hackett, C. J., B. A. Askonas, R. G. Webster, and K. van Wyke, Quantitation of influenza virus antigens on infected target cells and their recognition by cross-reactive cytotoxic T cells, *J. Exp. Med.* **151**:1014–1025 (1980).
56. Braciale, T. J., A. E. Lukacher, L. Morrison, V. L. Braciale, G. Smith, B. Moss, M.-J. Gething, and J. Sambrook, Influenza viral antigen recognition by class I and class II MHC restricted cytolytic T lymphocytes, in: *Options for the Control of Influenza* (A. P. Kendal and P. A. Patriarca, eds.), pp. 407–421, Liss, New York (1986).
57. McMichael, A. J., A. Townsend, C. A. Michie, F. M. Gotch, G. Smith, and B. Moss, Virus antigen recognition by influenza A virus specific cytotoxic T-cells, in: *Options for the Control of Influenza* (A. P. Kendal and P. A. Patriarca, eds.), pp. 445–451, Liss, New York (1986).
58. Askonas, B. A., and P. Pala, Cytotoxic T-cell repertoire: Individual variation in influenza A virus nucleoprotein recognition, in: *Options for the Control of Influenza* (A. P. Kendal and P. A. Patriarca, eds.), pp. 453–459, Liss, New York (1986).
59. Townsend, A. R. M., A. J. McMichael, N. P. Carter, J. A. Huddleston, and G. G. Brownlee, Cytotoxic T cell recognition of the influenza nucleoprotein and hemagglutinin expressed in transfected mouse L cells, *Cell* **39**:13–25 (1984).
60. Wraith, D. C., Induction of influenza A virus crossreactive cytotoxic T lymphocytes by purified viral proteins, in: *Options for the Control of Influenza* (A. P. Kendal and P. A. Patriarca, eds.), pp. 461–468, Liss, New York (1986).
61. Leung, K. N., and G. L. Ada, The effect of helper T cells on the primary in vitro production of delayed-type hypersensitivity to influenza virus, *J. Exp. Med.* **153**:1029–1043 (1981).
62. Katz, J. M., W. G. Laver, D. O. White, and E. M. Anders, Recognition of influenza virus hemagglutinin by subtype-specific and cross-reactive proliferative T cells: Contribution of $HA_1$ and $HA_2$ polypeptide chains, *J. Immunol.* **134**:616–622 (1985).
63. Liew, F. Y., Delayed type hypersensitivity to influenza virus, in: *Human Immunity to Viruses* (F. A. Ennis, ed.), pp. 163–175, Academic, New York (1983).
64. Leung, K. N., and G. L. Ada, Cells mediating delayed-type hypersensitivity in the lungs of mice infected with an influenza A virus, *Scand. J. Immunol.* **12**:393–400 (1980).
65. Buckley, C. E., III, M. J. Zitt, and T. R. Cate, Two categories of lymphocyte unresponsiveness to phytohemagglutinin, *Cell. Immunol.* **6**:140–148 (1973).

66. Casali, P., G. P. A. Rice, and M. B. A. Oldstone, Viruses disrupt functions of human lymphocytes: Effects of measles virus and influenza virus on lymphocyte-mediated killing and antibody production, *J. Exp. Med.* **159:**1322–1337 (1984).
67. Jarstrand, C., and J. Wasserman, Mitogen stimulation of lymphocytes from patients with epidemic influenza, *Scand. J. Infect. Dis.* **8:**7–11 (1976).
68. Kantzler, G. B., S. F. Lauteria, C. L. Cusumano, J. D. Lee, R. Ganguly, and R. H. Waldman, Immunosuppression during influenza virus infection, *Infect. Immun.* **10:**996–1002 (1974).
69. Reed, W. P., J. W. Olds, and A. L. Kisch, Decreased skin-test reactivity associated with influenza, *J. Infect. Dis.* **125:**398–402 (1972).
70. Roberts, N. J., Jr., Different effects of influenza virus, respiratory syncytial virus, and Sendai virus on human lymphocytes and macrophages, *Infect. Immun.* **35:**1142–1146 (1982).
71. Roberts, N. J., Jr., M. E. Diamond, R. G. Douglas, Jr., R. L. Simons, and R. T. Steigbigel, Mitogen responses and interferon production after exposure of human macrophages to infectious and inactivated influenza viruses, *J. Med. Virol.* **5:**17–23 (1980).
72. Roberts, N. J., Jr., and J. E. Nichols, Regulation of lymphocyte proliferation after influenza virus infection of human mononuclear leukocytes, *J. Med. Virol.* (in press).
73. Ettensohn, D. B., and N. J. Roberts, Jr., Influenza virus infection of human alveolar and blood-derived macrophages: Differences in accessory cell function and interferon production, *J. Infect. Dis.* **149:**942–949 (1984).
74. Kleinerman, E. S., C. A. Daniels, R. P. Polisson, and R. Snyderman, Effect of virus infection on the inflammatory response: Depression of macrophage accumulation in influenza-infected mice, *Am. J. Pathol.* **85:**373–382 (1976).
75. Warshauer, D., E. Goldstein, T. Akers, W. Lippert, and M. Kim, Effect of influenza viral infection on the ingestion and killing of bacteria by alveolar macrophages, *Am. Rev. Respir. Dis.* **115:**269–277 (1977).
76. Jakab, G. J., G. A. Warr, and M. E. Knight, Pulmonary and systemic defenses against challenge with *Staphylococcus aureus* in mice with pneumonia due to influenza A virus, *J. Infect. Dis.* **140:**105–108 (1979).
77. Nugent, K. M., and E. L. Pesanti, Effect of influenza infection on the phagocytic and bactericidal activities of pulmonary macrophages, *Infect. Immun.* **26:**651–657 (1979).
78. Kleinerman, E. S., R. Snyderman, and C. A. Daniels, Depressed monocyte chemotaxis during acute influenza infection, *Lancet* **2:**1063–1066 (1975).
79. Gardner, I. D., and J. W. M. Lawton, Depressed human monocyte function after influenza infection *in vitro*, *RES: J. Reticuloendothel. Soc.* **32:**443–448 (1982).
80. Kleinerman, E. S., R. Snyderman, and C. A. Daniels, Depression of human monocyte chemotaxis by herpes simplex and influenza viruses, *J. Immunol.* **113:**1562–1567 (1974).
81. Bell, D. M., N. J. Roberts, Jr., and C. B. Hall, Different antiviral spectra of human macrophage interferon activities, *Nature (Lond.)* **305:**319–321 (1983).
82. Roberts, N. J., Jr., R. G. Douglas, Jr., and R. T. Steigbigel, Interferon production by human macrophages, in: *Interferon: Properties and Clinical Uses* (A. Khan, N. O. Hill, and G. L. Dorn, eds.), pp. 85–93, Leland Fikes Foundation, Dallas, Texas (1980).
83. Roberts, N. J., Jr., A. H. Prill, and T. N. Mann, Interleukin 1 and interleukin 1 inhibitor production by human macrophages exposed to influenza virus or respiratory syncytial virus: Respiratory syncytial virus is a potent inducer of inhibitor activity, *J. Exp. Med.* **163:**511–519 (1986).
84. Watanabe, H., and J. S. Mackenzie, Spontaneous cytotoxicity against viral-infected cells: Effects of leucocytes and sera from older mice, *Cell. Immunol.* **70:**180–187 (1982).
85. Stein-Streilein, J., M. Bennett, D. Mann, and V. Kumar, Natural killer cells in mouse lung: Surface phenotype, target preference, and response to local influenza virus infection, *J. Immunol.* **131:**2699–2704 (1983).

86. Stein-Streilein, J., P. L. Witte, J. W. Streilein, and J. Guffee, Local cellular defenses in influenza-infected lungs, *Cell. Immunol.* **95**:234–246 (1985).
87. Ennis, F. A., A. Meagher, A. S. Beare, Q. Yi-Hua, D. Riley, G. Schwarz, G. C. Schild, and A. H. Rook, Interferon induction and increased natural killer-cell activity in influenza infections in man, *Lancet* **2**:891–893 (1981).
88. Greenberg, S. B., H. R. Six, S. Drake, and R Couch, Cell cytotoxicity due to specific influenza antibody production *in vitro* after recent influenza antigen stimulation, *Proc. Natl. Acad. Sci. USA* **76**:4622–4626 (1979).
89. Hashimoto, G., P. F. Wright, and D. T. Karzon, Ability of human cord blood lymphocytes to mediate antibody-dependent cellular cytotoxicity against influenza virus-infected cells, *Infect. Immun.* **42**:214–218 (1983).
90. Douglas, R. G., Jr., R. H. Alford, T. R. Cate, and R. B. Couch, The leukocyte response during viral respiratory illness in man, *Ann. Intern. Med.* **64**:521–530 (1966).
91. Faden, H., and P. Ogra, Neutrophils and antiviral defense, *Pediatr. Infect. Dis.* **5**:86–92 (1986).
92. Ruutu, P., A. Vaheri, and T. U. Kosunen, Depression of human neutrophil motility by influenza virus *in vitro*, *Scand. J. Immunol.* **6**:897–906 (1977).
93. Larson, H. E., and R. Blades, Impairment of human polymorphonuclear leucocyte function by influenza virus, *Lancet* **1**:283 (1976).
94. Schlesinger, J., C. Ernst, and L. Weinstein. Inhibition of human neutrophil chemotaxis by influenza virus, *Lancet* **1**:650–651 (1975).
95. Abramson, J. S., J. C. Lewis, D. S. Lyles K. A. Heller, E. L. Mills, and D. A. Bass, Inhibition of neutrophil lysosome–phagosome fusion associated with influenza virus infection in vitro: Role in depressed bactericidal activity, *J. Clin. Invest.* **69**:1393–1397 (1982).
96. Ruutu, P., Depression of rat neutrophil exudation and motility by influenza virus, *Scand. J. Immunol.* **6**:1113–1120 (1977).
97. Abramson, J. S., D. S. Lyles, K. A. Heller, and D. A. Bass, Influenza A virus-induced polymorphonuclear leukocyte dysfunction, *Infect. Immun.* **37**:794–799 (1982).
98. Busse, W. W., and J. M. Sosman, Altered luminol-dependent granulocyte chemiluminescence during an *in vitro* incubation with an influenza vaccine, *Am. Rev. Respir. Dis.* **123**:654–658 (1981).
99. Shult, P. A., E. C. Dick, K. A. Joiner, and W. W. Busse, Role of serum in stimulation of polymorphonuclear leukocyte, luminol-dependent chemiluminescence by influenza A, *Am. Rev. Respir. Dis.* **131**:267–272 (1985).
100. Mills, E. L., Y. Debets-Ossenkopp, H. A. Verbrugh, and J. Verhoef, Initiation of the respiratory burst of human neutrophils by influenza virus, *Infect. Immun.* **32**:1200–1205 (1981).
101. Loosli, C. G., Influenza and the interaction of viruses and bacteria in respiratory infections, *Medicine (Baltimore)* **52**:369–384 (1973).
102. Goldman, R., and N. Hogg, Enhanced susceptibility of virus-infected fibroblasts to cytostasis mediated by peritoneal exudate cells, *J. Immunol.* **121**:1657–1663 (1978).
103. Jao, R. L., E. F. Wheelock, and G. G. Jackson, Production of interferon in volunteers infected with Asian influenza, *J. Infect. Dis.* **121**:419–426 (1970).
104. Chonmaitree, T., N. J. Roberts, Jr., R. G. Douglas, Jr., C. B. Hall, and R. L. Simons, Interferon production by human mononuclear leukocytes: Differences between respiratory syncytial virus and influenza viruses, *Infect. Immun.* **32**:300–303 (1981).
105. Green, J. A., R. P. Charette, T-J. Yeh, and C. B. Smith, Serum interferon in humans with naturally acquired influenza, in: *Current Chemotherapy and Infectious Disease*, Vol. II (J. D. Nelson and C. Grassi, eds.), pp. 1332–1334, American Society for Microbiology, Washington, D. C. (1980).
106. Hall, C. B., R. G. Douglas, Jr., R. L. Simons, and J. M. Geiman, Interferon production in

children with respiratory syncytial, influenza, and parainfluenza virus infections, *J. Pediatr.* **93**:28–32 (1978).

107. McIntosh, K., Interferon in nasal secretions from infants with viral respiratory tract infections, *J. Pediatr.* **93**:33–36 (1978).

108. Wyde, P. R., M. R. Wilson, and T. R. Cate, Interferon production by leukocytes infiltrating the lungs of mice during primary influenza virus infection, *Infect. Immun.* **38**:1249–1255 (1982).

109. Burke, D. C., and A. Isaacs, Interferon: Relation to heterologous interference and lack of antigenicity, *Acta Virol. (Praha)* **4**:215–219 (1960).

110. Atkins, E., and W. C. Huang, Studies on the pathogenesis of fever with influenzal viruses. I. The appearance of an endogenous pyrogen in the blood following intravenous injection of virus, *J. Exp. Med.* **107**:383–401 (1958).

111. Atkins, E., and W. C. Huang, Studies on the pathogenesis of fever with influenzal viruses. II. The effects of endogenous pyrogen in normal and virus-tolerant recipients, *J. Exp. Med.* **107**:403–414 (1958).

112. Bennett, I. L., Jr., R. R. Wagner, and V. S. LeQuire, Pyrogenicity of influenza virus in rabbits, *Proc. Soc. Exp. Biol. Med.* **71**:132–133 (1949).

113. Jennings, S. T., D. B. Ettensohn, and N. J. Roberts, Jr., Influenza virus infection of human alveolar and peripheral blood-derived macrophages: Production of factors that alter fibroblast proliferation, *Am. Rev. Respir. Dis.* **130**:98–101 (1984).

114. Wilkinson, P. J., and R. Borland, Persistent infection of human lung cells with influenza virus, *Nature (Lond.)* **238**:153–155 (1972).

115. Ginsberg, H. S., and J. R. Blackmon, Reactions of influenza viruses with guinea pig polymorphonuclear leucocytes. I. Virus–cell interactions, *Virology* **2**:618–636 (1956).

116. Denman, A. M., B. Rager-Zisman, T. C. Merigan, and D. A. J. Tyrell, Replication or inactivation of different viruses by human lymphocyte preparations, *Infect. Immun.* **9**:373–376 (1974).

117. Hackeman, M. M. A., A. M. Denman, and D. A. J. Tyrrell, Inactivation of influenza virus by human lymphocytes, *Clin. Exp. Immunol.* **16**:583–591 (1974).

118. Roberts, N. J., Jr., and P. K. Horan, Expression of viral antigens after infection of human lymphocytes, monocytes, and macrophages with influenza virus, *J. Infect. Dis.* **151**:308–313 (1985).

119. Rodgers, B., and C. A. Mims, Interaction of influenza virus with mouse macrophages, *Infect. Immun.* **31**:751–757 (1981).

120. Rodgers, B. C., and C. A. Mims, Influenza virus replication in human alveolar macrophages, *J. Med. Virol.* **9**:177–184 (1982).

121. Wells, M. A., P. Albrecht, S. Daniel, and F. A. Ennis, Host defense mechanisms against influenza virus: Interaction of influenza virus with murine macrophages in vitro, *Infect. Immun.* **22**:758–762 (1978).

122. Zisman, B., and A. M. Denman, Inactivation of myxoviruses by lymphoid cells, *J. Gen. Virol.* **20**:211–233 (1973).

123. Wilson, A. B., D. N. Planterose, J. Nagington, J. R. Park, R. D. Barry, and R. R. A. Coombs, Influenza A antigens on human lymphocytes *in vitro* and probably *in vivo*, *Nature (Lond.)* **259**:582–584 (1976).

124. Butchko, G. M., R. B. Armstrong, and F. A. Ennis, Specificity studies on the proliferative response of thymus-derived lymphocytes to influenza viruses, *J. Immunol.* **121**:2381–2385 (1978).

125. Mock, D. J., F. Domurat, N. J. Roberts, Jr., E. E. Walsh, M. R. Licht, and P. Keng, Macrophages are required for influenza virus infection of human lymphocytes. *J. Clin. Invest.* **79**:620–624 (1987).

126. Brownson, J. M., B. W. J. Mahy, and B. L. Hazleman, Interaction of influenza A virus with human peripheral blood lymphocytes, *Infect. Immun.* **25**:749–756 (1979).

127. Merigan, T. C. Jr., and E. F. Wheelock, Virus–lymphocyte interactions, in: *Progress in Immunology* (B. Amos, ed.), pp. 1351–1353, Academic, New York (1971).

128. Woodruff, J. F., and J. J. Woodruff, T lymphocyte interaction with viruses and virus-infected tissues, *Prog. Med. Virol.* **19**:120–160 (1975).

129. Domurat, F., D. J. Mock, N. J. Roberts, Jr., and P. C. Keng, Proliferation alone does not increase human lymphocyte susceptibility to influenza virus infection, in: *Twenty-fifth Interscience Conference on Antimicrobial Agents and Chemotherapy, Minneapolis* (1985) (abst.).

130. Wyde, P. R., and T. R. Cate, Cellular changes in lungs of mice infected with influenza virus: Characterization of the cytotoxic responses, *Infect. Immun.* **22**:423–429 (1978).

131. Wyde, P. R., D. L. Peavy, and T. R. Cate, Morphological and cytochemical characterization of cells infiltrating mouse lungs after influenza infection, *Infect. Immun.* **21**:140–146 (1978).

132. Ada, G. L., K.-N. Leung, and H. Ertl, An analysis of effector T cell generation and function in mice exposed to influenza A or Sendai viruses, *Immunol. Rev.* **58**:5–24 (1981).

133. Liew, F. Y., and S. M. Russell, Inhibition of pathogenic effect of effector T cells by specific suppressor T cells during influenza virus infection in mice, *Nature (Lond.)* **304**:541–543 (1983).

134. Hurwitz, J. L., and C. J. Hackett, Influenza-specific suppression: Contribution of major viral proteins to the generation and function of T suppressor cells, *J. Immunol.* **135**:2134–2139 (1985).

135. Lipscomb, M. F., D. Yaekel-Houlihan, C. R. Lyons, R. R. Gleason, and J. Stein-Streilein, Persistence of influenza as an immunogen in pulmonary antigen-presenting cells, *Infect. Immun.* **42**:965–972 (1983).

136. Jakab, G. J., C. L. Astry, and G. A. Warr, Alveolitis induced by influenza virus, *Am. Rev. Respir. Dis.* **128**:730–739 (1983).

137. Owens, S. L., J. W. Osebold, and Y. C. Zee, Dynamics of B-lymphocytes in the lungs of mice exposed to aerosolized influenza virus, *Infect. Immun.* **33**:231–238 (1981).

138. Zee, Y. C., J. W. Osebold, and W. M. Dotson, Antibody responses and interferon titers in the respiratory tracts of mice after aerosolized exposure to influenza virus, *Infect. Immun.* **25**:202–207 (1979).

139. Astry, C. L., and G. J. Jakab, Influenza virus-induced immune complexes suppress alveolar macrophage phagocytosis, *J. Virol.* **50**:287–292 (1984).

140. Lindahl-Magnusson, P., P. Leary, and I. Gresser, Interferon inhibits DNA synthesis induced in mouse lymphocyte suspensions by phytohaemagglutinin or by allogeneic cells, *Nature (New Biol)* **237**:120–121 (1972).

141. Pacheco, D., R. Falcoff, L. Catinot, F. Floc'h, G. H. Werner, and E. Falcoff, Inhibitory effect of interferon on DNA and RNA synthesis in murine spleen cells stimulated by lectins, *Ann. Immunol.* **127C**:163–171 (1976).

142. Singer, S. H., P. Noguchi, and R. L. Kirschstein, Respiratory diseases in cyclophosphamide-treated mice. II. Decreased virulence of PR8 influenza virus, *Infect. Immun.* **5**:957–960 (1972).

143. Suzuki, F., J. Ohya, and N. Ishida, Effect of antilymphocyte serum on influenza virus infection, *Proc. Soc. Exp. Biol. Med.* **146**:78–84 (1974).

144. Abou-Donia, H., R. Jennings, and C. W. Potter, The spread and persistence of influenza viruses in normal and cyclophosphamide-treated mice, *J. Med. Virol.* **7**:251–262 (1981).

145. Reiss, C. S., and J. L. Schulman, Cellular immune responses of mice to influenza virus infection, *Cell. Immunol.* **56**:502–509 (1980).

146. Wells, M. A., F. A. Ennis, and P. Albrecht, Recovery from a viral respiratory infection. II. Passive transfer of immune spleen cells to mice with influenza pneumonia, *J. Immunol.* **126**:1042–1046 (1981).

147. Wells, M. A., P. Albrecht, and F. A. Ennis, Recovery from a viral respiratory infection. I.

Influenza pneumonia in normal and T-deficient mice, *J. Immunol.* **126:**1036–1041 (1981).

148. Ennis, F. A., M. A. Wells, G. M. Butchko, and P. Albrecht, Evidence that cytotoxic T cells are part of the host's response to influenza pneumonia, *J. Exp. Med.* **148:**1241–1250 (1978).

149. Yap, K. L., and G. L. Ada, The recovery of mice from influenza A virus infection: Adoptive transfer of immunity with influenza virus-specific cytotoxic T lymphocytes recognizing a common virion antigen, *Scand. J. Immunol.* **8:**413–420 (1978).

150. McLaren, C., and G. M. Butchko, Regional T- and B-cell responses in influenza-infected ferrets, *Infect. Immun.* **22:**189–194 (1978).

151. Sellers, T. F., Jr., J. Schulman, C. Bouvier, R. McCune, and E. D. Kilbourne, The influence of influenza virus infection on exogenous staphylococcal and endogenous murine bacterial infection of the bronchopulmonary tissues of mice, *J. Exp. Med.* **114:**237–255 (1961).

152. Gardner, I. D., Effect of influenza virus infection on susceptibility to bacteria in mice, *J. Infect. Dis.* **142:**704–707 (1980).

153. Giebink, G. S., and P. F. Wright, Different virulence of influenza A virus strains and susceptibility to pneumococcal otitis media in chinchillas, *Infect. Immun.* **41:**913–920 (1983).

154. Woodruff, J. F., and J. J. Woodruff, Lymphocyte receptors for myxoviruses and para-myxoviruses, *J. Immunol.* **112:**2176–2183 (1974).

155. Ennis, F. A., A. H. Rook, Q. Yi-Hua, G. C. Schild, D. Riley, R. Pratt, and C. W. Potter, HLA-restricted virus-specific cytotoxic T-lymphocyte responses to live and inactivated influenza vaccines, *Lancet* **2:**887–891 (1981).

156. Cunningham-Rundles, S., A. Brown, D. Gross, D. Braun, J. A. Hansen, R. A. Good, D. Armstrong, and B. Dupont, Association of HLA in immune response to influenza-A immunization, *Transplant. Proc.* **11:**1849–1852 (1979).

157. Spencer, M. J., J. D. Cherry, and P. I. Terasaki, HL-A antigens and antibody response after influenza A vaccination: Decreased response associated with HL-A type W16, *N. Engl. J. Med.* **294:**13–16 (1976).

158. Domurat, F., A. Nikaein, D. Mock, P. Keng, and N. J. Roberts, Jr., Demonstration and selective abrogation of class II HLA-DR-restricted activity in human mononuclear leukocytes exposed to respiratory syncytial virus, *Hum. Immunol.* **14:**164–165 (1985).

159. Lindenmann, J., E. Deuel, S. Fanconi, and O. Haller, Inborn resistance of mice to myxoviruses: Macrophages express phenotype *in vitro, J. Exp. Med.* **147:**531–540 (1978).

160. McLaren, C., and B. Pope, Macrophage dependency of *in vitro* B cell response to influenza virus antigens, *J. Immunol.* **125:**2679–2684 (1980).

# 17

# Paramyxoviruses

## RAIJA VAINIONPÄÄ and TIMO HYYPIÄ

## 1. BASIC PROPERTIES OF VIRUSES

### 1.1. Classification

The family Paramyxoviridae contains three genera: paramyxoviruses, morbilliviruses, and pneumoviruses[1] (Table I). They infect a large variety of mammals and birds, but the strains are very host specific. Classification in separate genera is based on differences in hemagglutinating and neuraminidase activities, as well as in morphology. The *Paramyxovirus* genus includes the human pathogens, mumps virus, parainfluenza virus types 1–4, and several animal pathogens, of which Sendai virus of mice, simian virus 5 (SV5) of monkey, and Newcastle disease virus (NDV) of birds are the best characterized. The genus *Morbillivirus* includes measles virus (MV) and three nonhuman viruses, canine distemper virus (CDV), rinderpest virus (RPV), and peste des petits ruminants virus (PPRV). Measles virus is not included in this review, because it has been presented in Chapter 18. Respiratory syncytial virus (RSV), bovine respiratory syncytial virus, and pneumonia virus of mice are members of the genus *Pneumovirus*. Throughout this chapter, the term paramyxovirus(es) is used to describe the whole family Paramyxoviridae.

### 1.2. Molecular Structure

Paramyxoviruses are pleomorphic enveloped RNA-viruses, with a particle size of about 150–300 nm. The genome is a single-stranded nonsegmented RNA molecule of negative polarity with a molecular weight of about $5–7 \times 10^6$ kDa, covered with nucleocapsid proteins. The helically coiled nucleocapsid structure is surrounded by a lipid-rich envelope of host cell origin with virus-

RAIJA VAINIONPÄÄ and TIMO HYYPIÄ • Department of Virology, University of Turku, SF-20520 Turku, Finland.

**TABLE I**
**Classification of the Family Paramyxoviridae**

| Virus | Activities | | Host |
|---|---|---|---|
| | Hemagglutination | Neuraminidase | |
| Paramyxoviruses | | | |
| Mumps virus | + | + | Humans |
| Parainfluenza virus | | | |
| type 1 | + | + | Humans |
| type 2 | + | + | Humans |
| type 3 | + | + | Humans |
| type 4a | + | + | Humans |
| type 4b | + | + | Humans |
| Sendai virus | + | + | Mouse |
| Simian virus-5 | + | + | Monkey |
| Newcastle disease virus | + | + | Birds |
| Morbilliviruses | | | |
| Measles virus | + | − | Humans |
| Canine distemper virus | − | − | Dog |
| Rinderpest virus | − | − | Cattle |
| Peste des Petits ruminants virus | − | − | Sheep |
| Pneumoviruses | | | |
| Respiratory syncytial virus | − | − | Humans |
| Bovine respiratory syncytial virus | − | − | Cattle |
| Pneumovirus of mice | + | − | Mouse |

specific glycoprotein spikes. The schematic model of a paramyxovirus is shown in Fig. 1. The nucleocapsid contains the RNA-dependent RNA polymerase activity. Viral replication takes place in the cytoplasm, and viruses are released by budding through the plasma membrane.

Paramyxovirus particles contain six virus-specific proteins and cellular actin, except RSV which has been reported to contain 7–8 virus-specific proteins. The characteristics of the structural proteins are summarized in Table II. The envelope spikes are formed by two different glycoproteins. The larger one contains the hemagglutinating and neuraminidase activities in the members of the *Paramyxovirus* genus and is therefore designated the HN protein. These activities are located at separate sites of the molecule.[2,3] The members of the *Morbillivirus* and the *Pneumovirus*, except measles virus and pneumonia virus of mice, probably do not contain these activities. The HN protein mediates virus–cell, attachment but the function of the neuraminidase activity in the replicative cycle is not completely understood. The HN protein is the major envelope protein[4] and is anchored to the membrane through its N-terminus.[5,6] It is also the major antigenic determinant of paramyxovirus and induces the production of neutralizing antibodies.[7] The gene of the larger glycoprotein of RSV is significantly smaller (918 nucleotides),[8] compared with the HN genes of the other paramyxoviruses (1595–1895 nucleotides).[5,6,9] The molecular

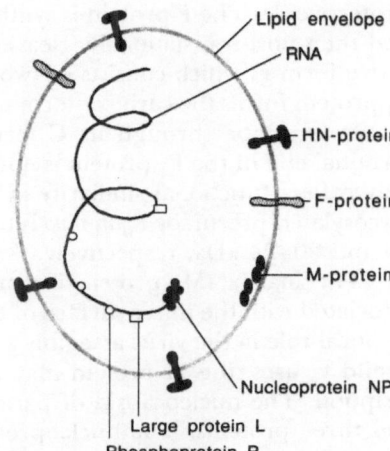

FIGURE 1. Schematic model of paramyxoviruses.

**FIGURE 1.** Schematic model of paramyxoviruses.

weight of the glycosylated form *in vivo* is comparable, however, to that of other paramyxoviruses, indicating its high carbohydrate content. In the native form, the HN protein occurs as a disulfide-bonded oligomer.[10] The fusion (F) glycoprotein is involved in hemolysis, cell fusion, and virus penetration into cells. Antibodies against F protein have also been shown *in vitro* to be highly

## TABLE II
### Structural Proteins of Paramyxoviruses

| Protein | Designation | Molecular weight *in vivo* (kDa) | Location | Function |
|---|---|---|---|---|
| Hemagglutinin | HN | 72–90 | Envelope | Attachment to host cell receptors |
| | | | | Hemagglutinating and neuraminidase activities |
| Fusion | $F_0-F(F_1F_2)$ | $F_0$ 60-70 | | |
| | | $F_1$ 40-60 | Envelope | Fusion, hemolysis, |
| | | $F_2$ 10-20 | | penetration |
| Nucleoprotein | NP | 43–70 | Nucleocapsid | Bound to RNA |
| Phosphoprotein | P | 34–70 | Nucleocapsid | Component of polymerase complex |
| Large | L | 180–200 | Nucleocapsid | Component of polymerase complex |
| Matrix | M | 30–36 | Inside virion | Assembly, regulation of transcription |
| Actin | | | | Host cell origin |

protective.[11] The F protein is synthesized as a precursor form, designated $F_0$, and then post-translationally cleaved by host cell proteases to the biologically active form F, which consists of two disulfide-linked subunits, $F_1$ and $F_2$. The $F_1$ protein forms the carboxy-terminal portion of the F protein and is anchored to the envelope through its C terminal. Considerable homology at the N-terminal end of the $F_1$ protein is noted for Sendai virus, SV5, and NDV, which may reflect functional similarity in this region. The molecular weight of the glycosylated precursor $F_0$ *in vivo* is 60–70 kDa, and the subunits $F_1$ and $F_2$ 40–60 and 20–24 kDa, respectively.[12–14]

The matrix (M) protein is nonglycosylated, strongly hydrophobic, and associated with the inner surface of the envelope. The M protein obviously has a critical role in the virus assembly and, based on the analogy to other negative strand viruses, the M protein may also participate in the regulation of transcription. The nucleocapsid of paramyxoviruses contains viral genomic RNA and three proteins. The nucleoprotein (NP) is the major protein of the nucleocapsid structure, and it covers the genomic RNA. High homology in amino acid sequences is present in NP between parainfluenza type 3 and Sendai viruses, as well as between measles and canine distemper viruses[15,16] but not between parainfluenza type 3 virus and RSV.[16] The phosphoprotein (P), another component of the nucleocapsid, has a role in the RNA-dependent RNA-polymerase activity. The large (L) protein is a minor component in the nucleocapsid structure, but it is assumed to be a crucial part in the RNA-dependent RNA-polymerase activity.[17] One or two nonstructural (C) virus-specific proteins are also synthesized in paramyxovirus-infected cells. The function(s) of these proteins is still unclear. In addition to the proteins mentioned above, some minor polypeptides are synthesized in infected cells, but their roles are unknown.

## 1.3. Replication of Paramyxoviruses

The first step in the infectious cycle is the attachment of the virus particle to the cellular receptors through the hemagglutinin protein. After the fusion of viral envelope with cellular membrane, mediated by the F protein, the viral nucleocapsid is released into the cytoplasm. The next event is the synthesis of viral messenger RNA (mRNA) molecules. Because eukaryotic cells do not contain RNA-dependent RNA-polymerase activity, virus nucleocapsid has to carry this activity into the cell, at least the viral proteins L and P are required for this activity. Virus-specific enzyme activity transcribes genome RNA to 6–10 individual mRNAs, which resemble eukaryotic messengers with their 3'-end poly-A-tails and 5'-end methylated cap structures. Viral mRNA molecules are translated to proteins on host cell ribosomes. The glycoproteins go through a complex maturation process during the transport through the Golgi apparatus to cellular membrane.[18,19]

The genome replication proceeds in two steps. The negative-strand RNA is transcribed to complementary-positive RNA and then, with this positive RNA as a template, the genomic negative-strand RNA is synthesized. The nucleoprotein immediately encapsidates the new genomic RNA. The L and P proteins also attach to the ribonucleoprotein structure.

The assembly process is not completely understood, but it is possible that the soluble M protein has a critical role in the budding process. The M protein may function as a bridge between virus-specific glycoproteins on cell membranes and nucleocapsid structures in the cytoplasm.

## 2. MODULATION OF IMMUNITY BY THE VIRUSES

### 2.1. Parainfluenza and Respiratory Syncytial Viruses

Parainfluenza viruses types 1–4 and RSV share several common features in their clinical manifestations, in pathogenesis, and in immune responses during infection. All are transmitted via direct contact or large droplets from respiratory secretions to the conjunctiva or nasal epithelium. The primary replication takes place in the epithelial cells of the upper respiratory tract. After an incubation period of 3–6 days, the virus is shed from the mucosal surface, and symptoms of upper respiratory tract infection develop. Croup or laryngotracheitis is the severe form of parainfluenza infections, whereas RSV is a common cause of bronchiolitis and pneumonia in infants and small children. Reinfections are common with all these viruses, but the clinical manifestations are usually less severe than during the primary infection. General immunosuppression is seldom observed, although the infections are relatively often associated with secondary bacterial infections and immunosuppressed patients tend to have prolonged and complicated forms of parainfluenza virus and RSV infections.

Several animal models have been used to study the pathogenesis of parainfluenza virus and RSV infections. One interesting model for RSV is the infection in cotton rats, used extensively by Prince et al.,[20] which will be presented here as an example. One day after intranasal infection of the animals, virus replicated in nose, trachea, and lungs. Duration of virus excretion was approximately 7 days with peak values on day 4. Viral antigen was detected by immunofluorescence in cells of the nasal epithelium and the luminal epithelial cells of the bronchi and bronchioles. Trachea and alveolar cells in the lungs were negative for RSV antigens. Serum-neutralizing antibodies developed by the ninth day of infection in parallel with the disappearance of virus shedding in the animals. In another study the active immunity and maternally transmitted immunity to RSV have been compared.[21] RSV-infected animals developed resistance to pulmonary reinfection that lasted at least 18 months, while the nasal resistance began to diminish by 8 months postinfection (p.i.). Transfer of lymphocytes from immune animals did not provide nasal or pulmonary resistance while complete or near-complete pulmonary resistance was seen in recipients of immune serum. However, minimal resistance was detected in the nose of these animals. Immune females were able to transfer antibody to their infants, with most of the antibody being transferred via colostrum and milk. Maternally transmitted immunity was more effective in the lungs than in the nose. It was recently shown that the passive immunity transferred by serum is complete or almost complete in the lungs, with a serum-neutralizing antibody titer of 1 : 380 or greater.[22] Walsh *et al.*[23] showed that passive immunization with circulating monoclonal antibody to either HN glycoprotein (gp90) or fusion polypeptide (VP70) reduces or prevents the growth of RSV in the lungs of infected cotton rats. Monoclonal antibody against the 44-kDa nucleoprotein does not have this effect and none of these antibodies is able to block the nasal infection.

In humans, most of the reports on the immune system and paramyxoviruses concern local and circulating antibodies. The importance of cell-mediated immunity *in vivo* in parainfluenza virus and RSV infections is largely unknown. The entry of these viruses occurs via the respiratory route, whereby the first specific defense mechanism is the barrier of local antibodies. Yanagihara and McIntosh[24] showed that in infants infected with parainfluenza virus type 1 or 2, a significant increase in nasopharyngeal secretory IgA is detected. However, there was a discordance in individual specimens between IgA and neutralizing activity because many secretions contained IgA but were not neutralizing, and vice versa. Similar results have been reported for RSV.[25] In spite of the observed discrepancy, the appearance of local antibodies correlates well with the disappearance of the viruses from the nasopharynx. The local antibodies disappear gradually and, 2 months after infection, they were detected in only 33% of RSV patients.[26] McIntosh[27] showed that the nasal secretions of RSV-infected children contain undetectable or low levels of interferon (IFN) activity in contrast to those infected with influenza A. Hall *et al.*[28] reported that IFN is detected in nasal washes in 5%, 30%, and 55% of children infected with RSV, parainfluenza, and influenza A viruses, respectively.

The appearance and levels of circulating antibodies to RSV have been analyzed by several groups. Meurman *et al.*[29] recently studied primary RSV infection in children, all of whom had an IgG antibody response; IgM class antibodies were detected in 73% of patients. Children under 6 months of age produced little or no IgM, while all patients aged 1–2 years produced RSV-specific IgM. IgA antibodies were detected in the serum of 77% of patients. IgM antibodies persisted from 2 to 10 weeks, and no IgA and low or undetectable levels of IgM were observed after 1 year in children with primary infection. Subsequent reinfection results in an accelerated antibody response (5–7 days) for all three Ig classes.[30] Detectable titers of IgG were found in 50% and IgA in 75% of colostrum samples.

In sera from children with primary RSV infections, the antibody response is mainly directed against polypeptides 30, 48, and 72 kDa, when studied by immunoprecipitation according to Vainionpää *et al.*[32] Ward *et al.*[33] described the strongest immune response in 6- to 12-month-old infants to be against VP41 and VPG48. An interesting phenomenon is the lack of antibodies against the larger glycoprotein G. On the contrary, the antibody level against the parainfluenza virus 3 HN polypeptide was consistently higher than the levels against the F protein in human sera as studied by Kasel *et al.*[35]

In a follow-up study, Fernald *et al.*[35] showed that neither humoral nor cell-mediated immunity (CMI) alone appears to act directly in modifying RSV infection and disease expression. These workers studied RSV-stimulated lymphocyte blastogenesis as the expression of CMI and showed that the mean stimulation index is near unity before the first infection and rises gradually during 3 years thereafter. Correspondingly Cranage and Gardner[36] observed that a great majority of children with RSV infection showed CMI approximately 1 year later. In addition, Scott *et al.*[31] reported significant proliferative response to RSV antigen in maternal, colostral, and cord blood cultures. The proportional number of peripheral blood mononuclear cells with the OKT8+ phenotype, so-called suppressor/cytotoxic cells, is decreased during convalence, but not during acute infection in patients with bronchiolitis.[37] Domurat *et al.*[38] reported that circulating mononuclear leukocytes from symptomatic children with RSV infections frequently express viral antigens. Viral antigens were detected more frequently in the cells of younger patients.

## 2.2. Mumps Virus

The pathogenesis of mumps virus differs significantly from that of RSV and parainfluenza viruses. Depression of delayed hypersensitivity is also clearly observed after administration of live attenuated mumps virus vaccine.[39,40] Although the primary replication of mumps virus also occurs on the mucosal surface, the clinical manifestation is not upper respiratory infection. Typically, parotitis is observed after an incubation period of about 18 days; the illness is often complicated with symptoms in the central nervous system (CNS), especially meningitis. Orchitis, carditis, and pancreatitis also are found in association with mumps infection. Viremia occurs during the course of the disease, and virus can be isolated from cerebrospinal fluid (CSF) in meningitis, and occasionally from urine.

Salivary IgA antibody is detected by day 4 p.i. in most cases during mumps infection; its appearance correlates with the disappearance of the virus.[41] Several groups have studied serum antibodies against mumps virus, mainly for diagnostic purposes. Anti-mumps IgG antibodies increase rapidly and reach maximum in about 3 weeks, while most patients have IgM antibodies from the second day p.i. with peak values within the first week.[42] The response in parotitis does not differ from that in meningitis/encephalitis, but relatively higher antibody titers are found in orchitis/epididymitis patients. In meningitis, the CSF IgG antibodies correlate well with titers in serum, whereas IgM antibodies do not show direct correlation, and some patients do not have mumps IgM in the CSF at all.[43] The inflammatory reaction in the CNS, characterized as mononuclear pleocytosis, Ig synthesis, and oligoclonal IgG, does not correlate with the clinical course of infection.[44] Antimumps Ig administration to mumps contacts is protective only when performed early after the outbreak of an epidemic.[45] However, application of antimumps IgG to patients within the first 5 days after onset seems to prevent the appearance of complications and reduces the duration of the disease.

Ilonen[46] showed that the lymphocytes of most of mumps-seropositive subjects respond to stimulation with inactivated mumps antigen. In individuals given inactivated mumps virus vaccine, the response to mumps antigen appeared by 20 weeks postvaccination and reached the levels found in seropositive subjects years after natural mumps infection.[47] No clear-cut correlation between antibody levels and blast transformation responses was observed. Fryden et al.[48] studied patients with mumps meningitis and showed that in four of five patients, the proliferation of CSF lymphocytes after mumps stimulation was higher than that of blood lymphocytes. The proliferation response was specific, indicating the occurrence of lymphocytes sensitized against mumps in the CNS during the acute infection.

Kreth et al.[49] tested cryopreserved lymphocytes from blood and CSF of 10 children with mumps meningitis in $^{51}$Cr-release assays against uninfected and mumps virus-infected phytohemagglutinin (PHA)-stimulated blasts. These workers showed that lymphocytes from all patients were specifically cytotoxic to autologous mumps virus-infected target cells, and the reaction was mediated by E-rosette-forming lymphocytes. The effector cells were detected over 2–3 weeks after the onset of meningitis. When peripheral lymphocytes from healthy blood donors were incubated with autologous mumps virus-infected stimulator cells maximal cytotoxicity was generated after 5–7 days in culture.[50] The cytotoxicity could be detected only in seropositive individuals. The major restriction antigens for cytotoxic T lymphocytes (CTL) were certain HLA B determinants, such as B18, B27, Bw35, Bw62, and Bw63. Tsutsumi et al.[51] showed that generation of cytotoxic activity is associated with a lymphoproliferative response to mumps virus.

## 3. IMMUNE REACTIVITY AND PARAMYXOVIRUSES

Interleukin-1 (IL-1) is a soluble stimulatory factor, produced by macrophages, influencing the proliferation of T lymphocytes. One of the most important properties of IL-1 is probably to induce interleukin-2 (IL-2) production. RSV-infected macrophages produce IL-1, but also IL-1 inhibitor. Activity of the latter in crude preparations was higher than the activity of IL-1.[52] The release of IL-2 has been reported also to be either markedly reduced or even negative in the presence of RSV-infected macrophages.[52,53] IL-2 is needed for induction of interferon-γ (IFN$_γ$), and RSV is known to be a poor inducer of IFN.[25,28,54] In contrast to RSV, mumps virus is known to induce IFN production[55] and mouse Paramyxovirus Sendai virus is a good inducer of IFN and has been used for the production of IFN for clinical use.[56] In addition, Sendai virus-infected mononuclear phagocytes produce cytotoxins that include tumor necrosis factor (TNF), as determined by monospecific antibodies.[57] NDV, known to induce high interferon titers, is able to inhibit the cytotoxic T-cell response against another virus and titers directly correlate with macrophage and natural killer (NK) cell activation.[58]

Several studies have focused on antibody-dependent cell mediated cytotoxicity (ADCC). Scott et al.[59] reported that ADCC was detected in vitro in the presence of anti-RSV antibodies and peripheral blood lymphocytes from healthy adults. IgG and IgA antibodies both in sera and in colostrum mediated this reaction. The ability to mediate cytotoxicity was connected with non-adherent effector cells. ADCC activity was also detected in respiratory secretions, and the development of ADCC correlated closely to IgG, IgM, and IgA titers.[60] The activity was detected 3 days after the onset of primary or secondary RSV infection in infants or young adults and peak levels were observed 14–29 days after the onset of illness.[61] However, no clear-cut correlation occurred between neutralizing antibodies and ADCC in vivo in children with primary or secondary infection or with live attenuated RSV vaccine infection.[62] ADCC antibodies rose more rapidly than neutralizing antibodies and fell more rapidly, too. RSV-infected cells became susceptible to ADCC 4–8 hr after infection indicating that ADCC can recognize infected cell membranes before the release of virus. IFN is not involved in the activity, at least not at the outset.[63]

Virus-dependent cellular cytotoxicity (VDCC) can be induced by treating peripheral blood lymphocytes with mumps or Sendai viruses. The effector cells can also be activated with purified viral glycoproteins to cytotoxicity against a variety of noninfected target cells. Alsheikhly et al.[64–66] showed with purified Sendai virus peplomers and with monoclonal antibodies against different mumps virus-specific proteins that HN protein is the only polypeptide involved in VDCC, although it seems that more than one serologically defined structure of HN protein is involved in activation. CTL are not involved in VDCC, which is also antibody independent. VDCC is not an immunologically specific defense mechanism, but this activity is generated more rapidly compared with many other immune functions, and it might have an important role as a primary defense mechanism. IFN does not seem to be involved in VDCC activity,[64,66] while it is involved in NK cell activation, which, in combination with VDCC, seems to form natural cytotoxicity against mumps virus. Human and mouse NK cells are known to bind well to mumps and Sendai virus-infected cells.[67,68]

## 4. IMMUNOPATHOGENESIS

It is typical that parainfluenza virus and RSV infections are restricted to the respiratory tract and that reinfections commonly occur. The symptoms of primary infection are more severe than during the reinfections that usually result in a common cold. Thus, immune protection is awakened, perhaps slowly, during these infections, but it is not always able to prevent the primary replication of the virus in the epithelial cells. The appearance of local neutralizing antibodies correlates well with the disappearance of virus and may be one of the most important mechanisms in preventing the illness. However, the circulating antibodies, and probably CMI as well, participate in preventing infections of the lower respiratory tract. There is a clear difference between these two viruses and mumps virus, which commonly features a viremia and viral spread to different organs throughout the body. One explanation may be in the different distribution of the receptors recognized by these viruses.

The immunopathogenesis of serious infections with RSV, including bronchiolitis, deserves special attention. These syndromes tend to occur early in childhood, during the time when the protective effect of transplacentally acquired antibodies is decreasing. Because the primary form of RSV disease is variable between individual children in this age group, the origin of differences has been studied with regard to virus strains, maternal immunity, and IgE production. The use of monoclonal antibodies against viral polypeptides has permitted the characterization of various strains of RSV.[69,70] RSV strains differ, especially in the epitopes of the G polypeptide. They have been isolated during epidemics in different years, and no evidence has yet been presented that they might differ in pathogenesis. In addition, the strain variation as detectable by neutralization tests in vitro did not necessarily relate to differences in cross-protection in vivo in animal studies.[22]

The role of maternal antibodies in the immunopathogenesis of RSV has largely been discussed, as the peak prevalence of severe infections occurs at the age at which these antibodies are still present. Maternal antibodies can modulate the disease by forming immune complexes. They may also permit the activation of the complement pathway and therefore lead to infected cell destruction or modulation of viral infection by a variety of mechanisms. Ward et al.[33] reported that higher levels of maternal antibodies against nucleoprotein significantly correlate to protection of small babies against RSV infection. During the infection of children under 1 year of age, antibodies against the nucleoprotein and the F protein were detected, but not antibody against the G glycoprotein. This is contrary, however, to the findings reported by Walsh et al.[23] in an animal model, which showed that monoclonal antibodies against the nucleoprotein do not protect against infection of the lungs. Further studies on the role of maternal antibodies are clearly needed.

Welliver et al.[71] studied the development of anti-RSV IgE in infants with various clinical syndromes. Anti-RSV IgE was present in most patients with wheezing but not significantly in others. Histamine was also detected significantly more often in the wheezing group, thus indicating that the IgE-mediated histamine release from mast cells could at least partially explain the pathogenesis in complicated RSV infections.

Immunopathogenic phenomena have been observed after vaccination with formalin-inactivated RSV and measles virus preparations but not with inactivated parainfluenza and mumps viruses.[72-75] The vaccinees developed atypical symptoms during the ensuing infection, and clinical disease was often more severe than during natural primary infection. The precise mechanism behind this phenomenon remains largely unknown. Murphy et al.[76] recently showed that formalin-inactivated RSV vaccine caused a good antibody response to the F protein but a very poor response to the G protein. This is the opposite of the results reported with formalin-inactivated measles virus vaccine.[77] In a recent study with formalin-inactivated RSV vaccine in cotton rats, Prince et al.[78] showed that the animals that have received vaccine develop an enhanced RSV disease with pulmonary lesions after intranasal challenge with live RSV. The manifestation is characterized primarily by increased infiltration of neutrophils into the alveoli and the intraalveolar and peripheral areas. The next influx of neutrophils appeared on day 4 p.i. These immunopathologic processes might be explained as type III and IV reactions, respectively. The vaccinated animals, had high levels of both anti-G and anti-F antibodies, however, and the effect of formalin could not be localized.

The live attenuated mumps vaccine has been in clinical use for several years but, for RSV, which is the most common cause of lower respiratory infections in infants and small children, attempts at prevention have not proved successful. Recently, a new approach with a genetically engineered vaccine has been described for protection against RSV.[79] It will be important to understand the role of maternal antibodies, the cause of reinfections, and the reason for modified infections caused by inactivated vaccines, before an effective vaccine for RSV can be successfully applied.

## 5. MECHANISMS OF IMMUNOSUPPRESSION

Lytic viral infection in immune cells, which cause cell death and decrease the number of viable cells, may result in direct suppression of immune responses. Immunosuppression can also reflect a disturbance in the balance between different activities of immunocompetent cells. Paramyxoviruses are known to infect cells involved in immune responses both *in vivo* and *in vitro*, but resting immune cells do not support detectable viral replication. Mitogen activation of lymphocytes, however, alters the nonproductive infection to a productive one with clear cytopathic effects and release of infectious virus. It has been reported with the animal paramyxoviruses, bovine rinderpest virus, and canine distemper virus that the ability to infect lymphocytes and macrophages correlates with virulence.[80,81]

Paramyxovirus infections have immunosuppressive features both *in vivo* and *in vitro*. Live attenuated mumps virus vaccine is known to cause depression in delayed hypersensitivity skin test.[39,40] Mumps virus infects B and T lymphocytes as well as monocytes and macrophages *in vitro* but clearly shows a preference to T lymphocytes.[82,83] Infectious virus is not produced in resting cells, indicating that no active viral replication occurs, whereas mitogen stimulation induces the production of infectious virus to high titers. RSV is known to depress *in vitro* lymphocyte transformation,[37,54] and it has been reported to infect T lymphocytes and monocytes and macrophages both *in vivo* and *in vitro*.[38] After stimulation with PHA, T lymphocytes supported viral replication to high titers. RSV infection *in vitro* caused an alteration in helper/suppressor T-lymphocyte ratios by decreasing the proportional number of T-helper cells (Th) and by increasing the number of suppressor cells (Ts). The decrease in numbers of Th may be the cause of the suppression in lymphocyte proliferation observed *in vitro*. Conversely, Welliver et al.[37] described a defect in Ts numbers and function in RSV patients with bronchiolitis during convalescence compared with patients with milder forms of illness. Canine distemper virus infection is observed to cause depression of cellular immunity in dogs, and probably two mechanisms are involved in suppression.[84] In previously immunized animals, the viral infection caused only transient depression of skin reactivity and *in vitro* responsiveness of lymphocytes, but in nonimmune dogs the infection caused prolonged depression of skin-test reactivity and lymphocyte stimulation. Similar suppression of CMI has later been reported in ferrets.[85]

Anti-RSV antibodies complexed with RSV antigens have an effect on oxidative metabolism of neutrophils by significantly increasing their chemiluminescence as determined by the luminol-dependent technique.[86] This response correlated directly to the amount of specific antibodies and antigens in the mixture. RSV antigens or specific antibodies alone did not have any effect on the induction of chemiluminescence by peripheral blood neutrophils. This indicates that the effect is specific for the antibody–antigen complex. Because induction was heat labile, some complement component(s) might be involved in induction. RSV-infected cells are able to activate complement by both classic and alternative pathways.[87] In addition, complement components are known to be present also in respiratory tract secretions.[26]

Sendai virus has been reported to cause dysfunction of phagocytic properties of mouse alveolar macrophages. The addition of immune serum resulted in a clear suppression of Fc-mediated phagocytic activity and this suppression was dose dependent.[88] Also, a defect in phagosome–lysosome fusion in alveolar macrophages from mice infected with Sendai virus has been reported to occur.[89] There are thus several individual observations on the role of paramyxoviruses in the immune suppression. However, the precise mechanism(s) by which paramyxoviruses can affect immunologic functions, especially in human infections, is still largely unknown.

# REFERENCES

1. Kingsbury, D. W., M. A. Bratt, P. W. Choppin, R. P. Hanson, Y. Hosaka, V. ter Meulen, E. Norrby, W. Plowright, R. Rott, and W. H. Wunner, Paramyxoviridae, *Intervirology* **10**:137–151 (1983).
2. Portner, A., The HN glycoprotein of Sendai virus: Analysis of site(s) involved in hemagglutinating and neuraminidase activities, *Virology* **115**:375–384 (1981).
3. Yewdell, J., and W. Gerhard, Delineation of four antigenic sites on a paramyxovirus glycoprotein via which monoclonal antibodies mediate distinct antiviral activities, *J. Immunol.* **128**:2670–2675 (1982).
4. Örvell, C., and E. Norrby, Immunologic properties of purified Sendai virus glycoproteins, *J. Immunol.* **119**:1882–1887 (1977).
5. Blumberg, B., C. Giorgi, L. Roux, R. Raju, P. Dowling, A. Chollet, and D. Kolakofsky. Sequence determination of the Sendai virus HN gene and its comparison to the influenza virus glycoproteins, *Cell* **41**:269–278 (1985).
6. Hiebert, S. W., R. G. Paterson, and R. A. Lamb, Hemagglutinin neuraminidase protein of the paramyxovirus simian virus 5: Nucleotide sequence of the mRNA predicts an N-terminal membrane anchor, *J. Virol.* **53**:1–6 (1985).

7. Örvell, C., and M. Grandien, The effects of monoclonal antibodies on biologic activities of structural proteins of Sendai virus, *J. Immunol.* **129**:2779–2787 (1982).
8. Wertz, G. W., P. L. Collins, Y. Huang, C. Gruber, S. Levine, and L. A. Ball, Nucleotide sequence of the G protein gene of human respiratory syncytial reveals an unusual type of viral membrane protein, *Proc. Natl. Acad. Sci. USA* **82**:4075–4079 (1985).
9. Elango, N., J. E. Copligan, R. C. Jambou, and S. Venkatesan, Human parainfluenza type 3 virus hemagglutinin–neuraminidase glycoprotein: Nucleotide sequence of mRNA and limited amino acid sequence of the purified protein, *J. Virol.* **57**:481–489 (1986).
10. Markwell, M. K., and C. F. Fox, Protein–protein interactions within paramyxoviruses identified by native disulfide bonding or reversible chemical cross-linking, *J. Virol.* **33**:152–166 (1980).
11. Löve, A., R. Rydbeck, G. Utter, C. Örvell, K. Kristensson, and E. Norrby. Monoclonal antibodies against the fusion protein are protective in necrotizing mumps meningoencephalitis, *J. Virol.* **58**:220–222 (1986).
12. Merz, D. C., A. C. Server, M. N. Waxham, and J. S. Wolinsky, Biosynthesis of mumps virus F glycoprotein: Non-fusing strains efficiently cleave the F glycoprotein precursor, *J. Gen. Virol.* **64**:1457–1467 (1983).
13. Paterson, R. G., T. J. R. Harris, and R. A. Lamb, Fusion protein of the paramyxovirus SV5: Nucleotide sequence of mRNA predicts a highly hydrophobic glycoprotein, *Proc. Natl. Acad. Sci. USA* **81**:6706–6710 (1984).
14. Jambou, R. C., N. Elango, and S. Venkatesan, Proteins associated with human parainfluenza virus type 3, *J. Virol.* **56**:298–302 (1985).
15. Rozenblatt, S., O. Eizenberg, R. Ben-Levy, V. Lavie, and W. J. Bellini, Sequence homology within the morbilliviruses, *J. Virol.* **53**:684–690 (1985).
16. Galinsky, M. S., M. A. Mink, D. M. Lambert, S. L. Wechsler, and M. W. Pons, Molecular cloning and sequence analysis of the human parainfluenza 3 virus RNA encoding the nucleocapsid protein, *Virology* **149**:139–151 (1986).
17. Buetti, E., and P. V. Choppin, The transcriptase complex of the Paramyxovirus SV5, *Virology* **82**:493–508 (1977).
18. Mottet, G., A. Portner, and L. Roux, Drastic immunoreactivity changes between the immature and mature forms of the Sendai virus HN and $F_0$ glycoproteins, *J. Virol.* **59**:132–141 (1986).
19. Waxham, M. N., D. C. Merz, and J. S. Wolinsky, Intracellular maturation of mumps virus hemagglutinin–neuraminidase glycoprotein: Conformational changes detected with monoclonal antibodies, *J. Virol.* **59**:392–400 (1986).
20. Prince, G. A., A. B. Jenson, R. L. Horswood, E. Camargo, and R. M. Chanock. The pathogenesis of respiratory syncytial virus infection in cotton rats, *Am. J. Pathol.* **93**:771–783 (1978).
21. Prince, G. A., R. L. Horswood, E. Camargo, D. Koening, and R. M. Chanock. Mechanism of immunity to respiratory syncytial virus in cotton rats, *Infect. Immun.* **42**:81–87 (1983).
22. Prince, G. A., R. L. Horswood, and R. M. Chanock, Quantitative aspects of passive immunity to respiratory syncytial virus infection in infant cotton rats, *J. Virol.* **55**:517–520 (1985).
23. Walsh, E. E., J. J. Schlesinger, and M. W. Brandriss, Protection from respiratory syncytial virus infection in cotton rats by passive transfer of monoclonal antibodies, *Infect. Immun.* **43**:765–758 (1984).
24. Yanagihara, R. and K. McIntosh. Secretory immunological response in infants and children to parainfluenza virus types 1 and 2, *Infect. Immun.* **30**:23–28 (1980).
25. McIntosh, K., H. B. Masters, I. Orr, R. K. Chao, and R. M. Barkin, The immunologic response to infection with respiratory syncytial virus in infants, *J. Infect. Dis.* **138**:24–32 (1978).
26. Kaul, T. N., R. C. Welliver, and P. L. Ogra. Appearance of complement components and

immunoglobulins on nasopharyngeal epithelial cells following naturally acquired infection with respiratory syncytial virus, *J. Med. Virol.* **9**:149–158 (1982).

27. McIntosh, K., Interferon in nasal secretions from infants with viral respiratory tract infections, *J. Pediatr.* **93**:33–36 (1978).

28. Hall, C. B., R. G. Douglas, R. L. Simons, and J. M. Geiman, Interferon production in children with respiratory syncytial, influenza, and parainfluenza virus infections, *J. Pediatr.* **93**:28–32 (1978).

29. Meurman, O., O. Ruuskanen, H. Sarkkinen, P. Hänninen, and P. Halonen, Immunoglobulin class-specific antibody response in respiratory syncytial virus infection measured by enzyme immunoassay, *J. Med. Virol.* **14**:67–72 (1984).

30. Welliver, R. C., T. N. Kaul, T. I. Putnam, M. Sun, B. S. Riddlesberger, and P. L. Ogra, The antibody response to primary and secondary infection with respiratory syncytial virus: Kinetics of class-specific responses, *J. Pediatr.* **96**:808–813 (1980).

31. Scott, R., M. Scott, and G. L. Toms, Cellular and antibody response to respiratory syncytial (RS) virus in human colostrum, maternal blood, and cord blood, *J. Med. Virol.* **8**:55–66 (1981).

32. Vainionpää, R., O. Meurman, and H. Sarkkinen, Antibody response to respiratory syncytial virus structural proteins in children with acute respiratory syncytial virus infection, *J. Virol.* **53**:976–979 (1985).

33. Ward, K. A., P. R. Lambden, M. M. Ogilvie, and P. J. Watt. Antibodies to respiratory syncytial virus polypeptides and their significance in human infection, *J. Gen. Virol.* **64**:1867–1876 (1983).

34. Kasel, J. A., A. L. Fran, W. A. Keitel, L. H. Taber, and W. P. Glezen, Acquistion of serum antibodies to specific viral glycoproteins of parainfluenza virus 3 in children, *J. Virol.* **52**:828–832 (1984).

35. Fernald, G. W., J. R. Almond, and F. W. Henderson. Cellular and humoral immunity in recurrent respiratory syncytial virus infections, *Pediatr. Res.* **17**:753–758 (1983).

36. Cranage, M. P., and P. S. Gardner, Systemic cell-mediated and antibody responses in infants with respiratory syncytial virus infections, *J. Med. Virol.* **5**:161–170 (1980).

37. Welliver, R. C., T. N. Kaul, M. Sun, and P. L. Ogra, Defective regulation of immune responses in respiratory syncytial virus infection, *J. Immunol.* **133**:1925–1930 (1984).

38. Domurat, F., N. J. Roberts, Jr, E. E. Walsh, and R. Dagan, Respiratory syncytial virus infection of human mononuclear leukocytes *in vitro* and *in vivo*, *J. Infect. Dis.* **152**:895–902 (1985).

39. Kupers, T. A., J. M. Petrich, A. W. Holloway, and J. W. Geme, Jr., Depression of tuberculin delayed hypersensitivity by live attenuated mumps virus, *J. Pediatr.* **76**:716–721 (1970).

40. Hall, C. B., and F. S. Kantor. Depression of established delayed hypersensitivity by mumps virus, *J. Immunol.* **108**:81–85 (1972).

41. Chiba, Y., K. Horino, M. Umetsu, Y. Wataya, S. Chiba, and T. Nakao, Virus excretion and antibody response in saliva in natural mumps, *Tohoku J. Exp. Med.* **111**:229–238 (1973).

42. Ukkonen, P., M-L. Granström, and K. Penttinen, Mumps-specific immunoglobulin M and G antibodies in natural mumps infection as measured by enzyme-linked immunosorbent assay, *J. Med. Virol.* **8**:131–142 (1981).

43. Ukkonen, P., M-L. Granström, J. Räsänen, E-M. Salonen, and K. Penttinen, Local production of mumps IgG and IgM antibodies in the cerebrospinal fluid of meningitis patients, *J. Med. Virol.* **8**:257–265 (1981).

44. Fryden, A., H. Link, and E. Norrby, Cerebrospinal fluid and serum immunoglobulins and antibody titers in mumps meningitis and aseptic meningitis of other etiology, *Infect. Immun.* **21**:852–861 (1978).

45. Copelovici, Y., D. Strulovici, A. L. Cristea, V. Tudor, and V. Armasu, Data on the

efficiency of specific anti mumps immunoglobulins in the prevention of mumps and of its complications, *Rev. Roum. Med. Virol.* **30**:171–177 (1979).

46. Ilonen, J., Lymphocyte blast transformation response of seropositive and seroegative subjects to herpes simplex, rubella, mumps and measles virus antigens, *Acta Pathol. Microbiol. Scand. Sect. C* **87**:151–157 (1979).

47. Ilonen, J., A. Salmi, K. Penttinen, and E. Herva, Lymphocyte blast transformation and antibody response after vaccination with inactivated mumps virus vaccine, *Acta Pathol. Microbiol. Scand. Sect. C* **89**:303–309 (1981).

48. Fryden, A., H. Link, and E. Möller, Demonstration of cerebrospinal fluid lymphocytes sensitized against virus antigens in mumps meningitis, *Acta Neurol. Scand.* **57**:396–404 (1978).

49. Kreth, H. W., L. Kress, H. G. Kress, H. F. Ott, and G. Eckert, Demonstration of primary cytotoxic T cells in venous blood and cerebrospinal fluid of children with mumps meningitis, *J. Immunol.* **128**:2411–2415 (1982).

50. Kress, H. G., and H. W. Kreth, HLA restriction of secondary mumps-specific cytotoxic T lymphocytes, *J. Immunol.* **129**:844–849 (1982).

51. Tsutsumi, H., Y. Chiba, W. Abo, S. Chiba, and T. Nakao, T-cell-mediated cytotoxic response to mumps virus in humans, *Infect. Immun.* **30**:129–134 (1980).

52. Roberts, Jr., N. J., A. H. Prill, and T. N. Mann, Interleukin 1 and interleukin 1 inhibitor production by human macrophages exposed to influenza virus or respiratory syncytial virus, *J. Exp. Med.* **163**:511–519 (1986).

53. Borysiewicz, L. K., P. Casali, B. Rogers, S. Morris, and J. G. P. Sissons, The immunosuppressive effects of measles virus on T cell function—Failure to affect IL-2 release of cytotoxic T cell activity *in vitro*, *Clin. Exp. Immunol.* **59**:29–36 (1985).

54. Roberts, N. J., Jr., Different effects of influenza virus, respiratory syncytial virus, and Sendai virus on human lymphocytes and macrophages, *Infect. Immun.* **35**:1142–1146 (1982).

55. Nakayama, T., Immune-specific production of gamma interferon in human lymphocyte cultures in response to mumps virus, *Infect. Immun.* **40**:486–492 (1983).

56. Dunnick, J. K., and G. J. Galasso, Clinical trials with exogenous interferon: Summary of a meeting, *J. Infect. Dis.* **139**:109–123 (1979).

57. Aderka, D., H. Holtmann, L. Toker, T. Hahn, D. Wallach. Tumor necrosis factor induction by Sendai virus, *J. Immunol.* **136**:2938–2942 (1986).

58. Brenan, M., and R. M. Zinkernagel, Influence of one virus infection on a second concurrent primary *in vivo* antiviral cytotoxic T-cell response, *Infect. Immun.* **41**:470–475 (1983).

59. Scott, R., M. O. De Landazuri, P. S. Gardner, and J. J. T. Owen, Human antibody-dependent cell-mediated cytotoxicity against target cells infected with respiratory syncytial virus, *Clin. Exp. Immunol.* **28**:19–26 (1977).

60. Cranage, M. P., P. S. Gardner, and K. McIntosh, *In vitro* cell-dependent lysis of respiratory syncytial virus-infected cells mediated by antibody from local respiratory secretions, *Clin. Exp. Immunol.* **43**:28–35 (1981).

61. Kaul, T. N., R. C. Welliver, and P. L. Ogra, Development of antibody-dependent cell-mediated cytotoxicity in the respiratory tract after natural infection with respiratory syncytial virus, *Infect. Immun.* **37**:492–498 (1982).

62. Meguro, H., M. Kervina, and P. F. Wright, Antibody-dependent cell-mediated cytotoxicity against cells infected with respiratory syncytial virus: Characterization of *in vitro* and *in vivo* properties, *J. Immunol.* **122**:2521–2526 (1979).

63. Alsheikhly, A. R., B. Wahlin, T. Andersson, and P. Perlmann, Virus-induced enhancement of lymphocyte-mediated antibody-dependent cytotoxicity (ADCC) *in vitro*, *J. Immunol.* **132**:2760–2766 (1984).

64. Alsheikhly, A., C. Örvell, B. Härfast, T. Andersson, P. Perlmann, and E. Norrby, Sendai-virus-induced cell-mediated cytotoxicity *in vitro*, *Scand. J. Immunol.* **17**:129–138 (1983).

65. Alsheikhly, A-R, T. Andersson, and P. Perlmann, Virus-mediated induction in human lymphocytes of antibody-independent cytotoxicity (ADCC) against natural killer-resistant tumor target cells, *Cell Immunol.* **88:**511–520 (1984).
66. Alsheikhly, A. R., C. Örvell, T. Andersson, and P. Perlmann, The role of serologically defined epitopes on mumps virus HN-glycoprotein in the induction of virus-dependent cell-mediated cytotoxicity, *Scand. J. Immunol.* **22:**529–538 (1985).
67. Härfast, B., T. Andersson, V. Stejskal, and P. Perlmann, Interactions between human lymphocytes and paramyxovirus-infected cells: Adsorption and cytotoxicity, *J. Immunol.* **118:**1132–1137 (1977).
68. Welsh, R., and L. A. Hallenbek, Effect of virus infections on target cell susceptibility to natural killer cell-mediated lysis, *J. Immunol.* **124:**2491–2497 (1980).
69. Andersson, L. J., J. C. Hierholzer, C. Tsou, R. M. Hendry, B. F. Fernie, Y. Stone, and K. McIntosh. Antigenic characterization of respiratory syncytial virus strains with monoclonal antibodies, *J. Infect. Dis.* **151:**626–633 (1985).
70. Mufson, M. A., C. Örvell, B. Rafnar, and E. Norrby, Two distinct subtypes of human respiratory syncytial virus, *J. Gen. Virol.* **66:**2111–2124 (1985).
71. Welliver, R. C., D. T. Wong, M. Sun, B. S. Middleton, Jr., R. S. Vaughan, and P. L. Ogra, The development of respiratory syncytial virus-specific IgE and the release of histamine in nasopharyngeal secretions after incubation, *N. Engl. J. Med.* **15:**841–846 (1981).
72. Kim, H. W., J. G. Canchola, C. D. Brandt, G. Pyles, R. M. Chanock, K. Jensen, and R. H. Parrott, Respiratory syncytial virus disease in infants despite prior administration of antigenic inactivated vaccine, *Am. J. Epidemiol.* **89:**422–434 (1969).
73. Fulginiti, V. A., J. J. Eller, O. F. Sieber, J. W. Joyner, M. Minamitani, and G. Meiklejohn, Respiratory virus immunization. I. A. field trial of two inactivated respiratoy virus vaccines: An aqueous trivalent parainfluenza virus vaccine and an alumprecipitated respiratory syncytial virus vaccine, *Am. J. Epidemiol.* **89:**435–448 (1967).
74. Chin, J., R. L. Magoffin, L. A. Shearer, J. H. Schieble, and E. H. Lennette, Field evaluation of a respiratory syncytial virus vaccine and a trivalent parainfluenza virus vaccine in a pediatric population, *Am. J. Epidemiol.* **89:**449–463 (1969).
75. Penttinen, K., E-P. Helle, and E. Norrby, Differences in antibody response induced by formaldehyde inactivated and live mumps vaccines, *Dev. Biol. St.* **43:**265–268 (1979).
76. Murphy, B. R., G. A. Prince, E. E. Walsh, H. W. Kim, R. H. Parrott, V. G. Hemming, W. J. Rodriguez, and R. M. Chanock, Dissociation between serum neutralizing and glycoprotein antibody responses of infants and children who received inactivated respiratory syncytial virus vaccine, *J. Clin. Microbiol.* **24:**197–202 (1986).
77. Norrby, E., G. Enders-Ruckle, and V. ter Meulen, Differences in the appearance of antibodies to structural components of measles virus after immunization with inactivated and live virus, *J. Infect. Dis.* **132:**262–269 (1975).
78. Prince, G. A., A. B. Jenson, V. G. Hemming, B. R. Murphy, E. E. Walsh, R. L. Horswood, and R. M. Chanock, Enchancement of respiratory syncytial virus pulmonary pathology in cotton rats by prior intramuscular inoculation of formalin-inactivated virus, *J. Virol.* **57:**721–728 (1986).
79. Elango, N., G. A. Prince, B. R. Murphy, S. Venkatesan, R. M. Chanock, and B. Moss, Resistance to human respiratory syncytial virus (RSV) infection induced by immunization of cotton rats with a recombinant vaccinia virus expressing the RSV G glycoprotein, *Proc. Natl. Acad. Sci. USA* **83:**1906–1910 (1986).
80. Appel, M. J. G., Reversion to virulence of attenuated canine distemper virus *in vivo* and *in vitro*, *J. Gen. Virol.* **41:**385–393 (1978).
81. Rossiter, P. B., and R. C. Wardley, The differential growth of virulent and avirulent strains of rinderpest virus in bovine lymphocytes and macrophages, *J. Gen. Virol.* **66:**969–975 (1985).
82. Duc-Nguyen, H., and W. Henle, Replication of mumps virus in human leukocyte cultures, *J. Bacteriol.* **92:**258–265 (1966).

83. Fleischer, B., and H. W. Kreth, Mumps virus replication in human lymphoid cell lines and in peripheral blood lymphocytes: Preference for T cells, *Infect. Immun.* **35**:25–31 (1982).

84. Mangi, R. J., T. P. Munyer, S. Krakowka, R. O. Jacoby, and F. S. Kantor, A canine distemper model of virus-induced anergy, *J. Infect. Dis.* **133**:556–563 (1976).

85. Kauffman, C. A., A. G. Bergman, and R. P. O'Connor, Distemper virus infection in ferrets: An animal model of immunosuppression, *Clin. Exp. Immunol.* **67**:617–625 (1982).

86. Kaul, T. N., H. Faden, and P. L. Ogra, Effects of respiratory syncytial virus and virus–antibody complexes on the oxidative metabolism of human neutrophils, *Infect. Immun.* **32**:649–654 (1981).

87. Smith, T. F., K. McIntosh, M. Fishaut, and P. M. Henson. Activation of complement by cells infected with respiratory syncytial virus, *Infect. Immun.* **33**:43–48 (1981).

88. Jakab, G. J., and G. A. Warr, Immune-enhanced phagocytic dysfunction in pulmonary macrophages infected with parainfluenza 1 (Sendai) virus[1–3], *Am. Rev. Respir. Dis.* **124**:575–581 (1981).

89. Silverberg, B. A., G. J. Jakab, R. G. Thompson, G. A. Warr, and K. S. Boo, Ultrastructural alterations in phagocytic functions of alveolar macrophages after parainfluenza virus infection, *J. Reticuloendothel. Soc.* **25**:405–416 (1979).

# 18

# Immunosuppression by Measles Virus

## PAOLO CASALI, MINORU NAKAMURA, and MICHAEL B. McCHESNEY

## 1. INTRODUCTION

The ability of measles virus to alter an expected immune (adaptive) response was first recognized by Clements von Pirquet in 1908. He observed that children who were tuberculin positive before contracting acute measles virus infection failed to mount specific skin responses to tuberculin during measles. Later reports confirmed von Pirquet's observation and extended his findings to other infectious agents.[1-8] Furthermore, during acute measles virus infection, humans may not make antibodies to tetanus toxoid or $H$ and $O$ antigens of *Salmonella typhi*. Lymphocytes harvested from persons undergoing acute measles virus infection respond poorly, *in vitro*, to a variety of mitogenic or antigenic stimuli and are deficient in producing chemotactic factors. Therefore, *in vivo* infection with measles virus can result in a weakened or abolished immune response and in susceptibility to concurrent infection with other viral or bacterial agents.[9,10-16] Indeed, it has been recognized that during measles virus infection, susceptibility to herpes simplex virus (HSV) increases[17] and pulmonary tuberculosis worsens.[18-20] By contrast, lipoid nephrosis, a disease that frequently responds to immune suppressive therapy, can dramatically improve in the presence of a concomitant measles virus infection.[21-23] The concept that measles virus can also impair the function of some cells involved in natural defense (nonadaptive) mechanisms stems from more recent findings and sug-

PAOLO CASALI and MINORU NAKAMURA • Laboratory of Oral Medicine, National Institute of Dental Research, National Institutes of Health, Bethesda, Maryland 20892.    MICHAEL B. McCHESNEY • Scripps Clinic and Research Foundation, La Jolla, California 92037.

gests the need for wider reconsideration of the *in vivo* immunosuppressive role of measles virus.[24] Even for children living in affluent Western societies, acute measles can lead to serious complications in the presence of important predisposing factors. Malnutrition or marginal malnutrition, which can be frequently observed in children in Third World countries, pose a significant threat to children infected with the virus. Indeed, in Africa, measles is often a serious disease, with high mortality and morbidity that can be largely ascribed to secondary viral or bacterial infections.[25]

This chapter reviews the experimental data available on the interaction between measles virus and the various subsets of human mononuclear cells. We focus on the understanding of mechanisms by which measles virus can modulate the specific immune response as well as the natural response of the body.

## 1.1. Characteristics and Basic Properties of Measles Virus

Measles virus is a member of the genus *Morbillivirus* in the family Paramyxoviridae. Canine distemper and rinderpest viruses are also members of this family. Unlike other paramyxoviruses, morbilliviruses lack any detectable neuraminidase activity. Structure and properties of this virus were recently reviewed by Norrby.[26] Virions are enveloped particles with a mean diameter of 150 nm and are morphologically indistinguishable from other paramyxoviruses. The viral particle contains a single-stranded RNA (250S) genome with negative polarity and six virion proteins (Fig. 1). Three of these proteins are internal, and the other three contribute to the formation of the viral envelope.[27–29] The three internal proteins complexed with the viral RNA include

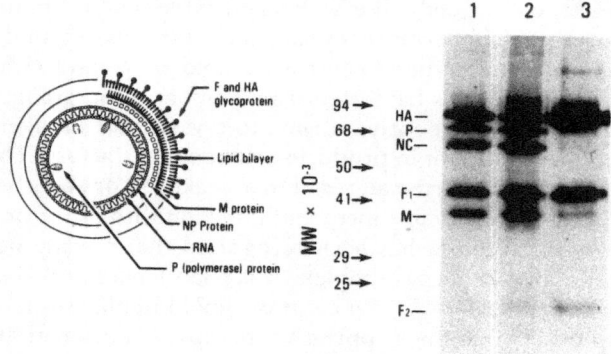

**FIGURE 1.** Schematic representation of a measles virion and autoradiogram of 10% sodium dodecyl sulfate–polyacrylamide gel electrophoresis (SDS–PAGE) of purified [³H]leucine metabolically labeled virions (lane 1). Passed through fraction from an agarose–lentil lectin column following application of NP-40 disrupted virions (lane 2). Purified HA and F (F₁ and F₂ moieties) glycoproteins eluted from the lentil lectin column by 0.1 M α-methylmannoside (lane 3). L protein is not visible in this preparation. F and HA glycoproteins are not distinguishable in electron micrographs. The F and HA glycoproteins appear to penetrate the lipid bilayer and may traverse the entire envelope. The M protein forms the inner layer of the envelope and maintains its structure and integrity. The actual arrangement of the NP protein in the nucleocapsid is unknown. (Modified from Casali *et al.*[112])

the phosphorylated nucleoprotein (NP), 60,000 daltons; the large protein (L), 180,000 daltons; and the phosphoprotein (P), 72,000 daltons. Recent cloning and sequencing of the entire phosphoprotein gene allowed for the identification of two open reading frames: a major coding for the P protein and a minor one coding for a previously unrecognized protein now designated C.[30] L and P proteins are only minor components and are likely part of the transcription complex. The three proteins associated with the envelope along with the lipid membrane comprise the matrix protein (M), 36,000-dalton; the hemagglutinin (HA), 79,000-dalton; and the fusion (F) 60,000-dalton peplomers. HA and F peplomers are glycosylated and expressed as transmembranous structures. HA is present mainly as a dimer in infected cells and virions. The F glycoprotein, as in other paramyxoviruses, is responsible for the cell-fusing activity of the virus and undergoes post-translational cleavage to give two disulfide-linked subunits, $F_1$ and $F_2$. Measles virus is atypical, however, in that all the carbohydrate in F is contained in the small $F_2$ subunit. HA and F glycoproteins protrude from the lipid envelope of the virion, structured in the form of stalks with terminal knobs, and mediate both the absorption of the virion to nucleated cells and erythrocytes and the fusion activity. After absorption to a cell, penetration of that cell, and replication (measles virus contains an internal RNA-dependent RNA polymerase), the virus matures at the surface of the infected cell by budding. Therefore, the virion envelope is constituted by the cytoplasmic cell membrane from which the virus derived. This implies that, besides virus-specific glycoproteins, other cell-related glycoproteins may occasionally be integrated into the virion envelope.

Measles virus replication in the host cell leads to the disruption of cellular cytoskeleton and the formation of either intracytoplasmic or intranuclear inclusion bodies, or both. The virus can also cause cell-fusion phenomena that may result in the formation of syncitia and cell death. In blood mononuclear cells, viral replication can occur in actively replicating or at least activated cells; little or no viral replication seems to take place in resting cells.

Immune elimination of infected cells and whole virions depends on recognition of virus-encoded cell-surface glycoproteins.[31–33] In cultures containing a source of complement, measles virus-infected cells can activate the alternative pathway,[34] leading to the deposition of C3 molecules on their own surfaces and lysis when anti-measles virus antibodies are present.[35] Measles virus-infected cells may also be lysed by immunoglobulin G (IgG) bound to viral glycoproteins expressed at the surface of infected cells and cytotoxic lymphocytes bearing receptors for the Fc portion of IgG,[36] by natural killer (NK) cells,[37] and by cytotoxic T lymphocytes (CTL).[38–40] In vitro generation of measles virus-specific CTL has resulted in effector cell lytic functions that are restricted by class II molecules of the major histocompatibility complex.[41,42]

## 1.2. Human Pathology Associated with Measles Virus Infection

In humans, infection with measles virus results most frequently in acute disease with widespread dissemination of the virus into epithelial membranes, small blood vessels, the lymphatic system, and often the central nervous system

(CNS). Generally, in otherwise healthy well-nourished persons, the acute infection resolves within a few weeks, without leaving any residual signs of disease. In malnourished and/or compromised persons with impaired natural and immune defense mechanisms, progressive infection and disease develop, with heavy lung, kidney, and CNS involvement, often leading to death.[25] In normal persons, more often than in immune-compromised hosts, measles infection can also result in symptomatic encephalitis. Although this outcome is observed in only 1 of 1000 infected patients, nonsymptomatic viremia and infection of the CNS are thought to occur in many self-resolving cases of acute measles infection. In persons who have had acute measles infection, imperfect clearance of the virus can result, later on in life, in a rare complication, such as subacute sclerosing panencephalitis (SSPE), due to persistent infection of the CNS. The CNS of patients with SSPE shows generalized perivascular inflammation, intranuclear inclusion bodies, and, in more advanced cases, demyelination. Indeed, measles virus antigens have been detected in nervous tissue by immunochemical methods[43–46] and more recently by DNA probe hybridization.[47] Evidence of persistence of measles virus has also been detected in mononuclear cells isolated from peripheral blood and tissues of patients with SSPE.[47–50] In addition, measles virus has been associated with other ailments of unknown etiology, such as multiple sclerosis[14,51–54] and Paget disease,[55,56] but conclusive evidence for its involvement as causative agent in these diseases has not been provided.

## 2. MODULATION OF ADAPTIVE (IMMUNE) AND NONADAPTIVE (NATURAL) RESPONSES BY MEASLES VIRUS

The immune responses mounted by the human body to a foreign antigen are articulated in an adaptive cascade of molecular and cellular associations that lead to the triggering and expression of specific cellular and humoral effector mechanisms. Following entry into the organism, a foreign component is trapped by accessory cells, i.e., monocytes/macrophages and dendritic cells, and is concentrated in lymphoid organs. Trapped antigen is processed by, and re-expressed at, the surface of such accessory cells or antigen-presenting cells (APC). APC present the processed antigen, in the context of their surface glycoproteins, to helper T cells (Th). Following such recognition, relevant Th clone(s) proliferate and release a number of lymphokines, including interleukin-2 (IL-2), B-cell growth factor (BCGF), and other soluble mediators. In the presence of APC and/or soluble antigen, these lymphokines drive the proliferation and differentiation of B lymphocytes to plasma cells (antibody-producing cells), as well as the proliferation and differentiation of the appropriate T precursor cells to CTL and delayed hypersensitivity ($T_D$) cells. Most antibodies and effector T cells generated from such cellular interactions will be specific for the antigen molecule(s) responsible for their generation. Therefore, antiviral antibodies will bind free virions and/or virus-encoded products expressed at the surface of infected cells. Similarly, specific CTL will bind and lyse infected cells expressing surface viral polypeptides in the context of class I, or in some cases class II, molecules. $T_D$ cells will release inflammatory mediators following recognition and binding of viral products passively or actively expressed in the context of surface class II molecules.

Measles virus can variously interfere with and modulate all cell functions involved in both adaptive and natural responses to *in vivo* invading microorganisms. Interference by measles virus with the mechanisms of activation, proliferation, and differentiation of various cell subsets leads to different patterns of alteration of the immune and natural responses.

## 2.1. Infection of Human Leukocytes by Measles Virus

In spite of the absolute peripheral blood lymphocytosis often observed during acute viral infection (e.g., infectious mononucleosis, mumps, rubella, infectious hepatitis), acute viral diseases in humans are generally characterized by varying degrees of leukopenia due to a drastic drop in the number of circulating polymorphonuclear (PMN) cells. By contrast, dramatic lymphopenia is an important component of the leukopenia associated with measles.[57,58] The reduction in lymphocyte count is mostly due to a paucity of T cells with little or no decrease in the number of B cells.[8,58] In subjects with measles, the relative distribution of helper T cells and cytotoxic/suppressor T cells is comparable to that present in healthy persons.[58]

The fact that the virus responsible for producing measles is associated with cells involved in human defense mechanisms was first shown by Papp,[59] who was able to induce an overt measles disease in a subject who had been given washed leukocytes from a donor with acute measles infection. Measles virus was later isolated from human leukocytes from subjects with acute measles.[60–62] The virus was found to be contained mainly in the mononuclear cell fraction.[63,64] An increase in the number of large lymphocytes can be observed during the acute phase of measles, and formation of giant cells containing measles virus antigens has been induced by phytohemagglutinin (PHA) stimulation of lymphocytes obtained from children up to several days after the onset of the rash.[65]

*In vitro*, measles virus can efficiently infect peripheral blood T and B lymphocytes and, to a lesser extent, monocytes from healthy adult donors.[64,66–68] Both lymphocytes and monocytes obtained from the umbilical cords of healthy newborns are more susceptible to infection with measles virus than are lymphocytes from adults.[64] Using an infectious centers assay and, later, DNA probe hybridization, it has been determined that T and B lymphocytes replicate measles virus more efficiently than do monocytes.[69] This finding confirms early data obtained by using immunofluorescence and electron microscopy techniques.[67,68]

Since the original observations of infection of lymphocytes by measles virus, it has been apparent that activated and dividing mononuclear cells replicate virus more efficiently than do resting cells.[64,67,70,71] The number of infectious centers produced after activation of mononuclear cells and subsequent infection with the virus depends on the modalities used to induce cellular activation. Indeed, the following activation stimuli allow for decreasing degrees of viral replication: (1) PHA; (2) allogeneic cells; (3) concanavalin A (Con A); and (4) pokeweed mitogen (PWM).[64]

It is now clear that infection of lymphocytes by virus does not require cell activation. Incubation of resting lymphocytes with virus results in a "silent infection"[24,40,69] with minimal or no expression of viral genes products at the cell surface and minimal or no release of virions. These "silently" infected lymphocytes can, however, be driven to produce large numbers of virions following activation with mitogens such as PHA, PWM, Con A, or specific antigens.[24,40] The comparative lack of viral replication in resting lymphocytes is likely due to a failure of the viral replication system after successful cell penetration by the virus. This silent infection of lymphocytes markedly differs from the "persistent" infection of mononuclear cells and other cell types (e.g., fibroblasts) by the same virus. Silently infected cells would not express easily detectable amounts of viral products on their own surfaces, while persistently infected cells variably express surface viral antigens. However, it is possible that such a clear-cut distinction would become inappropriate when better information about activity at the molecular level of the infected cell is gained.

The critical parameters of the silent infection by measles virus have been recently analyzed by the simultaneous application of immunofluorescence, immunoblotting, and nucleic acid hybridization techniques.[69] Indeed, whereas no detectable virions were released from infected resting lymphocytes, such cells were found to harbor many more virions than those used for their infection. Viral HA and NP polypeptides were consistently detected in trace amounts by immunoblotting; DNA probes showed the presence of both genomic and viral messenger RNA (mRNA). These data suggest that infection of resting lymphocytes by measles virus results in intracellular RNA replication and viral polypeptide production and assembly in the absence of an overt productive infectious cycle. Silent measles virus infection is therefore characterized by complete but limited viral replication. Subsequent mitogen stimulation allows a full productive infectious cycle to take place, indicating that maximal viral replication and release require mechanisms available only in activated cells.

## 2.2. Modulation of Antibody Production by Measles Virus

Whittle *et al.*[72] produced the first precise report that measles virus could suppress a specific antibody response. Children with acute measles, who were free from other manifest viral, bacterial, or parasitic infections, were unable to mount a normal antibody response to *H* and *O* antigens of *Salmonella typhi* and to tetanus toxoid following challenge with the relevant vaccines. However, total serum Ig concentration of these subjects was not different from that of healthy controls.[57,71] In agreement with these *in vivo* data, it was later shown that measles virus could negatively modulate IgM production *in vitro* in peripheral blood mononuclear cells (PBMC) cultured with PWM.[73] Like most *in vivo* naturally occurring antigens, including *Salmonella H* and *O* glycoproteins, PWM requires the presence of T lymphocytes and monocytes to activate B cells and drive them to proliferation and differentiation.[74,75] These initial observations were strengthened and expanded by Pelton et al.,[76] who found that measles virus can abrogate the production of not only IgM but of IgG as well. This negative effect on antibody production can also be exerted by influenza virus, but not by the laboratory strain (AD-169) of human cytomegalovirus (CMV)[24,77] (Table I). Infectious virus is required for this inhibitory activity because culturing of mononuclear cells with inactivated virus does not result in an abrogation of an antibody response.[24] In addition, in vitro antibody production is abolished only if PBMC are infected before, simultaneous to, or shortly after their activation by PWM.[24 73] Addition of virus to PBMC primed at least 3–4 days earlier results in little or no alteration of the amount of Ig accumulated at day 7 in the culture.[24] Similarly, in vitro antibody response to Diphtheria toxoid by tonsil cells can be suppressed by early infection with measles virus. However, infection of cultures already secreting antibody does not result in immunosuppression.[76] Measles virus suppression of Ig production in PBMC cultures is not the result of mere lytic events, since under the experimental conditions used viability of infected PBMC was not compromised after 7 days in culture,[24] although proliferation of at least some of the infected cells in culture seemed to be impaired.

Reconstitution experiments with purified cell subsets were designed to explore whether defective antibody production by infected PBMC is due to impairment of Th cell activity or to a direct effect of measles virus on B lymphocytes.[78,79] In vitro, T-cell and monocyte requirements for antibody production can be replaced by culture supernatants of stimulated T cells containing B-cell growth and differentiation factors.[75,80,81] In one experiment, purified B cells were cultured in T cell monocyte-conditioned medium (TCM)

## TABLE I
### Measles Virus and Influenza Virus but Not Human Cytomegalovirus Abrogate in Vitro T-Cell-Dependent Antibody Production by Human B Lymphocytes[a]

| Lymphocytes from donor | Nil | | Measles virus[b] Virion | | Measles virus[b] UV virion | | Influenza[c] Virion | | Influenza[c] UV virion | | HuCMV[d] Virion | | HuCMV[d] UV virion | |
|---|---|---|---|---|---|---|---|---|---|---|---|---|---|---|
| | IgM[e] | IgG[e] | IgM[e] | IgG[e] | IgM[e] | IgG[e] | IgM[e] | IgG[e] | IgM[e] | IgG[e] | IgM[e] | IgG[e] | IgM[e] | IgG[e] |
| A | 93 | 140 | 6 | 4 | 74 | 103 | 6 | 4 | 97 | 100 | 100 | 127 | — | 120 |
| B | 278 | 303 | 6 | 4 | 320 | 82 | 8 | 4 | 700 | 113 | 350 | 267 | 387 | 367 |
| C | 313 | 167 | 20 | 12 | 253 | 300 | 30 | 15 | 700 | 123 | 200 | 300 | 700 | — |
| D | 466 | 533 | 28 | 4 | —[f] | 200 | 30 | 4 | — | 16 | 360 | 517 | — | 533 |

[a] Human PBMC were infected with virus (MOI of 3), UV-inactivated virus or mock infected (nil), and then distributed in culture in presence of PWM. After 7 days, culture fluids were assessed for IgM and IgG concentration.
[b] Edmonston strain.
[c] A/WSN/33 strain (H1N1).
[d] AD-169 strain.
[e] Ig (ng) produced by $1 \times 10^5$ B lymphocytes. Mean of triplicate cultures.
[f] Not done.

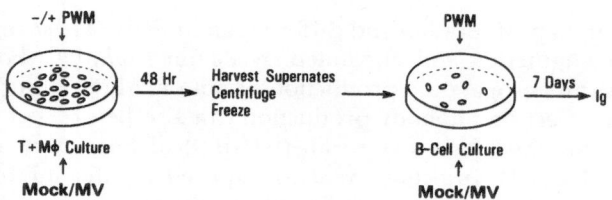

**FIGURE 2.** Effect of measles virus on T or B lymphocytes, or both. Method for preparing T-cell-conditioned medium (TCM) in the absence (Mock) or presence (MV) of measles virus, and its use in a pure B-cell PWM-driven proliferation assay in absence (Mock) or presence of virus (MV). (Modified from McChesney *et al.*[79])

produced by uninfected or infected T cells (Fig. 2) and PWM, following infection or mock infection with measles virus. The Ig produced was measured after 7 days of incubation. As shown in Table II, mock infected B cells, when cultured in TCM from stimulated T cells, secreted 10 times more IgG and six times more IgM than did control cultures. TCM from infected T cells were equally or more stimulatory than TCM from mock infected T cells for mock infected B cells. Ig secretion did not increase significantly above a minimum negative control level, whether infected B cells were cultured in TCM from mock infected or infected T cells, and for each TCM, the level of Ig secreted from infected B cells was suppressed compared with that from mock infected B cells.[78,79]

Because NK cells may modulate B-cell activity,[82] and given the fact that the only cell subpopulation contaminating the B-cell fractions used in the ex-

**TABLE II**
**Suppression by Measles Virus of PWM Driven Antibody Secretion Is Mediated by Infection of B Cells[a]**

|  | IgG (ng/ml) | | IgM (ng/ml) | |
|---|---|---|---|---|
| Medium | B-mock | B-MV | B-mock | B-MV |
| Medium, no PWM | 34[b] | 6 | 119 | 15 |
| TCM from |  |  |  |  |
| T-mock + mφ,[c] no PWM | 63 | 6 | 172 | 24 |
| T-mock + mφ + PWM | 369 | 12 | 610 | 82 |
| T-MV + mφ + PWM | 830 | 32 | 1256 | 128 |

[a]T cell-conditioned media (TCM) were prepared from 48-hr T-cell–monocyte (mφ) cultures in the presence or absence of PWM. T cells were mock infected (T-mock) or infected with measles virus (T-MV) (see Fig. 2). B cells, $1 \times 10^5$/well, were cultured in 50% TCM. PWM was present in all cultures. B cells were mock infected (B-mock) or infected with measles virus (B-MV). B-cell culture supernatants harvested after 7 days were assayed for IgG and IgM concentrations by an enzyme-linked immunosorbent assay (ELISA).
[b]Mean of triplicate cultures.
[c]mφ, macrophages.

periments described above was represented by some residual NK cells, a possible role of NK cells in mediating measles virus-induced suppression of Ig production has also been investigated.[78] As shown in Table III, although a high number of NK cells added to B cells may produce some decrease in the amount of Ig production, production of Ig is still reduced to baseline values in highly purified, NK cell free B lymphocyte preparations following infection with the virus. Therefore, measles virus suppression of *in vitro* PWM-driven Ig secretion is mediated by direct interaction with B lymphocytes, not through T lymphocytes, monocytes, or NK cells. Suppression of Ig production is not the result of a lytic event, although some alteration of cytoplasmic structures devoted to protein synthesis, glycosylation, and secretion is likely to occur in activated infected B lymphocytes. Short-term infection of T cells and monocytes by measles virus does not impair their ability to synthesize and secrete B-cell growth and differentiation factors. Measles virus can interfere with the early stages of B-cell differentiation, but it is unable to abrogate already established Ig secretory functions, nor does it likely interfere with the late stages of B-cell differentiation. Small resting B lymphocytes, whether memory B cells or virgin B cells, are activated *in vitro* and *in vivo* following crosslinking of the surface receptor molecules, and express surface receptors for proliferation and differentiation factors produced by T cells. These soluble factors can drive activated cells, but not resting B cells, through proliferation and eventual differentiation into plasma cells.[83] Measles virus-induced suppression of B-cell function would likely take place during cell activation or early proliferating stages.

As the data reviewed above clearly indicate, suppression of Ig synthesis by measles virus can result from direct infection of antibody-producing cell precursors (B lymphocytes), but this is probably not the only way in which the virus can alter Ig production. Indeed, measles virus is able to infect virtually every lympho-

## TABLE III
### Depletion and Reconstitution of NK Cells Does Not Affect Measles Virus Suppression of B-Cell Function

| B cells in culture[a] | IgG (ng/ml) | |
|---|---|---|
| | Mock | MV |
| 1 × 10⁵/well | 1412[b] | 49 |
| 1 × 10⁵/well plus NK cells, 5 × 10⁴/well | 1128 | 70 |
| 1 × 10⁵/well plus NK cells, 1 × 10⁵/well | 758 | 49 |

[a]B cells depleted of natural killer (NK) cells (by lysis with mouse monoclonal antibody B73.1 and complement), or B cells plus NK cells, were cultured in 50% TCM from stimulated T cells for 17 days. Culture supernatants were assayed for IgG by an enzyme-linked immunosorbent assay (ELISA). Cells were mock infected or infected with measles virus (MV) at initiation of culture.
[b]Mean of triplicate cultures.

reticular cell subpopulation, including T lymphocytes, monocytes/macrophages, and possibly dendritic cells. *In vivo* infection of proliferating Th cells involved in a humoral response to a given T-dependent antigen would eventually lead to cell lysis and depletion of the very same Th cells necessary for the production of the relevant antibody. As a result, no specific T help would be available for B cells and, for that matter, effector T-cell precursors. In addition, infection of mononuclear cells by measles virus *in vitro* results in efficient production of interferon-α (IFN$_\alpha$) and, to a lesser extent, IFN$_\gamma$.[37,84,85] In fact, remarkable concentrations of IFNs are present in intravascular and extravascular spaces during an *in vivo* viral infection.[86,87] IFN$_\alpha$ and IFN$_\gamma$ can efficiently suppress T lymphocyte, and probably B lymphocyte, proliferation in a number of systems involving mitogenic stimuli.[88] Furthermore, it has been clearly demonstrated that IFN$_\alpha$ can drastically reduce the antibody production by human PBMC cultured *in vitro* with PWM.[89]

The fact that IFNs can modulate *in vitro* lymphocyte proliferation, antibody production, and other functions suggests that IFNs may be responsible for at least some of the suppression of the antibody response observed *in vivo* during measles. Indeed, in one study in which seropositive subjects were vaccinated with attenuated measles virus vaccine, no traces of cell-free virus were found to be present in the circulation and circulating mononuclear cells did not carry the virus. Yet the response of these PBMC to mitogen and PPD was impaired,[90] suggesting a possible suppressive role for IFNs. Monocytes and macrophages may also play a direct role in the modulation of an antibody response *in vivo*. It has been demonstrated that human macrophages, i.e., monocytes allowed to adhere to tissue culture plastic for a given period, can suppress *in vitro* mitogen proliferation of syngeneic T cells and proliferation of T cells induced by stimulation with allogeneic irradiated B cells.[91] In this case, the inhibitory effect is not mediated through soluble cytokines but requires the direct macrophage–lymphocyte contact in order to be expressed.

On the basis of the data presented above, it could be postulated that measles virus modulates antibody production by (1) infection and direct alteration of B-cell biologic functions; (2) infection of Th cells and consequent abortion of the T help necessary for the proliferation and differentiation of specific B cells; and (3) infection of various mononuclear cells subsets, subsequent release of IFNs, and possibly other cytokines with inhibitory effect on Th and B lymphocytes. Following infection with measles virus *in vivo*, the resultant net modulation of antibody production will depend on whether all these mechanisms are operational at different times or simultaneously and at which degrees.

## 2.3. Modulation of Function of Amplifying T Cell and Effector T Cell by Measles Virus

The immune defect originally described by von Pirquet[92] in children with measles, i.e., lack of a cutaneous response to tuberculin, demonstrates that measles virus is capable of limiting the acquisition of some T-cell functions and the generation of effector $T_D$ cells. Indeed, impairment of various T-cell activities has been shown *in vivo* during measles and *in vitro* in lymphocytes from children with clinical disease. The same observation has been extended to antigens other than tuberculin purified protein derivative (PPD), such as *Candida* and streptococcal antigens, and dinitrochlorobenzene.[8,57,71] It has also been shown that the skin window migration of PMN is reduced in those patients.[93] The *in vivo* impairment of $T_D$ cell effector function is in agreement with numerous data obtained *in vitro* in lymphocytes from patients affected with acute measles. Indeed, the response of such lymphocytes to various mitogen (PHA, PWM, Con A), PPD, or other antigens (e.g., *Candida albicans*), as assessed by incorporation of [³H]thymidine by isolated PBMC, was found to be severely impaired.[7,8,57,58,90,94,95] Similarly, mononuclear cells from seropositive persons vaccinated with attenuated measles virus (Enders strain) did not proliferate in response to PHA and did not produce chemotactic factors when challenged with Con A or a number of antigens, including PPD and purified measles virus antigens.[90,96]

In agreement with the lack of response to various mitogens and antigens observed *in vitro* using PBMC from infected subjects, *in vitro* infection of mononuclear cells from healthy persons with measles virus invariably results in abrogation of their proliferative response (Fig. 3). Indeed, *in vitro* infected normal PBMC do not respond to PHA or PWM.[24,71,97−101] In addition, measles virus abrogates the *in vitro* normal PBMC proliferative response to PPD or allogeneic lymphocytes.[98,99] Live but not inactivated [ultraviolet (UV) light-inactivated or heat inactivated] virions can mediate these inhibitory activities. Analysis of cell size in culture shows that blast transformation is inhibited as well. The inhibitory effect of measles virus can take place only if PBMC cultures are

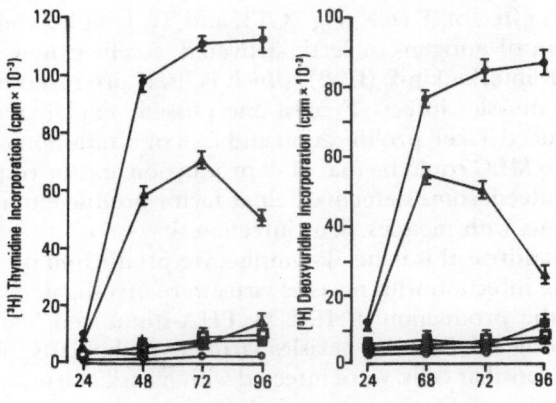

**FIGURE 3.** Measles virus inhibits human lymphocyte proliferation. Human PBMC (10⁵) were incubated with virus at MOI of 3 (■), UV-inactivated virus (▲), or were mock infected (●). Infected and uninfected lymphocytes were then cultured in the presence (filled symbols) or absence (open symbols) of PHA. [³H]Thymidine or [³H]uridine incorporation was measured between 0–24, 24–48, 48–72, and 72–96 hr. (Modified from Casali *et al.*[24])

infected before or soon after their challenge with mitogens, antigens, or allogeneic cells. Pretreatment of infectious virions with IgG anti-measles virus prevents the virus-induced inhibition of cell proliferation.[99] Measles virus abrogation of PBMC reactivity is likely due to a direct effect of the virions on prospective proliferating cells and not to soluble mediators and/or the presence of an accessory cell population, because (1) clear fluids from heavily infected cultures or $IFN_\alpha$ added to normal cells in quantities corresponding to those found in infected cultures cannot reproduce the inhibitory effect on normal lymphocytes otherwise observed in the presence of virus; and (2) virus can still abrogate the proliferation of PBMC cultures depleted of monocytes.[98,99]

Measles virus infects all T-lymphocyte populations[66,69,102] and the T4 + (Th) more efficiently than the T8 + (suppressor-Ts) subset.[102] Stimulation of purified infected T4 + lymphocytes with PWM results in a negligible degree of proliferation concomitant with a high level of virion release.[102] This is in agreement with earlier observations[73,99], showing that measles virus can abrogate an allogeneic response in mixed lymphocyte culture (MLC), in which lymphocytes proliferating in the first phases are almost exclusively Th cells. In vitro, [$^3$H]thymidine incorporation by mononuclear cells, as measured within the first 2–3 days following mitogen or antigeneic stimulation, is also due essentially to proliferation of Th lymphocytes. Therefore, measles virus would block the proliferation of Th cells which constitute an essential element in the initiation and amplification of any immune response. In the absence of a functional Th lymphocyte population, no cellular effector mechanisms can be triggered and amplified during an ongoing immune response. Given the pivotal role of T lymphocytes in the production of helper factors necessary for the recruitment and differentiation of effector T cells, as well as B cells, any reduction of their activity would in turn result in a defective expansion and differentiation of both effector T cells (e.g., CTL and $T_D$ lymphocytes) and B cells. The proliferation of antigen- or lectin-activated T cells is now known to depend entirely on interleukin-2 (IL-2), which is itself produced by activated T cells.[103,104] As measles infects T cells, one possible explanation for the reduced PHA-induced T-cell proliferation and lack of acquisition of cytotoxic T-cell activity in the MLC could be that IL-2 production and/or responsiveness is impaired.[105] Indeed, some defective helper factor production has been documented in patients with measles virus infection.[106]

In order to address this issue, T-lymphocyte production of, and response to, IL-2 following infection with measles virus were investigated by Borysewicz et al.[107] First, the production of IL-2 by PHA-stimulated lymphocytes was examined following infection by measles virus. Lymphoblasts, obtained from PHA-stimulated tonsillar cells, were infected with measles virus, washed free of virus, and then cultured for varying periods. Culture fluids from infected cells were not deficient in IL-2, as assessed by their ability to support proliferation of IL-2-dependent blasts (Fig. 4). Thus, measles virus-infected mononuclear cells are still capable of producing IL-2, which would therefore be available for driving proliferation of primed effector T lymphocytes as well as other helper T cells.

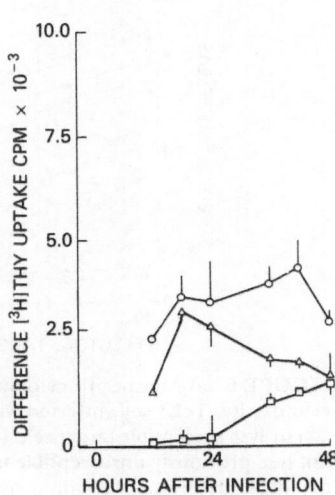

**FIGURE 4.** Measles virus infection does not abrogate IL-2 production in PHA-stimulated cultures. Tonsillar lymphocytes were pretreated with PHA and infected with measles virus (MOI, 5). Supernatants from cultures were harvested at various times and assayed on 6-day PHA blasts for their IL-2-like activity. Lymphocytes infected with measles in the presence of PHA (○) released IL-2 in similar amounts to uninfected PHA-pulsed lymphocytes (△). Cells infected with measles virus in absence of PHA (□) produced low levels of IL-2. (Modified from Borysiewicz et al.[107])

Next, the issue of whether infected T lymphocytes could still respond to IL-2 was addressed. The response to IL-2 of an established T-lymphocyte clone (Tpha) with the helper phenotype (Leu-3a+) was measured after infection with measles virus. Tpha cells, whose growth was partly dependent on exogenous IL-2, were infected with measles virus or inactivated virus, washed, and then cultured in the presence of an appropriate amount of IL-2. IL-2-dependent proliferation of Tpha lymphocytes was not affected by measles virus (Fig. 5). Similarly, measles virus did not impair responsiveness to IL-2 of lymphoblasts obtained 6 days after PHA activation of PBMC, nor did it affect IL-2-dependent proliferation of TcL1 lymphocytes, an allospecific cytotoxic T-

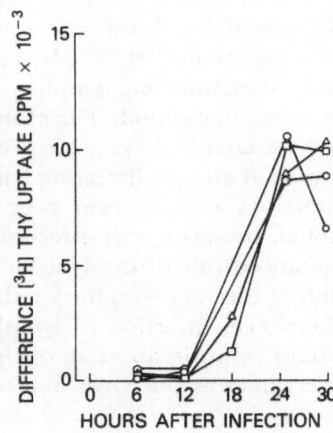

**FIGURE 5.** Established human helper T cells (clone Tpha, Leu3a phenotype) proliferate in response to IL-2 up to 30 hr following measles infection. Tpha cells were mock infected (○), infected with live (◇), UV-inactivated (△), or β-propiolactone inactivated (□) measles virus and cultured in presence of IL-2. Proliferation was estimated by a 4-hr pulse of [³H]thymidine and results expressed as the difference in [³H]thymidine uptake between IL-2 supplemented and medium alone cultures. At 30 hr, more than 80% of lymphocytes were viable in all the cultures and only live measles virus-infected cells expressed measles surface antigens (60%). (Modified from Borysiewicz et al.[107])

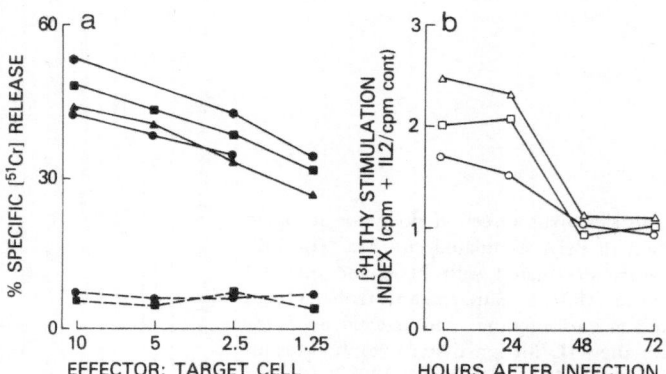

**FIGURE 6.** (a) Allospecific cytotoxic T cells (clone TcL1) continue to mediate antigen specific cytotoxicity. TcL1 cells infected with measles virus (MOI of 5) for 0 (●), 2 (▲), 24 (◆), or 48 (■) hr lyse susceptible target cells (PHA-stimulated PBL) in a 6-hr $^{51}$Cr-release assay. TcL1 do not lyse previously unsusceptible target cells (−−). (b) Measles virus-infected TcL1 (○) and 6-day PHA blasts (△, □) continue to respond by proliferation to IL-2 up to 26 hr postinfection. Viable lymphocytes infected with measles virus (48, 24, and 2 hr previously) and maintained in optimal concentrations of IL-2 were harvested, and $2 \times 10^4$ viable cells placed in medium alone or in medium supplemented with fresh IL-2 for a further 20 hr. [$^3$H]thymidine incorporation after a 4-hr pulse was estimated and results expressed as stimulation index (cpm in culture with IL-2/cpm in culture without IL-2). (Modified from Borysiewicz *et al.*[107])

cell clone generated in a mixed lymphocyte culture (Fig. 6b). In addition, these cytotoxic T lymphocytes (TcL1 clone) were still capable of killing appropriate allogeneic target cells following exposure to measles virus (Fig. 6a). Thus, measles virus-infected T lymphoblasts can still produce IL-2 and can proliferate in response to exogenous IL-2. Similarly, infected cytotoxic T-cell clones can still carry out their cytotoxic functions.

These findings do not appear to be at variance with the virus inhibition of T-lymphocyte proliferation and differentiation observable *in vitro* or with the depressed T-cell response occurring *in vivo* during measles or after vaccination with attenuated virus. Measles virus aborts T-lymphocyte effector functions only when infecting lymphocytes before or right after their priming by antigen or mitogenic stimuli. For example, the generation of allospecific CTL, but not their established cytotoxic activity, can be aborted by measles virus.[73] Thus, activated and proliferating lymphocytes, as well as T and B cells with acquired functions, do not seem to be functionally affected by the virus. It should be noted, however, that infection of actively proliferating cells is followed by an enhanced rate of cytolytic viral replication that may eventually lead to depletion of the very lymphocyte clones actively involved in a response. By contrast, short-term infection of lymphocytes before or at the time of their priming would result in abortion of their activation and proliferation with little or no overt microscopic virus-induced cell damage.

## 2.4. Modulation of Natural Killer Cell Activity by Measles Virus

Recently, considerable attention has focused on nonspecific host mechanisms of antiviral defense (natural, nonadaptive response) because of their potential to halt or reduce viral replication early in infection before the generation of specific antibodies or CTL. Examples of early antiviral mechanisms are are serum complement that can lyse some virions[107] and bind to infected cells in the absence of antibodies,[34] phagocytic cells, natural killer (NK) cells that lyse both infected and uninfected target cells,[109,110] and IFNs, which can inhibit viral replication[86,87] as well as induce new or enhance ongoing NK activity.[109–111]

The induction or enhancement of nonspecific cell-mediated cytotoxicity (CMC or NK activity) by viruses and/or IFNs has received much interest. Among other viruses, measles virus can produce dramatic enhancement of NK cell activity.[37,84,110] Both purified HA and F glycoproteins of measles virus either in soluble form or inserted in an artificial membrane enhance NK cell activity in a dose-dependent fashion.[83,111] The induction of nonspecific CMC by viral glycoproteins, either in the soluble state or inserted into artificial membranes, can be segregated from the CMC associated with whole virions in that glycoprotein-dependent CMC has faster kinetics and is not associated with release of IFNs[37,84] (Fig. 7). Accordingly, nonspecific CMC of measles-infected human fibroblasts or HeLa cells can be segregated into two modalities of expression. The first occurs early (within 2–4 hr) and in the absence of detectable IFN release. The second occurs later (8 hr) and is associated with extracellular IFN release. F(ab')$_2$ fragments to measles virus glycoproteins, but not F(ab')$_2$ fragment to human F(ab')$_2$ fragment, can abrogate the triggering of early CMC by purified glycoproteins in the fluid phase, suggesting that an antibody-dependent cellular cytotoxicity (ADCC) mechanism involving IgG anti-measles virus antibodies passively carried over by B or other lymphocytes

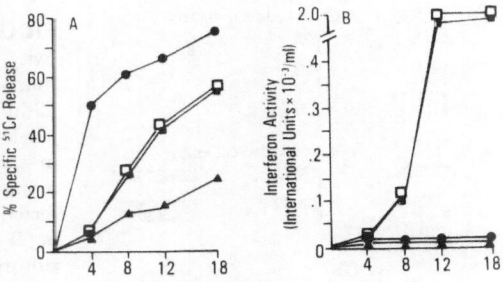

**FIGURE 7.** Measles virus glycoproteins and measles virions induce cell-mediated cytotoxicity with different modalities. Human PBL were cultured with $^{51}$Cr-labeled allogeneic human fibroblasts in presence of PBS (▲), UV-inactivated measles virions (■), measles virions (□), or purified measles virus glycoproteins (●). $^{51}$Cr released into the fluid phase was measured (A) and antiviral activity of the culture fluids titrated (B) at different times of incubation. Effector to target cell ratio was 100 : 1 in the mixed cultures. (Modified from Casali *et al.*[84])

is not operational in this killing process.[37] Conversely, anti-IFN$_\alpha$ antibody can abrogate the late IFN-associated CMC but not the early killing.[37]

After initial enhancement of the cytotoxic activity in NK cells, measles virus can silently infect the very lymphocytes and dramatically alter their function.[24,77] In a series of experiments using various NK susceptible target cells in presence or absence of exogenous IFN$_\alpha$, it was found that impairment of NK activity in human peripheral blood lymphocytes (PBL) by measles virus occurs as soon as 24 hr postinfection. In most cases, after 48 hr of incubation, NK cell activity is totally abrogated,[24] yet the viability of infected lymphocytes is not different from that of their uninfected counterparts. This abrogation of NK cell activity can be prevented by the addition of F(ab')$_2$ fragments to measles virus during the infection step. When equal numbers of PBL from three human donors were infected with measles virus at a multiplicity of infection (MOI) of 3, or were mock infected and then cultured under similar conditions, some decrease of cytotoxic activity in infected lymphocytes was already present at 24 hr of incubation (Fig. 8). PBL from donor 1, infected with measles virus, were roughly equivalent to uninfected PBL from the same donor in lysing K-562 cells, that is, measles virus-infected PBL, 61% specific $^{51}$Cr release; uninfected PBLs, 73% specific $^{51}$Cr release. After 48 and 72 hr of infection, killing by infected PBL of K562 decreased to 22% and 3% specific $^{51}$Cr release, respectively. By contrast, uninfected PBL showed no defect in killing at 48 and 72 hr, releasing 77% and 78% $^{51}$Cr, respectively. Equivalent lysis occurred with HLA-unmatched and -matched fibroblast targets. Impairment of NK cell activity by measles virus was even more overt when lysis of $^{51}$Cr

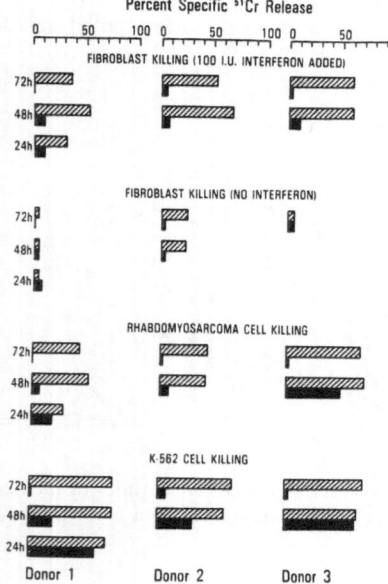

**FIGURE 8.** Measles virus inhibits NK cell activity. Measles virus was used to infect (MOI, 3) human PBL from three donors. At 24, 48, or 72 hr postinfection, PBL were added to $^{51}$Cr-labeled K-562 cells, rhabdomyosarcoma cells, or skin fibroblasts. Specific $^{51}$Cr release was monitored and compared with that of uninfected PBL. Skin fibroblasts were not HLA-A or B matched with donor PBL and were cultured in the absence or presence of 100 IU or IFN$_\alpha$. An effector to target cell ratio of 50 : 1 was used. Measles virus-infected lymphocytes (■); mock-infected lymphocytes (▨). (Modified from Casali *et al.*[24])

**TABLE IV**

**Measles Virus Inhibition of Cytotoxicity Mediated by Nonadherent OKT3⁻ Dr⁻ Lymphocytes (NK Cells)**

| Source of NK cells[a] | Virus culture | % viable cells | K-562 | Fibroblasts +PBS | Fibroblasts +IFNα[b] |
|---|---|---|---|---|---|
| Donor A | Nil | 99 | 53.1[c] | 4.3 | 35.0 |
| E:T, 3:1[d] | UV virus | 99 | 55.9 | 9.1 | 32.5 |
| (48 hr) | Virus | 99 | 9.1 | 5.1 | 3.5 |
| Donor B | Nil | 85 | 68.3 | 9.8 | 55.8 |
| E:T, 5:1 | UV virus | 88 | 27.1 | 15.9 | 33.1 |
| (72 hr) | Virus | 76 | 13.3 | 9.7 | 10.3 |

[a] PBL were depleted of monocytes and of OKT3 and HLA-DR-bearing cells by using appropriate mouse monoclonal antibodies and complement. These purified natural killer (NK) cells were <5% OKT3⁺, HLA-DR⁺, and >90% B73.1⁺ cells. NK cells from two donors were infected (MOI of 3) with measles virus or mock infected and handled identically. Cells were cultured for 48 or 72 hr and then assessed for their viability by trypan blue dye exclusion using at least 200 cells. Infected and noninfected NK cells were tested for their cytotoxic activity on [51Cr]-K-562 or [51Cr]fibroblasts as target cells, in the presence or absence of exogenous IFNα.
[b] Fibroblasts (adult skin, non-HLA matched) were either incubated with 100 IU/ml human IFMα present throughout the assay or left untreated.
[c] Percentage specific 51Cr release. Mean of triplicate cultures.
[d] Effector to target cell ratio.

fibroblasts in the presence of IFN-α was measured (Fig. 8). Decreasing efficiency in killing of rhabdomyosarcoma cells paralleled that of K-562 and fibroblast cells.

These data were confirmed using a purified NK cell population obtained by depletion of irrelevant cell subsets from PBL by appropriate monoclonal antibodies and complement[24] (Table IV). By contrast, data from parallel experiments using total PBL as the source of effector cells and rabbit antibody-coated P-815 cells as targets suggested that ADCC was only marginally affected by measles virus.[24] This finding contrasts with the now established concept that both NK cell activity and ADCC activity are functions expressed by the same non-T, non-B, Fc receptor⁺, DR⁻ cell subset.[113] The fact that the same infected cell would carry out one cytotoxic function but not the other certainly deserves further investigation. Modulation of nonspecific CMC by other human pathogen viruses, including influenza virus and cytomegalovirus (CMV) has also been investigated in experiments similar to those undertaken with measles virus.[24,77,114]

The mechanisms underlying the abrogation of NK cell functions by measles virus are still obscure. Killing of a target cell by an NK cell is an energy-dependent process requiring new protein synthesis and involving the release of cytotoxic granules or membranes perforating the cytoplasmic membrane of the target cell.[115] Thus, NK cell activity is not an established function, but rather an inducible one. Indeed, induction of NK activity by IFNs or infected target cells can be hampered by actinomycin D at a dose 10–100 times lower than that required for abrogation of ADCC[110] (P. Casali, unpublished observations). Within this context and according to the knowledge gained by investigation of the effect of measles virus on established cytotoxic T-cell clones (see Section 2.2), it may be argued that abrogation of NK cell activity by measles virus is due to some interference with the mechanisms involved in the early macromolecular events of cell activation, in the absence of any overt cell damage.

## 3. MECHANISMS OF IMMUNOSUPPRESSION AND SIGNIFICANCE OF VIRUS-INDUCED IMMUNOSUPPRESSION IN VIRAL PATHOGENESIS

Measles virus can infect and lyse virtually all mononuclear cell subsets; this may explain the severe lymphopenia observed *in vivo* during acute infection. However, disruption of infected immunocompetent cells is not the only mechanism by which the virus can produce severe immunosuppression; infection of lymphocytes often result in gross alteration of their function long before cell lysis (Fig. 9). *In vitro* models of infection suggest that measles virus can suppress the adaptive immune response as well as the natural nonadaptive host response. Impairment of nonadaptive first-line mechanisms of defense, such as NK cell activity, can be a major reason for increased susceptibility of subjects with acute measles to other viral infections.[17]

Measles virus infection, especially in association with other immunocompromising cofactors, can have devastating effects on effector functions of the adaptive response. This can take place through the direct infection of precursors of antibody-secreting cells and effector T cells or by negative modulation of Th lymphocyte function. Proliferation of Th cells is blocked by measles virus when infection is initiated before their priming. Some data, although conflicting, have been provided on the *in vivo* suppressive effect of the virus on this cell subset in both mouse and human subjects,[106,116] and showed a decreased production of lymphokines by Th cells necessary to drive B cells through proliferation and differentiation to plasma cells. Abrogation of Th cell proliferation and function can similarly affect the generation of effector T cells, such as CTL or $T_D$ cells. Monocytes and soluble mediators, e.g., IFNs, do not seem to be required for measles virus to carry out its modulatory activity on cells involved in nonadaptive or adaptive mechanisms of defense. However, data are available suggesting that mononuclear phagocytic cells, which can be activated during any viral infection, can modulate lymphocyte functions by direct contact[91] or through the biologic effect of IFNs.

Measles virus does not seem to interfere with replication of already proliferating lymphocytes, as shown by experiments involving infection of estab-

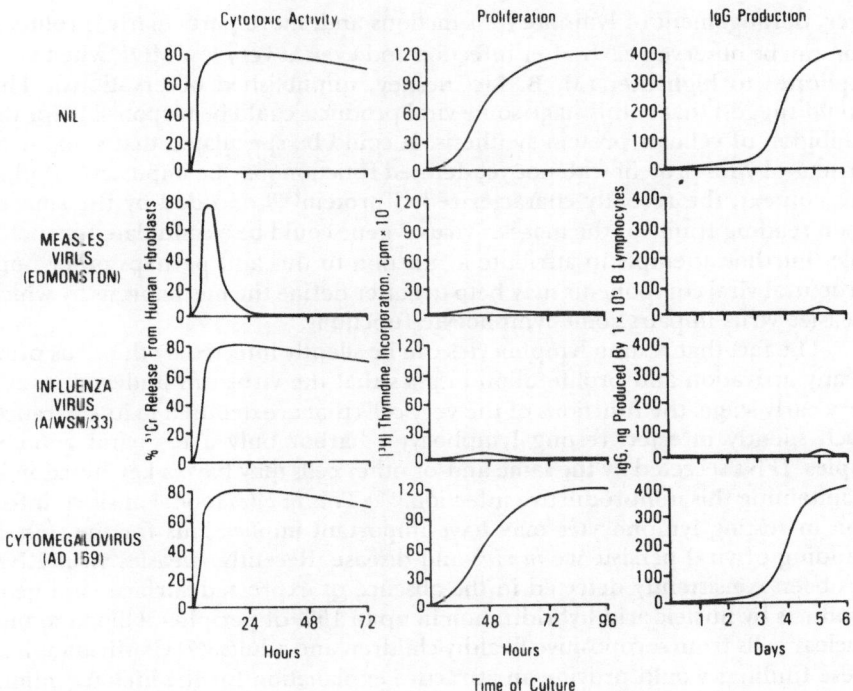

**FIGURE 9.** Infection of human lymphocytes with measles virus *in vitro* results in suppression of NK cell activity, lymphocyte proliferation, and antibody production. The effects of influenza virus and human cytomegalovirus on the same lymphocyte functions are shown for comparison. (Modified from Casali *et al.*[24])

lished T-helper or cytotoxic clones responding to IL-2. Thus, measles virus aborts the generation of inducible functions but not the expression of already established lymphocyte functions. Indeed, generation of cytotoxic lymphocytes is abrogated by the virus, but not the lytic activity of already established cytotoxic cell clones. Similarly, antibody production is abrogated by the virus when added to cell cultures before, simultaneous to, or soon after their priming, but not when added to B lymphocytes already activated that are proliferating and secreting antibody molecules. Therefore, viral inhibitory activity on lymphocytes is expressed at some early step in lymphocyte activation (i.e., blastogenesis and/or early cycling).

Inhibition of lymphocyte proliferation and other functions requires replicating viral particles and/or expression of some viral gene product(s). Replication of the virus is not connected with restriction of host cell metabolism. Yet, rapid impairment of lymphocyte function can be produced late in the infectious process by the exhaustive influence of intense virion replication. However, derangement of lymphocyte functions and, more particularly, proliferation can be observed early after infection and even at very low MOI, when virus replicates to high titer (M. B. McChesney, unpublished observations). This would suggest that at this stage some viral products could be responsible for the inhibition of cellular protein synthesis. It could be speculated that some nonstructural viral protein with not yet defined function may be important. Within this context, the recently characterized C protein[30] encoded by the smaller open reading frame of the measles virus P gene could be a candidate for such a role. Further attempts to attribute a function to this and perhaps other nonstructural viral components may help to better define the mechanisms by which measles virus impairs some lymphocyte functions.

The fact that resting lymphocytes can be silently infected by the virus prior to any activation and proliferation implies that the virus can undermine, at a very early stage, the functions of the very cells that are deputed to its clearance. Such silently infected resting lymphocytes harbor only a few viral genome copies. IFNs secreted by the same and/or other cells may have a key function in maintaining this nonproductive infection.[85] This *in vitro* model of silent infection in resting lymphocytes may have important implications for the understanding of viral persistence *in vivo* and disease. Recently, measles virus RNA has been consistently detected in the absence of expressed surface viral gene products by nucleic acid hybridization in up to 15% of peripheral blood mononuclear cells from seropositive healthy children and adults.[47] Confirmation of these findings would provide an attractive explanation for the lifelong immunity observed after acute measles.[13,14,26,117,118]

Activation of silently virus-infected lymphocytes *in vivo* or *in vitro* invariably leads to viral antigen expression and massive viral replication (productive infection), eventually resulting in cell lysis. Persistent infection can, however, occur in a minority of cells that normally succumb to a lytic infection, although this has not yet been produced in lymphocytes. A persistently infected cell displays abundant viral gene expression, overt production of infectious virus, and enhanced resistance to superinfection by the same virus. By contrast, a silently infected cell has few or no detectable viral antigens and does not generally produce infectious virions. Infection by measles virus of resting mononuclear cells should be considered silent only operationally, as viral gene products were recently demonstrated in 20–25% of such *in vitro* infected resting lymphocytes,[69] and minimal virion release can occasionally be observed (P. Casali, unpublished observations). Therefore, it is possible that in an activated lymphocyte, the abrogation of any function by measles virus is simply due to a transitory condition of functional cell damage. This may be a result of cytoplasmic and/or nuclear macromolecular alterations caused by the virus, occurring before detectable structural damage leading to cell lysis takes place.

In spite of the fact that measles virus can dramatically impair the expression of important functions of the host's adaptive as well as nonadaptive defense, full recovery within 6 weeks is the normal outcome of acute measles. Thus, in spite of virus-induced lymphopenia and immunosuppression, the residual potential of the immune system is still sufficient to handle the challenge brought to the body by other viral and/or bacterial pathogens during a measles infection. As already mentioned (Section 1), only in the presence of important predisposing factors, is the insult brought by measles virus to the immune system in many cases sufficient to allow severe secondary infection by other microbial agents to take place.

In view of the biologic effects of measles virus on the immune system and the above-mentioned data on the presence of virus in lymphocytes from healthy donors, the question arises whether, in an otherwise healthy host, the transitory immunosuppression by the virus could impair a radical clearance of the virus itself, leading in turn to persistent infection and pathology, as suggested by the impaired activity of specific CTL in patients with multiple sclerosis.[119] Although this possibility cannot be entirely ruled out, several findings indicate that persistent infection can be established even in the presence of fully operational immune effector mechanisms. Indeed, no impairment of immune function is observed in patients with SSPE or in healthy children, in whom reportedly up to 90% and 15%, respectively, of circulating lymphocytes harbor the virus.[47] Moreover, experimental data in vitro indicate that a strong immune response may paradoxically contribute to the establishment of persistent viral infection, as reviewed by Oldstone.[15]

A viral infection can become persistent when infected cells lack the virus-encoded surface molecules and can therefore escape the specific attack by the immune system. This can be accomplished in several ways. Measles virus can silently infect cells in absence of any virus-encoded glycoproteins expressed at the cell surface, making immune recognition of these cells impossible by specific antibodies, CTL, and nonadaptive cytotoxic cells (NK cells). For example, abundant viral genome has been detected in infected lymphocytes from patients with SSPE and in healthy subjects in the absence of surface expression of virus-encoded glycoproteins.[47] This pattern can be reproduced in lymphocytes from normal subjects by in vitro infection in selected conditions and in the absence of activating stimuli.[69] These silently infected cells will express virus-encoded surface gene products and replicate virus only following activation. Therefore, silently infected cells can harbor measles virus, still make it inaccessible to immune defense mechanisms. Even infected cells expressing abundant amounts of surface viral glycoprotein(s) and binding specific antibodies can escape immune elimination if lytic mechanisms, such as complement and/or killer cells, are not readily available. Indeed, surface HA and F are promptly modulated by bound antibodies and their reexpression cannot take place until anti-HA and F antibody molecules are present in the space surrounding the cell, leading to chronic modulation of these immunorecognitive molecules.[31,32,120] This may be one of the mechanisms by which persistent infection establishes during the last stages of acute measles, after a valid antiviral immune response has taken place. Alternatively, defective expression of virus-encoded molecules at the surface of an infected cell can result from synthetic defects intrinsic to the virus and/or the infected cells. Indeed, defective expression of HA glycoprotein has been found in lymphocytes from patients with acute measles.[69] Moreover, otherwise functional infecting and replicating virus may abort the budding process or release nonfunctional virions from infected cells due to the lack of certain cellular enzymes. In one study, infection of Daudi lymphoblastoid cells with Edmonston strain of measles virus resulted in lack of cleavage of fusion surface glycoprotein and virion maturation due to the lack of appropriate cellular protease.[121] Restriction of replication of infectious particles may also be dependent on extracellular factors, such as the presence of IFNs. IFNs secreted by the infected cells themselves can inhibit viral replication and release, which can be dramatically reversed following neutralization of IFN biologic activities by anti-IFN antibodies.[85] In other cases, escape of infected cells from the immunologic assault leading to viral persistence can result from selection of particular virus variants with abnormal expression of structural components and altered biologic properties, some of them greatly affecting the host's immune response. In humans, abnormal expression of measles virus components and variants has been observed in SSPE patients. The pathogenic virus is selected presumably within the patient, to be defective in production of M protein and possibly additional envelope components as well.[26,46,54,122–124] In these and other cases of viral persistence, virus spread probably occurs via cell-to-cell infection, with no or minimal presence of extracellular virus. Under these conditions, access to virus particles by immune effector mechanisms and complete virus clearance are unlikely events.

In conclusion, measles virus can silently infect resting blood mononuclear cells and induce functional impairment in absence of overt cellular damage. However, in most cases, activation of a resting infected cell leads to productive viral replication and eventually to a lytic event. Inhibition of lymphocyte function does require infectious particles and is due, in most cases, to direct action of the virus on lymphocytes in the absence of any modulatory effect by soluble mediators released by other activated and/or infected cells. Measles virus does not interfere with the proliferation of already dividing cells, nor does it inhibit the function of lymphocytes with an already established activity. *In vivo*, acute measles is associated with transitory, profound immunodepression. Silent infection of lymphocytes by the virus undermines the functions of the very cells adapted to its clearance and that of other microbic pathogens. However, the establishment of viral persistence so commonly observed after acute measles is unlikely due to the virus-induced transitory impairment of the immune functions. Rather, *in vivo* virus clearance can be impaired by various mechanisms intrinsic to the infected cells or to the virions themselves, or involving the action of the immune effector molecules or cells on infected target cells expressing viral gene products. Most recent data acquired by DNA hybridization suggest that the main reservoirs for viral persistence in healthy subjects following acute measles are blood mononuclear cells, lymphatic organs, and tissues of the CNS. Better knowledge of the mechanisms by which viral persistence is established would certainly shed some light on many patterns of human pathology for which a satisfactory explanation has not yet been identified.

ACKNOWLEDGMENTS.   The work on measles virus by Dr. Paolo Casali and Dr. Michael B. McChesney has been developed in the viral-immunobiology unit headed by Dr. Michael B. A. Oldstone, whom we thank for encouragement and support. We are grateful to Dr. Robert S. Fujinami and Dr. Abner Louis Notkins for helpful discussions. We thank Mrs. Eloise Mange for typing the manuscript.

# REFERENCES

1. Bloomfield, A. L., and J. Mateer, Changes in skin sensitiveness to tuberculin during epidemic influenza, *Am. Rev. Tuberc. Pulm. Dis.* **3**:166–168 (1919).
2. Westwater, J. S., Tuberculin allergy in acute infectious diseases: A study of the intracutaneous test, *Q. J. Med.* **4**:203–255 (1935).
3. Mellman, W. J., and R. Wilson, Depression of the tuberculin reaction by attenuated measles virus vaccine, *J. Lab. Clin. Med.* **61**:453–458 (1963).
4. Starr, S., and S. Berkovich, Effects of measles, gamma-globulin-modified measles and vaccine measles on the tuberculin test, *N. Engl. J. Med.* **270**:386–392 (1964).
5. Brody, A., and R. McAlister, Depression of tuberculin sensitivity following measles vaccination, *Am. Rev. Respir. Dis.* **90**:607–611 (1964).
6. Fireman, P., G. Friday, and G. Kumate, Effects of measles vaccine on immunological responsiveness, *Pediatrics* **43**:264–272 (1969).
7. Kantor, F. S., Infection, anergy and cell mediated immunity, *N. Engl. J. Med.* **292**:629–634 (1975).
8. Whittle, H. C., J. Dossetor, A. Odulojou, P. D. M. Bryce, and B. M. Greenwood, Cell mediated immunity during natural measles infection, *J. Clin. Invest.* **62**:678–684 (1978).
9. Notkins, A. L., S. E. Mergenhagen, and R. J. Howard, Effects of virus infections on the function of the immune system, *Annu. Rev. Microbiol.* **24**:525–538 (1970).
10. Finkel, A., and P. B. Dent, Virus–leukocyte interactions: Relationship to host resistance in virus infections in man, in: *Pathobiology Annual* (H. L. Loachimed), ed., pp. 47–70, Appleton-Century-Crofts, New York (1973).
11. Woodruff, J. E., and J. J. Woodruff, T. lymphocyte interaction with viruses and virus-infected tissues, *Prog. Med. Virol.* **19**:120–160 (1975).
12. Wheelock, E. F., and S. T. Toy, Participation of lymphocytes in viral infections, *Adv. Immunol.* **16**:123–184 (1975).
13. Black, F. L., Measles, in: *Viral Infections of Humans. Epidemiology and Control* (A. S. Evans, ed.), pp. 297–316, Wiley, London (1976).
14. Morgan, E. M., and F. Rapp, Measles virus and its associated diseases, *Bacteriol. Rev.* **41**:636–666 (1977).
15. Oldstone, M. B. A., Immunopathology of persistent viral infections, *Hosp. Pract.* **17**(12):61–72 (1982).
16. Oldstone, M. B. A., Virus can alter cell function without causing cell pathology: Disordered function leads to imbalance of homeostasis and disease, in: *Concepts in Viral Pathogenesis* (A. L. Notkins and M. B. A. Oldstone, eds.), pp. 269–276, Springer-Verlag, New York (1984).
17. Orren, A., A. Kipps, J. W. Moodie, D. W. Beatty, E. B. Dowdle, and J. P. McIntyre, Increased susceptibility to herpes simplex virus infections in children with acute measles, *Infect. Immun.* **31**:1–6 (1981).
18. Osler, W., *The Principles and Practice of Medicine*, Appleton, New York (1892).
19. Nalbant, J., The effect of contagious diseases on pulmonary tuberculosis and on the tuberculin reaction in children, *Am. Rev. Tuberc.* **36**:737–777 (1937).

20. Bech, V., Measles epidemics in Greenland, *Am. J. Dis. Child.* **103**:252–253 (1962).
21. Hutckins, G., and C. Janeway, Observations on the relationship of measles and remissions in the nephrotic syndrome, *Am. J. Dis. Child.* **73**:242–243 (1947).
22. Ooi, B. S., B. M. T. Chen, K. K. Tan, and O. T. Khoo, Longitudinal studies of lipoid nephrosis, *Arch. Intern. Med.* **130**:883–886 (1972).
23. Blumberg, R. W., and H. A. Cassady, Effect of measles on the nephrotic syndrome, *Am. J. Dis. Child.* **73**:151–166 (1974).
24. Casali, P., G. P. A. Rice, and M. B. A. Oldstone, Viruses disrupt functions of human lymphocytes: Effects of measles virus and influenza virus on lymphocyte mediated killing and antibody production, *J. Exp. Med.* **159**:1322–1337 (1984).
25. Morely, D., Severe measles in the tropics. I and II. *Br. Med. J.* **1**:297–300; 363–365 (1969).
26. Norrby, E., Measles, in: *Virology* (B. N. Fields, ed.), pp. 1305–1321, Raven, New York (1985).
27. Tyrrell, D. L. J., and E. Norrby, Structural polypeptides of measles virus, *J. Gen. Virol.* **39**:219–229 (1978).
28. Baczko, K., M. Billeter, and V. ter Meulen, Purification and molecular weight determination of measles virus genomic RNA, *J. Gen. Virol.* **64**:1409–1413 (1983).
29. Rima, B. K., The proteins of morbillivirus, *J. Gen. Virol.* **64**:1205–1219 (1983).
30. Bellini, W. J., G. Englund, S. Rozenblatt, H. Arnheiter, and C. D. Richardson, Measles virus P gene codes for two proteins, *J. Virol.* **53**:908–919 (1985).
31. Fujinami, R. S., and M. B. A. Oldstone, Alterations in the expression of measles virus polypeptides by antibody: Molecular events in antibody-induced antigenic modulation, *J. Immunol.* **125**:78–85 (1980).
32. Fujinami, R. S., J. G. P. Sissons, and M. B. A. Oldstone, Immune reactive measles virus polypeptides on the cell surface: Turnover and relationship of the glycoproteins to each other and to HLA determinants, *J. Immunol.* **127**:935–940 (1981).
33. Oldstone, M. B. A., R. S. Fujinami, A. Tishon, D. Finney, H. C. Powell, and P. W. Lampert, Mapping of the major histocompatibility complex and viral antigens on the plasma membrane of a measles virus infected cell, *Virology* **127**:426–437 (1983).
34. Sissons, J. G. P., M. B. A. Oldstone, and R. D. Shreiber, Antibody dependent activation of the alternative complement pathway by measles virus infected cells, *Proc. Natl. Acad. Sci. USA* **77**:559–562 (1980).
35. Sissons, J. G. P., R. D. Schreiber, L. H. Perrin, N. R. Cooper, M. B. A. Oldstone, and H. J. Muller-Eberhard, Lysis of measles virus infected cells by the purified cytolytic alternative complement pathway and antibody, *J. Exp. Med.* **150**:445–454 (1979).
36. Perrin, L. H., A. J. Tishon, and M. B. A. Oldstone, Immunological injury in measles virus infections. III. Presence and characterization of human cytotoxic lymphocytes, *J. Immunol.* **118**:282–290 (1977).
37. Casali, P., and M. B. A. Oldstone, Mechanisms of killing of measles virus-infected cells by human lymphocytes: Interferon associated and unassociated cell-mediated cytotoxicity, *Cell. Immunol.* **70**:330–344 (1982).
38. Kreth, H. W., V. ter Meulen, and G. Eckert, Demonstration of HLA restricted killer cells in patients with acute measles, *Med. Microb. Immunol.* **165**:203–214 (1979).
39. Wright, L. L., and N. L. Levy, Generation on infected fibroblasts of human T and non T lymphocytes with specific cytotoxicity, influenced by histocompatibility, against measles virus infected cells, *J. Immunol.* **122**:2379–2387 (1979).
40. Lucas, C. J., W. E. Biddison, D. L. Nelson, and S. Shaw, Killing of measles virus infected cells by human cytotoxic T cells, *Infect. Immun.* **38**:226–232 (1982).
41. Jacobson, S., J. R. Richert, W. E. Biddison, A. Satinsky, R. J. Hartzman, and H. F. McFarland, Measles virus specific T4+ human cytotoxic T cell clones are restricted by class II HLA antigens, *J. Immunol.* **133**:754–757 (1984).

42. Jacobson, S., G. T. Nepom, J. R. Richert, W. E. Biddison, and H. F. McFarland, Identification of specific HLA-DR2Ia molecules as a restriction element for measles virus-specific HLA class II-restricted cytotoxic T cell clones, *J. Exp. Med.* **161**:263–268 (1985).

43. Bouteille, M., C. Fontaine, C. Vedrenne, and J. Delarue, Sur un cas d'encéphalite subaigue à inclusions: étude anatomo-clinique et ultrastructurale, *Rev. Neurol.* **113**:454–458 (1965).

44. Connolly, J. H., I. V. Allen, L. J. Hurwitz, and J. H. D. Mollar, Measles-virus antibody and antigen in subacute sclerosing panencephalitis, *Lancet* **1**:542–544 (1967).

45. Payne, F. E., J. V. Baublis, and H. H. Itabashi, Isolation of measles virus from cell culture of brain from a patient with subacute sclerosing panencephalitis, *N. Engl. J. Med.* **281**:585–589 (1969).

46. Hall, W. W., and P. W. Choppin, Measles virus proteins in the brain tissue of patients with subacute sclerosing panencephalitis: Absence of the M protein, *N. Engl. J. Med.* **304**:1152–1155 (1981).

47. Fournier, J. G., M. Tardieu, P. Lebon, O. Robain, G. Ponsot, S. Rozenblatt, and M. Bouteille, Detection of measles virus RNA in lymphocytes from peripheral blood and brain perivascular infiltrates of patients with subacute sclerosing panencephalitis, *N. Engl. J. Med.* **313**:910–915 (1985).

48. Horta-Barbosa, L., R. Hamilton, B. Wittig, D. A. Fuccillo, J. L. Sever, and M. L. Vernon, Subacute sclerosing panencephalitis: Isolation of suppressed measles virus from lymphnode biopsies, *Science* **173**:840–841 (1971).

49. Wrzos, H. J., Z. Kulczycki, Z. Laskowski, D. Matacz, and W. J. Brzosko, Detection of measles virus antigen(s) in peripheral lymphocytes from patients with subacute sclerosing panencephalitis, *Arch. Virol.* **60**:291–297 (1979).

50. Robbins, S. J., H. Wrzos, A. L. Kline, R. B. Tenser, and F. Rapp, Rescue of cytopathic paramyxovirus from peripheral blood leukocytes in subacute sclerosing panencephalitis, *J. Infect. Dis.* **143**:396–403 (1981).

51. Norrby, E., Viral antibodies in multiple sclerosis, *Prog. Med. Virol.* **24**:1–39 (1978).

52. ter Meulen, V., and W. W. Hall, Slow virus infection of the nervous system: Virological, immunological and pathogenetic considerations, *J. Gen. Virol.* **41**:1–25 (1978).

53. Haase, A. T., P. Ventura, C. J. Gibbs, and W. W. Tourtellotte, Measles virus nucleotide sequences: Detection by hybridization in situ, *Science* **212**:672–675 (1981).

54. Norrby, E., T. A. Haase, K. P. Johnson, and C. Orvell, Persistent infections with paramyxovirus, in: *Medical Virology* (P. L. M. De la Maza and E. M. Peterson, eds.), pp. 217–236, Elsevier, New York (1982).

55. Rebel, A., M. Basle, A. Puoplard, S. Kouyoumdzian, R. Filmore, and A. Lepatezour, Viral antigens in osteoclasts from Paget's disease of bone, *Lancet* **2**:344–346 (1980).

56. Basle, M. F., J. G. Fournier, S. Rozenblatt, A. Rebel, and M. Bouitelle, Measles virus RNA detected in Paget's disease bone tissue by in situ hybridization, *J. Gen. Virol.* **67**:907–913 (1986).

57. Wesley, A., H. M. Coovadia, and L. Henderson, Immunological recovery after measles, *Clin. Exp. Immunol.* **32**:540–544 (1978).

58. Arneborn, P., and G. Biberfeld, T-lymphocyte subpopulations in relation to immunosuppression in measles and varicella, *Infect. Immun.* **39**:29–37 (1983).

59. Papp, K., Fixation du virus morbilleux aux leucocytes du sang des la période d'incubation de la maladie, *Bull. Acad. Med. Paris* **117**:46–51 (1937).

60. Berg, R. B., and M. S. Rosenthal, Propagation of measles in suspensions of human and monkey leucocytes, *Proc. Soc. Exp. Biol. Med.* **106**:581–585 (1961).

61. Gresser, I., and C. Chany, Isolation of measles virus from the washed leukocyte fraction of the blood, *Proc. Soc. Exp. Biol. Med.* **113**:695–698 (1963).

62. Gresser, I., and D. L. Lang, Relationships between viruses and leukocytes. *Prog. Med. Virol.* **8**:62–130 (1966).

63. Osunkoya, B. O., A. P. Cooke, O. Ayeni, and T. A. Adejumo, Studies on leukocyte cultures in measles. I. Lymphocyte transformation and giant cell formation in leukocyte cultures from clinical cases of measles, *Arch. Gesamte Virusforsch.* **44:**313–322 (1974).

64. Sullivan, J. L., D. W. Barry, S. J. Lucas, and P. Albrecht, Measles infection of human mononuclear cells. I. Acute infection of peripheral blood lymphocytes and monocytes, *J. Exp. Med.* **142:**773–784 (1975).

65. Osunkoya, B. O., A. R. Cooke, O. Ayeni, and T. A. Adejumo, Studies on leukocyte cultures in measles. II. Detection of measles virus antigen in human leukocytes by immunofluorescence, *Arch. Gesamte Virusforsch.* **44:**323–329 (1974).

66. Valdimarsson, H., G. Agnarsdottir, and P. J. Lachmann, Measles virus receptors on human T lymphocytes, *Nature (Lond.)* **255:**554–556 (1975).

67. Joseph, B. S., P. W. Lampert, and M. B. A. Oldstone, Replication and persistence of measles virus in defined subpopulations of human leukocytes, *J. Virol.* **16:**1638–1649 (1975).

68. Huddlestone, J. R., P. W. Lampert, and M. B. A. Oldstone, Virus–lymphocyte interactions: Infection of $T_G$ and $T_M$ subsets by measles virus, *Clin. Immunol. Immunopathol.* **15:**502–509 (1980).

69. Hyypiä, T., P. Korkiamaki, and R. Vainionpää, Replication of measles virus in human lymphocytes, *J. Exp. Med.* **161:**1261–1271 (1985).

70. Bloom, B. R., A. Senick, G. Stoner, G. Ju, M. Nowakowski, S. Kano, and L. Jimenez, Studies of the interactions between viruses and lymphocytes, *Cold Spring Harbor Symp. Quant. Biol.* **41:**73–83 (1977).

71. Lucas, C. J., J. C. Ubels-Postma, A. Rezee, and J. M. D. Galama, Activation of measles virus from silently infected human lymphocytes, *J. Exp. Med.* **148:**940–952 (1978).

72. Whittle, H. C., A. Bradley-Moore, A. Fleming, and B. M. Greenwood, Effects of measles on the immune response of Nigerian children, *Arch. Dis. Child.* **48:**753–756 (1973).

73. Galama, J. M. D., J. Ubels-Postma, A. Vos, and C. J. Lucas, Measles virus inhibits acquisition of lymphocyte functions but not established effector functions, *Cell. Immunol.* **50:**405–415 (1980).

74. Hirano, T., T. Kuritani, T. Kishimoto, and Y. Yamamura, *In vitro* immune response of human peripheral blood lymphocytes. I. The mechanism(s) involved in T cell helper functions in the pokeweed mitogen-induced differentiation and proliferation of B cells, *J. Immunol.* **119:**1235–1241 (1977).

75. Puck, J. M., and R. R. Rich, Regulatory interactions governing the proliferation of T cell subsets stimulated with pokeweed, *J. Immunol.* **132:**1106–1112 (1984).

76. Pelton, B. K., W. Hylton, and A. Denman, Selective immunosuppressive effects of measles virus infection, *Clin. Exp. Immunol.* **47:**19–26 (1982).

77. Casali, P., G. P. A. Rice, and M. B. A. Oldstone, Immune balance in the cytomegalovirus infected host, *Birth Defects* **20:**149–159 (1984).

78. McChesney, M. B., R. S. Fujinami, P. W. Lampert, and M. B. A. Oldstone, Viruses disrupt functions of human lymphocytes. II. Measles virus suppresses antibody production by acting on B lymphocytes, *J. Exp. Med.* **163:**1331–1336 (1986).

79. McChesney, M. B., R. S. Fujinami, and M. B. A. Oldstone, Virus induced suppression of immunoglobulin synthesis during measles virus infection is due to a direct effect on B lymphocytes, in: *The Biology of Negative Strand Viruses* (B. Mahy and D. Kolakofsky, eds.), pp. 298–303, Elsevier Science, New York (1987).

80. Schimpl, A., and E. Wecker, Replacement of T-cell function by a T-cell product, *Nature (Lond.)* **237:**15–17 (1972).

81. Howard, M., and W. E. Paul, Regulation of B-cell growth and differentiation by soluble factors, *Annu. Rev. Immunol.* **1:**307–334 (1984).

82. Tilden, A. B., T. Abo, and C. M. Balch, Suppressor cell function of human granular

lymphocytes identified by the HNK-1 (Leu 7) monoclonal antibody, *J. Immunol.* **130:**1171–1175 (1983).

83. Kehrl, J. H., A. Muraguchi, J. L. Butler, R J. M. Falkoff, and A. S. Fauci, Human B cell activation, proliferation and differentiation, *Immunol. Rev.* **78:**75–96 (1984).

84. Casali, P., J. G. P. Sissons, M. J. Buchmeier, and M. B. A. Oldstone, In vitro generation of human cytotoxic lymphocytes by virus: Viral glycoproteins induce nonspecific cell-mediated cytotoxicity without release of interferon, *J. Exp. Med.* **154:**840–855 (1981).

85. Jacobson, S., and H. F. McFarland, Measles virus persistence in human lymphocytes: A role for virus induced interferon, *J. Gen. Virol.* **63:**351–357 (1982).

86. Gresser, I., M. C. Tovey, M. T. Bandu, C. Maury, and D. Brontoy-Bogye, Role of interferon in the pathogenesis of virus disease in mice as demonstrated by the use of an anti-interferon serum. II. Study with herpes simplex, Moloney sarcoma, vesicular stomatitis, Newcastle disease and influenza viruses, *J. Exp. Med.* **144:**1316–1323 (1976).

87. Stewart, W. E. II, Mechanisms of antiviral actions in interferons, in: *The Interferon System* (W. E. Stewart, ed.), pp. 196–222, Springer-Verlag, New York (1979).

88. Kadish, A. S., F. A. Tansey, G. S. M. Yu, A. T. Doyle, and B. R. Bloom, Interferon as a mediator of human lymphocyte suppression, *J. Exp. Med.* **151:**637–650 (1980).

89. Harfast, B., J. R. Huddlestone, P. Casali, T. Merigan, and M. B. A. Oldstone, Interferon acts directly on human B lymphocytes to modulate immunoglobulin synthesis, *J. Immunol.* **127:**2146–2150 (1981).

90. Hirsch, R. L., F. Mokhtarian, D. E. Griffin, B. R. Brooks, J. Hess, and R. T. Johnson, Measles virus vaccination of measles seropositive individuals suppresses lymphocyte proliferation and chemotactic factor production, *Clin. Immunol. Immunopathol.* **21:**341–350 (1981).

91. Rinehart, J. J., M. Orser, and M. E. Kaplan, Human monocyte and macrophage modulation of lymphocyte proliferation, *Cell. Immunol.* **44:**131–143 (1979).

92. von Pirquet, C., Das Verhalten der kutanen Tuberculin-Reaktion während der Masern, *Dtsch. Med. Wochenschr.* **34:**1297–1300 (1908).

93. Anderson, R., A. R. Rabson, R. Sher, H. J. Koornhof, and D. Bact, Defective neutrophil motility in children with measles, *J. Pediatr.* **89:**27–32 (1976).

94. Smithwick, E. M., and S. Berkovich, In vitro suppression of the lymphocyte response to tuberculin by live measles virus, *Proc. Soc. Exp. Biol. Med.* **123:**276–278 (1966).

95. Finkel, A., and P. B. Dent, Abnormalities in lymphocyte proliferation in classical and atypical measles infection, *Cell. Immunol.* **6:**41–48 (1973).

96. Zweiman, B., D. Pappagianis, H. Maibach, and E. A. Hildreth, Effect of measles immunization on tuberculin hypersensitivity and in vitro lymphocyte reactivity, *Int. Arch. Allergy Appl. Immunol.* **40:**834–841 (1971).

97. Zweiman, B., and M. Miller, Effects of non-viable measles virus on proliferating human lymphocytes, *Int. Arch. Allergy* **46:**822–833 (1974).

98. Sullivan, J. L., D. W. Barry, P. Albrecht, and S. J. Lucas, Inhibition of lymphocyte stimulation by measles virus, *J. Immunol.* **114:**1458–1461 (1975).

99. Lucas, C. J., J. M. D. Galama, and J. Ubels-Postma, Measles virus-induced suppression of lymphocyte reactivity in vitro, *Cell. Immunol.* **32:**70–85 (1977).

100. Lucas, C. J., J. Ubels-Postma, J. M. D. Galama, and A. Rezee, Studies on the mechanism of measles-induced suppression of lymphocyte functions in vitro. Lack of a role for interferon and monocytes, *Cell. Immunol.* **37:**448–548 (1978).

101. Zweiman, B., R. P. Lisak, D. Waters, and H. Koprowski, Effects of purified measles virus components on proliferating human lymphocytes, *Cell. Immunol.* **47:**241–247 (1979).

102. Jacobson, S., and H. F. McFarland, Measles virus infection of human peripheral blood lymphocytes: Importance of the OKT4+ T-cell subset, in: *Non-segmented Negative Strand Viruses. Paramyxoviruses and Rhabdoviruses* (D. H. L. Bishop and R. W. Compans, eds.), pp. 435–442, Academic, New York (1984).

103. Smith, K. A., and F. W. Ruscetti, T growth factor and the culture of cloned functional T cells, *Adv. Immunol.* **31**:137–175 (1981).
104. Luger, T. A., J. S. Smolen, T. M. Chused, A. D. Steinberg, and J. J. Oppenheim, Human lymphocytes with either the OKT4 or OKT8 phenotype produce interleuken 2 in culture, *J. Clin. Invest.* **70**:470–473 (1982).
105. Wainberg, M. A., S. Vydelingum, and R. G. Margolese, Viral inhibition of lymphocyte mitogenesis: Interference with the synthesis of functionally active T cell growth factor (TCGF) activity and reversal of inhibition by the addition of the same, *J. Immunol.* **130**:2372–2378 (1983).
106. Joffe, M. I., and A. R. Rabson, Defective helper factor (LMF) production in patients with acute measles infection, *Clin. Immunol. Immunopathol.* **20**:215–223 (1981).
107. Borysiewicz, L. K., P. Casali, B. Rogers, S. Morris, and J. G. P. Sissons, The immunosuppressive effects of measles virus on T cell function. Failure to affect IL-2 release or cytotoxic T cell activity *in vitro*, *Clin. Exp. Immunol.* **59**:29–36 (1985).
108. Cooper, N. R., and M. R. Welsh, Antibody and complement dependent viral neutralization, *Springer Semin. Immunopathol.* **2**:285–310 (1979).
109. Trinchieri, G., and B. Perussia, Biology of disease: Human natural killer cells: Biologic and pathologic aspects, *Lab. Invest.* **50**:489–513 (1984).
110. Casali, P., and G. Trinchieri, Natural killer cells in viral infection, in: *Concepts in Viral Pathogenesis* (A. L. Notkins and M. B. A. Oldstone, eds., pp. 12–19, Springer-Verlag, New York (1984).
111. Trinchieri, G., and B. Perussia, Immune (γ) interferon: A pleiotropic lymphokine with multiple effects on cells of the adaptive and nonadaptive immune system, *Immunol. Today* **6**:131–136 (1985).
112. Casali, P., J. G. P. Sissons, R. S. Fujinami, and M. B. A. Oldstone, Purification of measles virus glycoproteins and their integration into artificial lipid membranes, *J. Gen. Virol.* **54**:161–171 (1981).
113. Perussia, B., S. Starr, S. Abraham, V. Fanning, and G. Trinchieri, Human natural killer cells analysed by B73.1, a monoclonal antibody blocking Fc-receptor function. I. Characterization of the lymphocyte subset reactive with B73.1, *J. Immunol.* **130**:2133–2141 (1983).
114. Rice, G. P. A., R. D. Schrier, P. Casali, and M. B. A. Oldstone, Cytomegalovirus and measles virus *in vitro* models of virus mediated immunosuppression, in: *Viral Mechanisms of Immunosuppression* (N. Gilmore and M. Wainberg, eds.), pp. 15–29, Liss, New York (1985).
115. Podack, E. R., and G. Dennert, The molecular mechanisms of lymphocyte mediated tumor cell lysis, *Immunol. Today* **21**:21–27 (1985).
116. McFarland, H. F., The effect of measles virus infection on T and B lymphocytes in the mouse. I. Suppression of helper cell activity, *J. Immunol.* **113**:1978–1983 (1974).
117. Lachmann, P. J., Immunopathology of measles, *Proc. R. Soc. Med.* **67**:1120–1122 (1974).
118. Lachmann, P., and D. K. Peters (eds.), *Clinical Aspects of Immunology*, Vol. II, Blackwell Scientific Publications, Oxford (1982).
119. Jacobson, S., M. Flerlage, and H. F. McFarland, Impaired measles virus specific cytotoxic T cell responses in multiple sclerosis, *J. Exp. Med.* **162**:839–850 (1985).
120. Lampert, P., B. S. Joseph, and M. B. A. Oldstone, Antibody-induced capping of measles virus antigens on plasma membrane studied by electron microscopy, *J. Virol.* **15**:1248–1255 (1975).
121. Fujinami, R. S., and M. B. A. Oldstone, Failure to cleave measles virus fusion protein in lymphoid cells: A possible mechanism for viral persistence in lymphocytes, *J. Exp. Med.* **154**:1489–1499 (1981).
122. Weschler, S. L., and Fields, B. N., Differences between intracellular polypeptides of measles and subacute sclerosing panencephalitis virus, *Nature (Lond.)* **272**:458–460 (1978).

123. Choppin, P. W., C. D. Richardson, D. C. Merz, W. W. Hall, and H. Scheid, The functions and inhibition of the membrane glycoproteins of paramyxoviruses and myxoviruses and the role of measles virus M protein in subacute sclerosing panencephalitis, *J. Infect. Dis.* **143**:352–363 (1981).
124. Carter, M. J., M. M. Willcocks, and V. ter Meulen, Defective translation of measles matrix protein in subacute sclerosing panencephalitis cell line, *Nature (Lond.)* **305**:153–155 (1983).

# Avian Retroviruses

## ROBERT W. STORMS and HENRY R. BOSE, JR.

## 1. BASIC PROPERTIES OF THE VIRUSES

Retrovirus particles are roughly spherical, about 100 nm in diameter, and have a dense inner core or nucleocapsid surrounded by an envelope with projecting glycoprotein spikes.[1] Although four basic subcategories of retrovirus particles have been identified, all avian retroviruses have C-type morphology.

Retroviruses contain single-stranded RNA genomes, and each particle contains two identical genomic RNAs joined at their 5′ ends. Each RNA subunit has a m7Gppp cap at the 5′ terminal and is polyadenylated at the 3′ end. Replication-competent retroviruses generally contain three cistrons: *gag*, *pol*, and *env*. The *gag* region encodes the structural proteins of the viral core. Homology within these polypeptides, the group-specific antigens, hence *gag*, is used as a means of classifying retroviruses. The core proteins are translated as a precursor polyprotein and are post-translationally processed to form the mature virion polypeptides. The *pol* region encodes an RNA-dependent DNA polymerase, or reverse transcriptase. This enzyme allows for the synthesis of the DNA provirus from the RNA genome.[2,3] The DNA provirus then integrates into the host chromosome and establishes itself as a heritable trait. The *env* region encodes the glycoproteins that become associated with the viral envelope. These proteins form spikes that protrude from the virion and that are involved in the attachment of the virion to specific receptors on the cell surface. The *env* gene products are translated from a spliced subgenomic messenger RNA (mRNA). As with *gag*, a precursor *env* polyprotein is synthesized and subsequently cleaved to form the mature *env* proteins. The *env* gene products are used to define virus subgroups. Differences within these glycoproteins are reflected by host range and by cell types susceptible to viral infection.

ROBERT W. STORMS and HENRY R. BOSE, JR. • Department of Microbiology, University of Texas at Austin, Austin, Texas 78712-1095.

Antibody directed against antigenic determinants of the *env* glycoproteins neutralizes the virus.

Those retroviruses capable of transforming cells *in vitro* and of rapidly inducing neoplastic disease *in vivo* are generally replication defective. These viruses depend on co-infection with a replication-competent helper virus to produce progeny.[4] Most replication-defective viral genomes contain deletions in all or part of the *gag*, *pol*, and *env* genes. In place of the structural coding sequences, these viruses have acquired sequences unrelated to the helper virus which are responsible for cell transformation.[4,5] The sequences are referred to as viral oncogenes and are related to unique sequences of host cellular DNA. The genetic structures of representative avian retroviruses are diagrammed in Figs. 1 and 2.

The avian retroviruses are composed of four groups,[1] which include (1) the avian sarcoma–leukosis virus complex (ASLV) of chickens; (2) the reticuloendotheliosis viruses (REV) of turkeys, ducks, and chickens; (3) the endogenous viruses native to some species of pheasants, partridges and quail; and (4) the lymphoproliferative disease virus of turkeys (Table I). Because studies on immunosuppression induced by avian retroviruses have focused on members of the ASLV group, in particular the acute leukemia viruses and their associated viruses, and on members of the REV group, only these viruses are discussed in depth.

## 1.1. Avian Sarcoma–Leukosis Complex

The ASLV is a large group of viruses defined on the basis of genetic and antigenic homologies to Rous-associated virus (RAV-O),[6] an endogenous retrovirus of chickens. On the basis of their envelope glycoproteins, the ASLV are divided into five subgroups: A, B, C, D, and E. The most common field isolates are of subgroups A and B.[7] The members of this group that induce acute transformation include sarcoma viruses and leukemia viruses. These viruses induce neoplasms after very brief latent periods and transform cells efficiently *in vitro*.[8] Because of deletions introduced into their structural genes upon acquisition of their transforming genes, all the avian acute leukemia

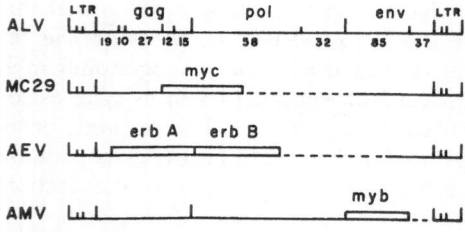

FIGURE 1. Genetic structures of the ASLV leukemia viruses. The proviral DNA genomes of the replication-defective retroviruses MC29, AEV, and AMV are shown in comparison to the general structure for the replication-competent ALV. The transduction of cellular genes (*myc*, *erb*A/*erb*B, or *myb*) into the viral genome has resulted in the deletion of viral structural genes essential for replication (dashed lines). Mature viral proteins encoded by ALV are indicated by their approximate molecular weights. The LTR (long terminal repeat) is the promoter that controls the expression of the viral genome.[5,102,103]

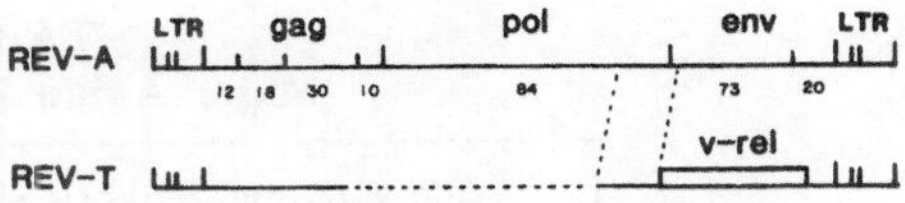

**FIGURE 2.** Genetic structures of the REV. The proviral DNA genome of REV-T is shown in comparison with its helper virus, REV-A. The insertion of *rel*, the cellular-derived oncogene, has resulted in extensive deletions (dashed line) within the REV-A genome. The mature viral proteins encoded by REV-A, as well as the other replication-competent REV, are indicated by their approximate molecular weights. The LTR, or long terminal repeat, is the viral promoter.[69,104]

viruses, as well as the replication-defective strains of Rous sarcoma virus (RSV), require helper viruses for replication. These replication-competent helper viruses are collectively referred to as the avian leukosis viruses (ALV), although individual isolates are named on the basis of their acutely oncogenic member (e.g., RAV is associated with RSV). The ALV replicate but fail to transform cells in culture[9]; however, with chronic infection, the ALV generally give rise to lymphoid neoplasms of the B-cell lineage; lymphoid leukosis is endemic within flocks of domestic chickens.[8]

Avian myeloblastosis viruses (AMV), avian myelocytomatosis viruses (e.g., MC29), and avian erythroblastosis viruses (AEV) are members of the ASLV group, which transform cells of the myeloid and erythroid series *in vitro*.[10] AMV and MC29 give rise to myeloid leukemias within 5–8 weeks following infection.[11,12] AEV induces a fatal erythroleukemia within 2 weeks after infection.[13] The helper viruses associated with these acutely transforming viruses have been designated myeloblastosis-associated viruses (MAV),[14] myelocytomatosis-associated viruses (MCAV),[15] and avian erythroblastosis-associated virus (AEAV).[16] AEAV is a subgroup B virus, while MAV and MCAV strains may be of either subgroup A or B specificities. The helper viruses induce a variety of disorders, including lymphoid leukosis, erythroblastosis, osteopetrosis, anemia, and nephroblastoma.[15,17,18] Osteopetrosis was initially characterized using field isolates of ALV that induced the disease with only relatively low incidence.[19,20] End-point cloning from a standard AMV stock allowed the isolation of a virus which induces a high incidence of osteopetrosis.[17] This virus, designated MAV-2(0), is of subgroup B specificity.

Many factors contribute to the relative ability of individual members of the ALV family to immunosuppress chickens. Early studies reported that ALV-infected birds exhibited either slight depressions[9,21] or no depression[22] in their ability to generate antibody against specific antigens. These studies were all conducted using ALV with subgroup A envelope specificity. It was later proposed[23] that humoral responsiveness during infections with subgroup A ALV was dependent on whether the strain of chicken used contained endogenous virus (RAV-0) within its genome. Similarly, small depressions in cellular immune responsiveness during infection by subgroup A ALV were dependent on the strain of virus and on the line of chicken used.[21] While these studies report depressions in immune responsiveness that are statistically significant, they are not profound. More recent studies have confirmed the relative immu-

**TABLE I**
**Major Avian Retroviruses**[a-c]

1. Avian sarcoma–leukosis viruses
   Leukosis
   　　Envelope subgroup A
   　　RAV-1
   　　RAV-3
   　　RAV-5
   　　RPL12
   　　HPRS1
   　　HPRS2
   　　ALV-F42
   　　MAV-1
   　　MCAV-A
   　　Envelope subgroup B
   　　RAV-2
   　　RAV-6
   　　MAV-2
   　　MCAV-B
   　　AEAV
   　　Envelope subgroup C
   　　RAV-7
   　　RAV-49
   　　Envelope subgroup D
   　　RAV-50
   　　Envelope subgroup E
   　　RAV-0
   　　RAV-60
   Sarcoma
   　　Envelope subgroup A
   　　SR-RSV-A
   　　PR-RSV-A
   　　EH-RSV
   　　RSV-29
   　　Envelope subgroup B
   　　SR-RSV-B
   　　PR-RSV-B
   　　HA-RSV
   　　Envelope subgroup C
   　　B77-RSV
   　　PR-RSV-C
   　　Envelope subgroup D
   　　SR-RSV-D
   　　CZ-RSV
   　　Envelope subgroup E
   　　SR-RSV-E
   　　PR-RSV-E
   　　Defective: BH-RSV, BS-RSV, FuSV, PRCII, PRCIV
   Myeloblastosis
   　　Defective: AMV-BAI/A, AMV-E26

**TABLE I** (*Continued*)

Myelocytomatosis
Defective: MC29, MH2, CMII, OK10
Erythroblastosis
Defective: AEV-ES4, AEV-R
2. Reticuloendotheliosis viruses
Reticuloendotheliosis-associated virus (REV-A)
Spleen necrosis virus (SNV)
Chick syncytial virus (CSV)
Duck infectious anemia virus (DIAV)
Defective: reticuloendotheliosis virus, strain T (REV-T)
3. Viruses of pheasants, quail, and partridges
Ring-necked pheasant virus (RPV)
RAV-61 (ring-necked pheasant)
Golden pheasant virus (GPV)
Amherst pheasant virus (APV)
4. Lymphoproliferative disease virus (LPDV)

---

[a]From Teich,[1] Calnek,[7] Purchase and Burmester,[8] Beard.[13]

[b]The major members of the avian sarcoma–leukosis complex, the reticuloendotheliosis viruses, and viruses of pheasants, partridges, and quail are listed. The lymphoproliferative disease virus of turkeys is the only member of its family. The viruses of pheasants, partridges, and quail have been identified primarily on the basis of genetic relation to RAV-O or by their ability to rescue replication-defective strains of RSV. Among the ASLV and REV, the term defective refers to replication defective. These defective viruses will assume the envelope specificity of their helper virus.

[c]*Abbreviations:* AEAV, avian erythroblastosis-associated virus; AEV, avian erythroblastosis virus, ALV, avian leukosis virus; AMV, avian myeloblastosis virus; BH-RSV, Bryan high titer RSV; BS-RSV, Bryan standard RSV; B77-RSV, Bratislava RSV; CZ-RSV, Carr-Zilber RSV; EH-RSV, Engelbreth-Holm RSV; FuSV, Fujinami sarcoma virus; HA-RSV, Harris RSV; HPRS, Houghton Poultry Research Station; MAV, myeloblastosis-associated virus; MCAV, myelocytomatosis-associated virus, PR-RSV, Prague RSV; PRC, Poultry Research Centre; RAV, Rous-associated virus; RPL, Regional Poultry Laboratory; RSV, Rous sarcoma virus; SR-RSV, Schmidt-Ruppin RSV. From Purchase *et al.*,[40] Fujita *et al.*,[105] Hanafusa *et al.*,[106] Chen and Vogt,[107] McDougall *et al.*[108]

---

nocompetence of birds infected with subgroup A ALV.[24–26] By contrast, subgroup B ALV has been demonstrated to immunosuppress infected birds severely.[27,28] This may be due, in part, to the susceptibility of macrophages to infection by subgroup B ASLV.[29,30]

## 1.2. Reticuloendotheliosis Viruses

The REV comprise a small group of genetically and antigenically distinct viruses.[31–33] REV strain T (REV-T) is a replication-defective acutely transforming virus that co-replicates with its helper, reticuloendotheliosis-associated virus (REV-A).[34] Oncogenic stocks of REV-T virus contain REV-A at a 100- to

1000-fold excess and are therefore referred to as REV-T(REV-A). Three other replication-competent members, duck infectious anemia virus (DIAV),[35] spleen necrosis virus (SNV),[36] and chick syncytial virus (CSV),[37] belong to this group and may serve as alternate helper viruses for REV-T.[38] The four nondefective REV share extensive sequence homology and may be distinguished from one another only on the basis of cross-neutralization tests and pathogenicity.[31,39] Infection by these viruses produces a variety of syndromes, including visceral reticuloendotheliosis, splenomegaly and spleen necrosis, nerve lesions, and B-cell lymphoma.[40,41] REV-T is perhaps the most virulent of all retroviruses and is invariably fatal in young chicks within 7–10 days postinfection.[42]

The replication-competent REV induce an extensive cytopathic effect in avian fibroblast cultures and lyse lymphoid cells *in vitro*.[34,43] Following the acute phase, the surviving cells establish a chronically infected culture.[44] All these viruses induce a transient immunosuppression in infected chickens.[45] REV-T, the replication-defective member, infects and transforms both avian fibroblasts and hematopoietic cells *in vitro*.[38,46] *In vitro* derived REV-T transformed nonvirus-producing hematopoietic cells induce lethal reticuloendotheliosis when injected into histocompatibly matched chickens.[47] Hematopoietic cells transformed by REV-T have lymphoblastoid morphology, express low levels of terminal-deoxynucleotidyl-transferase activity, and express B cell determinants; however, these cells fail to synthesize immunoglobulin (Ig) molecules.[47,48] Depending on the extent of their immunoglobulin gene rearrangements, the REV-T-transformed lymphoid cell lines fall into two major categories (J. M. Bishop and H. R. Bose, Jr., unpublished observations). Most of the lines exhibit rearrangements within both the heavy and light-chain Ig gene loci relative to embryonic tissues; however, a smaller number retain the embryonic configuration. None of these cell lines contains $\mu$ or $\lambda$ gene transcripts. Since these cells express B-cell membrane antigens, REV-T is infecting and transforming a very immature cell committed to B-cell differentiation. It is therefore proposed that transformation by REV-T blocks the differentiation of these cells, allowing for their uncontrolled proliferation and the development of a rapidly fatal disease.

## 2. MODULATION OF THE IMMUNE RESPONSE

### 2.1. Humoral Responses

The humoral response, as described here, refers to the production of antibody toward antigens and requires active communication among B cells, T cells, and macrophages.[49] Inactivation of any one or more of these cell types may alter humoral immune responsiveness.[50,51] It should be noted, however, that birds have an additional primary lymphoid organ not found in the mammalian species. This organ, the bursa of Fabricius, from which B cells derive their name, is involved in the differentiation of B cells.

Only limited information is available on the humoral response in birds infected with avian retroviruses. Birds infected with MAV-2(0) exhibit a depressed IgM and IgG response when assayed by direct and indirect hemolytic plaque assays.[27,52] Most osteopetrotic birds had visible decreases in the number of plaques generated as compared with normal birds; however, on occasion, low responses were also seen among infected chickens exhibiting no osteopetrosis upon necropsy. The depression in humoral response was dependent on the age of the bird at the time of infection and was most severe when embryos at the eleventh day of gestation were infected. In addition, serum antibody titers were assayed directly by agglutination and by enzyme-linked immunosorbent assay (ELISA).[52] Again, the responses were shown to be a function of age at the time of infection. The age-dependent depression in the humoral response corresponds to the pathology manifested in the spleen by MAV-2(0) infection. Birds infected as embryos show a total arrest of B-cell development, while chicks infected at hatching exhibit delayed splenic development.

Virus-specific antibody titers have been examined in both REV-A and MAV-2(0) infected birds.[53,54] Birds infected with the oncogenic REV-T(REV-A) stocks do not survive long enough to develop a humoral response; however, chickens infected with REV-A obtained by end-point dilution exhibit virus-specific antibodies 12–15 days after infection. The birds exhibiting this delayed humoral response are still unresponsive to T-cell mitogen stimulation. In studies on the anemia induced by MAV-2(0), a depressed antiviral humoral response is seen at the peak of the anemic episode, but an increasing serum titer is coincident with recovery and clearance of the virus. Furthermore, bursectomized birds, which fail to elicit a humoral response, are never able to survive the anemic episode.[55]

## 2.2. Cellular Responses

The cellular immune response is elicited and regulated by the cooperative interactions of various T cell subpopulations and macrophages.[56] The level of responsiveness of the cellular arm of the immune system may be demonstrated *in vitro* by the polyclonal stimulation of T cells to blastogenesis by mitogens or by mixed lymphocyte reactions (MLR).[57–59] Phytohemagglutinin (PHA) and concanavalin A (Con A) are T-cell specific mitogens and require cooperation between T cells and macrophages to induce their proliferative responses.[60,61]

Infection of chicks with REV-T(REV-A), or with REV-A alone, eliminates the ability of T lymphocytes to respond to mitogens.[45,62] The suppression can be detected within 3 days after REV-A infection, and the lack of proliferation is not due to a loss of lymphocyte viability.[63,64] In addition, spleen cell preparations from REV-A-infected birds exhibit a depressed MLR when mixed with allogeneic spleen cells.[53] Similarly, rejection of allogeneic skin grafts in REV-A-infected birds exhibit a depressed MLR when mixed with allogeneic spleen cells.[53] Similarly, rejection of allogeneic skin grafts in REV-A-infected birds is delayed.[53]

Infections by MAV-2(0), other related subgroup B ALV helper viruses, or the acute leukemia viruses of the ASLV group result in depressed responses to mitogens.[27,28] However, with MAV-2(0) infections, the responsiveness to mitogens is compartmentalized. At the outset of any gross pathologic disturbances, the response to Con A by splenic lymphocytes is generally severely suppressed, whereas the depression of Con A blastogenesis seen with peripheral blood lymphocytes (PBL) is nominal until late in the disease process.[27,65]

The studies discussed above indicate a generalized suppression of the cellular response within REV and MAV-2(0)-infected birds. The development of delayed hypersensitivity (DTH) has been studied in MAV-2(0) infections.[52] Compared with normal birds, MAV-2(0)-infected chickens were suppressed in their DTH response elicited against human gammaglobulin. The ability of REV-infected birds to elicit virus-specific cellular responses has also been investigated. The activity of cytotoxic T lymphocytes (CTL) was assayed *in vitro* with lymphocytes from birds infected with REV-T-transformed lymphoid cells.[53] Chickens infected with low dosages showed normal mitogen responsiveness, displayed no visible lymphomatosis, and exhibited a detectable CTL response specific for the REV-T-transformed cells. With increased dosage, the birds' responsiveness to mitogens was eliminated, and no CTL activity was detected. These birds displayed visceral lesions characteristic of reticuloendotheliosis.

In both REV and MAV-2(0) infections, the first insight into the precise lesion in the immune system was provided by mixing lymphocyte preparations from the immunosuppressed birds with mitogen-stimulated uninfected lymphocytes from age-matched syngeneic chickens. (Syngeneic here refers only to the B locus, the major histocompatibility locus of chickens.) In REV infections, the spleen cells from immunosuppressed birds were capable of eliminating the normal mitogen responsiveness of uninfected birds.[62,66] This finding suggested the presence of a suppressor cell population in the spleens of infected chickens. By contrast, the blastogenic response of MAV-2(0)-infected lymphocytes was restored in the presence of normal cell populations.[65] Adherent cells from either the spleen or the peripheral blood of uninfected birds were found capable of complementing the suppressed lymphocytes. These adherent populations were morphologically macrophage-like cells, possessed Fc receptors, participated in Fc-dependent phagocytosis, and displayed nonspecific esterase activity. Adherent cells obtained from infected birds displayed these same characteristics. Furthermore, the adherent cells were present in equivalent numbers in spleen and PBL preparations from both infected and uninfected birds, and peritoneal exudate cells were elicited in similar numbers, regardless of infection. These characteristics implicate a defect in relatively differentiated macrophages in MAV-2(0)-infected chickens. Normal macrophages complemented the blastogenesis of lymphocytes from both transiently (anemic) and persistently (osteopetrotic) immunosuppressed chickens.

A decrease in the production of interleukin-2 (IL-2), or T-cell growth factor, has been implicated in MAV-2(0)-induced suppression.[67] IL-2 production is decreased in spleen cell preparations from osteopetrotic birds. Furthermore, the addition of exogenous IL-2 resulted in a marked increase in Con A-induced blastogenesis. These studies, however, failed to duplicate the restoration of the mitogenic response seen by Price and Smith[65] upon the addition of normal macrophages. This could simply be a reflection of different stages of the progression of the disease. Alternatively, the restoration of lymphocyte blastogenic responses to mitogens by normal macrophages could be specific within the strain of bird. Boni-Schnetzler et al.[67] used a strain that has a genetically determined low responsiveness to Con A.[68] Normal macrophages from this strain may be unable to restore the suppression induced by MAV-2(0).

## 3. MECHANISMS OF IMMUNOSUPPRESSION

Infection by REV-T(REV-A) or end-point-purified REV-A virus, or inoculation with REV-T-transformed lymphoid cells, results in the induction or activation of a suppressor cell population.[45,62,66] The PHA response of normal uninfected spleen cells may be suppressed by the addition of lymphocyte preparations from immunosuppressed REV-infected birds. In these studies, the normal cells are preincubated with PHA to permit induction. The suppressive cells from age-matched syngeneic birds are then added. The suppressor cells are tentatively classified as a T cell. The removal of glass-adherent populations[63] or the elimination of macrophages by toxic lysosomotropic agents (R. W. Storms and H. R. Bose, Jr., unpublished observations) does not affect the suppressor activity of spleen cells from infected birds. The suppressor cell may be separated from granulocytes on the basis of density in Percoll gradients (M. T. Crakes and H. R. Bose, Jr., unpublished observations). The depletion of B cells from suppressor cell preparations with specific anti-Ig sera, likewise, has no significant effect on the capacity of the preparations to depress the normal response to PHA (M. T. Crakes and H. R. Bose, Jr., unpublished observations). Furthermore, the depletion of T cells from these same preparations corresponds with a decrease in the suppressive capacity.

The inhibition of blastogenesis is a contact-mediated event.[63] Normal cells induced with PHA were separated from the suppressor cell population by a membrane within a double-diffusion chamber. The membrane was of sufficient pore size (0.4 μm) to permit the free exchange of soluble factors. If soluble factors are involved in the inhibition, they alone are not sufficient to suppress the normal response to PHA. Furthermore, the cellular communication involved in suppression is mediated by trypsin-sensitive protein(s) on the surface of the suppressor cell.[64]

Lymphocytes from the spleens of REV-T(REV-A)-infected birds fail to respond to PHA as early as 3 days after infection. By 6 days postinfection, these cells are capable of inhibiting the PHA response of normal cells.[64] The inhibition is seen with normal cells present at a 20-fold excess over the spleen cells of infected birds. This suggests that the suppressor cell is capable of interaction with multiple target cells.

Infection by immunosuppressive doses of REV-T-transformed lymphoid cells inhibits the ability of the host to elicit specific cytotoxic cells against the invading tumor.[53] However, the cytotoxic effector function of cells present at the time of infection appears to remain normal.[64] Normal and REV-T(REV-A)-infected birds were compared on their ability to respond to allogeneic targets. Spleen cells from these animals were induced with PHA, then co-incubated with $^{51}Cr$-labeled chicken red blood cells (RBCs), and the release of label was assayed. The results indicated that the cytotoxic function was not impaired by REV infection. This would argue that the suppressor cell inhibits the blastogenic capacities of lymphocytes but does not interfere with the viability or certain effector functions of these cells.

The mechanism by which the suppressor cell is induced is unclear. The suppressor is presumed to be host derived rather than tumor or virus derived, as the addition of REV-T-transformed lymphoid cells or virus preparations is not sufficient to inhibit the blastogenic response of normal cells *in vitro*.[64] The suppressor cells are not productively infected with REV-A, as suppressive populations of spleen cells from REV-T(REV-A)-infected birds are not sensitive to complement-mediated lysis in the presence of REV-specific antiserum.[64] The activation of the suppressor cell population does, however, require active viral replication. The injection of ultraviolet (UV)-inactivated REV-T(REV-A) preparations into birds does not induce immunosuppression.[64]

REV-T-transformed nonvirus-producing (NP) lymphoid cells have been isolated that do not contain REV-A proviral sequences[69] and that do not contain detectable virus-specified proteins by radioimmunoprecipitation techniques.[47] These NP cells are capable of inducing or activating the suppressor population.[45,53] The suppressor cells induced by NP cells and those induced by REV-A virus act by a contact-mediated mechanism, express T-cell determinants, and have a similar density[63] (M. T. Crakes and H. R. Bose, Jr., unpublished observations). Furthermore, the suppressor cells induced in each infection appear with the same kinetics relative to the manifestation of the infection. It is not known whether the means by which the suppressor cell is induced is the same in both infections. Since REV-T has deleted most of the *pol* and *env* cistrons,[69–71] the only viral gene that could be involved would be the *gag* gene, which is largely intact in REV-T.[69]

The transmembranal envelope proteins (pl5E or p20E) of several retroviruses have been found to be responsible for the ability of these viruses to immunosuppress their hosts.[72–77] Indeed, these retroviruses contain a common sequence within their transmembranal proteins that has been found to account for at least part of the immunosuppressive ability of these transmembranal proteins.[78] While REV-A does share this region of homology with other immunosuppressive retroviruses,[71,79] there are important distinctions between REV-A and these viruses: (1) unlike the other viruses, REV-A particles do not perturb the blastogenic responses of T-cell mitogen-stimulated normal lymphocytes *in vitro;* and (2) with the exception of Moloney murine leukemia virus,[80,81] none of the other viruses that process this immunosuppressive sequence has been demonstrated to induce a suppressor cell. It is possible that the p20E protein of REV-A is involved in the induction of the suppressor cell and that this induction is too complex to be duplicated *in vitro*.

REV-T does not encode the viral transmembranal protein[70,71] thus, expression of this protein cannot explain the ability of NP lymphoid cells to induce the suppressor population. If the p20E protein is involved in the induction of the suppressor cell by REV-A, either REV-T transformed NP cells activate the suppressor by a different mechanism or they have induced the expression of a cellular protein which functions in the place of the viral transmembranal protein. Several murine and human tumor cell lines have been shown to express a protein of approximately 20 kDa that shares antigenic determinants with the transmembranal protein of immunosuppressive retroviruses.[83–85] Furthermore, this protein is expressed by normal human lymphocytes when they are stimulated by T-cell mitogens.[85] It has been proposed that this protein downregulates proliferative immune responses and that it works within an immunoregulatory network with IL-2.[77] The induction of this protein within REV-T-transformed NP cells could theoretically be achieved by transformation, i.e., by active cell division or aberrant expression.

Infection by MAV-2(0), and most probably the related subgroup B ASLV, does not induce or activate a suppressor cell population[30,65] but rather appears to induce a disturbance in the production of interleukins.[67] The pathways for T-cell induction by interleukins are not well defined in chickens; however, if they parallel mammalian pathways,[86,87] some speculations on the mechanism of immunosuppression by MAV-2(0) may be made. T cells from MAV-2(0)-infected birds are receptive to exogenous IL-2.[67] Furthermore, PBL and spleen cell preparations from MAV-2(0)-infected animals produce decreased amounts of IL-2 as compared with uninfected birds. Therefore, any interaction in the pathway before, or including, the release of IL-2 may be defective. The cell type(s) that induce IL-2-producing cells may be totally functional or present but unresponsive to IL-1. Likewise, nothing is known of the levels of IL-1 in infected chickens.

Using T-cell mitogen assays, Price and Smith[65] demonstrated a defect in macrophage populations of MAV-2(0)-infected birds. The macrophages from infected birds appear to be intact and to perform normal phagocytic functions. Furthermore, spleen cell preparations appear to bind Con A at normal, or above-normal, levels. However, the addition of normal macrophages to MAV-2(0)-infected lymphocytes restores mitogenic responses. It would appear likely that the defect in macrophages in MAV-2(0)-infected birds would be at the level of macrophage interaction with T cells.

## 4. SIGNIFICANCE OF IMMUNOMODULATION IN VIRAL PATHOGENESIS

The pathogeneses of avian retroviral disease states are complex problems involving a myriad of host–parasite interactions. The most intensively studied disease processes of these viruses has been the induction of neoplastic or hyperplastic disorders. In the natural transmission of the acute leukemia viruses, infection by a replication-defective oncogenic virus is accompanied by the simultaneous infection of the host with a replication-competent helper virus; the presence of a viremia has several obvious points of advantage for the progression of the tumor mass. Virus production allows for the active recruitment of newly transformed cells into the tumor mass. Furthermore, many helper viruses by themselves compromise the immune status of the host, thereby providing for the unrestricted progression of the tumor. The REV system provides an excellent model for the study of this phenomenon. Infection by REV-T(REV-A) results in a rapidly fatal leukemia characterized by severe hepatosplenomegaly with either lymphoproliferative or necrotic lesions.[42,88] Death invariably occurs within 7–10 days.[88,89] The helper virus, REV-A, has been demonstrated to be responsible for the anemia, peripheral nerve lesions, paralysis, and severe runting[34,62,90] first observed after the inoculation of chicks with reduced concentrations of REV-T(REV-A).[91]

The lethal dose for an REV-T-transformed virus-producing cell line may be several logs lower than that required for a nonvirus-producing cell line.[45,53] In both tumor transfers, the development of reticuloendotheliosis is accompanied by immunosuppression. Furthermore, there is a temporal correlation between the accumulation of the tumor mass within the visceral organs and the appearance of the suppressed state. In the absence of immunosuppression, the host mounts a response with CTL specific for REV-T-transformed cells. The NP cell lines by themselves are immunosuppressive to the host. However, the presence of REV-A in infections by virus-producing lines may work synergistically toward this suppression. Furthermore, in the disease produced by viral infection, in which the number of tumor cells would initially be low, the suppression induced by REV-A may be considered a major contributory factor in the progression of the disease state. The immunosuppression induced by REV-A is transient.[45,64] The suppression is detectable within 3 days post-infection and is maintained for a period of 2–3 weeks. In infections caused by REV-T(REV-A), death occurs well within this time frame.

In addition to their role in oncogenesis, the helper viruses of both the REV and ASLV groups are responsible for a variety of pathologic disturbances. The MAV-2(0) virus system has allowed for the dissection of two disorders common to the ALV: osteopetrosis and anemia. Osteopetrosis and the immunosuppression induced by MAV-2(0) infection may both arise from a generalized dysfunction in macrophage-like cells.

Virally induced osteopetrosis in chickens is considered a hyperplasia rather than a neoplasm.[20] Smith and Ivanyi[92] proposed that the osteopetrosis induced by MAV-2(0) results from an impairment in bone-resorptive function. The osteoblast proliferation characteristic of the disease is seen as the result of a malfunction in osteoclast populations, leading to an imbalance in the cellular homeostasis of the bone. The functions of osteoclasts in bone resorption are dependent on at least two soluble factors: osteoclast-activating factor (OAF)[93] and IL-1.[94] OAF is produced by lymphocytes by a monocyte-dependent mechanism, while IL-1 appears to act primarily on osteoblasts, which, in turn, stimulate osteoclasts in their bone-resorptive functions by a contact-mediated mechanism. Thus, a lack of bone resorption could result from a failure to produce or respond to either OAF or IL-1 or from a failure of either of two separate cell interactions. Furthermore, it is proposed that the osteoclast precursor is a macrophage-like cell[95,96] and is the primary target population in MAV-2(0)-induced osteopetrosis.[92] Thus, osteopetrosis could be the result of direct viral infection of the macrophage-like osteoclast precursor.

Immunosuppression in MAV-2(0) infections is apparently the result of a macrophage defect, possibly involving a macrophage–lymphocyte interaction.[65,67] MAV-2(0) induces a transient suppression in anemic birds but a persistent one in osteopetrotic birds.[27,65] Susceptibility to osteopetrosis correlates with the stage of embryogenesis, when seeding of the primary lymphoid organs begins.[92] By contrast, anemia is most prominent when MAV-2(0) is injected into older chicks (e.g., 8 days old).[54] Thus, the differences in immunosuppression (transient versus persistent) may be a function of lymphoid development. This idea is borne out by the fact that osteopetrotic birds display lymphoid involution,[52,97] possibly resulting from an impairment in the proliferation of follicular cells in the bursa and thymus.[52,92] Furthermore, avian lymphoid cells and osteoclasts share a hematogenic origin.[98,99] It is proposed that the MAV-2(0) target cell is a progenitor of both lymphoid cells and osteoclasts.[92]

Osteopetrosis and immunosuppression represent manifestations of the same viral infection. In both cases, a defect in macrophage or macrophage-like populations has been implicated. It is of interest to note that, in both cases, a defect on the part of macrophages to communicate with lymphocytes could be involved. The precise involvement of the virus in the macrophage defect remains unclear. The susceptibility of macrophages to infection within some strains of birds has been demonstrated to be linked with the subgroup B envelope specificity.[29,30] Considering the short latent period in the induction of osteopetrosis or immunosuppression (as demonstrated in anemia), it would seem unlikely that a cellular gene could be induced by an integration-specific mechanism; however, cellular gene expression could be influenced by the activity of a viral protein provided *in trans*. Furthermore, in recent studies involving subgroup E ALV, osteopetrosis resulted from osteoblasts in which persistent viral DNA synthesis was occurring aberrantly from newly synthesized viral RNA.[100] The ability to confer osteopetrosis by these subgroup E viruses is believed to involve the *pol* gene, while the severity of osteopetrosis is dependent on the strength of the viral promoter.[101] There is still no evidence to support a similar mechanism for MAV-2(0)-induced osteopetrosis.

ACKNOWLEDGMENTS.   The work on immunosuppression by avian reticuloendotheliosis was supported by U.S. Public Health Service grants Ca-26169 and Ca-33192 from the National Cancer Institute. Bob Storms is a predoctoral trainee supported by U.S. Public Health Service grant Ca-09182 from the National Cancer Institute.

# REFERENCES

1. Teich, N., Taxonomy of retroviruses, in: *RNA Tumor Viruses, Molecular Biology of Tumor Viruses*, 2nd ed. (R. Weiss, N. Teich, H. Varmus, and J. Coffin, eds.), pp. 25–209, Cold Spring Harbor Laboratory, Cold Spring Harbor, New York (1982).

2. Temin, H. M., and D. Baltimore, RNA-directed DNA synthesis and RNA tumor viruses, *Adv. Virus Res.* **17**:129–186 (1972).

3. Varmus, H., and R. Swanstrom, Replication of retroviruses, in: *RNA Tumor Viruses, Molecular Biology of Tumor Viruses*, 2nd ed. (R. Weiss, N. Teich, H. Varmus, and J. Coffin, eds.), pp. 369–512, Cold Spring Harbor Laboratory, Cold Spring Harbor, New York (1982).

4. Graf, T., and H. Beug, Avian Leukemia viruses. Interaction with their target cells *in vivo* and *in vitro*, *Biochim. Biophys. Acta* **512**:269–299 (1978).

5. Bishop, J. M., Cellular oncogenes and retroviruses, *Annu. Rev. Biochem.* **52**:301–354 (1983).

6. Vogt, P. K., and R. R. Friis, An avian leukosis virus related to RSV(O): Properties and evidence for helper activity, *Virology* **43**:223–234 (1971).

7. Calnek, B. W., Lymphoid leukosis virus: A survey of commercial breeding flocks for genetic resistance and incidence of embryo infection, *Avian Dis.* **12**:104–111 (1968).

8. Purchase, H. G., and B. R. Burmester, Leukosis/sarcoma group, in: *Diseases of Poultry* (M. S. Hofstad, B. W. Calnek, C. F. Helmboldt, W. M. Reid, and H. W. Yoder, Jr., eds.), pp. 418–468, Iowa State University Press, Ames, Iowa (1978).

9. Peterson, R. D. A., H. G. Purchase, B. R. Burmester, M. D. Cooper, and R. A. Good, Relationships among visceral lymphomatosis, bursa of Fabricius, and bursa-dependent lymphoid tissue of the chicken, *J. Natl. Cancer Inst.* **36**:585–598 (1966).

10. Beug, H., A. von Kirckbach, G. Doederlein, J. F. Conscience, and T. Graf, Chicken hematopoietic cells transformed by seven strains of defective acute leukemia viruses display three distinct phenotypes of differentiation, *Cell* **18**:375–390 (1979).

11. Mlandenov, Z., U. Heine, D. Beard, and J. W. Beard, Strain MC29 avian leukosis virus. Myelocytoma, endothelioma, and renal growth: Pathomorphological and ultrastructural aspects, *J. Natl. Cancer Inst.* **38**:251–258 (1967).

12. Burmester, B. R., M. A. Gross, W. G. Walter, and A. K. Fontes, Pathogenicity of a viral strain (RPL12) causing avian visceral lymphomatosis and related neoplasms. II. Host–virus interactions affecting response, *J. Natl. Cancer Inst.* **22**:103–127 (1959).

13. Beard, J. W., Avian virus growths and their etiological agents, in: *Advances in Cancer Research* (A. Haddow and S. Weinhouse, eds.), pp. 1–127, Academic, New York (1963).

14. Moscovici, C., and P. K. Vogt, Effect of genetic cellular resistance on cell transformation and virus replication in chicken hematopoietic cultures infected with avian myeloblastosis virus (BAI/A), *Virology* **35**:487–497 (1968).

15. Langlois, A. J., L. Veprek, D. Beard, R. B. Fritz, and J. W. Beard, Isolation of a non-focus forming virus from stock MC29 by endpoint dilution, *Cancer Res.* **31**:1010–1016 (1971).

16. Ishizuki, R., and T. Shimuzi, Heterogeneity of strain R avian (erythroblastosis) virus, *Cancer Res.* **30**:2827–2831 (1970).

17. Smith, R. E., and C. Moscovici, The oncogenic effects of nontransforming viruses from avian myeloblastosis virus, *Cancer Res.* **29**:1356–1366 (1969).

18. Graf, T., B. Royer-Pokora, G. E. Schubert, and H. Beug, Evidence for the multiple oncogenic potential of cloned leukemia virus: *In vitro* and *in vivo* studies with avian erythroblastosis virus, *Virology* **71**:423–433 (1976).

19. Holmes, J. R., Experimental transmission of avian osteopetrosis, *J. Comp. Pathol.* **68**:439–443 (1958).

20. Sanger, V. L., T. N. Fredrickson, C. C. Merrill, and B. R. Burmester, Pathogenesis of osteopetrosis in chickens, *Am. J. Vet. Res.* **27**:1735–1744 (1966).
21. Purchase, H. G., R. C. Chubb, and P. M. Biggs, Effect of lymphoid leukosis and Marek's disease on the immunological responsiveness of the chicken, *J. Natl. Cancer Inst.* **40**:583–592 (1968).
22. Dent, P. B., M. D. Cooper, L. M. Payne, J. J. Solomon, B. R. Burmester, and R. A. Good, Pathogenesis of avian lymphoid leukosis. II. Immunological reactivity during lymphomagenesis, *J. Natl. Cancer Inst.* **41**:391–401 (1968).
23. Crittenden, L. B., Exogenous and endogenous leukosis virus genes—A review, *Avian Pathol.* **10**:101–112 (1981).
24. Granlund, D. J., and R. W. Loan, Effect of lymphoid leukosis virus infection on the cell immune capacity of the chicken, *J. Natl. Cancer Inst.* **52**:1373–1374 (1974).
25. Fadly, A. M., L. F. Lee, and L. D. Bacon, Immunocompetence of chickens during early and tumorigenic stages of Rous-associated virus-1 infection, *Infect. Immun.* **37**:1156–1161 (1982).
26. Meyers, P., G. D. Ritts, and D. R. Johnson, Phytohemagglutinin-induced leukocyte blastogenesis in normal and avian leukosis virus infected chickens, *Cell. Immunol.* **27**:140–146 (1976).
27. Smith, R. E., and L. J. van Eldik, Characterization of the immunosuppression accompanying virus-induced avian osteopetrosis, *Infect. Immun.* **22**:452–461 (1978).
28. Rup, B. J., J. D. Hoelzer, and H. R. Bose, Jr., Helper viruses associated with avian acute leukemia viruses inhibit the cellular immune response, *Virology* **116**:61–71 (1982).
29. Gazzolo, L., M. G. Moscovici, and C. Moscovici, Replication of avian sarcoma viruses in chicken macrophages, *Virology* **58**:514–525 (1974).
30. Gazzolo, L., M. G. Moscovici, C Moscovici, and P. K. Vogt, Susceptibility and resistance of chicken macrophages to avian RNA tumor viruses, *Virology* **67**:553–565 (1975).
31. Kang, C-Y., and H. M. Temin, Lack of sequence homology among RNAs of avian leukosis-sarcoma viruses, reticuloendotheliosis viruses and chicken endogenous RNA-directed DNA polymerase activity, *J. Virol.* **12**:1314–1324 (1973).
32. Maldonado, R. L., and H. R. Bose, Jr., Relationship of reticuloendotheliosis virus to the avian tumor viruses: Nucleic acid and polypeptide composition, *J. Virol.* **11**:741–747 (1973).
33. Maldonado, R. L., and H. R. Bose, Jr., Group-specific antigen shared by the members of the reticuloendotheliosis virus complex, *J. Virol.* **17**:983–990 (1976).
34. Hoelzer, J. D., R. B. Franklin, and H. R. Bose, Jr., Transformation by reticuloendotheliosis virus: Development of a focus assay and isolation of a nontransforming virus, *Virology* **93**:20–30 (1979).
35. Ludford, G. G., H. G. Purchase, and H. W. Cox, Duck infectious anemia virus associated with *Plasmodium lophurea*, *Exp. Parasitol.* **31**:29–30 (1972).
36. Trager, W. A., New virus of ducks interfering with development of malaria parasites (*Plasmodium lophurea*), *Proc. Soc. Exp. Biol. Med.* **101**:578–582 (1959).
37. Cook, M. K., Cultivation of a filterable agent associated with Marek's disease, *J. Natl. Cancer Inst.* **43**:203–213 (1969).
38. Hoelzer, J. D., R. B. Lewis, C. R. Wasmuth, and H. R. Bose, Jr., Hematopoietic cell transformation by reticuloendotheliosis virus: Characterization of the genetic defect, *Virology* **100**:462–474 (1980).
39. Purchase, H. G., and R. L. Witter, The reticuloendotheliosis viruses, *Curr. Top. Microbiol. Immunol.* **71**:103–124 (1975).
40. Purchase, H. G., C. Ludford, K. Nazerian, and H. W. Cox, A new group of oncogenic viruses: Reticuloendotheliosis, chick syncytial, duck infectious anemia and spleen necrosis viruses, *J. Natl. Cancer Inst.* **51**:489–499 (1973).

41. Witter, R. L., and L. B. Crittenden, Lymphomas resembling lymphoid leukosis in chickens inoculated with reticuloendotheliosis virus, *Int. J. Cancer* **23**:673–678 (1979).

42. Sevoian, M. R., N. Larose, and D. M. Chamberlain, Avian lymphomatoses VI: A virus of unusual potency and pathogenicity, *Avian Dis.* **8**:336–347 (1964).

43. Temin, H. M., and V. K. Kassner, Replication of reticuloendotheliosis virus in cell culture: Acute infection, *J. Virol.* **13**:291–297 (1974).

44. Temin, H. M., and V. K. Kassner, Replication of reticuloendotheliosis virus in cell culture: Chronic infection, *J. Gen. Virol.* **27**:267–274 (1975).

45. Rup, B. J., J. L. Spence, J. D. Hoelzer, R. B. Lewis, C. R. Carpenter, A. S. Rubin, and H. R. Bose, Jr., Immunosuppression induced by avian reticuloendotheliosis virus: Mechanism of induction of the suppressor cell, *J. Immunol.* **123**:1362–1370 (1979).

46. Franklin, R. B., C-Y. Kang, M. K. Wan, and H. R. Bose, Jr., Transformation of chick embryo fibroblasts by reticuloendotheliosis virus, *Virology* **83**:313–321 (1977).

47. Lewis, R. B., J. McClure, B. Rup, D. W. Niesel, R. F. Garry, J. D. Hoelzer, K. Nazerian, and H. R. Bose, Jr., Avian reticuloendotheliosis virus: Identification of the hematopoietic target cell for transformation, *Cell* **25**:421–431 (1981).

48. Beug, H., H. Muller, S. Grieser, G. Doederlein, and T. Graf, Hematopoietic cells transformed *in vitro* by REV-T avian reticuloendotheliosis virus express characteristics of very immature lymphoid cells, *Virology* **115**:295–309 (1981).

49. Singer, A., and R. J. Hodes, Mechanisms of T cell-B cell interaction, *Annu. Rev. Immunol.* **1**:211–241 (1983).

50. Evans, G., and J. Ivanyi, Antibody synthesis by chicken spleen cell *in vitro*. I. Requirements of B cells at various stages after immunization for T cell, macrophages and antigen, *Eur. J. Immunol.* **5**:747–752 (1975).

51. McArthur, W. P., D. G. Gilmour, and G. J. Thorbecke, Immunocompetent cell in the chicken. II. Synergism between thymus and either bursa or bone marrow cells in the humoral response to sheep erythrocytes, *Cell. Immunol.* **8**:103–111 (1973).

52. Hirota, Y., M-T. Martin, M. Viljanen, P. Toivanen, and R. M. Franklin, Immunopathology of chickens infected *in ovo* and at hatching with the avian osteopetrosis virus MAV-2(0), *Eur. J. Immunol.* **10**:929–936 (1980).

53. Walker, M. H., B. J. Rup, A. S. Rubin, and H. R. Bose, Jr., Specificity in the immunosuppression induced by avian reticuloendotheliosis virus, *Infect. Immun.* **40**:225–235 (1983).

54. Paterson, R. W., and R. E. Smith, Characterization of anemia induced by avian osteopetrosis virus, *Infect. Immun.* **22**:891–900 (1978).

55. Price, J. A., and R. E. Smith, Influence of bursectomy on bone growth and anemia induced by avian osteopetrosis viruses, *Cancer Res.* **41**:752–759 (1981).

56. Green, D. R., P. M. Flood, and R. K. Gershon, Immunoregulatory T-cell pathways, *Annu. Rev. Immunol.* **1**:439–463 (1983).

57. Naspilz, C. K., and M. Richter, The action of phytohemagglutinin *in vivo* and *in vitro*, *Prog. Allergy* **12**:1–85 (1968).

58. Miggiano, V. C., I. Bergen, and J. R. L. Pink, The mixed lymphocyte reaction in chickens: Evidence for control by the major histocompatibility complex, *Eur. J. Immunol.* **4**:397–401 (1974).

59. Hovi, T., J. Suni, L. Hortling, and A. Vaheri, Stimulation of chicken lymphocytes by T- and B-cell mitogens, *Cell. Immunol.* **39**:70–78 (1978).

60. Greaves, M. F., I. M. Roitt, and N. E. Rose, Effect of bursectomy and thymectomy on the responses of chicken peripheral blood lymphocytes to phytohemagglutinin, *Nature (Lond.)* **220**:293–295 (1968).

61. Toivanen, P., and A. Toivanen, Selective activation of chicken T lymphocytes by concanavalin A, *J. Immunol.* **111**:1602–1603 (1973).

62. Scofield, V. L., and H. R. Bose, Jr., Depression of the mitogen response in spleen cells

from reticuloendotheliosis virus-infected chickens and their suppressive effect on normal lymphocyte response. *J. Immunol.* **120**:1321–1325 (1978).

63. Carpenter, C. R., H. R. Bose, Jr., and A. S. Rubin, Contact-mediated suppression of mitogen-induced responsiveness by spleen cells in reticuloendotheliosis virus-induced tumorigenesis, *Cell. Immunol.* **33**:392–401 (1977).

64. Carpenter, C. R., A. S. Rubin, and H. R. Bose, Jr., Suppression of the mitogen-stimulated blastogenic response during reticuloendotheliosis virus-induced tumorigenesis: Investigations into the mechanism of action of the suppressor cells, *J. Immunol.* **120**:1313–1320 (1978).

65. Price, J. A., and R. E. Smith, Inhibition of concanavalin A response during osteopetrosis virus infection, *Cancer Res.* **42**:3617–3624 (1982).

66. Carpenter, C. R., K. E. Kempf, H. R. Bose, Jr., and A. S. Rubin, Characterization of the interaction of reticuloendotheliosis virus with the avian lymphoid system, *Cell. Immunol.* **39**:309–315 (1978).

67. Boni-Schnetzler, M., J. Boni, and R. M. Franklin, Reversal of T-cell unresponsiveness by T-cell-conditioned medium in retrovirus MAV-2(0)-induced immunosuppression in chickens, *Cancer Res.* **45**:4871–4875 (1985).

68. Pink, J. R., and O. Vainio, Genetic control of the response of chicken T-lymphocytes to concanavalin A: Cellular localisation of the low responder defect, *Eur. J. Immunol.* **13**:571–575 (1983).

69. Rice, N. R., R. R. Hiebsch, M. A. Gonda, H. R. Bose, Jr., and R. V. Gilden, Genome of reticuloendotheliosis virus: Characterization by use of cloned provirus DNA, *J. Virol.* **42**:237–252 (1982).

70. Stephens, R. M., N. R. Rice, R. R. Heibsch, H. R. Bose, Jr., and R. V. Gilden, Nucleotide sequence of v-*rel:* The oncogene of reticuloendotheliosis virus, *Proc. Natl. Acad. Sci. USA* **80**:6229–6233 (1983).

71. Wilhelmsen, K. C., K. Eggleton, and H. M. Temin, Nucleic acid sequences of the oncogene v-*rel* in reticuloendotheliosis virus strain T and its cellular homolog, the proto-oncogene c-*rel, J. Virol.* **52**:172–182 (1984).

72. Mathes, L. E., R. G. Olsen, L. C. Hebebrand, E. A. Hover, and J. P. Schaller, Abrogation of lymphocyte blastogenesis by a feline leukaemia virus protein, *Nature (Lond.)* **274**:687–689 (1978).

73. Mathes, L. E., R. G. Olsen, L. C. Hebebrand, E. A. Hoover, J. P. Schaller, P. W. Adams, and W. S. Nichols, Immunosuppressive properties of a virion polypeptide, a 15,000-dalton protein, from feline leukemia virus, *Cancer Res.* **39**:950–955 (1979).

74. Hebebrand, L. C., R. G. Olsen, L. E. Mathes, and W. S. Nichols, Inhibition of human lymphocyte mitogen and antigen response by a 15,000 dalton protein from feline leukemia virus, *Cancer Res.* **39**:443–447 (1979).

75. Cianciolo, G. J., T. J. Matthews, D. P. Bolognesi, and R. Synderman, Macrophage accumulation in mice is inhibited by low molecular weight products from murine leukemia viruses, *J. Immunol.* **124**:2900–2905 (1980).

76. Copelan, E. A., J. J. Rinehart, M. Lewis, L. Mathes, R. Olsen, and A. Sangone, The mechanism of retrovirus suppression of human T cell proliferation *in vitro, J. Immunol.* **131**:2017–2020 (1983).

77. Snyderman, R., and G. J. Cianciolo, Immunosuppressive activity of the retroviral envelope protein p15E and its possible relationship to neoplasia, *Immunol. Today* **5**:240–244 (1984).

78. Cianciolo, G. J., T. D. Copeland, S. Oroszlan, and R. Snyderman, Inhibition of lymphocyte proliferation by synthetic peptide homologous to retroviral envelope proteins, *Science* **230**:453–455 (1985).

79. Sonigo, P., C. Barker, E. Hunter, and S. Wain-Hobson, Nucleotide sequence of Mason-Pfizer monkey virus: An immunosuppressive D-type retrovirus, *Cell* **45**:375–385 (1986).

80. Cerny, J., K. D. Grinwich, and R. A. Stiller, Immunosuppression by spleen cells from Moloney leukemia. III. Evidence for a suppressor cell that is not the leukemic, virus-producing cell, *J. Immunol.* **119**:1097–1081 (1977).
81. Caulfield, M. J., and J. Cerny, Cell interactions in leukemia-associated immunosuppression: Suppression of thymus-independent antibody responses by leukemia spleen cells (Moloney) *in vitro* is mediated by normal T cells, *J. Immunol.* **124**:255–260 (1980).
82. Cianciolo, G. J., J. Hunter, J. Silva, J. S. Haskill, and R. Snyderman, Inhibitors of monocyte responses to chemotaxins are present in human cancerous effusion and react with monoclonal antibodies to the p15E protein of retroviruses, *J. Clin. Invest.* **68**:831–844 (1981).
83. Cianciolo, G. J., M. E. Lostrom, M. Tam, and R. Snyderman, Murine malignant cells synthesize a 19,000-dalton protein that is physiochemically and antigenically related to the immunosuppressive retroviral protein, p15E, *J. Exp. Med.* **158**:885–900 (1983).
84. Hershey, P., C. Bindon, M. Czerniecki, A. Spurling, J. Wass, and W. H. McCarthy, Inhibition of interleukin 2 production by factors released from tumor cells, *J. Immunol.* **131**:2837–2842 (1983).
85. Cianciolo, G. J., D. Phipps, and R. Snyderman, Human malignant and mitogen-transformed cells contain retroviral p15E-related antigen, *J. Exp. Med.* **159**:964–969 (1984).
86. Smith, K. A., Interleukin-2, *Annu. Rev. Immunol.* **2**:319–333 (1984).
87. Durum, S. K., J. A. Schmidt, and J. J. Oppenheim, Interleukin-1: An immunological perspective, *Annu. Rev. Immunol.* **3**:263–287 (1985).
88. Theilen, G. H., R. F. Zeigel, and M. J. Twiehaus, Biological studies with RE virus (strain T) that induces reticuloendotheliosis in turkeys, chickens and japanese quail, *J. Natl. Cancer Inst.* **37**:731–743 (1966).
89. Bose, H. R., Jr., and A. S. Levine, Replication of reticuloendotheliosis virus (strain T) in chicken embryo cell culture, *J. Virol.* **1**:1117–1121 (1967).
90. Garry, R. F., G. M. Shackleford, L. J. Berry, and H. R. Bose, Jr., Inhibition of hepatic phosphoenolpyruvate carboxykinase by avian reticuloendotheliosis virus, *Cancer Res.* **45**:5020–5026 (1985).
91. Witter, R. L., H. G. Purchase, and G. H. Burgoyne, Peripheral nerve lesions similar to those of Marek's disease in chickens inoculated with reticuloendotheliosis virus, *J. Natl. Cancer Inst.* **45**:567–577 (1970).
92. Smith, R. E., and J. Ivanyi, Pathogenesis of virus-induced osteopetrosis in the chicken, *J. Immunol.* **125**:523–530 (1980).
93. Yoneda, T., and G. R. Mundy, Monocytes regulate osteoclastactivating factor production by releasing prostaglandins, *J. Exp. Med.* **150**:338–350 (1979).
94. Thomson, B. M., J. Saklatvala, and T. J. Chambers, Osteoblasts mediate interleukin-1 stimulation of bone resorption by rat osteoclasts, *J. Exp. Med.* **164**:104–112 (1986).
95. Marks, S. C., Jr., Studies on the mechanism of spleen cell cure for osteopetrosis in *ia* rats: Appearance of osteoblasts with ruffled borders, *Am. J. Anat.* **151**:119–123 (1978).
96. Marks, S. C., Jr., and G. B. Schneider, Evidence for a relationship between lymphoid cells and osteoclasts: Bone resorption restored in *ia* (osteopetrotic) rats by lymphocytes, monocytes, and macrophages from a normal litter mate, *Am. J. Anat.* **152**:331–341 (1978).
97. Banes, A. J., and R. E. Smith, Biological characterization of avian osteopetrosis, *Infect. Immun.* **16**:876–884 (1977).
98. Moore, M. A. S., and J. J. T. Owen, Experimental studies on the development of the bursa of Fabricius, *Dev. Biol.* **14**:40–51 (1966).
99. Jotereau, F. V., and N. M. le Douarin, The development relationship between osteocytes and osteoclasts: A study using the quail-chick nuclear marker in endochondral ossification, *Dev. Biol.* **63**:253–265 (1978).
100. Robinson, H. L., and B. D. Miles, Avian leukosis virus-induced osteopetrosis is associated with the persistent synthesis of viral DNA, *Virology* **141**:130–143 (1985).

101. Shank, P. R., P. J. Schatz, L. M. Jensen, P. N. Tsichlis, J. M. Coffin, and H. L. Robinson, Sequences in the *gag-pol*-5' *env* region of avian leukosis viruses confer the ability to induce osteopetrosis, *Virology* **145:**94–104 (1985).

102. Coffin, J., Structure of the retroviral genome, in: *RNA Tumor Viruses, Molecular Biology of Tumor Viruses,* 2nd ed. (R. Weiss, N. Teich. H. Varmus, and J. Coffin, eds.), pp. 261–368, Cold Spring Harbor Laboratory, Cold Spring Harbor, New York, (1982).

103. Linial, M., and D. Blair, Genetics of retroviruses, in: *RNA Tumor Viruses, Molecular Biology of Tumor Viruses,* 2nd ed. (R. Weiss, N. Teich, H. Varmus, and J. Coffin, eds.), pp. 649–783, Cold Spring Harbor Laboratory, Cold Spring Harbor, New York, (1982).

104. Tsai, W-P., T. D. Copeland, and S. Oroszlan, Purification and chemical and immunological characterization of avian reticuloendotheliosis virus *gag*-gene-encoded structural proteins, *Virology* **140:**289–312 (1985).

105. Fujita, D. J., Y. C. Chen, R. R. Friis, and P. K. Vogt, RNA tumor viruses of pheasants: Characterization of avian leukosis subgroups F and G, *Virology* **60:**558–571 (1974).

106. Hanafusa, T., H. Hanafusa, C. E. Metroka, W. S. Hayward, C. W. Rettenmier, R. C. Sawyer, R. M. Dougherty, and H. S. Di Stefano, Pheasant viruses: A new class of ribodeoxyviruses, *Proc. Natl. Acad. Sci. USA* **73:**1333–1337 (1976).

107. Chen, Y. C., and P. K. Vogt, Endogenous leukosis viruses in the avian family *Phasianidae,* *Virology* **76:**740–750 (1977).

108. McDougall, J. S., P. M. Biggs, R. W. Shilleto, and B. S. Milne, Lymphoproliferative disease of turkeys. II. Experimental transmission and aetiology, *Avian Pathol.* **7:**141–155 (1978).

# 20

# Nonhuman Mammalian Retroviruses

## STEVEN SPECTER, MAURO BENDINELLI, and HERMAN FRIEDMAN

## 1. BASIC PROPERTIES

It is beyond the scope of this chapter to present a comprehensive discussion of the biology of mammalian retroviruses. This has been reviewed effectively elsewhere.[1,2] The brief overview provided here focuses on the aspects of retrovirus biology that may assist in understanding the effects of these viruses on immunity.

Retroviridae is the family classification for all RNA viruses that replicate by way of a DNA intermediate, which can become integrated into the host cell genome. This includes the subfamilies Oncovirinae, the tumorigenic retroviruses; Lentivirinae, "slow" or chronic degenerative disease-causing viruses; and Spumavirinae, foamy and syncytial viruses (Table I). Mammalian retroviruses are a homogeneous group regarding general chemical composition, genetic organization, method of replication, and structure. However, they differ drastically in the spectrum of pathogenic diseases they cause. The oncogenic retroviruses are classified according to morphology, characteristics of maturation, and natural host species. They are further subdivided based on antigenic markers, experimental host range, sensitivity to cross-interference, and other biologic or pathogenic properties. These viruses generally cause leukemia/lymphomas or fibrosarcomas, although other tumor types may result. Three genuses comprise this subfamily and are designated C, B, and D (Table I).

STEVEN SPECTER and HERMAN FRIEDMAN • Department of Medical Microbiology and Immunology, College of Medicine, University of South Florida, Tampa, Florida 33612-4799. MAURO BENDINELLI • Department of Biomedicine, University of Pisa, I-56100 Pisa, Italy.

## TABLE I
### Major Species of Nonhuman Mammalian Retroviruses

Subfamily Oncovirinae
  Genus oncovirus C
    Murine leukemia viruses
      Friend leukemia virus (FLV)
      Moloney leukemia virus (MLV)
      Radiation leukemia virus (RadLV)
      Rauscher leukemia virus
    Murine sarcoma viruses
      Moloney sarcoma virus (MSV)
      Harvey sarcoma virus (HSV)
      Kirsten sarcoma virus (KSV)
    Feline leukemia and sarcoma viruses
      (FeLV)
    Primate leukemia and sarcoma viruses
      Simian sarcoma virus (SSV)
      Gibbon ape leukemia virus
      Baboon endogenous virus
    Other species
      Cattle
      Rat
      Hamster
  Genus oncovirus B
    Mouse mammary tumor virus (MMTV)
    Viruses of baboon, guinea pigs, possibly
      others
  Genus oncovirus D
    Mason–Pfizer monkey virus (MPMV)
      and related viruses
    Squirrel monkey virus
    Guinea pig virus

Subfamily Lentivirinae
  Visna maedi
  Progressive pneumonia virus
  Caprine arthritis encephalitis virus (CAEV)
  Equine infectious anemia virus (EIAV)
  Simian immune deficiency virus (SIV)
  Candidate feline and bovine lentiviruses
Subfamily Spumavirinae
  Simian foamy virus
  Canine foamy virus
  Bovine syncytial virus
  Feline syncytial virus
  Hamster syncytial virus

The virion is approximately of 100-nm diameter and consists of an icosahedral nucleocapsid surrounded by a lipid envelope. The genome is a single-stranded RNA of positive polarity. Generally, each particle contains two genomic copies. Genome replication occurs in the nucleus with nucleocapsid formation in the cytoplasm. Maturation takes place at the cytoplasmic membrane from which mature particles obtain their envelope as they are released from the cell by budding. The genome is linear, weighs $5–7\times10^6$ daltons, and consists of three known genes, *gag*, *pol*, and *env*, in the order 5' to 3'. Another major feature retroviruses have in common is the *pol* gene product, an RNA-dependent DNA-polymerase (reverse transcriptase). Type C oncoviruses share a centrally located dense nucleocapsid, whereas the type B nucleocapsid is eccentric. Both have lipid envelopes with glycoprotein spikes. Type D particles have a nucleocapsid intermediate located between B- and C-type particles and have less prominent spikes on their envelope.

Mammalian retroviruses can be further described as ecotropic (growth only on cells of species of origin), xenotropic (efficient growth only in cells other than the species of origin), or amphitropic (growth in both types of cells). Retroviral genomes are frequently found in the form of proviruses stably integrated into the host DNA, whose level of expression can vary greatly, depending on the host's genetics and epigenetic factors. Nothing is known about the immunomodulating properties of endogenous retroviruses in their own host. Their general significance within the context of the host's pathophysiology is also unknown. They may be directly associated with oncogenesis, but this is not common. Recombinants between endogenous and exogenous retroviruses have been implicated, however, in the genesis of tumors induced by exogenous retroviruses. Recently, a viral preparation derived from a strain of radiation leukemia virus (RadLV) of mice and containing a mixture of ecotropic, xenotropic, and recombinant viruses was shown to cause a strikingly rapid and profound suppression of all measured aspects of cellular and humoral immunity, as well as polyclonal lymphocyte activation.[3] The role of the endogenous and recombinant retroviruses in the genesis of immunosuppression appears to be worthy of further investigation. Exogenous retroviruses spread within their natural host species horizontally or vertically in a manner similar to that of other viruses. Genetically, cell transformation is believed to be due to the introduction of informational (oncogenes) or regulatory sequences into the host cell genome, while much uncertainty still exists with regard to the biochemical changes responsible for the transformed phenotype. The role of constitutive and adaptive host resistance factors in retrovirus-induced oncogenesis has been extensively studied. The general consensus is that immune mechanisms are crucial in determining not only host resistance to the infecting virus but also tumor development and progression.[1,2]

# 2. MODULATION OF IMMUNITY BY RETROVIRUSES

## 2.1. Introduction

Modulation of immune responses by mammalian retroviruses has been recognized for more than a quarter century. Several reviews covering this subject have been published.[4-9] Therefore, this chapter highlights material covered in these reviews, focusing on more recent findings and the mechanisms of immune suppression.

## 2.2. Murine Leukemia/Sarcoma Viruses

### 2.2.1. Humoral Immunity

Alteration of antibody responsiveness by murine retroviruses was first documented during the 1960s.[10-12] At least eight different murine oncornaviruses have been recorded to inhibit primary and/or secondary antibody production.[4-9] The most extensive studies have been performed with the Friend leukemia virus (FLV). Ceglowski and Friedman, as well as others, have reported decreased antibody production using both serum antibody titers and enumeration of antibody plaque-forming cells due to this murine retrovirus.[13,14] Inhibition of antibody production against sheep red blood cells (SRBC) as well as protein, bacterial, and viral antigens has been reported.[4-9,12,15,16] In addition, total lymphocyte counts and the ability of B lymphocytes to home to the spleen were reduced.[17-19] As the disease progresses, the depression of antibody formation becomes more severe. Electron microscopic examination of plasma cells from FLV-infected mice reveals the presence of C-type particles. This is seen even in antibody-producing cells, indicating that infection of plasma cells as well as antibody synthesis are not mutually exclusive.[20,21] Rather, it appears that failure of precursor B cells to develop into plasma cells causes the depressed response. Whether this is due to direct infection of these cells or effects of FLV on other cells (T lymphocytes and/or macrophages) is not entirely clear (see Section 3.2). FLV induces rapid erythroblastosis and erythroleukemia in the spleen. However, many of the FLV effects outlined above occur before significant histopathologic changes take place and are also seen following single infection with the helper virus alone, which is otherwise asymptomatic for several months postinfection.[16,22] This finding indicates that viral infection per se is responsible for the development of immunosuppressive symptoms.

Accordingly, *in vitro* studies have indicated that both infected spleen cells and cell-free virus are capable of inhibiting the generation of primary antibody responses by normal lymphoid cells.[23-25] Virus had to be added at the initiation of the culture or shortly thereafter, confirming that only early events in antibody formation are inhibited by the virus.

### 2.2.2. Cell-Mediated Host Defenses

*2.2.2.a. T Lymphocytes.* Extensive studies have been performed concerning cellular immune responses in FLV infection, including T-lymphocyte function, macrophage activity, natural killer (NK) cell function, and lymphokine production. The earliest studies examined functions associated with T cells. Friedman and co-workers as well as others reported defective allograft rejection, tumor rejection, delayed hypersensitivity and contact sensitivity responses.[26-28] Lymphocyte-mediated cytotoxicity and production of migration inhibitory factor were also noted to be impaired.[27,29] These studies showed that the decreased activity was located in cells that did not adhere to plastic surfaces, indicating that the FLV effect was on the T lymphocytes rather than on macrophages, which act as accessory cells. Population studies have indicated a drastically reduced number of Thy 1.2 + cells in the spleens of FLV-infected mice with an increased proportion in the lymph nodes early in infection. After 3 weeks of infection, cells expressing B- and T-lymphocyte markers were decreased in the spleen and lymph nodes.[30,31] More recently, Kitagawa and co-workers demonstrated that both Lyt 1 + and Lyt 2 + cell numbers decrease to zero in the spleen of susceptible mice infected with FLV.[32] However, there is only a transient number decrease in T cell numbers of both subsets in the lymph nodes. Murine leukemia viruses have also been shown to enhance susceptibility to bacteria and to oncogenic and nononcogenic viruses.[33–35] This is most likely another reflection of defective cellular immunity.

2.2.2.b. *Macrophages.* Macrophages are vital protagonists of immunity, they serve as accessory cells to T- and B-lymphocyte function and as effector cells. Their function in phagocytosis of foreign particles and tumor cell killing is a first line of the host defenses. Phagocytosis of SRBC and carbon particles has been demonstrated to be severely inhibited in FLV infection[36–38] and in other retrovirus infections of mice.[39] More importantly, macrophage function may be vital to general immune responsiveness in murine retrovirus infections.

Kirchner *et al.*[40] reported that Moloney virus-induced immunosuppression is associated with suppressor macrophages. Likewise, Moody *et al.*[41] demonstrated that suppressor macrophages mediate inhibition of NK cell activity in FLV infection (D. J. Moody, S. Specter, and H. Friedman, unpublished observations). Suppressor macrophages have also been described in several other retrovirus infections as well as in animals having retrovirus-induced tumors.[42] Bendinelli *et al.*[43] and Specter *et al.*[44] reported that the addition of normal macrophages to FLV-infected spleen cells in culture could reverse suppression of B-lymphocyte function, although others failed to restore *in vivo* antibody responses with normal macrophages.[45] Thus, macrophage functions, as accessory cells and effector cells, are severely depressed in murine retrovirus infection, while suppressor function is intact or even enhanced.

2.2.2.c. *Natural Killer Cells/Antibody-Dependent Cellular Cytotoxicity.* NK cell activity is also believed to be important in the first line of host defense against intracellular pathogens and tumor cells. NK function was shown to be severely compromised in FLV-infected mice.[41] Furthermore, it was demonstrated *in vitro* that spleen cells from FLV-infected mice inhibited the NK activity of normal spleen cells. This suppression was attributed to splenic cells that were nonspecific esterase positive and nonadherent to nylon wool or plastic, indicating that they were macrophages. When these macrophages from FLV-infected mice were incubated with normal spleen cells in the presence of indomethacin, suppression did not occur, suggesting that a cyclo-oxygenase product (prostaglandin E2?) was the mediator of this effect (D. J. Moody, S. Specter, and H. Friedman, unpublished observations). Cell binding between NK cells and target cells is normal. Thus, the suppressive effects attributable to FLV infection were on the lytic phase of NK activity.[41] Antibody-dependent cellular cytotoxicity (ADCC) was also shown to be suppressed by FLV infection.[46]

2.2.2.d. *Lymphokine/Monokine Activity.* Earliest studies on the production of soluble factors in murine retrovirus infection measured the ability of antigen-sensitized lymphocytes to produce migration inhibitory factor and interferon (IFN).[27,47] Both factors were shown to be depressed in AKR leukemia and FLV infection.[27,47] More recent studies have focused on interleukin production. Butler *et al.*[48] reported that serum from *Escherichia coli* lipopolysaccharide (LPS)-treated mice contained a factor that reversed the suppression of antibody formation. Although not identified by these investigators, this was suspected of being interleukin-1 (IL-1). However, supplementing cultures of FLV-infected spleen cells with exogenous IL-1 failed to restore antibody responsiveness to any significant respect.[49] Subsequently, IL-1 production has been investigated in FLV-infected mice[50] (D. Matteucci and M. Bendinelli, unpublished observations). These mice produced normal or elevated levels of IL-1. Moreover, thymocytes from FLV-infected mice responded normally to the addition of exogenous IL-1.

Unlike the observations with IL-1, interleukin-2 (IL-2) production by spleen cells is markedly decreased during FLV infection. Even when T cells are separated from tumor cells and macrophages using a Percoll gradient, the suppression was detected.[50] This indicates that decreased IL-2 activity is not due to a dilution of producer cells by tumor cells. It is interesting to speculate that the failure to produce IL-2 may be related to the increase in IL-1 levels. If IL-2 is involved, through a feedback circuit, in controlling IL-1 production, failure to produce normal amounts of IL-2 could lead to an increase in IL-1 levels. Furthermore, it is possible that the failure to produce cytokines may be responsible for many of the defective immune responses seen in murine retrovirus infections. Studies are in progress in the FLV model to determine the importance of these cytokines in mediating the suppression or responsiveness in these infected mice.

## 2.3. Feline Retroviruses

Feline leukemia virus (FeLV) is a type C retrovirus first described to induce immune hyporesponsiveness by Perryman *et al.*[51] Unlike the murine viruses, there was no notable effect on antibody production, but later studies have shown suppressed antibody responses to defined oligopeptides.[52] The spectrum of immune defects noted in FeLV infection has recently been reviewed.[9,53] Studies have reported inhibition of mitogen and antigen induced T lymphocyte proliferation, a decrease in cell mediated lympholysis and lymphokine production. In addition, reduced mobility of lymphocyte membrane receptors for concanavalin A (Con A) has been reported, as well as hypocomplementemia, thymic atrophy, and depletion of paracortical zones of lymph nodes.[54] A major consequence is that cats that have a persistent FeLV viremia develop fatal secondary illnesses, mostly autoimmune or infectious diseases.

Reports have indicated that *in vitro* ultraviolet (UV)-inactivated FeLV also is capable of suppressive activity for feline immunoglobulin (Ig) and IFN synthesis.[55,56] Hebebrand *et al.*[57] reported a decrease in cellular immunity by a 15,000-dalton envelope protein (p15E) from FeLV. FeLV p15E also inhibits neutrophil function *in vitro*[58] and T-cell proliferation in response to IL-2.[55] Although IL-2-dependent cells continued to absorb IL-2, they were not stimulated to grow in the presence of inactivated virus or p15E. The mechanism by which p15E causes these effects is unclear.

Recently, the immunosuppressive properties of FeLV have been attributed to a replication defective mutant that arises during the course of infection.[59] Moreover, a new highly T lymphotropic retrovirus has been isolated from FeLV-seronegative cats and tentatively designed feline T-lymphotropic lentivirus. Experimentally infected kittens develop generalized lymphadenopathy, fever and leukopenia.[60] So far, its effects on immune responses have not been described.

## 2.4. Bovine Retroviruses

Inhibition of primary antibody responses to administered antigens, especially the IgM responses has been reported in cows with lymphocytic leukemia.[61,62] Although no virus was isolated from these animals, it is now established that an exogenous type C retrovirus is the etiologic agent for such tumors. Lymph node cells from the tumor-bearing cow were unable to respond to an immunization with SRBC. However, the cow did produce antibodies to an antigen, E. coli, to which it was previously exposed. Thus, it appears that suppression affected development of immunity to new antigens but did not alter immunologic memory. It is interesting to note that bovine leukemia virus causes a persistent lymphocytosis (PL) in infected cattle.[63] This is composed mainly of B lymphocytes (63–67% of peripheral blood lymphocytes versus 18–28% in normal controls). B lymphocytes are the preferred target for bovine leukemia virus replication. When lymphosarcoma develops in these animals with PL, there is a concomitant loss of B lymphocytes.[63] This may be responsible for the reduced ability to produce antibody seen by Celer et al.[61] Bovine leukemia virus is infectious to rabbits and induces leukemia in such animals.[64] This may be a useful model if the virus can be shown to also induce immunosuppression, since bovine leukemia has a serious economic impact on an important food source.

A recently described virus of cattle that causes persistent lymphocytosis, lymphadenopathy, lesions of the CNS, progressive weakness, and emaciation was characterized as a candidate lentivirus. Since it shares structural, serologic, and genetic features with HIV, the name bovine immunodeficiency virus has been suggested.[65]

## 2.5. Ovine/Caprine Retroviruses

Known sheep- and goat-infecting retroviruses are not associated with oncogenesis but cause slow progressive diseases. These viruses are classified as lentiviruses. They have recently gained popularity as animal models for human immunodeficiency virus (HIV) pathogenesis, since this too is a lentivirus.[66] The mechanisms whereby these viruses circumvent the host's immune system and persist are under close scrutiny. They appear to include restricted replication within immunocompetent cells, low susceptibility to neutralization by antibody, and an unusual frequency of mutation in the env gene. The latter mechanism permits the emergence of antigenic variants in the infected host that are resistant to neutralization by antibody formed against the infecting virus. The new variant is then free to replicate until the immune system can mount a primary response against it (antigenic shift).

The two viruses discussed here are the closely related visna maedi virus of sheep and caprine arthritis encephalitis virus (CAEV). Both viruses result in an insidious onset of a multiorgan disease. The disease progresses slowly affecting the CNS, lungs, and/or joints. Emaciation, respiratory distress, chronic arthritis, or paralysis result, leading to death within a few months.[66] These viruses have been most thoroughly studied immunologically in their interaction with cells of the monocyte/macrophage lineage.[66–74]

Viral replication occurs in mature macrophages. If monocytes are infected in culture and prevented from maturing into macrophages, replication is inhibited.[67,68] *In vitro* studies using sheep peripheral blood monocytes have shown that the entire life cycle of the visna maedi virus, including attachment to the cell, viral gene expression, and virion assembly, is dependent on the level of maturation/differentiation of monocytic cells.[74] CAEV replication restriction in such cells was recently attributed to a CAEV-specific IFN-like substance produced by T lymphocytes stimulated by virus-infected cells.[73] This unique IFN might also be responsible for the enhanced Ia antigen expression in macrophage-like cells seen in inflammatory lesions of infected lambs.[72] No infection in lymphocytes by ruminant lentiviruses has been observed either in inflamed tissues or in virus-inoculated cultures of blood leukocytes.[73] Although CAEV replicates in the mature macrophage, it does not inhibit macrophage functions, such as expression of Fc and complement receptors, phagocytosis of Latex particles, and production of hydrolytic enzymes.[69] Unlike most oncogenic retroviruses and HIV, the ovine/caprine lentiviruses do not appear to cause a great deal of immunosuppression in their natural host. However, in mice infected with visna virus before antigenic challenge, both humoral and cellular immunity were inhibited.[75] This suppressive activity could be abrogated by inactivation of the virus unlike that seen with FeLV, indicating a different mechanism of immunosuppression.

## 2.6. Equine Retroviruses

Equine infectious anemia virus (EIAV) is a widespread insect-transmitted lentivirus infection that is usually asymptomatic. Symptomatic infections are characterized by recurrent episodes of fever, anemia, weight loss, ataxia, and lymphadenopathy. Acute infection is characterized by viremia. In chronic disease, viruses are present only during bouts of illness. Hypergammaglobulinemia, thrombocytopenia, and lymphoid necrosis are important characteristics of the disease. Viral antigens can be detected in virtually all tissues of infected hosts, the monocytic cells being the predominant population of infected cells.[76] Alteration of selected immune responses has been reported in EIAV infected horses, but information on this is still limited.[77–81] Henson and McGuire showed that cell-mediated immunity is involved in disease pathogenesis, since immunosuppressants limited disease symptoms.[82] Recently, a transforming retrovirus was isolated from a cell line established from an equine derived fibrosarcoma.[83] The pathogenic manifestations are yet to be extensively studied, including its interaction with the immune system.

## 2.7. Nonhuman Primate Retroviruses

The retroviruses that infect nonhuman primates fall into the genuses of oncoviruses C, B, and D as well as the Spumavirinae represented by the simian foamy virus. The recognition of the retroviral etiology of acquired immune deficiency syndrome (AIDS) has spurred considerable interest in the immunosuppressive properties of nonhuman primate retroviruses as possible models for studying the pathogenetic mechanisms of the human disease and developing therapeutic and preventive treatments.

That simian retroviruses cause immunosuppression was first recognized while studying the oncogenic properties of the Mason–Pfizer monkey virus (MPMV), a type D retrovirus specifically isolated from a rhesus female with spontaneous breast carcinoma. Most rhesus monkeys inoculated with the virus developed thymic atrophy, lymphoid depletion, and a wasting syndrome with resulting infections, instead of the expected tumors.[84] The significance of this observation was realized much later, however.

More recently, additional immunosuppressive retroviruses were isolated from captive monkeys affected by a recently described disease, sometimes referred to as simian AIDS (SAIDS). The disease strikes rhesus (*Macaca mulata*) and other *Macaca* monkeys and resembles human AIDS not only clinically but also in terms of immunology and histopathology. Clinical features include generalized lymphadenopathy and splenomegaly, skin-test anergy, neutropenia and lymphopenia, weight loss, persistent diarrhea, and opportunistic infections by a wide spectrum of agents similar to those seen in AIDS. Retoperitoneal fibromatosis also occurs in affected monkeys. Consistent immunologic abnormalities include a reduced T-helper/T-suppressor ratio in circulating lymphocytes and a strikingly diminished proliferative response to mitogens, the latter being partially reversible by exogenous IL-2.[9,85]

From macaques with SAIDS, retroviruses have repeatedly been isolated that reproduce the disease upon inoculation into susceptible monkeys. The experimentally induced disease closely resembles natural pathology but is often more pronounced, probably because of the large viral inoculum and of the youth of injected animals. Some of the retroviruses isolated from SAIDS have type D properties. The relationship of different type D isolates to each other and to MPMV is being actively investigated. They appear to be closely related but present differences, especially in the *env* region, and it is not clear whether they should be considered different members or variants of a single virus.[86–88] It is noteworthy that the same problem exists for HIV isolates. It is also uncertain to what extent other retroviruses, such as the simian immune deficiency virus (SIV), may contribute to the SAIDS cases from which type D viruses have been isolated.[9] However, SAIDS has been induced with a molecular clone of a type D retrovirus[89] and prevented by immunization with formalin-killed virus.[90]

From captive monkeys with SAIDS, a simian lentivirus more closely related to HIV than were the viruses discussed above has also been isolated.[91] The virus was initially called simian T-lymphotropic virus (STLV-3) but is now generally referred to as SIV. Several similar isolates were subsequently obtained from diseased macaques and healthy African green monkeys, mangabeys, and other monkeys.[92,93] Serologic evidence indicates that SIV and/or related viruses circulate widely among free-living monkeys in Africa.[94]

Simian immune-deficiency virus is morphologically indistinguishable from HIV and appears to be related to this virus by the antigenicity of its proteins (especially internal ones) and by a number of important biologic properties, including tropism for CD4-bearing lymphocytes.[9] The complete genomes of selected SIV isolates have been cloned and sequenced. The results indicate that SIV is genetically closer to HIV-2, the agent of AIDS in West Africa, than to HIV-1. However, the divergence of SIV and HIV-2 sequences is greater than those among different HIV-1 isolates.[95] A virus even more closely related to SIV has been reported to have been isolated from healthy humans in Africa, HTLV-4. However, sequence analysis has shown that it is almost identical to SIV, calling into question the independent origin of these viruses.[96,97]

In experimentally infected rhesus macaques, SIV readily induces a spectrum of changes that remarkably parallel those induced by HIV in humans, including encephalopathy. By contrast, natural or experimental SIV infection of other monkeys has not yet been associated with disease. Although chimpanzees can be infected with HIV, they do not appear to develop AIDS-like diseases. Thus, SIV infection of macaques is considered the animal model more closely related to AIDS.[9]

Those mentioned above are not the only retroviruses of nonhuman primates that appear to possess immunosuppressive properties. The foamy viruses cause a persistent infection of peripheral blood leukocytes even in the presence of neutralizing antibodies. However, little pathogenesis is associated with infection. The name "foamy" comes from the production of multinucleated vacuolated cell cultures. Modulation of immunity by the primate retrovirus was first reported in infected rabbits.[98] The virus inhibited cellular immune function, as measured by decreased thymidine uptake following phytohemagglutinin (PHA) stimulation and suppression of IFN production. Antibody responses to SRBC were unaffected by the virus. Weislow *et al.*[99] showed that baboon endogenous virus, a group C retrovirus, was capable of depressing PHA-induced blastogenesis in baboons as well as nonprimate (mouse) and human cells. Two fractions from a Sephacryl S200 column purification of the virus exhibited this suppressive activity. One fraction was approximately 45,000 daltons and the other of uncertain size. Neither fraction's activity was altered by trypsin treatment or neutralized by antibody to the virus. Thus, Weislow *et al.*[99] suggest that this substance may not be a protein. In similar studies, Denner *et al.*[100] demonstrated that MPMV also shared this suppressive activity.

# 3. MECHANISMS OF IMMUNOSUPPRESSION

## 3.1. General

The ability of mammalian retroviruses to suppress humoral and cell-mediated immune responses has been extensively investigated and reported. The mechanisms by which this group of viruses suppresses immune functions are varied, including the direct infection of effector cells, suppressor cells, and factors. Although these are clearly described, the subcellular mechanisms that are altered by these various viruses are unclear. The following section attempts to put these mechanisms into perspective and indicates how they may bring about suppression.

## 3.2. Viral Replication within Cells of the Immune System

The range of cell types that support the replication of mammalian retroviruses is extensive and is often unrelated to the type of tumor generated. RadLV seems to multiply only in T lymphocytes,[7,8] while other murine leukemias and feline, bovine, and caprine retroviruses may multiply in B and T lymphocytes, as well as macrophages.[8,53,63,67] The ability to replicate within cells does not necessarily indicate that immunologic function in these cells will be impaired.

Friend leukemia virus can be detected in antibody-producing cells.[20,21] Thus, infection of immunologically competent B cells does not alter antibody secretion. The infection of precursor B cells, however, prevents the development of antibody-producing cells.[101] This was shown convincingly in cell-transfer studies using thymocytes from normal mice and bone marrow from FLV-infected mice, and vice versa. Only when normal bone marrow was used could the cells produce antibody. However, these experiments did not confirm that infection per se was responsible for the suppression. FLV-induced immunosuppression is a complex phenomenon probably involving many immunocompetent cell types, a conclusion consistent with the documented ability of this virus to infect several such cells.[9,16] The possibility that suppressor cells or factors could be involved was not eliminated. Likewise, the ability of murine leukemia viruses to replicate within macrophages may affect their function, but often this appears to be mediated through suppressor cells or factors rather than direct damage to the infected cells.[40,41,102,103] Infection of lymphocytes by FeLV does not appear to be an important event in the immunosuppression associated with this virus.[54,57] This is further substantiated by the observation that suppression can be induced in the absence of live virus. With the lentiviruses causing ovine and caprine infections, persistent infection of cells of the monocyte/macrophage lineage is necessary for progression of virus replication, and these cells must be differentiated into macrophages for replication to occur.[68,71,74] However, the infection by these viruses did not alter macrophage functions such as phagocytosis and receptor expression.

In other systems, however, such as FeLV and the regressing strain of FLV, phagocytosis and other easily tested functions of macrophages were not impaired, yet a close correlation between productive infection of these cells and progression of retrovirus induced leukemia and decreased lymphocyte function has been noted.[42] This might indicate that macrophages represent a key cell for the amplification of the infection or that other as yet undefined macrophage functions important for the host's ability to deal with the infection are inhibited by the internal presence of the virus. Defects in the ability to process or present antigen by accessory cells of retrovirus-infected hosts have been suggested.[42]

Viral replication within T lymphocytes and possibly other immunocompetent cells seems to have a more direct role in immunosuppression caused by AIDS and SAIDS viruses. Much, however, remains to be learned on the basic mechanisms of these diseases. The speculation that the immune defects observed in AIDS are brought about by more complex mechanisms than by direct viral damage to the infected T cells is a recurrent one in the current medical literature. Macrophages have been reported to replicate HIV very effectively.[104]

It seems reasonable to conclude that the relationship between direct infection of lymphoid cells by mammalian retroviruses and their ability to induce immune suppression is not a consistent one. Some viruses infect lymphoid cells with evidence of direct suppression, while others can suppress lymphoid cells without infecting the altered cell. It seems apparent that within this virus family, the mechanisms of induction of suppression are quite diverse.

## 3.3. Virion-Associated Immunosuppressive Products

In a number of cases, in vitro suppression of varying types of immune functions has been observed using inactivated viruses, either by UV irradiation or by other procedures.[35,55,56] Wainberg and Israel[105] also reported that nonviral membrane vesicles that approximate virus in size showed a similar activity. This finding led these investigators to postulate a direct perturbation of lymphocytes or accessory cells by physical interaction with retroviral particles.

It is now generally accepted, however, that retroviruses may possess virion components endowed per se with the ability to modulate immunocompetent cell responsiveness. Several murine retroviruses have been shown to be mitogenic for lymphocytes and, at least in one case the major envelope glycoprotein (gp70) was seen to exert a similar effect.[106] Another virion protein which has been shown to inhibit immune functions is p15E. This envelope protein has been reported to inhibit a number of functions of lymphocytes, monocytes, and macrophages. For example, p15E of FeLV has been noted to inhibit production of $IFN_\gamma$ and IL-2.[56,107] It may be that depression of the production of these mediators results in the defective cell mediated immune functions seen in FeLV infection. This effect does not appear to be related to blocking of IL-2 receptors or to destruction of IL-2. The molecular mechanism that p15E alters resulting in decreased production of these lymphokines remains to be determined. A region of this protein is highly conserved among murine and feline retroviruses. Interestingly, in recent reports, Cianciolo *et al.* demonstrated that a peptide synthesized to correspond to a portion of this region of homology inhibits the alloantigen-stimulated proliferation of lymphocytes and the proliferation of an IL-2-dependent T-cell line and NK cell activity and inactivates IL-1.[108–110] This implies that the immunosuppressive activity of p15E is attributable to its conserved region.

These findings also imply that, at least theoretically, retroviruses do not need to be produced within the immune system to effect significant immunomodulation. Virus or viral products might inhibit immune cell functions wherever they interact in sufficient concentration to generate an effect. Thus, at least some retroviruses could truly be viewed as self-replicating immunosuppressive factors.

### 3.4. Suppressor Cells

The appearance of cells with suppressive activity is well documented in mammals infected by retroviruses. In cocultivation experiments, lymphoid cells of infected animals were found capable of suppressing the normal responsiveness of uninfected lymphoid cells. The responses adversely affected by cocultivation include lectin- and antigen-stimulated lymphocyte blastogenesis, antibody and interleukin production, T-cell cytotoxicity, and NK activity.

The cells responsible for such suppression have most often been described as macrophages. Indeed, activation or recruitment of suppressor macrophages appears to be a frequent occurrence in retrovirus-infected mammals. The list of retroviral infections that have been shown to induce suppressor macrophages includes dimethylbenzanthracene-induced leukemia virus,[111] Moloney leukemia virus,[112] mouse mammary tumor virus (MMTV),[113] and FLV.[41] Most often, such suppressor macrophages have been detected in overtly tumorous mice. It is therefore not clear whether they are activated or recruited directly by the viral infection or by the induced tumors.

While retrovirus-specific suppressor T cells have often been observed in retrovirus-infected mammals[114] and also in uninfected animals,[115] there are no reports of nonspecific T-suppressor cells that might contribute to the generalized immunosuppressive states induced by these viruses. Recently, however, the immunosuppressive effects of a RadLV-derived murine retroviral complex have been shown to depend on the availability of functional T cells.[116] There are also reports that mammalian retroviruses, such as FLV and FeLV, might damage T-suppressor cells or their precursors.[117,118]

In mice infected with MMTV[119] and FLV,[120] B lymphocytes were found that showed suppressive activity. In the latter study, the cells that inhibited antibody responsiveness of lymphoid cell cultures *in vitro* were characterized as B lymphocytes and, at later stages of infection, as neoplastic cells. The suppressing effect of such cells was blocked by FLV-specific antiviral antibody, suggesting that it might simply be due to the release of suppressive virus or viral products. The possibility that B lymphocytes may function as suppressor cells is increasingly being recognized, however.

### 3.5. Role of Nonviral Soluble Factors

Soluble mediators are now known to be implicated in the correct functioning and regulation of the immune system. It seems likely that imbalances in the physiology of such factors are involved in generating immunosuppression by mammalian retroviruses. Several defects in the production of or responsiveness to IL-2 or $IFN_\gamma$ have been reported in AIDS patients.[121] Information on this matter in nonhuman mammalian retroviruses is still scanty.

One distinct possibility is that mediators that exert a helper or amplifying effect in immune responses are produced in lower amounts or are consumed at a higher rate in retrovirus-infected as compared with normal animals. Mice infected with FLV were found to have reduced levels of thymic factors in serum,[122] although the significance of this is uncertain. IL-1 production or activity did not appear to be altered during FLV infection.[49,50] By contrast, IL-2 secretion has been shown to be inhibited by FLV infection *in vivo*[50] and by UV-inactivated FeLV *in vitro*.[107] The latter virus was also shown to inhibit IFN production *in vitro*.[56] Lymphocyte sensitivity to IL-2 was also shown to be reduced by FeLV.[55] Although limited, this evidence indicates that disturbances in the production of IL-2 and other lymphokines could play a central role in retrovirus-induced immunosuppression.

Production of suppressor factors or factors that interfere with the activity of mediators that enhance immune responsiveness has been reported in only a few circumstances. In studies by Moody *et al.*,[41] suppression of NK function by FLV-infected cells *in vitro* could be reversed by indomethacin, suggesting that suppression might be due to prostaglandins. In addition, Ceglowski *et al.*[36] reported a soluble factor from FLV-infected mice that suppresses cellular immune function but not humoral immunity. The factor was heat stable (56° for 30 min), resistant to nucleases and neuraminidase, but sensitive to trypsin digestion. Antibody that neutralizes FLV infectivity was ineffective against this suppressive factor, suggesting that it might be host generated. Ceglowski and colleagues did not identify the cellular source of the factor.

Retroviruses have been shown to induce IFN in infected hosts.[123] Different IFNs have different effects on immune functions, depending on timing, dosage, and chemical composition. The consequences to immune responsiveness of the mixtures of IFNs produced in response to retrovirus infection are still impossible to predict. The demonstration of abnormal IFNs during some retrovirus infections adds to this complexity.[72,121] The production of IL-1 inhibitors by virus-infected cells has been reported.[124] It is not known whether such factors are produced in retroviral infections, although a synthetic peptide homologous to the envelope proteins of retroviruses has been shown to inactivate IL-1.[110]

Thus, taken together, the above information indicates that nonhuman mammalian retrovirus-induced suppression may be the result of (1) direct infection of lymphoid cells, (2) modulation of immunocompetent cells by viral components, (3) induction of suppressor cells or factors, or (4) a combination of these factors.

## 4. SIGNIFICANCE OF VIRUS-INDUCED IMMUNOSUPPRESSION

The ability of retroviruses to suppress immunity would seem a highly advantageous evolutionary development with regard to the nature of infections associated with this virus family. These infections lead to viral persistence, which may be associated with the transformation of cells (oncoviruses) or not (foamy viruses, lentiviruses, and some oncoviruses). Because of the chronic nature of these infections, the capacity to suppress immunity might be critical to survival. A major complication of the host–virus interaction is that the generalized immunosuppression that often results frequently leads to opportunistic malignancies and/or infections that are fatal. Death is not a direct retrovirus-induced phenomenon, suggesting that immunosuppression is a successful adaption that permits long-term viral replication.

It is still unclear what effect virus-induced immunosuppression has on establishing the primary infection. This is generally seen as detrimental to the host. However, it is well known that immunopathologic processes significantly contribute to damage to host tissues. Thus, host-mediated immune suppression may be an adaptation by the host to reduce immunopathologic lesions.

Nevertheless, it is now generally appreciated that retrovirus-induced immunodeficiencies represent a continuous threat to life because they can pave the way to serious secondary illness of fungal, protozoan, bacterial, and viral origin as well as to neoplasia. Devastating opportunistic infections and tumors, similar to those of human AIDS, are observed in retrovirus-infected monkeys and cats. Indeed, it has been stated that in free-living cats, FeLV causes a higher lethality by functioning as an immunosuppressant than by acting as an oncogenic agent.[56] In other mammals, there are no data dealing with naturally occurring superinfections, but an increased susceptibility to challenge with various infective agents and experimental tumors has been documented in mice infected with a number of retroviruses.[33–35] Thus, it is generally believed that treatments capable of preventing or alleviating the immunosuppressive manifestations associated with retroviral infections would be of significant benefit to retrovirus-infected hosts.

### Prevention and Treatment

Prevention or treatment for retroviral disease of nonhuman mammals is a relatively undeveloped area. There are no active or passive preventatives, i.e., no vaccines or immunoglobulins. Vaccine development, especially for feline leukemia, is being investigated but still with only limited success.[125] Development of antiviral drugs is also being pursued but again with limited success. Drugs that inhibit retroviral replication *in vitro* have been identified. More recently, selected drugs have been used *in vivo* with encouraging results.[126–128] Despite the astonishingly rapid progress in our understanding of the molecular biology of retroviruses, however, hopes to control retroviral infections with effective antiviral treatments in the near future are meager.

The implications of retrovirus-induced immunosuppression in this context are not entirely clear. Genovesi *et al.*[129] attributed much of the therapeutic effects exerted by passive immunotherapy with xenogeneic antisera of FLV-infected mice to a reduction of the immunosuppressive burden constituted by the infecting virus. Moreover, leukemogenic and immunosuppressive properties were closely associated in clones of the murine RadLV.[130] However, the impact of retrovirus-induced immunodepression on the other pathology directly caused by the virus (i.e., tumors) is essentially unknown.

Initial attempts to treat retroviral infections were done with nonspecific immunostimulants dating back to the early 1970s, when there were reported improvements in retrovirus-induced leukemias and mammary tumors of mice after treatment with bacillus Calmette–Guérin (BCG).[131,132] These studies were followed by reports of partial or total success with *Corynebacterium parvum* and *granulosum* preparations and a host of other nonspecific immunostimulants (as reviewed by Bendinelli.)[42] In certain instances, however, the same substance ameliorated or aggravated the course of disease, depending on such factors as dose, route of inoculation, timing, and genetics of the host. Likewise, a recent study suggesting that tuftsin can suppress Friend virus-induced leukemia was highly time and dose dependent.[133]

The results of the above studies can best be summarized by stating that aspecific immunostimulants can modulate the outcome of retroviral infections but that the direction of modulation depends on too many variables to be consistently predictable. In any case, reduced incidence and severity of retrovirus-induced diseases and increased host survival were observed only when the immunostimulants were given prior to infection. No immunopotentiating treatment has thus been substantiated as an effective therapeutic for practical use in retrovirus induced disease.

Viewed from the present context, a major limitation of these studies is that it is not known whether the beneficial effects of nonspecific stimulants were due to reversal of the immunosuppressive action of the infecting virus or to other mechanisms. Most of these studies have looked only at the development and progression of the virus-induced neoplasia as a parameter of treatment efficacy, which can hardly be considered a direct indicator of the host's immune responsiveness. In fact, most immunostimulants used in such experiments were complex biologic response modifiers endowed with so many activities that it is not possible to be certain that the observed effects were due to their immunostimulatory activity.

More recently, Day and co-workers[134] reported effective treatment of

FeLV leukemia by *in vivo* administration of staphylococcal protein A (SPA). More than one half the cats treated had a positive response, and many were in complete remission after several months. These animals showed no evidence of leukemia upon histologic examination.[134] The mechanisms by which SPA reverses disease and the accompanying immunosuppression is not yet determined. Earlier studies using SPA *in vitro* or for extracorporeal therapy were based on the removal of immune complexes.[135] However, this therapy did not result in the same beneficial results as *in vivo* therapy. Thus, it has been hypothesized that *in vivo*, besides removing immune complexes, the SPA therapy is effective due to its immunostimulatory action.[134]

In addition to the use of nonspecific immunostimulants, others have been investigating the use of lymphokines. Supplementation with exogenous IFN or interleukins has been shown *in vitro* to reverse some inhibited immune responses, but results are conflicting, so that no conclusions can be drawn.

## 5. CONCLUSIONS

Considerable information is presently available on the cellular and subcellular events responsible for retrovirus-induced immune deficiencies in selected nonhuman mammalian systems. One of the lessons learned is that the pathways leading to immune deficiency are multiple and that their relative importance may vary, depending not only on the virus–host system being considered, but also on variables such as the duration of infection and the immunologic situation of the host at the time of infection. Although there are no studies dealing specifically with this aspect, it seems probable that the secondary infections that afflict retrovirus-infected hosts contribute significantly to overall immunosuppression, representing another source of variation.

Nevertheless, certain common features are apparent in most systems that have been attentively investigated. Among these are findings that show impairment of macrophage functions to be a frequent occurrence and probably an important mechanism in the genesis of immune deficiency. Another unifying concept is that many immunosuppressive retroviruses possess structural components endowed with a striking ability to modulate selected immune functions. Thus, it seems likely that continuing efforts to unravel the mechanisms of suppression by retroviruses of lower mammals and nonhuman primates will lead also to a better understanding of AIDS and related diseases.

Although fraught with many difficulties, attempts to reverse retrovirus-induced immune deficiencies in nonhuman primate and lower animals warrant pursuing because they might help generate the basic knowledge needed to develop effective immune reconstruction treatments to be used in clinical situations. Several clinical trials of selected immune response modifiers have attempted to reconstitute the immune functions of AIDS patients and to prevent progression in HIV-infected persons. Other studies are currently under way in the United States and elsewhere. Unfortunately, this empirical approach— justified as it may have been, given the severity of the disease—has met with limited success, if any. It seems worthwhile to go back to animal models and to undergo a systematic rigidly controlled search for treatments capable of counteracting the immunosuppressive action of retroviruses or at least of leading to an effective and balanced potentiation of the host's antimicrobial resistance mechanisms, which might reduce the incidence and severity of opportunistic infections.

ACKNOWLEDGMENTS.    During the preparation of this work, M. B.'s laboratory was supported by grants from the Italian Research Council, Special Project Oncology, and from the Italian Ministry of Public Education.

# REFERENCES

1. Weiss, R., N. Teich, N. Varmus, and J. Coffin (eds.), *RNA Tumor Viruses: Molecular Biology of Tumor Viruses*, 2nd ed., Cold Spring Harbor Laboratory, Cold Spring Harbor, New York (1985).
2. Lowy, D. R., Transformation and oncogenesis: Retroviruses, in: *Virology* (B. N. Fields, D. M. Knipe, R. M. Chanock, J. L. Melnick, B. Roizman, and R. E. Shope, eds.), pp. 235–263, Raven, New York (1985).
3. Mosier, D. E., R. A. Yetter, and H. C. Morse III, Retroviral induction of acute lymphoproliferative disease and profound immunosuppression in adult C57BL/6 mice, *J. Exp. Med.* **161:**766–784 (1985).
4. Notkins, A. L., S. E. Mergenhagen, and R. J. Howard, Effect of virus infections on the function of the immune system, *Annu. Rev. Microbiol.* **24:**525–538 (1970).
5. Dent, P. B., Immunodepression by oncogenic viruses, *Prog. Med. Virol.* **14:**1–35 (1972).
6. Dent, P. B., Immunodepression by oncogenic viruses, in: *The Immune System and Infectious Diseases, Fourth International Convocation of Immunology, Buffalo, N.Y.* pp. 95–107, Karger, Basel (1975).
7. Specter, S., and H. Friedman, Viruses and the immune response, *Pharmacol. Ther. A* **2:**595–622 (1978).
8. Bendinelli, M., D. Matteucci, and H. Friedman, Retrovirus-induced acquired immunodeficiencies, *Adv. Cancer Res.* **45:**125–181 (1985).
9. Desrosiers, R. C., and N. L. Letvin, Animal models for acquired immunodeficiency syndrome, *Rev. Infect. Dis.* **9:**438–446 (1987).
10. Peterson, R. D. A., R. Hendrickson, and R. A. Good, Reduced antibody forming capacity during the incubation period of passage A leukemia in C3H mice, *Proc. Soc. Exp. Biol. Med.* **114:**517–520 (1966).
11. Odaka, T., H. Ishii, K. Yamaura, and T. Yamamoto, Inhibitory effect of Friend leukemia virus infection on the antibody formation to sheep erythrocytes in mice, *Jpn. J. Exp. Med.* **36:**277–290 (1967).
12. Salaman, M. H., and N. Wedderburn, The immunodepressive effects of Friend virus, *Immunology* **10:**445–448 (1966).
13. Ceglowski, W. S., and H. Friedman, Suppression of the primary antibody plaque response of mice following infection with Friend disease virus, *Proc. Soc. Exp. Biol. Med.* **126:**662–666 (1967).
14. Ceglowski, W. S., and H. Friedman, Immunosuppression by leukemia viruses. I. Effect of Friend disease virus on cellular and humoral hemolysin response of mice to a primary immunization with sheep erythrocytes, *J. Immunol.* **101:**594–604 (1968).
15. Hirano, S., W. S. Ceglowski, J. L. Allen, and H. Friedman, Effect of Friend leukemia

virus on antibody forming cells to a bacterial antigen, *J. Natl. Cancer Inst.* **43**:1337–1345 (1969).

16. Morrison, R. D., J. Nishio, and B. Chesebro, Influence of the murine MHC (H-2) on Friend leukemia virus-induced immunosuppression, *J. Exp. Med.* **163**:301–314 (1986).

17. Bainbridge, D. R., and M. Bendinelli, Circulation of lymphoid cells in mice infected with Friend leukemia virus, *J. Natl. Cancer Inst.* **49**:773–781 (1972).

18. Friedman, H., and W. S. Ceglowski, Virus tumorigenesis and immunity: Influence of immunostimulation and immunodepression, in: *The Role of Immunologic Factors in Viral and Oncogenic Processes* (R. F. Beers, R. C. Tilghman, and E. G. Bassett, eds.), pp. 187–210, Johns Hopkins University Press, Baltimore (1974).

19. Zatz, M. M., A. White, and A. L. Goldstein, Lymphocyte populations of AKR/J mice. II. Effect of leukemogenesis on migration patterns, response to PHA and expression of theta antigen, *J. Immunol.* **111**:1519–1525 (1973).

20. Koo, G. C., W. S. Ceglowski, and H. Friedman, Immunosuppression by leukemia viruses. V. Ultrastructural studies of antibody forming spleens of mice infected with Friend leukemia virus, *J. Immunol.* **106**:799–814 (1971).

21. Koo, G. C., W. S. Ceglowski, and H. Friedman, Immunosuppression by leukemia viruses. VI. Ultrastructure of individual antibody-forming cells in the spleens of Friend leukemia virus-infected mice, *J. Immunol.* **106**:815–830 (1971).

22. Bendinelli, M., and L. Nardini, Immunodepression by Rowson-Parr virus in mice. II. Effect of Rowson-Parr virus infection on the antibody response to sheep red blood cells *in vivo* and *in vitro*, *Infect. Immun.* **7**:160–166 (1978).

23. Kateley, J. R., I. Kamo, G. Kaplan, and H. Friedman, Suppressive effect of leukemia virus-infected lymphoid cells on *in vitro* immunization of normal splenocytes, *J. Natl. Cancer Inst.* **53**:1371–1378 (1974).

24. Weislow, O. S., and E. F. Wheelock, Depression of humoral immunity to sheep erythrocytes *in vitro* by Friend virus leukemic spleen cells: Induction of resistance by statolon, *J. Immunol.* **114**:211–215 (1975).

25. Specter, S., N. Patel, and H. Friedman, Immunosuppression induced *in vitro* by cell-free extracts of Friend leukemia virus infected splenocytes, *J. Natl. Cancer Inst.* **56**:143–147 (1976).

26. Friedman, H., H. Melnick, L. Mills, and W. S. Ceglowski, Depressed allograft immunity in leukemia virus infected mice, *Transplant. Proc.* **5**:981–986 (1973).

27. Mortensen, R. F., W. S. Ceglowski, and H. Friedman, Leukemia virus induced immunosuppression. IX. Depression of delayed hypersensitivity and MIF production after infection of mice with Friend leukemia virus, *J. Immunol.* **111**:1810–1819 (1973).

28. Deodhar, S. D., and T. Chiang, Immunosuppression with Friend virus in an allogeneic murine tumor system, *Fed. Proc.* **29**:560 (1970).

29. Mortensen, R. F., W. S. Ceglowski, and H. Friedman, Leukemia virus induced immunosuppression. X. Depression of T cell-mediated cytotoxicity after infection with Friend leukemia virus, *J. Immunol.* **112**:2077–2086 (1973).

30. Kateley, J. R., J. Holderbach, and H. Friedman, Leukemia virus induced alteration of lymphocyte Ig surface receptors and 'capping' response of mouse spleen and lymph node cells, *J. Natl. Cancer Inst.* **53**:1135–1141 (1974).

31. Farber, P., S. Specter, and H. Friedman, Leukemia virus induced immune suppression. Scanning electron microscopy of infected spleen cells, *Science* **160**:467–471 (1975).

32. Kitagawa, M., O. Matsubara, and T. Kasuga, Dynamics of lymphocytic subpopulations in Friend leukemia virus-induced leukemia, *Cancer Res.* **46**:3034–339 (1986).

33. Glendhill, A. W., Enhancement of the pathogenicity of mouse hepatitis virus (MHV) by prior infection of mice with certain leukemia agents, *Br. J. Cancer* **15**:531–538 (1961).

34. Chirigos, M. A., K. Perk, W. Turner, B. Burka, and M. Gomez, Increased oncogenicity of the murine sarcoma virus (Moloney) by co-infection with murine leukemia virus, *Cancer Res.* **28**:1055–1063 (1968).

35. Specter, S., F. Basolo, and M. Bendinelli, Retroviruses as immunosuppressive agents, in: *The Nature, Cellular, and Biochemical Basis and Management of Immunodeficiencies* (R. A. Good and E. Lindenlaub, eds.), pp. 128–145, Schattauer Verlag, Stuttgart, (1987).

36. Ceglowski, W. S., G. U. Labadie, and A. A. Mascio, Effects of leukemia viruses on cellular immune responses, in: *Tumor Virus Infections and Immunity* (R. L. Crowell, H. Friedman, and J. Prier, eds.), pp. 165–173, University Park Press, Baltimore (1976).

37. Old, L., B. Benacerraf, D. A. Clark, and M. Goldsmith, The reticuloendothelial system and the neoplastic process, *Ann. NY Acad. Sci.* **88:**264–280 (1960).

38. Wheelock, E. F., S. T. Toy, O. S. Weislow, and M. H. Levy, Restored immune and non-immune functions in Friend virus leukemic mice treated with statolon, *Prog. Exp. Tumor Res.* **19:**369–390 (1974).

39. Seidel, H. J., and W. Nothdurft, The phagocytic activity of the reticuloendothelial system of mice infected with Rauscher leukemia virus, *J. Reticuloendothel. Soc.* **19:**173–181 (1976).

40. Kirchner, H., T. M. Chused, R. B. Herberman, H. T. Holden, and H. Lavrin, Evidence of suppressor cell activity in spleens of mice bearing primary tumors induced by Moloney sarcoma virus, *J. Exp. Med.* **139:**1473–1478 (1974).

41. Moody, D. J., S. Specter, M. Bendinelli, and H. Friedman, Suppression of natural killer cell activity by Friend murine leukemia virus, *J. Natl. Cancer Inst.* **72:**1349–1356 (1984).

42. Bendinelli, M., The reticuloendothelial system in infection with RNA tumor viruses, in: *The Reticuloendothelial System: A Comprehensive Treatise,* Vol. 10: *Infection* (M. Escobar and J. P. Utz, eds.), pp. 297–347, Plenum, New York (1988).

43. Bendinelli, M., G. S. Kaplan, and H. Friedman, Reversal of leukemia virus induced immunosuppression *in vitro:* Role of peritoneal exudate macrophages, *J. Natl. Cancer Inst.* **55:**1425–1432 (1975).

44. Specter, S., N. Patel, and H. Friedman, Peritoneal exudate cell induced restoration of antibody formation by leukemia virus suppressed spleen cell cultures, *Proc. Soc. Exp. Biol. Med.* **151:**163–167 (1976).

45. Ceglowski, W. S., and H. Friedman, Failure of peritoneal exudate macrophages to reverse immunologic impairment induced by Friend leukemia virus, *Proc. Soc. Exp. Biol. Med.* **148:**808–811 (1975).

46. Genovesi, E. V., D. Livnat, and J. J. Collins, Immunotherapy of murine leukemia. VII. Prevention of Friend leukemia virus induced immunosuppression by passive serum therapy, *Int. J. Cancer* **30:**609–624 (1982).

47. Demaeyer-Guignard, J., Mouse leukemia: Depression of serum interferon production, *Science* **177:**797–799 (1972).

48. Butler, R. C., J. M. Frier, M. S. Chapekar, M. O. Graham, and H. Friedman, Role of antibody response helper factors in immunosuppressive effects of Friend leukemia virus, *Infect. Immun.* **39:**1260–1264 (1983).

49. Bendinelli, M., D. Matteucci, A. M. Giangregorio, and P. G. Conaldi, Restoration of antibody responsiveness by endotoxin in retrovirus-immunosuppressed mice: Role of macrophages, in: *Immunobiology and Immunopharmacology of Bacterial Endotoxins* (A. Szentivanyi and H. Friedman, eds.), pp. 465–478, Plenum, New York (1986).

50. Lopez-Cepero, M., S. Specter, D. Matteucci, H. Friedman, and M. Bendinelli, Altered interleukin production during Friend leukemia virus infection. *Proc. Soc. Exp. Biol. Med.* **188:**353–363 (1988).

51. Perryman, L. E., E. A. Hoover, and D. S. Yohr, Immunologic reactivity of the cat: Immunosuppression in experimental feline leukemia, *J. Natl. Cancer Inst.* **49:**1357–1365 (1972).

52. Wernicke, D., Z. Trainin, H. Ungar-Waron, and M. Essex, Humoral immune response of asymptomatic cats naturally infected with feline leukemia virus, *J. Virol.* **60:**669–673 (1986).

53. Hardy, W. D., and M. Essex, FeLV-induced feline acquired immune deficiency syndrome. A model for human AIDS, *Prog. Allergy* **37**:353–376 (1986).
54. Pack, F. D., and W. L. Chapman, Light and electron microscopic evaluation of thymuses from feline leukemia virus-infected kittens, *Exp. Pathol.* **18**:96–110 (1980).
55. Orosz, C. G., N. E. Zinn, R. G. Olsen, and L. E. Mathes, Retrovirus mediated immunosuppression. I. FeLV-UV and specific FeLV proteins alter T lymphocyte behavior by inducing hyporesponsiveness to lymphokines, *J. Immunol.* **134**:3396–3403 (1985).
56. Yasuda, M., R. A. Good, and N. K. Day, Influence of inactivated feline retrovirus on feline alpha interferon and immunoglobulin production, *Clin. Exp. Immunol.* **69**:240–245 (1987).
57. Hebebrand, L. C., R. G. Olsen, L. E. Mathes, and W. S. Nichols, Inhibition of human lymphocyte mitogen and antigen response by a 15,000 dalton protein from feline leukemia virus, *Cancer Res.* **39**:443–447 (1979).
58. Lafrado, L. J., M. G. Lewis, L. E. Mathes, and R. G. Olsen, Suppression of *in vitro* neutrophil function by feline leukemia virus (FeLV) and purified FeLV-p15, *J. Gen. Virol.* **68**:507–513 (1987).
59. Mullins, J. I., C. S. Chen, and E. A. Hoover, Disease-specific and tissue-specific production of unintegrated feline leukemia virus variant DNA in feline AIDS, *Nature (Lon.)* **319**:333–336 (1986).
60. Pedersen, N. C., E. W. Ho, M. L. Brown, and J. K. Yamamoto, Isolation of a T lymphotropic virus from domestic cats with an immunodeficiency-like syndrome, *Science* **235**:790–793 (1987).
61. Celer, V., L. Cerny, E. Jelinkova, and P. Nedbal, Absence of primary immune response in acute bovine lymphatic leukemia. I. Occurrence of natural antibodies against *E. coli*, *Neoplasma* **18**:523–528 (1971).
62. Trainin, Z., H. Ungar-Waron, R. Mairon. A. Barma, and M. Sela, IgG and IgM antibodies in normal and leukaemic cattle, *J. Comp. Pathol.* **86**:571–580 (1976).
63. Thorne, R. M., P. Gupta, S. J. Kenyon, and J. F. Ferrer, Evidence that the spontaneous blastogenesis of lymphocytes from bovine leukemia virus-infected cattle is viral antigen specific, *Infect. Immun.* **34**:84–89 (1981).
64. Burny, A., C. Bruck, Y. Cleuter, D. Couez, J. Deschamps, J. Chysdael, D. Gregoire, R. Kettmann, M. Mammerickx, G. Marbaix, and D. Portetelle, Bovine leukemia virus: A new model of leukemogenesis, in: *Advances in Viral Oncology*, Vol. 5: *Viruses as the Causative Agents of Naturally Occurring Tumors* (G. Klein, ed.), pp. 35–56, Raven, New York (1985).
65. Gonda, M. A., M. J. Braun, S. G. Carter, T. A. Kost, J. W., Jr. Bess, L. O. Arthur, and M. J. Van Der Marten, Characterization and molecular cloning of a bovine lentivirus related to human immunodeficiency virus, *Nature (Lond.)* **330**:388–391 (1987).
66. Haase, A. T., Pathogenesis of lentivirus infections, *Nature (Lond.)* **322**:130–136 (1986).
67. Narayan, O., J. S. Wolinsky, J. E. Clements, J. D. Strandberg, D. E. Griffin, and L. C. Cork, Slow virus replication: The role of macrophages in the persistence and expression of visna viruses of sheep and goats, *J. Gen. Virol.* **53**:345–356 (1982).
68. Anderson, L. W., P. Klevjer-Anderson, and H. D. Liggitt, Susceptibility of blood-derived monocytes and macrophages to caprine arthritis encephalitis virus, *Infect. Immun.* **41**:837–840 (1983).
69. Narayan, O., S. Kennedy-Stoskopf, D. Sheffer, D. E. Griffin, and J. E. Clements, Activation of caprine arthritis–encephalitis virus expression during maturation of monocytes to macrophages, *Infect. Immun.* **41**:67–73 (1983).
70. Gendelman, H. E., O. Narayan, S. Kennedy-Stoskopf, J. E. Clements, and G. H. Pezeshkour, Slow virus–macrophage interactions: Characterization of a transformed cell line of sheep alveolar macrophages that express a marker for susceptibility to ovine–caprine lentivirus infections, *Lab. Invest.* **51**:547–555 (1984).

71. Gendelman, H. E., O. Narayan, S. Molineaux, J. E. Clements, and Z. Ghotbi, Slow, persistent replication of lentiviruses: Role of tissue macrophages and macrophage precursors in bone marrow, *Proc. Natl. Acad. Sci. USA* **82**:7086–7090 (1985).
72. Kennedy, P. G. E., O. Narayan, Z. Ghotbi, J. Hopkins, H. E. Gendelman, and J. E. Clements, Persistent expression of Ia antigen and viral genome in visna-maedi virus induced inflammatory cells: Possible role of lentivirus-induced interferon, *J. Exp. Med.* **162**:1970–1982 (1985).
73. Narayan, O., D. Sheffer, J. E. Clements, and G. Tennekoon, Restricted replication of lentiviruses: Visna viruses induce a unique interferon during interaction between lymphocytes and infected macrophages, *J. Exp. Med.* **162**:1954–1969 (1985).
74. Gendelman, H. E., O. Narayan, S. Kennedy-Stoskopf, P. G. E. Kennedy, Z. Ghotbi, J. E. Clements, J. Stanley, and G. Pezeshkpour, Tropism of sheep lentiviruses for monocytes: Susceptibility to infection and virus gene expression increase during maturation of monocytes to macrophages, *J. Virol.* **58**:67–74 (1986).
75. Svennerholm, B., O. Strannegard, and E. Lycke, Immune reactivity of visna virus-inoculated mice, *Infect. Immun.* **20**:412–417 (1978).
76. Salinovich, O., S. L. Payne, R. C. Montelaro, K. A. Hussain, C. J. Issel, and K. L. Schnorr, Rapid emergence of novel antigenic and genetic variants of equine infectious anemia virus during persistent infection, *J. Virol.* **57**:71–80 (1986).
77. Banks, K. L., and J. B. Henson, Quantitation of immunoglobulin-bearing lymphocytes and lymphocyte response to mitogens in horses persistently infected by equine infectious anemia virus, *Infect. Immun.* **8**:679–682 (1973).
78. Henson, J. B., and T. C. McGuire, Equine infectious anemia, *Prog. Med. Virol.* **18**:143–159 (1974).
79. McGuire, T. C., Suppression of synthesis of an IgG subclass in a persistent viral infection, *Immunology* **30**:17–24 (1976).
80. Fujimiya, Y., L. E. Perryman, and T. B. Crawford, Leukocyte cytotoxicity in a persistent virus infection: Presence of direct cytotoxicity but absence of antibody-dependent cellular cytotoxicity in horses infected with equine infectious anemia virus, *Infect. Immun.* **24**:628–636 (1979).
81. Cheevers, W. R., and T. C. McGuire, Equine infectious anemia virus: Immunopathogenesis and persistence, *Rev. Infect. Dis.* **7**:84–150 (1985).
82. Henson, J. B., and T. C. McGuire, Immunopathology of equine infectious anemia, *Am. J. Clin. Pathol.* **56**:306–314 (1971).
83. Teich, N., Taxonomy of retroviruses, in: *RNA Tumor Viruses: Molecular Biology of Tumor Viruses,* 2nd ed. (R. Weiss, N. Teich, H. Varmus, and J. Coffin, eds.), pp. 1–16, Cold Spring Harbor Laboratory, Cold Spring Harbor, New York (1985).
84. Fine, D. L., J. C. Landon, R. J. Pienta, M. T. Kubicek, M. J. Valerio, W. G. Loeb, and H. C. Chopra, Responses of infant rhesus monkeys to inoculation with Mason–Pfizer monkey virus material, *J. Natl. Cancer Inst.* **54**:651–658 (1975).
85. Gardner, M. B., and P. A. Marx, Simian acquired immunodeficiency syndrome, in: *Advances in Viral Oncology,* Vol. 5: *Viruses as the Causative Agents of Naturally Occurring Tumors* (G. Klein, ed.), pp. 57–82, Raven, New York (1985).
86. Power, M. D., P. A. Marx, M. L. Bryant, M. B. Gardner, P. J. Barr, and P. A. Luciw, Nucleotide sequence of RSV-1, a type D simian acquired immune deficiency retrovirus, *Science* **231**:1567–1572 (1986).
87. Sonigo, P., C. Barker, E. Hunter, and S. Wain-Hobson, Nucleotide sequence of Mason–Pfizer monkey virus: An immunosuppressive type D retrovirus, *Cell* **45**:375–385 (1986).
88. Thayer, R. M., M. D. Power, M. L. Bryant, M. B. Gardner, P. J. Barr, and P. A. Luciw, Sequence relationship of type D retroviruses which cause simian acquired immunodeficiency syndrome, *Virology* **157**:317–330 (1987).
89. Heidecker, G., N. W. Lerche, L. J. Lowenstine, A. A. Lackner, K. G. Osborn, M. G.

Gardner, and P. A. Marx, Induction of simian acquired immune deficiency syndrome (SAIDS) with a molecular clone of a type D SAIDS virus, *J. Virol.* **61**:3066–3071 (1987).

90. Marx, P. A., N. C. Pedersen, N. W. Lerche, K. G. Osborn, L. J. Lowenstine, A. A. Lackner, D. H. Maul, H.-S. Kwang, J. D. Kluge, C. P. Zaiss, V. Sharpe, A. P. Spinner, A. C. Allison, and M. B. Gardner, Prevention of simian acquired immune deficiency syndrome with a formalin-inactivated type D retrovirus vaccine, *J. Virol.* **60**:431–435 (1986).

91. Daniel, M. D., N. L. Letvin, N. W. King, M. Kannagi, P. K. Schgal, R. D. Hunt, P. J. Kanki, M. Essex, and R. C. Desrosiers, Isolation of T-lymphotropic retrovirus related to HTLVIII/LAV from wild-caught African green monkeys, *Science* **230**:951–954 (1985).

92. Murphey-Corb, M., L. N. Martin, S. R. S. Rangan, G. B. Baskin, B. J. Gormus, R. H. Wolf, W. A. Andes, M. West, and R. C. Montelaro, Isolation of an HTLV-III-related retrovirus from macaques with simian AIDS and its possible origin in asymptomatic mangabeys, *Nature (Lond.)* **321**:435–437 (1986).

93. Fultz, P. N., H. M. McClure, D. C. Anderson, R. B. Swenson, R. Anand, and A. Srinivasan, Isolation of a T-lymphotropic retrovirus from naturally infected mangabeys monkeys (*Cercocebus atys*), *Proc. Soc. Acad. Sci. USA* **83**:5286–5290 (1986).

94. Benveniste, R. E., L. O. Arthur, C.-C. Tsai, R. Sowder, T. D. Copeland, L. E. Henderson, and S. Oroszlans, Isolation of a lentivirus from a macaque with lymphoma: Comparison with HTLV-III/LAV and lentiviruses, *J. Virol.* **60**:483–490 (1986).

95. Chakrabarti, L., M. Guyader, M. Alizon, M. D. Daniel, R. C. Desrosiers, P. Tiollais, and P. Sonigo, Sequence of simian immunodeficiency virus from macaque and its relationship to other human and simian retroviruses, *Nature (Lond.)* **328**:543–546 (1987).

96. Hahn, B. H., L. I. Kong, S.-W. Lee, P. Kumar, M. E. Taylor, S. K. Arya, and G. M. Shaw, Relation of HTLV-4 to simian and human immunodeficiency-associated viruses, *Nature (Lond.)* **330**:184–187 (1987).

97. Kornfeld, H., N. Riedel, G. A. Viglianti, V. Hirsh, and J. I. Mullins, Cloning of HTLV-4 and its relation to simian and human immunodeficiency viruses, *Nature (Lond.)* **326**:610–613 (1987).

98. Hooks, J. J., and B. Detrick-Hooks, Simian foamy virus-induced immunosuppression in rabbits, *J. Gen. Virol.* **44**:383–390 (1979).

99. Weislow, O. S., O. U. Fisher, Jr., D. R. Twardzik, A. Hellman, and A. K. Fowler, Depression of mitogen-induced lymphocyte blastogenesis by baboon endogenous retrovirus-associated components, *Proc. Soc. Exp. Biol. Med.* **166**:522–527 (1981).

100. Denner, J., V. Wunderlich, and D. Bierwolf, Suppression of human lymphocyte mitogen response by proteins of the type D retrovirus PMFV. *Int. J. Cancer* **37**:311–316 (1986).

101. Ceglowski, W. S., and H. Friedman, Immunosuppression by leukemia viruses. IV. Effect of Friend leukemia virus on antibody precursors as assessed by cell transfer studies, *J. Immunol.* **105**:1406–1415 (1970).

102. Israel, E., B. Beiss, and M. A. Wainberg, Viral abrogation of lymphocyte mitogenesis: Induction of a soluble factor inhibitory to cellular proliferation, *Immunology* **40**:77–85 (1980).

103. Israel, E., and M. A. Wainberg, Viral inhibition of lymphocyte mitogenesis: The role of macrophages as primary targets of virus–cell interaction, *J. Reticuloendothel. Soc.* **29**:105–116 (1981).

104. Nicholson, J. K. A., G. D. Cross, C. S. Callaway, and J. S. McDougal. In vitro infection of human monocytes with human T lymphotropic virus type III/Lymphadenopathy-associated virus, *J. Immunol.* **137**:323–329 (1986).

105. Wainberg, M. A. and E. Israel, Viral inhibition of lymphocyte mitogenesis. I. Evidence for the nonspecificity of the effect, *J. Immunol.* **124**:64–70 (1980).

106. Bubbers, J. E., J. H. Elder, and F. J. Dixon, Stimulation of murine lymphocytes by Rauscher leukemia virus *in vitro, J. Immunol.* **124**:388–394 (1980).

107. Orosz, C. G., N. E. Zinn, R. G. Olsen, and L. E. Mathes, Retrovirus-mediated immunosuppression. II. FeLV-UV alters *in vitro* murine T lymphocyte behavior by reversibly impairing lymphokine secretion, *J. Immunol.* **135**:583–590 (1985).

108. Schmidt, D. M., N. K. Sidhu, G. J. Cianciolo, and R. Snyderman, Recombinant hydrophilic region of murine retroviral protein p15E inhibits stimulated T-lymphocyte proliferation, *Proc. Natl. Acad. Sci. USA* **84**:7290–7294 (1987).

109. Harris, D. T., G. J. Cianciolo, R. Snyderman, S. Argov, and H. S. Koren, Inhibition of human natural killer cell activity by a snythetic peptide homologous to a conserved region in the retroviral protein, p15E, *J. Immunol.* **138**:889–894 (1987).

110. Kleinerman, E. S., L. B. Lachman, R. D. Knowles, R. Snyderman, and G. J. Cianciolo, A synthetic peptide homologous to the envelope proteins of retroviruses inhibits monocyte-mediated killing by inactivating interleukin 1, *J. Immunol.* **139**:2329–2337 (1987).

111. Roder, J. C., L. Tyler, J. K. Ball, and S. K. Singhal, The immunocompetence of tumor T cells and their role in generalized immunosuppression and immunostimulation following inoculation of dimethylbenzanthracene-induced leukemia virus in mice, *Cell. Immunol.* **36**:128–142 (1978).

112. Grinwich, K. D., T. S. Alexander, and J. Cerny, Properties of murine leukemia-associated suppressor cells. I. Preferential suppression of thymus-dependent antibody responses and the requirement for syngenicity in the K region of the H-2 complex, *J. Immunol.* **122**:1108–1115 (1979).

113. Tagliabue, A., D. Boraschi, and J. L. McCoy, Development of cell-mediated antiviral immunity and macrophage activation in C3H/HeN mice infected with mouse mammary tumor virus, *J. Immunol.* **124**:2203–2208 (1980).

114. Yefenof, E., and J. Ben-David, Suppressor and reactive lymphocytes in radiation leukemia virus-induced leukemogenesis, *Cancer Immunol. Immunother.* **16**:48–52 (1983).

115. Tilkin, A.-F., B. Begue, E. Gomard, and J.-P. Levy, Natural suppressor cell inhibiting T killer responses against retroviruses: A model for self-tolerance, *J. Immunol.* **134**:2779–2782 (1985).

116. Mosier, D. E., R. A. Yetter, and H. C. Morse, Functional T lymphocytes are required for a murine retrovirus-induced immunodeficiency disease (MAIDS), *J. Exp. Med.* **165**:1738–1742 (1987).

117. Garaci, E., G. Migliorati, T. Jezzi, A. Bartocci, L. Gioia, C. Rinaldi, and E. Bonmassar, Impairment of *in vitro* generation of cytotoxic or T suppressor lymphocytes by Friend leukemia virus infection in mice, *Int. J. Cancer* **28**:367–373 (1981).

118. Stiff, M. and R. Olsen, Loss of short-lived suppressive function of peripheral leukocytes in feline retrovirus-infected cats, *J. Clin. Lab. Immunol.* **7**:133–138 (1982).

119. Rudczynski, A. B., and R. F. Mortensen, Suppressor cells in mice with murine mammary tumor virus-induced mammary tumors. I. Inhibition of mitogen-induced lymphocyte stimulation, *J. Natl. Cancer Inst.* **60**:205–211 (1978).

120. Bendinelli, M., D. Matteucci, A. Toniolo, and H. Friedman, Suppression of *in vitro* antibody response by spleen cells of mice infected with Friend-associated lymphatic leukemia virus, *Infect. Immun.* **24**:1–6 (1979).

121. Spickett, G. P., and A. G. Dalgleish, Cellular immunology of HIV-infection, *Clin. Exp. Immunol.* **71**:1–7 (1987).

122. Tonietti, G., G. B. Rossi, V. Gobbo, L. Accinni, A. Ranucci, F. Titti, M. G. Premrov, and E. Garaci, Effects of *in vivo* Friend leukemia virus infection on levels of serum thymic factors and on selected T-cell functions in mice, *Cancer Res.* **43**:4355–4363 (1983).

123. Blank, K. J., and D. M. Murasko, Induction of interferon in AKR mice by various murine leukaemia viruses, *Nature (Lond.)* **283**:494–495 (1980).

124. Rodgers, B. C., D. M. Scott, J. Mundin, and J. G. P. Sissons, Monocyte-derived inhibitor of interleukin 1 induced by human cytomegalovirus, *J. Virol.* **55**:527–532 (1985).

125. Hardy, W. D., Jr., Feline retroviruses, in: *Advances in Viral Oncology*, Vol. 5: *Viruses as the*

*Causative Agents of Naturally Occurring Tumors* (G. Klein, ed.), pp. 1–34, Raven, New York (1985).

126. Ruprecht, R. M., L. D. Rossoni, W. A. Haseltine, and S. Broder, Suppression of retroviral propagation and disease by suramin in murine systems, *Proc. Natl. Acad. Sci. USA* **82**:7733–7737 (1985).

127. Mitsuya, H., and S. Broder, Strategies for antiviral therapies in AIDS, *Nature (Lond.)* **325**:773–778 (1987).

128. Mitsuya, H., R. F. Jarret, M. Matsukura, F. Di Marzio Veronese, A. L. DeVico, M. G. Sarngadharan, D. G. Johns, M. S. Reitz, and S. Broder, Longterm inhibition of human T-lymphotropic virus type III/Lymphadenopathy-associated virus (human immunodeficiency virus) DNA synthesis and RNA expression in T cells protected by 2′,3′-dideoxynucleosides *in vitro, Proc. Natl. Acad. Sci. USA* **84**:2033–2037 (1987).

129. Genovesi, E. V., D. Livnat, and J. J. Collins, Immunotherapy of murine leukemia. VIII. Efficacy of passive serum therapy of Friend leukemia virus-induced disease in immunocompromised mice, *J. Natl. Cancer Inst.* **70**:311–322 (1983).

130. David, Y. B., M. Kotler, and E. Yefenof, A highly leukemogenic radiation leukemia virus isolate is a thymotropic, immunosuppressive retrovirus with a unique RNA structure, *Int. J. Cancer* **39**:495–497 (1987).

131. Larson, C. L., R. E. Baker, R. N. Ushijima, H. B. Baker, and C. Gillespie, Immunotherapy of Friend disease in mice employing viable BCG vaccine, *Proc. Soc. Exp. Biol. Med.* **140**:700–702 (1972).

132. Sklaroff, D. M., S. J. DeCourcy, Jr., S. Specter, and H. Friedman, Suppressed development of mammary tumorigenesis in RIII mice treated neonatally with BCG, *Proc. Soc. Exp. Biol. Med.* **158**:235–237 (1978).

133. Wleklik, M., S. B. Levy, M. Luczak, and V. A. Najjar, Suppression of Friend virus-induced leukaemia in mice by tuftsin, *J. Gen. Virol.* **67**:2001–2004 (1986).

134. Liu, W. T., R. A. Good, L. Q. Trang, R. W. Engleman, and N. K. Day, Remission of leukemia and loss of feline leukemia virus in cats injected with *Staphylococcus* protein A: Association with increased circulating interferon and complement-dependent antibody, *Proc. Natl. Acad. Sci. USA* **81**:6471–6475 (1984).

135. Snyder, H. W., Jr., M. C. Singhal, W. D. Hardy, Jr., and F. R. Jones, Clearance of feline leukemia virus from persistently infected pet cats treated by extracorporeal immunoadsorption is correlated with an enhanced antibody response to FeLV gp 70, *J. Immunol.* **132**:1538–1543 (1984).

# 21

# Implications for Immunotherapy of Viral Infections

## MAYRA LOPEZ-CEPERO, STEVEN SPECTER, and JOHN HADDEN

## 1. INTRODUCTION

That host-defense mechanisms control viral infection and eventually limit the spread of disease has been recognized for many years. Preceding modern medicine, Chinese physicians in the eleventh century observed that the inhalation of smallpox crusts prevented the subsequent occurrence of disease. Later, in the eighteenth century in England, variolation was practiced by Lady Montagu as a primitive form of immunization to protect against smallpox.

The future of modern immunology was ensured when Edward Jenner made the surprising discovery that inoculation with cowpox crusts protects humans against smallpox. Eighty years later, further development of preventive immunization was made possible by Louis Pasteur, who prepared the first inactivated vaccine for rabies virus.

During the twentieth century, numerous successes have been achieved by performing specific immunoprophylaxis with vaccines. The development and administration of live attenuated vaccines for mumps, poliomyelitis, rubella (German measles), rubeola (measles), and smallpox has significantly reduced the morbidity and mortality of these diseases. Inactivated vaccines are also commonly used for hepatitis B virus (HBV), influenza, poliomyelitis,

MAYRA LOPEZ-CEPERO and STEVEN SPECTER • Department of Medical Microbiology and Immunology, College of Medicine, University of South Florida, Tampa, Florida 33612-4799.     JOHN HADDEN • Departments of Internal Medicine and Medical Microbiology and Immunology, College of Medicine. University of South Florida, Tampa, Florida 33612-4799.

and rabies. In addition, new vaccines for cytomegalovirus (CMV), herpes simplex virus (HSV), varicella–zoster virus (VZV), respiratory syncytial virus (RSV), and rotavirus, among others, are under clinical trial or current investigation.

Vaccines, however, are only a prophylactic approach to immunological intervention against viral infections. A second type of intervention in antiviral treatment has been the use of immunoglobulin (Ig) preparations. Although these preparations are generally administered to patients after exposure to a virus, they are still a form of prophylaxis, since they are effective only shortly after exposure and before the onset of clinical disease (see Section 2.2).

There is a lack of successful chemotherapeutic antiviral agents for therapy of established infections, with the exception of (1) nucleic acid analogues such as acycloguanosine, arabinoside a (Ara-a) arabinoside c (Ara-C), and 5'-iododeoxyuridine (IDU), which are used in the treatment of HSV infections; (2) methisazone for smallpox; and (3) amantadine for influenza A. This has led to a new approach in therapy for viral pathogens—immunostimulation—using biologic molecules, such as biologic response modifiers (BRM) and/or drugs that increase the host's defenses rather than directly killing the pathogen. The treatment of infections with immunostimulating substances is called the pro-host approach.[1]

These BRM and drugs have a relative specificity for one or another effector cell of the immune system permitting approaches to the manipulation of selected components of the immune system. The cells affected are T lymphocytes,[2–5] macrophages,[6,7] and natural killer (NK) cells, considered the principal immune cell types involved in protection against viral infections.[8–10]

Since 1957, when Isaacs and Lindenmann first reported the presence of interferon (IFN), a host factor induced by viral infection that interfered with further viral replications, new advances in the understanding of host-defense mechanisms have been achieved. Products of the host-immune system such as IFNs produced during viral infections by leukocytes ($IFN_\alpha$), fibroblasts ($IFN_\beta$), or T lymphocytes ($IFN_\gamma$) are able to enhance the activity of other populations of immune cells important in the primary host defense against viral infections. This includes macrophages and NK cells. For example, $IFN_\gamma$, $IFN_\beta$, and mainly $IFN_\gamma$ have been reported to enhance the cytotoxic activity of NK cells against virus-infected cells.[11] Another lymphokine produced by activated T lymphocytes, interleukin-2 (IL-2), as well as $IFN_\gamma$ is able to enhance the cytotoxic activity of NK cells against virus-infected cells.[8,9]

The development and function of T lymphocytes is under the regulation of thymic hormones, IL-1 and IL-2. Thymic hormones [$thymosin_{\alpha1}$, thymic humoral factor (THF), thymopoietin, and thymulin (FTS)], in addition to IL-2, promote the maturation of immature thymocytes. These will be very important in the therapy of T-cell-deficient persons for the potentiation of this cell population. The decrease in susceptibility to viral pathogens is highly dependent on functioning T lymphocytes.[12] A number of drugs capable of mimicking the activity of the thymus (i.e., thymomimetic drugs) act directly to induce pro-thymocyte differentiation and promote T-cell functions, including IL-2 pro-

**TABLE I**
**Immunomodulatory Substances Used as Antiviral Agents**

| Microbial products | Products of the immune system | Drugs |
|---|---|---|
| Protozoa | Immunoglobulins | Levamisole |
| Gram-negative bacteria | Thymic hormones | Vitamins A and C |
| Lipopolysaccharides | Transfer factor | Tilorone |
| Gram-positive bacteria | Interferons and interferon inducers | Pyran |
| Peptidoglycans | Interleukin 1 | Inosiplex (isoprinosine) |
| Muramyl dipeptide | Interleukin 2 | NPT 15392 |
| Mycobacteria | | |

duction, which is crucial in the immunologic cascade that develops against pathogens. Thymomimetic drugs, such as isoprinosine and NPT 15392, are also able, like IFN and IL-2, to increase the activity of NK cells against virus-infected cells.[10,13] Prohost drugs such as isoprinosine have been tested in clinical trials against viral diseases, such as herpesviruses, influenza, and measles. Levamisole, which is indirectly a thymomimetic drug because it promotes *in vivo* the appearance of thymic hormonelike substances, is another example of the new approach to immunotherapy.

The prohost approach to therapy of viral infections is popularized by the observations that virus-induced immune suppression is a common feature of such infections, as is evident from the preceding chapters of this volume. Because immunosuppression, although often transient, is so common in viral infections, it is believed to be important in viral pathogenesis. Immunotherapy therefore attacks this central defect and has the potential to lead to rapid recovery from severe consequences of viral infections.

The mechanisms by which immunotherapeutic substances induce prohost responses against viral infections are the focus of this chapter. The reader is referred to previous reviews for a summary of information not presented here on other immunostimulatory therapeutic agents (Table I), for which there is little new information.[14,15]

## 2. IMMUNOSTIMULATORY AGENTS

### 2.1. Bacteria and Their Products as Immunomodulators

A wide variety of microorganisms, including bacteria, have been studied in terms of immunomodulatory activity.[16] Adjuvants based on the use of Mycobacteria in water-and-oil emulsions have extensively been employed to stimulate immune responses to weakly immunogenic substances. However, it is now recognized that the ability to enhance immunoresponsiveness is not limited to Mycobacteria.[17]

## 2.1.1. Gram-Positive Bacteria

Many of the cell constituents of gram-positive bacteria have been shown to be important as immunomodulator molecules, inducing inflammatory responses. For example, protein A from *Staphylococcus aureus* shows nonspecific immunostimulatory activities and stimulates lymphocyte blastogenic responses.[18] Staphylococcal, streptococcal, and pneumococcal components such as teichoic acids and peptidoglycans are immunostimulatory in their action. For example, they have been shown to increase macrophage-mediated cell cytotoxicity.[19]

Mycobacteria have been used for decades as immunoenhancing agents. It has been known for a number of years that Mycobacteria can exert protective effects against viral infections. For example, the parenteral administration of live bacille Calmette-Guérin (BCG) has been shown to enhance resistance of mice to a variety of viruses, including HSV-1 and HSV-2 and influenza A[20] and Friend leukemia virus (FLV), as shown by Larson *et al.*[21]

The immunoenhancing material of BCG proved to be peptidoglycan, and the minimal adjuvant structure was found to be *N*-acetyl-muramyl-L-alanyl-D-isoglutamine or muramyl dipeptide (MDP).[22] Since MDP has been the most thoroughly investigated of the gram-positive bacterial products, we focus on this material.

The MDP used experimentally is a synthetic copy of part of the peptidoglycan moiety of the cell wall of many bacteria. Originally derived from Mycobacteria, it has a spectrum of biologic and specifically immunologic activities similar to that of the whole microorganisms but devoid of their harmful effects.

Muramyl dipeptides have been derived that are modifications of the original MDP in an effort to enhance activity. This search has proved successful in some cases and still continues.[23] These have been shown to exert adjuvant activity when associated with a number of antigens including bacterial, parasitic, and viral vaccines, among others.[24] Influenza, HSV, and HBV vaccines are part of the list of viral antigens that become more effective when associated with MDP. For example, in the case of the influenza vaccine, antibody responses are higher when MDP is combined with the vaccine, which otherwise is poorly immunogenic.[25] Thompson *et al.*[26] reported that MDP enhanced the immunogenicity of formalinized whole HSV. In view of these results, it can be expected that MDP will be of practical importance in the field of conventional and synthetic vaccines for humans and animals.

Regarding the route of administration of these compounds, several investigators showed that in addition to the parenteral routes, the oral and nasal routes are effective for immunoenhancement and permit direct stimulation of the secretory immune mechanisms.[27−30] Therefore, it can be expected that in combination with vital antigens, locally administered MDP will be used to increase the production of neutralizing antiviral secretory antibodies.

Muramyl dipeptides have no direct antiviral effects,[31] so it can be inferred that they act by modulating the host's mechanisms of defense. MDPs do not seem to alter the production of IFN induced by viruses.[31] The activation of cellular mechanisms by MDP can also account for part of their antiviral effects. MDPs are inducers of monokines such as IL-1[32] and endogenous pyrogens, which increase body temperature and thereby counteract viruses. Also, MDPs are enhancers of NK cell activity,[33] antibody-dependent cell cytotoxicity (ADCC),[34] and macrophage cytotoxicity,[35] which are important mechanisms of defense against viral infections.

Thus, the role of MDP in the antiviral strategy is seen in a dual aspect: one deals with their adjuvant effects and the other with their activity on the effectors of nonspecific resistance. This second aspect involves the stimulation of the host nonspecific defenses that will lead to recovery from the viral infection. It is not clear whether the beneficial effects of MDP also involve the ability to reduce the immunodepression induced by the virus.

### 2.1.2. Gram-Negative Bacterial Endotoxins

Endotoxins or lipopolysaccarides (LPS) are associated with the external cell envelope of all gram-negative bacteria. The polysaccharide is linked to the lipid A moiety, which is responsible for the toxic manifestations of endotoxins. Both lipid A and the polysaccharide are important in immune alterations.[36] Endotoxins are T-independent antigens, are mitogenic for murine B lymphocytes, and stimulate polyclonal antibody formation. Another major role of endotoxins as immunoregulatory molecules is the stimulation of macrophages, with the consequent release of soluble mediators such as IL-1. This stimulates other cell types to produce a cascade of cytokines such as IL-2 that can enhance or facilitate immune responses.

The beneficial effects of endotoxin on the immune response are clearly seen in the retrovirus-induced leukemia in mice caused by FLV.[37,38] FLV has also been used as a model for studying immunosuppression resulting from retrovirus infection *in vivo* and *in vitro*. Spleen cells from FLV-infected mice show a marked immunosuppression early after exposure to the virus both *in vivo* and *in vitro*. The antibody responsiveness in these mice is markedly diminished. It has been shown that the addition of bacterial endotoxins to these depressed spleen cell cultures from FLV-infected mice results in marked enhancement of the immune response. The effect may be due to the activation of macrophages in the spleen cultures, since the addition of normal mouse spleen cells and/or normal adherent splenocytes, as well as peritoneal exudate cells, has resulted in a marked enhancement of the immune response by FLV-infected spleen cell cultures.[39] However, other effects of LPS may contribute to the recovery of antibody responsiveness.[37,38] Beneficial effects in other experimental viral infections also have been reported.[40–42]

## 2.2. Products of the Immune System

### 2.2.1. Immunoglobulins

The need for effective treatment of viral diseases requires the use of products of the immune system. Gammaglobulins have been used for decades to transfer passive immunity. This has been performed using pooled gammaglobulins as a general prophylactic measure to protect healthy people (or immunodeficient patients) in outbreaks of hepatitis and for other viral diseases. For certain viruses, specific Ig preparations have been used as a prophylactic, either before exposure to a virus in epidemic settings or for protection shortly after exposure. Specific Igs are available for HBV, rabies virus, vaccina virus, and VZV and have been reviewed elsewhere.[15]

While the Ig prevent infection by enhancing the immune clearance of the virus, they may also be important in precluding the immunodepressive effects of the virus. More recently, the approach to the use Ig has been reassessed. For example, it has been found that the route of administration affects the efficacy of these preparations when used for prophylactic therapy against viral infections. Stiehm[43] showed that intramuscular (i.m.) IgG preparations, in the prevention and modification of viral infections, were less effective than intravenous (i.v.) preparations. This may be because i.v. IgG can be given in higher doses than can i.m. Several clinical trials with i.v. IgG focus on severe viral infections. For example, in leukemic children, varicella may be associated with serious complications (e.g., visceral dissemination) or bacterial superinfections. These patients have been protected by passive immunization with the specific varicella–zoster immune globulin (VZIG), which has been shown to decrease the incidence and severity of varicella in these high-risk patients.[44] CMV as well as HSV infections cause serious problems in immunocompromised patients, such as organ transplant patients. Studies by Winston *et al.*[45] showed that prevention and modification of the disease can be achieved by passive immunization with i.v. Ig. Winston's study showed the effect of a polyvalent Ig preparation containing antibodies against CMV in a group of 18 bone marrow transplant patients. These patients received i.v. IgG before and once every week after transplantation for 17 weeks. The incidence of CMV infection was similar in control and experimental groups; however, the symptomatic disease, including CMV pneumonia, was less frequent in patients who received the IgG.

As these studies suggest, the prevention and therapy of VZV and CMV infections could be achieved by the use of i.v. IgG. The problem that still needs to be solved is whether hyperimmune sera are superior to polyvalent IgG, which would warrant a search for donor plasma containing high antibody titers. It is important to note in this age of concern about transmission of AIDS that the use of Ig has not been implicated in the transmission of infectious agents. Recent epidemiologic reports show that the use of these preparations does not transmit the HTLV-III/LAV infection; therefore, the current indications for their clinical use should not be changed on the basis of such a concern.[46]

## 2.2.2. *Immunity Induced by Viral Vaccines*

The control of viral diseases relies mainly on the use of preventive vaccines. The first goal in the establishment of antiviral vaccines is the ability to generate an immune response against the virus, characterized by neutralizing antibodies and cellular immunity. The attachment of the virus to the host's cell membrane is indispensable for the subsequent penetration of the viral genome into the cell. Therefore, vaccines that generate an immune response that neutralizes viral antigens involved in the attachment step are effective in the prevention of these diseases. For example, HBV has an antigen on its envelope known as $HB_sAg$, which interacts with the hepatocyte membrane. Without it, the viral genome cannot penetrate into the hepatocyte. Therefore, the antibodies against $HB_sAg$ are protective, since they neutralize the attachment of the HBV.[47−49]

In a similar fashion, the envelope glycoprotein antigens of influenza virus, neuraminidase and hemagglutinin, which interact with the epithelial cells of the upper respiratory tract, can be neutralized by antineuraminidase and antihemagglutinin antibodies that will inhibit the attachment, penetration, and infection by the influenza virus of epithelial cells. However, not all the attachment antigens of viruses are known, and some are difficult to isolate. In these cases, intact killed or live attenuated virus is used in the preparation of vaccines.

The control of HAV is one of the unsolved sociosanitary problems of many parts of the world. However, the recent propagation of high-titer virus in monkey fetus kidney cells and in human embryonic lung fibroblasts provides the basis for the presupposition that the preparation of an HAV vaccine is closer to development.[50]

Infection with CMV is one of the most important causes of disease in immunocompromised persons. *In utero* it can cause fetal death or serious neurologic and hepatic damage, while in adults it causes fatal pneumonitis or inhibition of organ graft acceptance. The availability of a vaccine for the prevention of CMV infections is limited to clinical trials.

Respiratory syncytial virus causes lower respiratory tract infections that are often fatal in small children. Buynak *et al.*[51] reported the use of attenuated live virus vaccines that can be injected parenterally and induce the appearance of neutralizing antibodies in 75–85% of tested subjects. Another clinical trial showed that the percentage of responders was even higher when the complement-fixing antibodies were considered. Specific antibodies belonging to the IgA class were found also in nasal secretions.[52] All subjects who seroconverted to anti-RSV antibody were protected against natural RSV infection.

One must also be aware that viral vaccines may be immunosuppressive. Transient immune suppression has been demonstrated with attenuated vaccines for measles, polio, rubella, vaccinia, and yellow fever.[53] Thus, care must be taken not to administer vaccines with this potential to any patient who is already immunologically compromised.

### 2.2.3. Thymic Hormones

Thymic hormones are immunologically active peptides produced by the thymus. A number of such hormones have been extracted. Crude preparations have been prepared by boiling the thymus and isolating the active portions. Thymosin fraction V is one such preparation that contains more than 35 different peptides. Another such preparation is thymostimulin. From such preparations, isolated hormones have been purified. Examples are thymopoietin, thymosin-1, and thymulin (formerly factor thymique sérique, or FTS). These three hormones have been shown to be derived from thymic epithelial cells and to circulate in the blood. Interestingly, as with thymus weight itself, their activities in the blood are highest in the first 10 years of life, declining to low levels in the third and fourth decades. Because of their effects on the immune system, thymic hormones have been implicated for the treatment of severe viral infections in immunocompetent hosts. Among these, thymic humoral factor (THF) has been the most extensively studied.

Thymic humoral factor is an acidic peptide derived from calf thymus. Among its biologic properties is the reconstitution of impaired immune function of neonatally thymectomized mice and stimulation of colony-forming capacity in the bone marrow of these animals. THF also enhances antitumor killer function and mitogen-driven proliferation, as well as IL-2 production by mouse spleen cells. It also raises the competence of T cells of normal mice to participate in graft-versus-host (GVH) and mixed-lymphocyte (MLR) reactions. Therefore, THF apparently has an immunomodulatory function, permitting a return to normal of impaired immune balance.[54]

The first evidence of the antiviral effect of THF was reported by Rager-Zisman *et al.*[55] in Sendai virus-infected mice. These investigators found a significant number of the animals injected with a lethal dose of Sendai virus to be protected by the administration of THF at the time of infection, as compared with untreated infected controls. In humans, THF induces an increase in the proliferation of peripheral blood lymphocytes challenged *in vitro* with VZV antigen.[56] Thus, THF appears to reverse the virus-induced immune depression.

These findings, in addition to biologic properties, make THF suitable for treating viral disease. The mechanism of THF action is unknown. However, it is postulated that this hormone repairs the adverse effects caused by the infection, either recruiting more T cells or by restoring the ability of the infected lymphocytes to produce soluble products such as IL-2, which are necessary for an intact immune response.[54]

Thymic humoral factor does not induce IFN activity in murine spleen cells. Thus, IFN is not a major mechanism of THF-induced repair. Therefore, it is further postulated that the antiviral effect of THF is mediated via enhancement and clonal expansion of the pool of cytotoxic T cells (CTL) by such soluble factors as IL-2. These CTL are responsible for the specific antiviral immunity. Trainin *et al.*[57] demonstrated that the targets for the activity of THF are early and late T-lymphocyte precursors that are driven to proliferation and differentiation.

Therefore, THF is a powerful inducer of T-cell proliferation and functional maturation of remaining T-cell precursors in immunosuppressed patients. These can result in a significant cell-mediated antiviral effect that is applicable to host recovery from infection. Since these reports do not indicate any side effects of THF, it should receive serious consideration as an agent for the treatment of viral diseases.

## 2.2.4. Interleukin-1

Interleukin-1 is a cellular mediator produced by macrophages and other cells that has a multiplicity of effects on immunologic and inflammatory reactions.[58] Besides stimulating thymocyte proliferation, IL-1 activates B cells, induces neutrophilia, triggers the production of acute-phase proteins, induces fever, and stimulates bone resorption.[59,60] IL-1 has been shown to have no direct effect on the generation of NK cell activity, but it does so indirectly by augmenting the effects of IFN and IL-2 on NK cell cytotoxicity.[61] Dinarello and co-workers[62] showed that the treatment of target cells with IL-1 enhanced their binding to NK cells, resulting in increased cytotoxicity. The importance of this interleukin in the enhancement of first-line defenses against tumor cells is further supported by the observation that monocytes from human peripheral blood show considerable cytotoxicity against tumor cells when treated with IL-1. Onozaki et al.[63] showed that IL-1 released by macrophages and monocytes plays an important role in the host defense against neoplastic cells by acting on monocytes as an autostimulating factor, maintaining high spontaneous levels of cytotoxic activity rather than inducing new effectors.

There are no reports of IL-1 production defects in viral infections. Thus, the administration of exogenous IL-1 probably would not compensate for any virus-induced immunodepression. Thus, it seems unlikely that IL-1, which is able to maintain and enhance first-line defenses active against virus-infected cells, would be suggested for preventing or limiting viral infections.

## 2.2.5. Interleukin-2

Interleukin-2 is a glycosylated peptide released from antigen- or mitogen-stimulated T lymphocytes, that functions to mediate a switch in T cells from late $G_1$ into the proliferative phases of the cell cycle.[64] IL-2 induces also the maturation and proliferation of T-cell precursors.[65]

Interleukin-2 has biologic activities on other cell populations, such as NK cells. In studies performed by Rook et al.,[66] IL-2 was shown to enhance the NK cell and CTL activities against CMV-infected targets in control subjects as well as in patients with AIDS. These results indicate that IL-2 can substantially potentiate the cytotoxic effector functions of peripheral blood leukocytes from normal as well as immunosuppressed patients. It is also known that IL-2 induces the production of tumor necrosis factor ($TNF_\alpha$ and $TNF_\beta$) in peripheral blood mononuclear cells. This activity of IL-2 is enhanced by $IFN_\gamma$. Therefore, it is suggested that IL-2 as well as $IFN_\gamma$-induced tumor cell destruction is mediated by $TNF_\alpha$ and $TNF_\beta$ (lymphotoxin).[67] Recently, in studies by Rosenberg,[68] IL-2 was demonstrated to activate mouse spleen cells as well as human peripheral blood leukocytes in vitro. These lymphokine-activated killer cells (LAK) in vitro kill tumor cells but not normal cells in in vivo. Studies by Conlon et al.[69] show that IL-2 therapy appears to restore the in vivo responsiveness of immunosuppressed recipients to allogeneic tumor cell challenge. The two previous studies, although leaving many questions to be answered, have established a new approach to immunotherapy, the regulation of immune responses, and the treatment of immune deficiencies. Although these have been applied only to tumor systems to date, their application for severe viral infections should follow because of the similarity of antitumor and antiviral immunity.[70]

### 2.2.6. Interferon

Human IFN are classified into three antigenically distinct groups designated $IFN_\alpha$, $IFN_\beta$, and $IFN_\gamma$. $IFN_\alpha$ is produced mainly by lymphoreticular cells and is induced by viruses, tumor cells, bacteria, and B-cell mitogens. It is encoded by a multigene family consisting of at least 13 nonallelic and 8 allelic members.[71,72] $IFN_\beta$ is produced mainly by fibroblasts and epithelial cells during viral infections. $IFN_\gamma$ is produced mainly by T cells stimulated by specific antigens or mitogens. $IFN_\beta$ and $IFN_\gamma$ are each encoded by a single gene.[73,74]

Interferon and the immune system have many interactions; therefore, IFN may be considered one of the regulators of immunity.[75] The IFN system is induced during most viral infections. There are two arms in the IFN system: the afferent arm and the efferent arm. During the former, production of IFN proteins are induced; in the latter, effector cells are activated by the IFN proteins, leading to viral and cellular growth inhibition.

The induction of the IFN system by viral infections leads to the production of different types of IFN by at least four major pathways. Viral infection of macrophages derepresses many of the genes for $IFN_\alpha$ produced in the first few hours, which can diffuse throughout the body.[76] The second pathway for the IFN induction is the stimulation of B lymphocytes by viral membranes or virus-infected cells, leading to the production of $IFN_\alpha$. The third pathway is the induction of $IFN_\gamma$ by T lymphocytes. The inducing stimulus is either the foreign viral antigen to which the T cell is sensitized or the mitogens identified on viruses, as well as bacteria and parasites.[77] The fourth pathway is the induction of $IFN_\beta$ in human cells or $IFN_\alpha/IFN_\beta$ in murine cells by virus infection of epithelial or fibroblastic cells. $IFN_\beta$ and $IFN_\gamma$, in contrast to $IFN_\alpha$, tend to remain localized at the site of production.

The production of IFN by the cells of the immune system is only part of the simultaneous induction of monokines and lymphokines that initiate the inflammatory and immune responses.[75] Besides the antiviral action, the IFN effects include antitumor action, immunoregulatory action, cell growth inhibition, alteration of cell membranes, macrophage activation, enhancement of cytotoxicity of lymphocytes and NK cells, influence on subsequent production of IFN, and hormonelike activation of cells.

The immunoregulatory activities of IFN are now well characterized. A number of different immune responses have been shown to be affected by IFN, including highly purified recombinant DNA-derived subtypes and analogues. Of the immune functions affected by IFN, NK cell activation has received the most attention, probably because of the apparent dependence of NK cell maturation and activation on IFN. Studies by Gresser and co-workers[78] showed that purified $IFN_\alpha$ and $IFN_\beta$ augment the NK cytolytic activity after *in vivo* and *in vitro* administration. Anti-IFN antibodies, which neutralize IFN, were shown to block the NK enhancement seen both in mice[78] and in humans.[79]

Inducers of IFN, such as Newcastle disease virus,[78] polyinosinic-poly-citidylic acid,[80,81] and tumor-derived or virus-infected cells,[82] also enhance NK activity in humans, rats, and mice. Studies by Targan and Dorey[83] and Timonen *et al.*[84] showed that IFN stimulates NK activity in three different ways: (1) increasing the number of NK cells available to bind to the targets, (2) stimulating the kinetics of the lysis, and (3) increasing the ability of NK cells to recycle and kill more than one target cell.

Interferon is also known to augment monocyte and macrophage activities, as well as the number of Fc receptors on such cells. Fc receptor enhancement on K cells is probably the basis for the IFN-induced increase of ADCC. In addition to stimulation of direct cytotoxic activity of macrophages and CTL, IFN may enhance the effects of antiviral and antitumor cytotoxic mechanisms by inducing increased neoantigen presentation on cells. Antigen presentation by macrophages is mediated in association with histocompatibility antigens, the expression of which is augmented by IFN treatment.[85]

That IFN can be used as immunotherapeutic agents is indicated by the work of Murray *et al.*,[6] who showed that alveolar macrophages from AIDS patients can readily be activated by soluble T-cell products and $IFN_\gamma$. T cells from these patients fail to generate $IFN_\gamma$, but their peripheral blood and tissue macrophages respond normally to the direct activating effect of this molecule. It seems reasonable to proceed to evaluate $IFN_\gamma$ as an immunotherapeutic agent in AIDS patients with opportunistic infections. Such therapeutic trials would be especially appropriate for patients infected with intracellular pathogens against which host defenses are thought to require an intact T-cell-dependent macrophage-mediated response for control and eradication.

The use of $IFN_\alpha/IFN_\beta$ treatment has been effectively shown against infections by papovaviruses and respiratory viruses such as coronavirus, rhinovirus, and influenza virus type A.[86] IFN treatment of laryngeal papillomas induced by human papillomaviruses causes the regression and degeneration of these tumors in 60–80% of all cases treated.[87–90]

Nevertheless, the mechanism(s) by which IFN protects against viruses and tumors is not fully understood. However, several mechanisms have been proposed for the explanation of this phenomenon, which include (1) the antiviral and cell growth inhibitory properties of IFN, (2) the enhancing effect of IFN in the cytocidal and tumoricidal activities of NK cells and macrophages, and (3) the production of tumoricidal factors that may control and reduce tumor cell growth.[91]

It is probable that a combination of two or more of these mechanisms contributes in the effective elimination of tumors and viral infections.

## 2.3. Antiviral Immunostimulatory Drugs

While a variety of drugs have been tried as immunostimulatory agents in the therapy of viral infections,[14] two have been selected for review here. Both drugs, levamisole and isoprinosine, have been demonstrated to have some value in restoring defective cellular immune responses and in aiding in the recovery from viral infections in experimental models and clinical trials.

### 2.3.1. Levamisole

Levamisole is a potent anthelminthic drug, known since 1966. It has been found useful as an immunomodulating drug since 1971. It is active mainly on T cells but also affects phagocytes. It restores to a normal range the depressed functions of T cells, macrophages, and neutrophils. It enhances cellular levels of cyclic guanosine monophosphate, which is believed to be responsible for this activity. Its use results in the maturation of precursor T lymphocytes, acting like a thymic hormone; i.e., it is thymomimetic in action. While levamisole is not an antiviral agent by itself, the most reliable results and the main fields of employment of levamisole in clinical trials have been in recurrent herpetic infections (labialis, progenitalis, and keratitis), recurrent upper respiratory tract infections, and opportunistic infections in cancer and immunosuppressed patients.[91] The use of levamisole is still of practical value in recurrent infections as a prophylactic drug despite its various side effects, which include skin rashes, febrile illness, metallic taste, gastrointestinal upset, anxiety, and neutropenia.[92,93] Its use is obviously limited to severe or life-threatening infections.

### 2.3.2. Isoprinosine

Isoprinosine has been shown to have beneficial effects in ameliorating the symptoms of infections with HSV, CMV, subacute sclerosing panencephalitis (SSPE), rhinovirus, and in some cases of influenza virus.[94] Isoprinosine is a synthetic immunomodulatory agent. Some of its immunomodulating properties include enhancement of *in vitro* function of T cells and macrophages. It induces the appearance of T-cell markers and enhances the lymphocyte response to mitogens. This property appears to be due to the synthesis of IL-2. *In vivo* it also increases T-cell functions and macrophage activities and increases antibody formation. Isoprinosine has been shown to restore T-cell function and the lymphocyte response to mitogens in immunosuppressed cancer patients following radiotherapy. It also potentiates the antiviral and antitumor activity of IFN.[95]

Studies by Hersey and Edwards[10] showed that isoprinosine increases NK activity. The mechanism of the effect of isoprinosine on NK activity may be linked to the release of mediators known to stimulate NK activity, i.e., IFN and IL-2. These workers also showed that isoprinosine potentiates the release of

IL-1 and IL-2 in response to stimulation by mitogens *in vitro* and modulates suppressor cell activity.

Tsang *et al.*[5] demonstrated the capacity of isoprinosine to restore, at least partially, some aspects of the depressed cellular functions associated with AIDS. These investigators suggest the potential clinical use of this agent in the treatment of high-risk patients and in patients with mild symptoms of AIDS.[5] The cellular mechanism(s) of action of isoprinosine in enhancing proliferative responses remain obscure. It seems that isoprinosine facilitates, in the early phases of blastogenesis, the processing or the presentation of antigens to lymphocytes. Alternatively, a mitogenic helper factor may be released by the effector cells during the early hours of isoprinosine stimulation, generating T-cell growth factors (IL-2), thereby enhancing the immune response. Isoprinosine is a nontoxic synthetic compound, but some of its side effects may include nausea or a transient rise in serum and urinary uric acid with no resultant sequelae.

## 3. CONCLUSIONS

Viruses, in general, cause suppression of the host-immune system, most frequently affecting cellular immune responses. Suppression, in many cases, is manifested as a cell dysfunction, i.e., a decrease in T lymphocytes or NK cells and macrophage activities and/or inhibition of production of soluble mediators such as interleukins and IFNs. A variety of immunomodulatory agents have been described that are being used to restore these deficient immune functions in experimental models or clinical trials.

The restoration of T-cell functions seems to be attainable using thymic hormones or thymomimetic drugs. IL-2 enhances the activity of NK cells and activates LAK cells; recently, a new role has been attributed to IL-2 as an inducer of prothymocyte differentiation.[96] IFNs are well known as antiviral agents and potent immunostimulants. They enhance nonspecifically the microbicidal and cytocidal capacities of macrophages and NK cells, respectively. Among the bacterial products, MDP, the smallest active immunostimulatory component of the mycobacterial cell wall, have adjuvant effects and enhance macrophage microbicidal and tumoricidal activities. It seems likely that combined immunotherapy, making use of bacterial products such as MDP, thymic hormones, thymomimetic drugs, interleukins (mainly IL-2), and other cytokines such as IFN may prove useful in the treatment of immune deficiencies and in the prevention and control of viral infections.

However, nonspecific immunotherapy should be used with the precaution that the immune system is an homeostatic system and that any alteration may lead to an imbalance that might increase host pathologic processes. In fact, there are reports of immunotherapeutic regimens that enhance the pathologic processes of neoplasia and viral infections. This has been reported for endotoxins,[37] levamisole,[97,98] thymic hormones,[97] isoprinosine[97] and bacterial preparations.[99,100] Thus, it is necessary to have an understanding of the dose, route of administration, and temporal relationship between time of infection and therapeutic intervention, since these may contribute to whether the reversal of suppression is beneficial or detrimental. In this regard, immunosuppression is not necessarily a detrimental response. Immunosuppression may be an important mechanism by which the host reduces immunopathologic processes that develop following many viral infections. Unfortunately, previous experience indicates that our current knowledge will not help us predict when this will occur. Only empirical testing has led to an answer. A better understanding of virus-induced pathogenesis is an important prerequisite to the use of such therapeutic approaches in clinical medicine.

# REFERENCES

1. Hadden, J. W., C. Lopez, R. J. O'Reilly, and E. M. Hadden, Levamisole and inosiplex: Antiviral agents with immunopotentiating action, *Ann. NY Acad. Sci.* **284**:139–152 (1977).
2. Wainberg, M. A., S. Vydelingum, and R. G. Margolese, Viral inhibition of lymphocyte mitogenesis: Interference with the synthesis of functionally active T cell growth factor (TCGF) activity and reversal of inhibition by the addition of same, *J. Immunol.* **130**:2372–2378 (1983).
3. Specter, S., and J. W. Hadden, New approaches to immunotherapy: Thymomimetic drugs, *Springer Semin. Immunopathol.* **61**:1–10 (1985).
4. Tsang, P., F. Lew, G. O'Brien, I. J. Selikoff, and J. G. Bekesi, Immunopotentiation of impaired lymphocyte functions *in vitro* by isoprinosine in prodromal subjects and AIDS patients, *Int. J. Immunopharmacol.* **7**:511–514 (1985).
5. Tsang, P. H., K. Tangnavard, S. Solomon, and J. G. Bekesi, Modulation of T and B lymphocyte functions by isoprinosine in homosexual subjects with prodromata and in patients with acquired immune deficiency syndrome (AIDS), *J. Clin. Immunol.* **4**:469–477 (1984).
6. Murray, H. W., R. A. Gellene, D. M. Libby, C. D. Rothermel, and B. Y. Rubin, Activation of tissue macrophages from AIDS patients: In vitro response of AIDS alveolar macrophages to lymphokines and interferon gamma, *J. Immunol.* **135**:2374–2377 (1985).
7. Koff, W. C., I. J. Fidler, S. D. Showalter, M. K. Chakraborty, B. Hampar, L. M. Ceccorulli, and E. S. Kleinerman, Human monocytes activated by immunomodulators in liposomes lyse herpesvirus infected but not normal cells, *Science* **224**:1007–1008 (1984).
8. Henney, C. S., K. Kuribayashi, D. E. Kern, and S. Gillis, Interleukin 2 augments natural killer cell activity, *Nature (Lond.)* **291**:335–338 (1981).
9. Kuribayashi, K., S. Gillis, D. E. Kern, and C. S. Henney, Murine NK cell cultures: Effects of interleukin 2 and interferon on cell growth and cytotoxic reactivity, *J. Immunol.* **126**:2321–2327 (1981).
10. Hersey, P., and A. Edwards, Effect of isoprinosine on natural killer cell activity of blood mononuclear cells *in vitro* and *in vivo*, *Int. J. Immunopharmacol.* **6**:315–320 (1984).
11. Welsh, R. M., Do natural killer cells play a role in virus infections?, *Antiviral Res.* **1**:5–12 (1981).
12. Fiorilli, M. M. Crescenzi, and F. Aiuti, Thymic hormone therapy of viral infections, *Riv. Immunol. Immunofarmacol.* **5**:72–73 (1985).

13. Hadden, J. W., E. M. Hadden, T. Spira, R. Settineri, L. Simon, and A. Giner-Sorolla, Effects of NPT 15392 *in vitro* on human leukocyte functions, *Int. J. Immunopharmacol.* **4:**235–242 (1982).

14. Friedman, H., and S. Specter, Immunotherapy and immunoregulation, in: *Chemotherapy of Viral Infections* (P. E. Came and L. A. Caliguiri, eds.), pp. 313–330, Springer-Verlag, Berlin, (1982).

15. Hadden, J. W., F. Sorice, and J. L. Touraine (eds.), Immunotherapy and viral diseases, *Riv. Immunol. Immunofarmacol.* **5:**51–52 (1985).

16. Lagrange, P. H. Immunomodulation of bacterial infection, in: *Advances in Immunopharmacology*, Vol. II (J. W. Hadden, L. Chedid P. Dukor, F. Spreafico, and D. Wulloughby, (eds.), pp. 195–204, Pergamon London (1983).

17. Werner, G. H., Immunopotentiating substances with antiviral activity, *Pharmacol. Ther.* **6:**235–273 (1979).

18. Liu, W. T., R. A. Good, L. Q. Trang, R. W. Engleman, and N. K. Day, Remission of leukemia and loss of feline leukemia virus in cats injected with *Staphylococcus* protein Ai Association with increased circulating interferon and complement-dependent antibody, *Proc. Natl. Acad. Sci. USA* **81:**6471–6475 (1984).

19. Friedman, H., T. Klein, and R. C. Butler, Bacterial antigens as immunomodulators, in: *Immunomodulation* (H. H. Fudenberg, H. D. Whitten, and F. Ambrogi, (eds.), pp. 209–228, Plenum, New York (1982).

20. Suenaga, T., T. Okuyama, I. Yoshida, and M. Azuma, Effect of *Mycobacterium tuberculosis* BCG infection on the resistance of mice to Ectromelia virus infection: Participation of interferon in enhanced resistance, *Infect. Immun.* **20:**312–314 (1978).

21. Larson, C. L., R. E. Baker, R. N. Ushijima, M. Baker, and C. A. Gillespie, Immunotherapy of Friend disease in mice employing viable BCG vaccine, *Proc. Soc. Exp. Biol. Med.* **140:**700–702 (1972).

22. Ellouz, F., A. Adam, R. Ciorbaru, and E. Lederer, Minimal structural requirements for adjuvant activity of bacterial peptidoglycan derivatives, *Biochem. Biophys. Res. Commun.* **59:**1317–1325 (1974).

23. Chedid, L., Adjuvants for vaccines, in: *Advances in Immunopharmacology*, Vol. II (J. W. Hadden, L. Chedid, P. Dukor, F. Spreafico, and D. Willoughby, (eds.), pp. 401–406, Pergamon, London (1983).

24. Audibert, F., M. Jolivet, and C. Carelli, Use of MDP with diphteric or other synthetic oligopeptides as a model for totally synthetic vaccines, in: *Advances in Immunopharmacology*, Vol. II (J. W. Hadden, L. Chedid, P. Dukor, F. Smeafico, and D. Willoughby (eds.), pp. 429–434, Pergamon, London, (1983).

25. Webster, R. G., W. P. Glezen, C. Hanroun, and W. G. Laner, Potentiation of the immune response to influenza virus subunit vaccines, *J. Immunol.* **119:**2073–2077 (1977).

26. Thomson, T. A., J. Hilfenhaus, H. Moser, and P. S. Morahan, Comparison of effects of adjuvants on efficacy of virion envelope herpes simplex virus vaccine against labial infection of BALB/c mice, *Infect. Immun.* **41:**556–562 (1983).

27. Chedid, L., F. Audibert, P. Lefrancier, J. Choay, and E. Lederer, Modulation of the immune response by a synthetic adjuvant and analogs, *Proc. Natl. Acad. Sci. USA* **73:**2472–2475 (1976).

28. Butler, J. E., H. B. Richerson, P. A. Swanson, W. C. Kopp, and M. T. Suelzer, The influence of muramyl dipeptide on the secretory immune response, *Ann. NY Acad. Sci.* **109:**669–687 (1983).

29. Genco, R. J., R. Linzer, and R. T. Evans, Effect of adjuvants on orally administered antigens, *Ann. NY Acad. Sci.* **109:**650–663 (1983).

30. Taubman, M. A., J. L. Ebersole, D. J. Smith, and W. Stack, Adjuvants for secretory immune responses, *Ann. NY Acad. Sci.* **109:**637–649 (1983).

31. Morin, A., B. Chorley, and L. Chedid, Muramyl peptides in the antiviral strategy, *Riv. Immunol. Immunofarmacol.* **5**:74–79 (1985).

32. Damais, C., G. Riveau, M. Parant, N. Gerota, and L. Chedid, Production of lymphocyte activating factor in the absence of endogenous pyrogen by rabbit or human leukocytes stimulated by MDP derivative, *Int. J. Immunopharmacol.* **5**:403–410 (1983).

33. Sharma, S. D., V. Tsai, J. L. Krahenbuhl, and J. S. Remington, Augmentation of mouse natural killer cell activity by muramyl depeptide and its analogs, *Cell. Immunol.* **62**:101–109 (1981).

34. Leclerc, C., D. Juy, E. Bourgeois, and L. Chedid, *In vivo* regulation of humoral and cellular immune responses of mice by a synthetic adjuvant, N-acetyl-muramyl-L-alanyl-D-isoglutamine. Muramyl dipeptide from MDP, *Cell Immunol.* **45**:199–206 (1979).

35. Fidler, I. J., S. Sone, W. E. Fogler, and Z. L. Barnes, Eradication of spontaneous metastases and activation of alveolar macrophages by intravenous injection of liposomes containing muramyl dipeptide, *Proc. Natl. Acad. Sci. USA* **78**:1680–1684 (1981).

36. Nowotny, A., (Ed.) *Beneficial Effects of Endotoxins*, Plenum, New York (1983).

37. Bendinelli, M., D. Mattencci, A. M. Giangregorio, and P. G. Conaldi, Restoration of antibody responsiveness by endotoxin in retrovirus immunodepressed mice: Role of macrophages, in: *Immunobiology and Immunopharmacology of Bacterial Endotoxins* (A. Szentivanyi and H. Friedman, eds.), pp. 465–478, Plenum, New York (1986).

38. Friedman, H., S. Specter, and R. C. Butler, Stimulation of immunomodulatory factors by bacterial endotoxins and non-toxic polysaccharides, in: *Beneficial Effects of Endotoxins* (A. Nowotny, ed.), pp. 273–282, Plenum, New York (1983).

39. Specter, S., M. Bendinelli, W. S. Ceglowski, and H. Friedman. Macrophage-induced reversal of immunosuppression by leukemia viruses. Macrophage functions in immunity, *Fed. Proc.* **37**:97–101 (1978).

40. Finklestein, R. A., Alteration of susceptibility of embryonated eggs to Newcastle disease virus by *Escherichia coli* and endotoxin, *Proc. Soc. Exp. Biol. Med.* **106**:481–484 (1961).

41. Gledhill, A. W., Sparing effect of serum from mice treated with endotoxin upon certain murine virus diseases, *Nature (Lond.)* **183**:185–186 (1959).

42. Wagner, R. R., R. M. Snyder, E. W. Hook, and C. N. Luttrell. Effect of bacterial endotoxin on resistance of mice to viral encephalitides, *J. Immunol.* **83**:87–98 (1959).

43. Stiehm, E. R., Standard and special human immune serum globulins as therapeutic agents, *Pediatrics* **63**:301–319 (1979).

44. Orenstein, W. A., D. L. Heymann, R. J. Ellis, R. L. Rosenberg, J. Nakano, N. A. Halsey, G. D. Overturf, G. F. Hayden, and J. J. Witte, Prophylaxis of varicella in high-risk children: Dose response effect of zoster immune globulin, *J. Pediatr.* **98**:368–373 (1981).

45. Winston, D. J., H. G. Winston, C-H. Lin, M. D. Budinger, R. E. Champlin, R. P. Gale, Intravenous immunoglobulin for modification of cytomegalovirus infections associated with bone marrow transplantation, *Am. J. Med.* **76**:238–133 (1984).

46. Centers for Disease Control, *MMWR* **35**:231–233 (1986).

47. McAleer, W. J., H. Z. Markus, D. E. Wampler, E. B. Buynak, W. J. Miller, R. E. Weibel, A. A. McLean, and M. R. Hilleman, Vaccine against human hepatitis B virus prepared from antigen derived from human hepatoma cells in culture (41801), *Proc. Soc. Exp. Med.* **175**:314–319 (1984).

48. Valenzuela, P., A. Medina, W. J. Rutter, G. Ammerer, and B. D. Hall, Synthesis and assembly of hepatitis B virus surface antigen particles in yeast, *Nature (Lond.)* **298**:347–350 (1982).

49. McAleer, W. J., E. B. Buynak, R. Z. Maigetter, D. E. Wampler, W. J. Miller, and M. R. Hilleman, Human hepatitis B vaccine from recombinant yeast, *Nature (Lond.)* **307**:178–180 (1984).

50. Ticehurst, J. R., Hepatitis A virus: Clones, cultures and vaccines, *Semin. Liver Dis.* **6**:46–55 (1986).

51. Buynak, E. B., R. E. Weibel, A. A. McLean, and M. R. Hilleman. Live respiratory

syncytial virus vaccine administered parenterally (40112), *Proc. Soc. Exp. Biol. Med.* **195**:636–642 (1978).

52. Buynak, E. B., R. E. Weibel, A. J. Carlson, A. A. McLean, and M. R. Hilleman. Futher investigations of live respiratory syncytial virus vaccine administered parenterally (40433), *Proc. Soc. Exp. Biol. Med.* **160**:272–277 (1979).

53. Bendinelli, M., Immunomodulaton in viral infections: Virus or infection-induced? in: *Immunomodulation: New Frontiers and Advances* (H. H. Fugenberg, H. D. Whitten, and F. Ambrogi, eds.), pp. 161–195, Plenum, New York (1984).

54. Handzel, Z. T., Y. Burnstein, B. Rager-Zisman, and N. Trainin, The effects of thymic humoral factor (THF) on viral infection in humans, *Riv. Immunol. Immunofarmacol.* **5**:68–71 (1985).

55. Rager-Zisman, R. B., A. Harish, V. Rotter, Y. Yakir, and N. Trainin, Antiviral effects of THF, in: *Advances in Allergy and Immunology* A. Oehling, F. J. Dison, and H. G. Kunkel, eds.), pp. 25–37, Academic, New York (1980).

56. Trainin, N., V. Rotter, Y. Yakir, and R. Leve, Biochemical and biological properties of THF in animal and human models, *Ann. NY Acad. Sci.* **332**:9–22 (1979).

57. Trainin, N., M. Pecht, and Z. T. Handzel, Thymic humoral factor (THF), *Immunol. Today* **4**:16–17 (1983).

58. Dinarello, C. A., Interleukin 1, *Rev. Infect. Dis.* **6**:52–95 (1984).

59. Durum, S. K., J. A. Schmidt, and J. J. Oppenheim, IL-1: An Immunological perspective, *Ann. Rev. Immunol.* **3**:263–87 (1985).

60. Gowen, M., and G. R. Mundy, Actions of recombinant interleukin 1, interleukin 2 and interferon-gamma on bone resorption *in vitro, J. Immunol.* **136**:2478–2482 (1986).

61. Dempsey, R. A., C. A. Dinarello, J. W. Mier, L. J. Rosenwasser, M. Aylegrelta, T. E. Brown, and D. R. Parkinson, The differential effects of human leukocyte pyrogen, lymphocyte activating factor, T cell growth factor and interferon on human NK activity, *J. Immunol.* **129**:2504–2510 (1982).

62. Herman, J., C. A. Dinarello, M. C. Kero, and A. R. Rabson, The role of interleukin 1 in tumor NK cell interactions, correction of defective NK cell activity in cancer patients by treating target cells with interleukin 1, *J. Immunol.* **4**:2882–2886 (1985).

63. Onozaski, K. K. Matshushima, E. S. Kleinerman, T. Saito, and J. J. Oppenheim, Role of interleukin 1 in promoting human monocyte mediated tumor cytotoxicity, *J. Immunol.* **135**:314–320 (1985).

64. Smith, K. A., Interleukin 2, *Annu. Rev. Immunol.* **2**:319–333 (1984).

65. Erard, F., P. Corthesy, M. Nabholz, J. W. Lowenthal, P. Z., G. Plaetinck, and H. R. McDonald, Interleukin 2 is both necessary and sufficient for the growth and differentiation of lectin-stimulated cytolytic T lymphocyte precursors, *J. Immunol.* **3**:1644–1652 (1985).

66. Rook, A. H., H. Masur, H. Clifford, W. Frederick, T. Kasahara, A. M. Macher, J. Y. Djeu, J. F. Manischewitz, L. Jackson, A. S. Fauci, and G. V. Quinnan, Interleukin 2 enhances the depressed natural killer and cytomegalovirus-specific cytotoxic activities of lymphocytes from patients with the acquired immune deficiency syndrome, *J. Clin. Invest.* **72**:398–403 (1983).

67. Nedwin, G. E., L. P. Svedersky, T. S. Bringman, M. A. Palladino, and D. V. Goeddel, Effect of interleukin 2, interferon-gamma, and mitogens on the production of tumor necrosis factors alpha and beta, *J. Immunol.* **135**:2492–2497 (1985).

68. Rosenberg, S. Lymphokine-activated killer cells: A new approach to immunotherapy of cancer, *J. Natl. Cancer Inst.* **75**:595–603 (1985).

69. Conlon, P. J., T. L. Washkewicz, D. Y. Mochizuki, K. L. Urdal, S. Gillis, and C. S. Henney, The treatment of induced immune deficiency with interleukin 2, *Immunol. Lett.* **10**:307–314 (1985).

70. Weinberg, A. Acute genital infection in guinea pigs: Effect of recombinant interleukin 2 (rIL-2) on Herpes simplex virus-2, *J. Infect. Dis.* **154**:134–140 (1986).

71. Goeddel, D. V., D. W. Leung, T. J. Duel, M. Gross, R. M. Lawn, R. McCandliss, P. H. Seeburg, A. Ullrich, E. Yelverton, and P. W. Gray, The structure of eight distinct clone human leukocyte interferon c-DNAs, *Nature (Lond.)* **290:**30–27 (1981).

72. Nagata, S., N. Mantei, and C. Weissman, The structure of one of the eight or more distinct chromosomal genes for human interferon-alpha, *Nature (Lond.)* **287:**401–408 (1981).

73. Gray, P. W., and D. V. Goeddel, Structure of the human immune interferon gene, *Nature (Lond.)* **298:**859–863 (1982).

74. Ohno, S., and T. Taniguehi, Structure of a chromosomal gene for human interferon-beta, *Proc. Natl. Acad. Sci. USA* **78:**5305–5309 (1981).

75. Friedman, R. A., and S. N. Vogel, Interferon with special emphasis on the immune system, *Adv. Immunol.* **34:**97–133 (1983).

76. Stewart, W. E. II, *The Interferon System,* Springer-Verlag, New York (1979).

77. Baron, S., V. Howie, M. Langford, E. M. McDonald, G. J. Stanton, J. Reitmeyer, and D. A. Wleigent, Induction of interferon by bacteria, protozoa and viruses, Defensive role, *Tex. Rep. Biol. Med.* **41:**150–157 (1982).

78. Orn, A., H. Wigzell, A. Senik, and I. Gresser, Enhanced natural killer cell activity in mice injected with interferon and interferon inducers, *Nature (Lond.)* **273:**759–761 (1978).

79. Herberman, R. R., J. R. Ortaldo, and G. D. Bonnard, Augmentation by interferon of human natural and antibody dependent cell mediated cytotoxicity, *Nature (Lond.)* **277:**221–223 (1979).

80. Oehler, J. R., L. R. Lindsay, M. E. Nunn, H. T. Holden, and R. B. Herberman, Natural cell mediated cytotoxicity in rats. II. *In vivo* augmentation of natural killer cell activity, *Int. J. Cancer* **21:**210–220 (1978).

81. Oehler, J. R., and R. B. Herberman, Natural cell mediated cytotoxicity in rats. III. Effects of immunopharmacologic treatments on natural reactivity and on reactivity augmented by polyinosinic-polycytidylic acid, *Int. J. Cancer,* **21:**221–229 (1978).

82. Santoli, D., G. Trinchieri, and H. Koprowski, Cell mediated cytotoxicity against virus infected target cells in humans, *J. Immunol.* **121:**532–538 (1978).

83. Targan, S., and F. Dorey, Interferon activation of prespontaneous killer cells (Pre-SK) and alteration in kinetics of lysis of both pre-SK and active SK cells, *J. Immunol.* **124:**2157–2161 (1980).

84. Timonen, T., J. R. Ortaldo, and R. B. Herberman, Analysis by a single cell cytotoxicity assay of natural killer cell frequencies among human large granular lymphocytes of the effects of interferon on their activity, *J. Immunol.* **128:**2514–2521 (1982).

85. Stebbing, N., Analogs of human interferons, *Riv. Immunol. Immunofarmacol.* **5:**64–67 (1985).

86. Tyrrell, D. A. J., Interferon in the prevention and treatment of respiratory virus infections, *Interferon Syst.* **24:**243–246 (1985).

87. Haglund, S., P. G. Lundquist, K. Cantell, and H. Strander, Interferon therapy in juvenile laryngeal papillomatosis, *Arch. Otolaryngol.* **107:**327–332 (1981).

88. Bomholt, A., Interferon therapy for laryngeal papillomatosis in adults, *Arch. Otolaryngol.* **109:**550–552 (1983).

89. Schonfeld, A., A. Schattner, M. Crespi, H. Levani, J. Shoham, S. Nitke, D. Wallach, T. Han, O. Yarden, T. Doerner, and M. Revel, Intramuscular human interferon-beta injections in treatment of condylomata acuminata, *Lancet* **1:**1038–1041 (1984).

90. Gresser, I., How does interferon inhibit tumor growth? *Interferon* **6:**93–126 (1985).

91. Hadden, J. W., Immunomodulators in the immunotherapy of cancer and other diseases, *TIPS* **3:**191–194 (1982).

92. DelGiacco, G. S., Levamisole, *Riv. Immunol. Immunofarmacol.* **5:**88–91 (1985).

93. DelGiacco, G. S., F. Locci, L. Cengiarotti, E. Chessa, A. DiTucci, G. Meloni, E. Montaldo, M. Pautasso, G. Piludu, and M. C. Piras, The present status of clinical use of levamisole,

in: *Immunomodulation* (H. H. Fudenberg, H. D. Whitten, and F. Ambrogi, eds.), pp. 303–310, Plenum, New York (1984).

94. Hadden, J. W., C. Lopez, R. J. O'Reilly, and E. M. Hadden, Levamisole and inosiplex: Antiviral agents with immunopotentiating action, *Ann. NY Acad. Sci.* **284**:139–152 (1977).

95. Talal, N., and J. Hadden, Hormones, immunomodulatory drugs, and autoimmunity, in: *Handbook of Inflammation*, Vol. 5: *The Pharmacology of Inflammation* (I. L. Bonta, M. A. Bray, M. J. Parnham, eds.), pp. 355–369, Elsevier, New York (1985).

96. Hadden, J. W., S. Specter, A. Galy, J-L. Touraine, and E. M. Hadden, Thymic hormones, interleukins, endotoxin, and thymomimetic drugs in T lymphocyte ontogeny, in: *Advances in Immunopharmacology*, Vol. III (J. W. Hadden, L. Chedid, P. Dukor, F. Spreafico, and D. Willoughby, eds.), pp. 87–95, Pergamon, Elmsford, New York (1986).

97. Matteuci, D., A. Toniolo, P. G. Conaldi, F. Basolo, Z. Gori, and M. Bendinelli, Systemic lymphoid atrophy in coxsackie-B-3-infected mice: Effects of virus and immunopotentiating agents, *J. Infect. Dis.* **151**:1100–1108 (1985).

98. Gudvangen, R. J., P. S. Duffey, R. E. Paque, and C. J. Gauntt, Levamisole exacerbates coxasacki-B-3-induced murine myocarditis, *Infect. Immun.* **41**:1157–1165 (1983).

99. Meade, B. D., Altered mononuclear phagocyte function in mice treated with lymphocytosis promoting factor of *Bordetella pertussis*, *Dev. Biol. Std.* **61**:63–78 (1985).

100. Hewlett, E., Depression of delayed hypersensitivity responses in patients with pertussis, *Dev. Biol. Std.* **61**:241–247 (1985).

# 22

# Conclusions and Prospects

## MAURO BENDINELLI, STEVEN SPECTER, and HERMAN FRIEDMAN

Previous chapters of this multiauthored volume have dealt with the interactions that individual viruses or groups of viruses establish with the host's immune system and with the functional consequences of these interactions. Rather than simply recapitulating data and concepts already developed, this chapter focuses on future goals of research in this area. The interest here is provocative thought. Thus, speculations are presented especially in regard to the clinical implications of the problem.

## 1. PHENOMENOLOGY

The amount of information available concerning the immunosuppressive activity of different viruses is uneven, but a great deal of phenomenology has been described. As a result of extensive investigation, we now know that (1) virtually no acute systemic viral infection is devoid of effects on the ability of the immune system to respond normally to heterologous immunogens. In addition to those covered extensively in this book, viruses for which there is very little information but that are known to be endowed with at least some immunomomodulatory activity include Norwalk agent,[1] canine, and murine parvoviruses,[2,3] African swine fever virus,[4] and others; (2) enhancement of selected immune responses is occasionally observed, but immunosuppressive changes are largely predominant (Table I); and (3) as dramatically exemplified by patients with the acquired immune deficiency syndrome (AIDS), but clearly evident in many other viral infections as well, this immune deficit can be

MAURO BENDINELLI • Department of Biomedicine, University of Pisa, I-56100 Pisa, Italy.    STEVEN SPECTER and HERMAN FRIEDMAN • Department of Medical Microbiology and Immunology, College of Medicine, University of South Florida, Tampa, Florida 33612-4799.

## TABLE I
### Alterations of Immune Effector Functions, as Measured with Heterologous Stimuli, Observed in Virus-Infected Hosts

| Parameter | Viruses inducing[a] | |
|---|---|---|
| | Enhancement | Depression |
| Antibody responsiveness | Few | Many |
| Immunoglobulin level in serum | Many | Some |
| Antibody-dependent hypersensitivity | — | Few |
| Antibody-dependent cellular cytotoxicity | — | Few |
| Circulating autoantibody | Several | Some |
| Lymphocyte circulation | Some | Several |
| Delayed hypersensitivity reactions | Several | — |
| Contact sensitivity | — | Several |
| T-cell-mediated cytotoxicity | — | Several |
| Skin allograft rejection | — | Several |
| Graft-versus-host reaction | — | Some |
| Immunologic maturation | — | Some |
| Autoimmune lesions | Some | Few |
| Tolerance induction | Few | Some |
| Clearance of foreign particles from blood | Some | Some |
| Lymphocyte trapping in spleen | Few | Few |
| Antigen trapping by spleen | Few | — |
| Interferon responsiveness to inducers | — | Several |
| Natural killer cell activity | Many | Some |

[a]Reported differences may reflect the frequency with which various parameters have been examined.

paralleled by an increased susceptibility to superinfections that may have considerable clinical significance.

There still are, however, aspects that are in need of much scrutiny. Most experimental work on viral immunosuppression has been done by infecting otherwise immunologically normal hosts. For example, mouse models of infection have almost invariably employed genetically homogeneous otherwise healthy young adult animals often bred under specific pathogen-free conditions. This has the obvious advantage of minimizing compounding variables but can hardly reproduce the clinical setting, in which the extent and consequences of virus-induced immunomodulation might be quite diverse, depending on many disposing factors, such as genetic makeup, extreme ages, nutritional deficiencies or excesses, hormonal imbalances, and the concomitance of underlying pathology, including superimposed infections and neoplasia. Studies concerned with defining the genetic regulation of susceptibility to viral immunosuppression are few, but there are indications that the matter is quite complex. In mice, for example, the same H-2 haplotype can have markedly different effects on immunosuppression, depending on the infecting virus.[5] Individuals whose immune system is already not completely functional due to physiologic, pathologic, or iatrogenic reasons have an increased risk of developing more severe viral diseases. It seems logical to assume that in such individuals the impact of viral infections on residual immune competence is particularly severe.[6] Recent observations that cytomegalovirus (CMV) infection of immunosuppressed allograft recipients can result in chronic T-cell inversion[7] indicate that exploration of these aspects might be particularly rewarding. A multifactor hypothesis for the etiology of AIDS has been the subject of much speculation, but there are virtually no experimental data to support the contention.

Another aspect deserving further scrutiny is the effect of localized viral infections and, more generally, of primary viral replication during the early stages of systemic infections, on the functioning of immunologic effectors and anti-infective defenses operating locally at the site of infection and in its proximity. Judging from influenza, the only localized viral infection for which there is sufficiently detailed information (see Chapter 16), it seems likely that substantial changes of such effectors of immunity take place. These changes might contribute to the successful initiation of infection, its progression, or the facilitation of local superinfections. Investigation of this problem has been hampered by the limited knowledge of immune mechanisms that operate in infected tissues and organs and by the shortage of specific assays to assess their activity.

A third important aspect that requires further phenomenologic analysis, both in the clinical setting and in experimental models, are the long-term consequences of persistent and latent viral infections on immunity. Although persistence of viruses such as lymphocytic choriomeningitis, lactic dehydrogenase, and chronic leukemia viruses has been associated with significant changes of immunocompetent cells and immunomodulation,[8-10] at this point it would appear that immunologic perturbations associated with chronic viral infection are far less dramatic than observed in most acute infections. However, many viruses that can linger more or less dormantly in lymphoreticular tissues might serve as a cause of cumulative injury to immunocompetent cells. This may result from bursting out of the virus from time to time or otherwise and may lead to clinically significant immunologic deficits as well as other immunopathology. Determination of the involvement of persistence in marginal or slowly progressing immune deficiencies remains as a promising avenue of research. Given the present pace of progress in virologic and immunologic technologies, many difficulties that still prevent a clear definition of these aspects might soon be overcome.

## 2. PATHOPHYSIOLOGY

Understanding the mechanisms whereby viruses immunosuppress their hosts remains a major challenge, because even in the infections most extensively investigated a precise definition of the cellular and subcellular pathways that culminate in the immunosuppressed state are not completely understood. Viruses are associated with pathology of the immune system much more frequently than are other infectious agents, be they bacterial, fungal, protozoan, or metazoan. Even benign infections, such as a rhinoviral common cold, are associated with transient changes in the immunologic profile of patients.[11]

As frequently suggested in the preceding chapters, most often this seems to be related to the fact that viruses are strictly intracellular parasites and that viruses encounter immunocytes very early after entrance into the host. Apparently, both lymphocytes and macrophages are among the body's cells which produce abundant interferon (IFN), but this does not appear to be sufficient to protect them from infection. Several viruses selectively infect B cells, T cells, or macrophages, and many replicate in such cells as in others (Table II). The tropism of many viruses for macrophages and specific lymphocyte subsets, the range of interactions that viruses can establish with such cells, and their functional consequences have only just begun to be appreciated.

Virus-immunocompetent cell encounter may result in the virus being eliminated without apparent consequences or in serious damage to the cells, but most often results in a split victory, whereby the acute viral infection is kept under partial check and the cells remain overtly or latently infected, often for prolonged periods or forever. Lymphocytes and macrophages have often been likened to Trojan horses; because of their high mobility, they are supposed to convey viruses throughout the organism and protect them from immune surveillance. The reasons that viruses so frequently and easily lodge and persist in lymphoreticular tissues are not clear. Cellular activation is often a prerequisite for efficient viral replication in lymphoid and monocytic cells.[12–14] It seems possible that the virus itself causes an activation of lymphoreticular cells, by classic immunologic mechanisms or otherwise, and this allows for the continuous replenishment/recruitment of permissive cells. Numerous additional sophisticated explanations have been suggested, including the clonal distribution

### TABLE II
### A Summary of Viruses Known to Replicate within Macrophages and Lymphocytes

| Virus | Monocytes/ macrophages | Lymphocytes[a] |
|---|---|---|
| Adenoviruses | − | L |
| Arenaviruses | | |
| Lymphocytic choriomeningitis | + | B |
| Coronaviruses | | |
| Mouse hepatitis virus | + | L |
| Enteroviruses | | |
| Coxsackie B | − | L* |
| Echovirus | − | L* |
| Poliovirus | + | L* |
| Hepatitis B virus | − | B,T |
| Herpesviruses | | |
| Bovine herpesvirus | + | N.D. |
| Cytomegalovirus | | |
| Human | + | B |

**TABLE II** (*Continued*)

| Virus | Monocytes/ macrophages | Lymphocytes[a] |
|---|---|---|
| Mouse | + | L |
| Epstein–Barr | − | B |
| Equine herpesvirus | + | B,T |
| Guinea pig herpes-like virus | + | B,T |
| Herpes simplex | + | B*,T* |
| Herpesvirus sylvilagus | − | B,T |
| Human herpesvirus 6 | + | B,T |
| Infectious laryngotracheitis | + | N.D. |
| Marek disease | − | B,T |
| Mouse thymic virus | − | T |
| Pseudorabies virus | + | L |
| Varicella–zoster | + | ? |
| Influenza | + | L* |
| Papovaviruses | | |
| BK virus | − | B,T |
| B lymphotropic papovavirus | − | B |
| Paramyxoviruses | | |
| Measles | + | B,T* |
| Mumps | + | B*,T* |
| Parainfluenza | + | B,T |
| Respiratory syncytial virus | + | T* |
| Poxviruses | | |
| Ectromelia | + | N.D. |
| Leporipoxviruses (fibroma/myxoma) | ? | B,T |
| Vaccinia | + | L* |
| Variola | ? | ? |
| Reoviruses | + | T (newborn mice) |
| Retroviruses | | |
| Avian | + | L |
| Bovine | − | B |
| Caprine/ovine | +* | N.D. |
| Equine | + | L |
| Feline | + | L |
| Human | + | B*,T* |
| Murine | + | B*,T* |
| Simian | ? | T |
| Rhabdoviruses | | |
| Vesicular stomatitis virus | + | T* |
| Togaviruses | | |
| Dengue | + | B,T* |
| Rubella | + | T |
| Yellow fever | + | T |

[a]B, B lymphocytes; T, T lymphocytes; L, undefined lymphocyte population; *, replicate preferentially in activated cells; ?, replication reported but not confirmed; +, positive; −, negative information; N.D., no data.

of molecules expressed on the surface of lymphocytes and the fact that macrophages can be entered by viruses by means other than plasma membrane receptor binding (sometimes with the help of virus-specific antibodies or complement), but none has been proved with certainty,[15,16] leaving enormous matter for future research.

The events linking the invasion of immunocompetent cells to pathology of the immune system now are being addressed increasingly in molecular terms. Apparently, direct cytopathology is neither sufficient nor necessary. Even in AIDS, in which the only evidence of a cythopathic effect *in vivo* is the description of giant cells in the brain of infected individuals, it seems unlikely that the immunodeficiency is the result of direct progressive destruction of T4 cells, because *in situ* hybridization and other sensitive techniques indicate that less than 1 in 10,000 lymphocytes are replicating the virus at any given time, and *in vitro* studies support this view. Indeed, a number of alternative ingenious hypotheses are being envisaged to explain the T-lymphocyte depletion observed in this disease, for example, autoimmune destruction of viral receptor-expressing cells by anti-idiotypical antibody or by antiviral antibody following passive absorption of viral proteins produced in other cells.[14] In certain instances, some satisfactory answers have emerged. For example, virion proteins and glycoproteins mitogenic for B cells and that have other immunomodulatory activities are increasingly being described regarding infections due to several viruses, including human immune deficiency virus (HIV).[17,18] Analysis of the functionally active part of such molecules has also been initiated and at least in one case, the functional part has been synthesized.[19,20] Clearly, identifying the viral molecule(s) and fragment(s) thereof that are responsible for the immunosuppressive effects may lead not only to a better understanding of viral immunosuppression but may permit the delineation of rational therapeutic approaches, the development of better-designed (subunit) vaccines, and possibly provide new clinically useful immunosuppressive products as well. Interestingly, virus-related immunosuppressive products have been detected in certain tumors.[21] Molecular biology has now provided the necessary tools to address these important aspects.

Immune functions and antimicrobial defenses in general result from the coordinate collaboration of many cell types, classes, and subclasses, the interactions of which are mediated by physical contact and soluble factors and regulated by suppressor cell circuits that may be either specific or nonspecific in activity. While these avenues are explored, it should not be forgotten that in such a complex network, even minor virus-induced modifications of the cell surface or alterations in the synthesis and response to soluble mediators are bound to reflect on other cells, generating multiple cascade effects that can ultimately overwhelm physiologic check-and-balance mechanisms and lead to immune hyporeactivity and dysfunction. As discussed in previous chapters, the activation of suppressor cells is a frequent occurrence in viral infections. Moreover, in certain viral infections, profound involution of lymphoid organs has been noted in the absence of detectable viral replication in such organs, and the intervention of autoreactive phenomena has been invoked.[22] Further investigation of these aspects may be rewarding, as mechanisms that regulate immune function are progressively better understood. This is particularly true of the suppressor lymphokines, which are at different stages of characterization.

Within this context, it is worth recalling that IFN and other substances released by virus-infected cells not only have direct immunoregulatory effects but may also induce cells, which normally do not, to express immunoregulatory relevant molecules such as class I[23] and class II[24] major histocompatibility complex antigens. Increased numbers of Ia-like-positive cells have been detected in patients infected with viruses as diverse as mumps[25] and HTLV-I.[26] Lymphocytes have also been noted to change surface phenotype as a consequence of viral infection.[27] Lymphocytotoxic substances are frequently found in the serum of patients during the acute phase of viral infections, but nothing is known about their genesis and function. They may be either cytotoxic or cytostatic, depending on their concentration and the nature of the target cells. Furthermore, lymphotoxic, IFN-inhibitory, and more generally immunomodulatory substances have been detected in several virus-infected cell cultures and hosts.[28−31] However, the full extent to which this increasingly wide range of soluble mediators contributes to immunosuppression remains to be established. Also to be defined is the role that the formation of immune complexes and the consequent activation of complement may have in the genesis of immunosuppression. These are presumably physiologic events in antigen clearance but, when occurring on a large scale due to the self-replicating nature of viral antigens and in proximity to immunocompetent cells, might provoke significant perturbation of immune homeostasis.[32]

As a final comment on pathophysiology, we would like to mention again that present knowledge of mechanisms of immunosuppression derives mainly from studying animals, especially rodents, experimentally infected with very large doses of laboratory passaged viruses, usually by routes that do not necessarily reflect the natural disease process. Recent studies show that viral variants and the passage history of the virus can affect the ability of a virus to immunosuppress.[33] Future experimental studies should examine whether the mechanisms of immunosuppression vary depending not only on the infecting virus but on the many variables that influence infections as well. At the single-cell level, this possibility is exemplified by repeated observations showing that in vitro virus−host lymphocyte balance is sensitive both to external influences and to the physiologic state of these cells. Recently added examples are observations showing that herpes simplex virus (HSV) and HIV replication are enhanced by interleukin-2 (IL-2)[13] and prostaglandin $E_2$ ($PGE_2$).[34] At the organism level, it is suggested by recent findings that in irradiated reconstituted mice retroviruses immunosuppress by mechanisms at least partially different from those in normal mice.[35]

# 3. BIOLOGIC SIGNIFICANCE

There is no factual basis for an attempt to guess the significance of viral immunosuppression in the biologic cycle of viruses. We wish nevertheless to touch briefly on a couple of points. In order to be able to be perpetuated in nature, a virus must rapidly adapt to each new situation in the evolutionary continuum as the host undergoes a modification that can impede the biologic cycle of the virus. Indeed, the close evolutive parallelism that exists between viruses and their hosts finds increasing supportive evidence as light is shed on the mechanisms of viral replication and infection. It also seems likely that the ability possessed by many viruses to dodge the host's immune system is a result of this evolutive adherence. Although this important instrument of defense has been shaped over millions of years to increase resistance against infectious agents, viruses are able to avoid or resist its action effectively, for the time needed to replicate and diffuse to new hosts, and often for much longer periods. The strategies used by different viruses to avoid the action of the immune system seems to be quite varied, and each virus seems to employ those more suited to its general properties and to the characteristics of infection.[15] The only characteristic that appears to be common to most, if not all, viruses is the ability to immunosuppress. This suggests that viral immunosuppression is an essential element in the economy of viral infections, possibly a prerequisite for other, more sophisticated, mechanisms of escape from the host's defenses to engage in action. A possible example of obligatory mutualism between a virus and a host mediated by the ability of the former to abate the host's defense mechanisms has been described in an insect.[36] Viral immunosuppression has also been suggested to be important in the collaboration between helper and defective viruses *in vivo*.[37]

Teleologically, one might argue that the transient nature of viral immunosuppression, as observed in most cases, is in the best interest of the virus, since this permits both survival of the virus and survival of the host. The tendency of viruses to reach an equilibrium with the host is proved by the high frequency of persistent viral infection, wherein the host becomes a potential reservoir for the spread of the virus. Regardless of the argument taken, it is apparent that there is a balance to be achieved between the ability of the virus to depress host defenses and the ability of the host to generate a response to the virus capable of leading to recovery. Fortunately, in most instances, the balance favors the host, but far too often it favors the virus, with severe or fatal consequences for the host. Achieving an understanding of the circumstances under which the virus is favored and developing methods to tip the balance in favor of the host are major goals of this field of study.

## 4. EFFECTS ON THE EVOLUTION OF THE INDUCING INFECTION

Evidence concerning the effects of immune suppression on the evolution of the inducing infection is limited. In virus-immunosuppressed hosts, responses against heterologous antigens are usually scarcely impaired if antigen is given at the outset of infection. Since exposure to antigens of the infecting virus occurs before the immunosuppressed state is fully established, this has been interpreted as evidence that virus-specific responses might be affected

only marginally if at all. In most viral infections, antiviral antibody and effectors of cell-mediated immunity are readily demonstrable. Stimulation by
viral antigens continues, however, for the duration of infection and responses
to different viral epitopes become detectable at different times, making it quite
possible that overall immunity mounted against the infecting virus is lower
than it would be in the absence of viral immunosuppression. In fact, we cannot
know what the antiviral response would be like if the viral infection was not
accompanied by immunomodulatory effects. Further investigation of these
aspects is clearly warranted. In these studies, attention should again be paid to
passage history of the virus and other variables that, as recently suggested,[33]
might affect the ability to suppress immune responses to homologous antigens.

All things considered, the infected organism does not appear to cope very
effectively with viral infections. Complete elimination of the infecting virus is
infrequent. Theoretically, viral immunosuppression may influence the course
of the inducing infection in several ways. For example, it might (1) limit the
host's ability to block viral spread from the primary site(s) of replication to
target organs; (2) lengthen the duration of the acute unchecked phase of viral
growth; (3) be a prerequisite for the development of virus-specific T-suppressor (Ts) cells and more generally for specific unresponsiveness known to
occur in several viral infections; and (4) facilitate the establishment of viral
persistence. Although currently available data neither prove nor disprove such
possibilities, a correlation between these parameters of infection and viral immunosuppression has often been suggested. For example, a common feature
of heart transplant patients with primary or secondary CMV infections is a
large increase of Ts cells and an inversion of the T-helper/Ts ratio that lasts up
to 3 years. In a recent study, chronic excretion of the virus was found to
correlate with persistence of such changes.[38]

## 5. EFFECTS ON PATHOGENESIS BY THE INDUCING VIRUS

Theoretically, the overall impact of viral immunosuppression on the severity and extent of pathogenesis caused by the inducing virus may be either
detrimental or beneficial to the host, depending on the mechanisms whereby
the disease induced by the infecting virus is generated (Table III). When the
disease is sustained mainly by direct viral damage to cells and tissues (e.g., due
to cytolysis), viral immunosuppression could lead to an aggravation of damage
by affecting the host's ability to mount protective immunity against the virus,
thereby enhancing its replication and hampering or delaying its clearance.
Interestingly, in recent studies, lymphocytes from patients susceptible to frequent recurrences of HSV-induced lesions proved most susceptible to inhibition of proliferative responsiveness by exposure to HSV *in vitro* and were less
efficiently restored by exogenous IL-2 than were cells from patients subject to
infrequent recurrences.[39] The same seems to be true for viral diseases sustained by the proliferation of cells bearing antigens readily recognized by the

**TABLE III**
**Clinical Implications of Viral Immunosuppression**

| | Human infections with | | Animal models |
| --- | --- | --- | --- |
| | HIV | Other viruses | |
| For disease | | | |
| Facilitates spread of infecting virus | ? | ? | Possibly |
| Prolongs infection | ? | ? | ? |
| Enhances virus-induced damage | ? | ? | Possibly |
| Decreases virus-induced immunopathology | ? | ? | Possibly |
| Facilitates secondary infections | Yes | Yes | Yes |
| Facilitates tumor development/progression | Yes | ? | Yes |
| Causes long-term sequelae | | | |
|   Growth retardation | Possibly | ? | Yes |
|   Apparently idiopathic immune deficiencies | Yes | Possibly | ? |
|   Autoimmunity | Possibly | ? | Possibly |
|   Others | ? | ? | ? |
| For treatment | | | |
| Justifies use of immunopotentiating drugs | No[a] | No | No |

[a]Except for full-blown AIDS.

host's immune system, such as in virus-induced hyperplastic or neoplastic growth. Indeed, a correlation between viral immunosuppression and progression of virus-induced tumors has been noted. Examples are tumors caused by Aleutian disease virus of mink[40] and leukemias caused by oncogenic retroviruses of mammals and birds.[10] It is clear, however, that documented information in this area is extremely sketchy.

Immunopathology appears to play a key role in the pathogenesis of many viral infections.[41] In viral infections in which immunopathogenetic mechanisms are essential determinants of disease, turning down of immune responses might have an adaptive value because it might interfere with the mechanisms whereby cell and tissue damage is generated. Although the possibility that viral immunosuppression is beneficial to the host remains essentially speculative, indirect support comes from clinical observations that the nephrotic syndrome, a disease that frequently responds to immunosuppressive therapy, and allergic manifestations can undergo remission in the course of measles.[42] Supportive evidence comes also from findings showing that the administration of immunopotentiating drugs can exacerbate virus-induced pathology in certain animal models of infection.

Thus, although limited, available information clearly indicates that generalizations on these issues are not possible. Each viral infection should be examined and evaluated separately with regard to the detrimental or beneficial effects of immune suppression.

## 6. EFFECTS ON RESISTANCE TO OTHER INFECTIONS

Acquired immune-deficiency syndrome is the prototype example of a viral infection that predisposes to secondary opportunistic infections. However, long before AIDS was recognized, clinical practice had shown that patients with, or convalescent from, viral infections presented an increased susceptibility to superinfections and to the reactivation of latent infections. For example, following measles there may be an increased susceptibility to bacteria and viruses for periods of 1 year or more.[43] It is generally accepted that in patients with influenza, the diminution of resistance is usually short-lived and is most evident locally as enhanced incidence and gravity of bacterial pneumonia. It is recognized that CMV infection predisposes graft recipients to superinfections that may represent a serious threat to life.[44] Furthermore, patients hospitalized with specific infectious diseases in an area of Japan in which HTLV-1 is endemic were found to have an almost threefold incidence of antibody to that virus than did the general population,[45] suggesting that HTLV-1-induced immunosuppression may have considerable public health significance. Natural viral infections of animals known to predispose their hosts to secondary infections include feline leukemia virus, canine parvovirus, and bovine herpesviruses.[46] In many such infections, the viral attack would cause minor symptoms and lesions and then resolve within a short period, if secondary bacterial, fungal, or protozoal infection did not occur.

While there seems to be little doubt that the enhanced susceptibility to superinfections associated with viral infections is due to the underlying virus-induced immunosuppression, much remains to be learned on the relative importance of the various immunologic dysfunctions caused by the virus in determining such increased susceptibility. Thus, for example, the relative contribution of damage to the adaptive and natural effectors of antimicrobial defenses remains to be established. The recognition that HIV replicates very effectively in cells of the monocyte–macrophage series[47,48] and in altered B lymphocytes[49] has led to a proposal that pathogenesis leading to AIDS requires persistant infection in macrophages followed by subsequent infection of lymphocytes and neural cells.

In conclusion, the study of viral immunosuppression in animal models should always be completed by the assessment of resistance to challenge with a battery of superinfecting agents known to take advantage of defects in different branches of immunity. This would not only prove the significance of the observed changes in immunologic parameters but would possibly lead to a better correlation between such changes and increased susceptibility to selected agents as well.

## 7. EFFECTS ON RESISTANCE TO TUMORS

One of the hallmarks of progressive HIV infection is the development of Kaposi sarcomas and other neoplasms that are usually very rare and less virulent in the normal population. Many such tumors have a suspected viral etiology; it is therefore possible that their enhanced incidence is another aspect of the reduced resistance to superimposed infections discussed earlier. Apart from anecdotal evidence, there are no indications that other viral infections of humans are related to subsequent neoplasia.

The few experimental studies in animals have been performed with retroviruses and indicate that both virus-induced and transplanted tumor growth is facilitated.[10] Clearly, it is an important area worth further investigation.

## 8. IMPLICATIONS FOR THERAPY OF VIRAL DISEASES

This aspect has been extensively covered in Chapter 21. Here, we wish to emphasize the need for great prudence in suggesting the use of immunopotentiating and immunorestorative agents for the treatment of viral infections. The low number of efficient antiviral drugs available has led to consideration of substances that nonspecifically stimulate the immune system as potentially useful therapeutic tools (in certain instances rushing them into clinical use). Superficially, the immunosuppressed state so often associated with viral infection might be considered a further rationale for this kind of treatment. However, the information available is far from encouraging in this direction. The reasons are manifold, but three are stressed here.

1. The limited knowledge we still have of the contribution of immunopathology to viral diseases, especially of humans:[41] Our understanding of viral pathogenesis is insufficient to predict the effects immunopotentiating treatments can have on specific viral diseases and their sequelae. In properly controlled experimental systems, immunopotentiation has given contradictory results. For example, while a thymic hormone increased the survival rate of mice infected with mengovirus,[50] the administration of immunopotentiating agents to mice infected with a similarly cytolytic virus, coxsackievirus $B_3$, resulted in exacerbation of cardiac damage.[51]
2. The fact that *in vitro* many viruses replicate more efficiently in stimulated than in resting lymphocytes and macrophages and most immunopotentiating agents are mitogenic for lymphocytes and/or activate macrophages. Thus, an unwanted result of immunopotentiating treatment might be more extensive replication of the virus within the lymphoreticular tissue. This might, for example, facilitate the establishment of viral persistence.
3. The possibility of paradoxical effects on the functioning of the immune system: Interestingly, immunopotentiating agents were found to enhance the lymphoid depletion caused by coxsackievirus $B_3$ infection of mice.[22]

In conclusion, far more experimental knowledge of the immunobiology and immunopathology of each viral infection is needed before immunostimulants can be safely used for the treatment or prevention of viral diseases. Empirical use of such substances is justified only for the patient whose prognosis is very poor. It should be mentioned in this context that, so far, the use of immunostimulants in patients with AIDS has given unsatisfactory results and that concern has often been raised that treating patients with minor symptoms of HIV infection with IL-2, IFN, or other drugs with lymphocyte stimulatory activity in the absence of a specific antiviral therapy might be hazardous because it could facilitate progression of the disease. Treatment of full-blown AIDS with leukocyte transfusions or bone marrow grafts has proved unsuccessful because grafted cells were rapidly overwhelmed by infection.[52]

## 9. POSSIBLE ROLE OF VIRUSES IN THE GENESIS OF IDIOPATHIC IMMUNE DEFICIENCIES

A viral etiology has repeatedly been proposed for some persistent immune deficiences of unknown origin. Cases of congenital hypogammaglobulinemia have, for example, been tentatively considered sequelae of congenital rubella or other intrauterine infections.[53] The importance that persistent viral infections might have in the genesis of such syndromes has attracted little experimental attention, however. Viral persistence often involves low-level viral replication in lymphoid tissues and, on a chronic basis even subtle changes might lead to alterations no longer compensated by balance mechanisms. Moreover, it has recently been emphasized that viruses can produce disorders of specialized cells and systems also in the absence of the familiar footprints of viral replication.[54] The effects of viral persistence on the functioning of the immune system have been little explored even in animal models. A role for viral immunosuppression in the genesis of the runting syndromes associated with congenital or perinatal viral infections of animals has, however, often been postulated. Recently, a similar failure to thrive was described in children congenitally infected with HIV.[55] In humans, the existence of a chronic paucisymptomatic pathology associated with immunologic disorders and due to Epstein–Barr virus (EBV) infection is slowly emerging.[56] Furthermore, a number of viruses, including EBV, hepatitis B, and human parvovirus, have been linked to bone marrow aplasia.[57]

Often immune-deficient hosts show altered functions of the existing immune apparatus as well as a lack of certain immunologic effector mechanisms. In idiopathic immune deficiencies, the incidence of allergic, autoimmune, and collagen diseases is high, suggesting that impaired and disordered functioning of the immune system is strongly intermingled. It has often been thought that viruses may trigger autoimmune diseases, a suggestion that is finding some interesting experimental basis.[16,58] Some among the most immunosuppressive viruses of animals and humans also cause polyclonal activation of B cells;[18] such stimulation of B cells could potentially lead to autoimmune phenomena if tolerance is broken. It is also possible that viral infections act synergistically with other pathologic or physiologic causes of reduced immune responsiveness. In mice persistently infected with parainfluenza or CMV, the long-term effects on immunity were dependent on cofactors, such as the animal's age and diet.[59,60] Viewed from this standpoint, viral immunosuppression might also be considered as a possible factor in immune senescence.

Transient immunosuppression is difficult to evaluate immunologically. Nevertheless, it is everyday experience in clinical practice that most persons in certain periods of their life become more susceptible to infections. For example, they become more apt to be affected by minor respiratory illness and tend to recover slowly from such infections. For these elusive maladies, a viral origin seems a likely possibility. Examples of such infections include Kawasaki disease and chronic fatigue syndrome. The recent discovery of HIV encourages the search for novel immunosuppressive viruses that might be implicated in the genesis of such illnesses. Most interestingly, in recent experiments, serial transfer of lymphocytes pre-exposed to allogeneic cells resulted in the establishment of an infectious form of immune deficiency that appeared to involve a viruslike agent.[61]

In conclusion, although there is little more than educated guessing in the study of virus-induced immunosuppression, a rich harvest of research is to be expected, that can only be achieved through an active cooperation between biologists and clinicians. It is ironic that a lentivirus (HIV) is responsible for the acceleration of activity in this discipline.

# REFERENCES

1. Dolin, R., R. C. Reichman, and A. S. Fauci, Lymphocyte populations in acute viral gastroenteritis, *Infect. Immun.* **14**:422–428 (1976).
2. Norley, S. G., and R. C. Wardley, Investigation of porcine natural-killer cell activity with reference to African swine-fever virus infection, *Immunology* **49**:593–597 (1983).
3. Olsen, C. G., M. I. Stiff, and R. G. Olsen, Comparison of the blastogenic response of peripheral blood lymphocytes from canine parvovirus-positive and -negative outbred dogs. *Vet. Immunol. Immunopathol.* **6**:285–290 (1984).
4. Engers, H. D., J. A. Louis, R. H. Zubler, and B. Hirt, Inhibition of T cell-mediated functions by MVM(i), a parvovirus closely related to minute virus of mice, *J. Immunol.* **127**:2280–2285 (1981).
5. Morrison, R. P., J. Nishio, and B. Chesebro, Influence of the murine MHC (H-2) on Friend leukemia virus-induced immunosuppression, *J. Exp. Med.* **163**:301–314 (1986).
6. Rinaldo, C. R., Jr., R. L. DeBiasio, W. H. Hamoudi, B. Rabin, M. Liebert, and T. R. Hakala, Effect of herpesvirus infections on T-lymphocyte subpopulations and blastogenic responses in renal transplant recipients receiving cyclosporine, *Clin. Immunol. Immunopathol.* **38**:357–366 (1986).
7. Maher, P., C. M. O'Toole, T. G. Wreghitt, D. J. Spiegelhalter, and T. A. H. English, Cytomegalovirus infection in cardiac transplant recipients associated with chronic T cell subset ratio inversion with expansion of a Leu7+ Ts-c+ subset, *Clin. Exp. Immunol.* **62**:515–524 (1985).

8. Oldstone, M. B. A., A. Tishon, J. M. Chiller, W. O. Weigle, and F. J. Dixon, Effect of chronic viral infection on the immune system. I. Comparison of the immune responsiveness of mice chronically infected with LCM virus with that of noninfected mice, *J. Immunol.* **110**:1268–1278 (1973).

9. Inada, T., and C. A. Mims, Mouse Ia antigens are receptors for lactate dehydrogenase virus, *Nature (Lond.)* **309**:59–61 (1984).

10. Bendinelli, M., D. Matteucci, and H. Friedman, Retrovirus-induced acquired immunodeficiencies, *Adv. Cancer Res.* **45**:125–181 (1985).

11. Levandowski, R. A., D. W. Ou, and G. G. Jackson, Acute-phase decrease of T lymphocyte subsets in rhinovirus infection, *J. Infect. Dis.* **153**:743–748 (1986).

12. Haase, A. T., Pathogenesis of lentivirus infections. *Nature (Lond.)* **322**:130–136 (1986).

13. Braun, R. W., and H. Kirchner, T lymphocytes activated by interleukin 2 alone acquire permissiveness for replication of herpes simplex virus, *Eur. J. Immunol.* **16**:709–711 (1986).

14. Fauci, A. S., The human immunodeficiency virus. Infectivity and mechanisms of pathogenicity, *Science* **239**:617–622 (1988).

15. Mims, C. A., and D. O. White, *Viral Pathogenesis and Immunology*, Blackwell, Oxford, 1984.

16. Notkins, A. L, and M. B. A. Oldstone, *Concepts in Viral Pathogenesis*, Springer-Verlag, New York, 1984.

17. Fries, L. F., H. M. Friedman, G. H. Cohen, R. J. Eisenberg, C. H. Hammer, and M. M. Frank, Glycoprotein C of herpes simplex virus 1 is an inhibitor of the complement cascade, *J. Virol.* **137**:1636–1641 (1986).

18. Pahwa, S., R. Pahwa, C. Saxinger, R. C. Gallo, and R. A. Good, Influence of the human T-lymphotropic virus/lymphadenopathy associated virus on functions of human lymphocytes: Evidence for immunosupressive effects and polyclonal B-cell activation by banded viral preparations, *Proc. Natl. Acad. Sci. USA* **82**:8198–8202 (1985).

19. Cianciolo, G. J., T. D. Copelan, S. Oroszlan, and R. Snyderman, Inhibition of lymphocyte proliferation by a synthetic peptide homologous to retroviral envelope proteins, *Science* **230**:433–455 (1985).

20. Harrell, R. A., G. J. Cianciolo, T. D. Copelan, S. Oroszlan, and R. Snyderman, Suppression of the respiratory burst of human monocytes by a synthetic peptide homologous to envelope proteins of human and animal retroviruses, *J. Immunol.* **136**:3517–3520 (1986).

21. Balm, A. J., I. B. Tan, H. A. Drexhage, R. J. Scheper, B. M. von Blomberg, M. de Haan, and G. B. Snow, Head and neck carcinomas contain retroviral-p15(E) related factors which inhibit monocyte migration, in: *Fifth Joint Meeting of the American Society of Head and Neck Surgery and Society of Head and Neck Surgeons, Dorado, Puerto Rico, May 5–8, 1985*, p. 68.

22. Matteucci, D., A. Toniolo, P. G. Conaldi, F Basolo, and M. Bendinelli, Systemic lymphoid atrophy in coxsackievirus B3-infected mice: Effects of virus and immunopotentiating agents, *J. Infect. Dis.* **151**:1100–1108 (1985).

23. Suzumura, A., E. Lavi, S. R. Weiss, and D. H. Silberg, Coronavirus infection induces H-2 antigen expression on oligodendrocytes and astrocytes, *Science* **232**:991–993 (1986).

24. Todd, I., R. Pujol-Borrell, L. J. Hammond, G. F. Bottazzo, and M. Feldman, Interferon-gamma induces HLA-DR expression in thyroid epithelium, *Clin. Exp. Immunol.* **61**:265–273 (1985).

25. Nagai, H., T. Morishima, Y. Morishima, S Isomura, and S. Suzuki, Local T cell subsets in mumps meningitis, *Arch. Dis. Child.* **58**:927–935 (1983).

26. Morishima, Y., K. Ohya, T. Morishima, K. Nishikawa, and T. Fukuda, Immunological studies on adult T cell leukaemia virus (ATLV) carriers, *Clin. Exp. Immunol.* **64**:457–464 (1986).

27. Sing, G. K., and H. M. Garnett, The effects of human cytomegalovirus challenge *in vitro* on subpopulations of T cells from seronegative donors, *J. Med. Virol.* **14**:363–371 (1984).

28. Rodgers, B. C., D. M. Scott, J. Mundin, and Y. G. P. Sissons, Monocyte-derived inhibitor of interleukin 1 induced by human cytomegalovirus, *J. Virol.* **55**:527–532 (1985).

29. Sheridan, J. F., M. Beck, L. Aurelian, and M. Radowsky, Immunity to herpes simplex virus: Virus reactivation modulates lymphokine activity, *J. Infect. Dis.* **152**:449–456 (1985).

30. Blazar, B. A., L. M. Lutton, and M. Strome, Immunomodulatory activity in supernatants from EBV-immortalized lymphocytes, *Cancer Immunol. Immunother.* **22**:62–67 (1986).

31. Roberts, N. J. Jr., A. H. Prill, and T. N. Mann, Interleukin 1 and interleukin 1 inhibitor production by human macrophages exposed to influenza virus or respiratory syncytial virus: Respiratory syncytial virus is a potent inducer of inhibitor activity. *J. Exp. Med.* **163**:511–519 (1986).

32. McDonald, T. L. Blocking of cell-mediated immunity to Moloney murine sarcoma virus-transformed cells by lactate dehydrogenase virus-antibody complex, *J. Natl. Cancer Inst.* **70**:493–497 (1983).

33. Schrier, R. D., G. P. A. Rice, and M. B. A. Oldstone, Suppression of natural killer cell activity and T cell proliferation by fresh isolates of human cytomegalovirus, *J. Infect. Dis.* **153**:1084–1091 (1986).

34. Kuno, S., R. Ueno, O. Hayaishi, H. Nakashima, and S. Harada, Prostaglandin $E_2$, a seminal constituent, facilitates the replication of acquired immune deficiency syndrome virus *in vitro*, *Proc. Natl. Acad. Sci. USA* **83**:3487–3490 (1986).

35. Toniolo, A., D. Matteucci, P. G. Conaldi, and M. Bendinelli, Virus-induced immunodeficiency. Antibody responsiveness of MuLV-infected spleen cells following transfer into irradiated mice, *Med. Microbiol. Immunol.* **173**:197–206 (1986).

36. Edson, K. M., S. B. Vinson, D. B. Stoltz, and M. D. Summers, Virus in a parasitoid wasp: Suppression of the cellular immune response in the parasitoid's host, *Science* **211**:583–584 (1981).

37. Bendinelli, M., and L. Nardini, Immunodepression by Rowson–Parr virus in mice. II. Effect of Rowson–Parr virus infection on the antibody response to sheep red cells *in vivo* and *in vitro*, *Infect. Immun.* **7**:160–166 (1973).

38. O'Toole, C. M., J. J. Gray, P. Maher, and T. G. Wreghitt, Persistent excretion of cytomegalovirus in heart transplant patients correlates with inversion of the ratio of T helper/T suppressor-cytotoxic cells, *J. Infect. Dis.* **153**:1160–1162 (1986).

39. Wainberg, M. A., J. D. Portnoy, B. Clecner, S. Hubschman, J. Lagacé-Simard, N. Rabinovitch, Z. Remer, and J. Mendelson, Viral inhibition of lymphocyte proliferative responsiveness in patients suffering from recurrent lesions caused by herpes simplex virus, *J. Infect. Dis.* **152**:441–448 (1985).

40. Hwan, S., and B. N. Wilkie, Mitogen- and viral antigen-induced transformation of lymphocytes from normal mink and from mink with progressive or nonprogressive Aleutian disease, *Infect. Immun.* **34**:111–114 (1981).

41. Sissons, J. G. P., and L. K. Borysiewicz, Viral immunopathology, *Brit. Med. Bull.* **41**:34–40 (1985).

42. Smithwick, E. M., Common viruses as immunosuppressive agents, in: *The Nature, Cellular, and Biochemical Basis and Management of Immunodeficiencies*, (R. A. Good and E. Lindenlaub, (eds.), pp. 115–126, Schattauer Verlag, Stuttgart, 1986.

43. Coovadia, H. M., Recent advances in the understanding of measles, *South Afr. J. Hosp. Med.* **6**:143–148 (1980).

44. Ho, M., *Cytomegalovirus. Biology and Infection*, Plenum, New York, 1982.

45. Essex, M. E., M. F. McLane, N. Tachibana, D. P. Francis, and T. Lee, Seroepidemiology of human T-cell leukemia virus in relation to immunosuppression and the acquired immunodeficiency syndrome, in: *Human T-Cell Leukemia/Lymphoma Virus* (R. C. Gallo, M. E. Essex, and L. Gross, (eds.), pp. 355–362, Cold Spring Harbor Laboratory, Cold Spring Harbor, New York (1984).

46. Bielefeldt Ohmann, H., and L. A. Babiuk, Viral infections in domestic animals as models for studies of viral immunology and pathogenesis, *J. Gen. Virol.* **66**:1–25 (1986).
47. Nicholson, J. K. A., G. D. Cross, C. S. Callaway, and J. S. McDougal, In vitro infection of human monocytes with human T lymphotropic virus type III/lymphadenopathy-associated virus, *J. Immunol.* **137**:323–329 (1986).
48. Gartner, S., P. Markovitz, D. M. Markovitz, M. H. Kaplan, R. C. Gallo, and M. Popovic, The role of mononuclear phagocytes in HTLV-III/LAV infection, *Science* **233**:215–219 (1986).
49. Ablashi, D. V., E. A. Hunter, K. L. Ablashi, S. Z. Salahuddin, P. D. Markham, B. Kramarsky, and R. C. Gallo, Chronic HTLV-III infection in EBV-genome containing B-cell lines, in: *Abstracts of the Second International Symposium on Epstein–Barr Virus and Association of Malignant Diseases*, p. 115 (1986).
50. Klein, A. S., R. Fixler, and J. Shoham, Antiviral activity of a thymic factor in experimental viral infections. I. Thymic hormonal effect on survival, interferon production and NK cell activity in mengo virus-infected mice, *J. Immunol.* **132**:3159–3163 (1984).
51. Gudvangen, R. J., P. S. Duffey, R. E. Paque, and C. J. Gauntt, Levamisole exacerbates coxsackievirus B3-induced murine myocarditis, *Infect. Immun.* **41**:1157–1165 (1983).
52. Lane, H. C., H. Masur, D. L. Longo, H. G Harvey, A. H. Rook, G. V. Quinnan, R. G. Steis, A. Macher, G. Whalen, L. C. Edgar, and A. S. Fauci, Partial immune reconstitution in a patient with acquired immunodeficiency syndrome, *N. Engl. J. Med.* **311**:1099–1103 (1984).
53. Hayward, A. R., Immunodeficiency, in: *Immunological Aspects of Infectious Diseases* (G. Dick, ed.), pp. 151–200, MTP Press, Lancaster, (1979).
54. Oldstone, M. B. A., An old nemesis in new clothing: Viruses playing new tricks by causing cytopathology in the absence of cytolysis, *J. Infect. Dis.* **152**:665–667 (1985).
55. Pahwa, S., M. Kaplan, S. Fikrig, R. Pahwa, M. G. Sarngadharan, M. Popovic, and R. C. Gallo, Spectrum of human T cell lymphotropic virus type III infection in children. *JAMA* **255**:2299–2305 (1986).
56. Rickinson, A. B., Chronic, symptomatic Epstein–Barr virus infections, *Immunol. Today* **7**:13–14 (1986).
57. Chorba, T., P. Coccia, R. C. Halman, P. Tattersall, L. J. Anderson, J. Sudman, N. S. Young, E. Kurczynski, U. M. Saarinen, R. Moir, D. N. Lawrence, J. M. Jason, and E. Evatt, The role of parvovirus B19 in aplastic crisis and erythema infectiousum (fifth disease), *J. Infect. Dis.* **154**:383–393 (1986).
58. Wolfgram, L. J., K. W. Beisel, and N. R. Rose, Heart-specific autoantibodies following murine coxsackievirus B3 myocarditis, *J. Exp. Med.* **161**:1112–1121 (1985).
59. Kay, M. M. B., Long term subclinical effects of parainfluenza (Sendai) infection on immune cells of aging mice, *Proc. Soc. Exp. Biol. Med.* **158**:326–331 (1978).
60. Cruz, J. R., and J. L. Waner. Effect of concurrent cytomegaloviral infection and undernutrition on the growth and immune response of mice, *Infect. Immun.* **21**:436–441 (1978).
61. Kim, B. S., and K. M. Hui, Induction of infectious immunodeficiency in BALB/c mice by serial transfer of lymphocytes immune to alloantigens, *J. Immunol.* **135**:255–260 (1985).

# Index

459